Functional Polymer Composites with Nanoclays

RSC Smart Materials

Series Editors:
Professor Hans-Jörg Schneider, *Saarland University, Germany*
Professor Mohsen Shahinpoor, *University of Maine, USA*

Titles in this Series:
1: Janus Particle Synthesis, Self-Assembly and Applications
2: Smart Materials for Drug Delivery: Volume 1
3: Smart Materials for Drug Delivery: Volume 2
4: Materials Design Inspired by Nature
5: Responsive Photonic Nanostructures: Smart Nanoscale Optical Materials
6: Magnetorheology: Advances and Applications
7: Functional Nanometer-Sized Clusters of Transition Metals: Synthesis, Properties and Applications
8: Mechanochromic Fluorescent Materials: Phenomena, Materials and Applications
9: Cell Surface Engineering: Fabrication of Functional Nanoshells
10: Biointerfaces: Where Material Meets Biology
11: Semiconductor Nanowires: From Next-Generation Electronics to Sustainable Energy
12: Supramolecular Materials for Opto-Electronics
13: Photocured Materials
14: Chemoresponsive Materials: Stimulation by Chemical and Biological Signals
15: Functional Metallosupramolecular Materials
16: Bio-Synthetic Hybrid Materials and Bionanoparticles: A Biological Chemical Approach Towards Material Science
17: Ionic Polymer Metal Composites (IPMCs): Smart Multi-Functional Materials and Artificial Muscles Volume 1
18: Ionic Polymer Metal Composites (IPMCs): Smart Multi-Functional Materials and Artificial Muscles Volume 2
19: Conducting Polymers: Bioinspired Intelligent Materials and Devices
20: Smart Materials for Advanced Environmental Applications
21: Self-cleaning Coatings: Structure, Fabrication and Application
22: Functional Polymer Composites with Nanoclays

How to obtain future titles on publication:
A standing order plan is available for this series. A standing order will bring delivery of each new volume immediately on publication.

For further information please contact:
Book Sales Department, Royal Society of Chemistry, Thomas Graham House, Science Park, Milton Road, Cambridge, CB4 0WF, UK
Telephone: +44 (0)1223 420066, Fax: +44 (0)1223 420247
Email: booksales@rsc.org
Visit our website at www.rsc.org/books

Functional Polymer Composites with Nanoclays

Edited by

Yuri Lvov
Louisiana Tech University, Ruston, LA, USA
Email: ylvov@coes.latech.edu

Baochun Guo
South China University of Technology, Guangzhou, China
Email: psbcguo@scut.edu.cn

Rawil F. Fakhrullin
Kazan Federal University, Kazan, Republic of Tatarstan, Russian Federation
Email: kazanbio@gmail.com

ROYAL SOCIETY OF CHEMISTRY

RSC Smart Materials No. 22

Print ISBN: 978-1-78262-422-6
PDF eISBN: 978-1-78262-672-5
EPUB eISBN: 978-1-78262-904-7
ISSN: 2046-0066

A catalogue record for this book is available from the British Library

Published by the Royal Society of Chemistry,
Thomas Graham House, Science Park, Milton Road,
Cambridge CB4 0WF, UK

Registered Charity Number 207890

For further information see our web site at www.rsc.org

Printed in the United Kingdom by CPI Group (UK) Ltd, Croydon, CR0 4YY, UK

Foreword

Over the last fifteen years, increased knowledge about the science of materials and their organization has kept researchers scrambling to keep up with an ever increasing demand for engineered materials of higher and higher specifications. Nanotechnology and the incorporation of clay–polymer composite materials are ushering in a new age of materials with an impact on society that cannot be underestimated. Clay–polymer composite materials, exemplified by layered nacre-like composites as well as other similar biomimetic materials, can resolve numerous problems in the areas of energy production and storage, electronics and sensors, and medicine, to name a few. The industrial impact of many nanoclay composites, especially as a reinforcement component of common plastics, has already been realized. As one example of that, the global nanoclay market size is estimated to have exceeded $800 million in 2014. At the same time, we see only a fraction of the potential impact of nanoscale clay materials, as evidenced by many chapters in this book. One could expect that future advances in the use of nanoclay composites in self-healing materials, nanoscale sensors, cardiovascular stents, tissue engineering, neural–electronic interfaces, information-nano-bio integration, and nano-information biological-cognitive systems could radically change the lives of humankind.

The central goal of this fascinating research effort is to discover how to create materials that combine the requisite properties, including both mechanical (that is, strength, stiffness, and toughness) with so called functional properties (such as transparency, corrosion resilience, flame retardancy, *etc*). However, only recently has it been understood how such composites can be prepared with control of the nanoscale organization. The design of organized clay–polymeric nanocomposites demands not just exfoliation but well defined clay nanoparticles. An important role in composite processes belongs not only to the material composition, but also to the clay particle shape and

RSC Smart Materials No. 22
Functional Polymer Composites with Nanoclays
Edited by Yuri Lvov, Baochun Guo and Rawil F. Fakhrullin

Published by the Royal Society of Chemistry, www.rsc.org

size as well. Clay nanoparticles have been used to prepare sheets as thin as 1 nm and a few hundred nm wide (such as kaolin and montmorillonite), and their organized multilayers made them very strong and non-gas-permeable. Recently, an important development was reached for elongated tubule and fibrous clay particles such as halloysite, imogolite, sepiolite, and others, and this book is concerned mostly with these clay types. With natural availability, abundant deposits, low-cost, simple purification, and fibrous or hollow tubular nanostructures, clay nanocomposites have distinct advantages over synthetic nanomaterials like carbon or metal oxide nanotubes.

Halloysite tubule nanoclays represent an emerging type of nanomaterial with functional properties comparable to carbon nanotubes but are considerably less expensive. Their applied use began in high performance ceramics, as reinforcements for polymers, controlled release of active agents for composites with flame-retardant, anticorrosion, antimicrobial, and antioxidant properties. Their applications have broadened into biological areas including biomimetic materials such as bone cements and dentist formulation tissue scaffolds. Drug delivery systems, cosmetics, catalyst immobilization, specific adsorbents, and mesoporous materials for water purification can also be made from nanoclays, as exemplified by the multiple studies described in this book. Fibrous clays provided significant reinforcement properties to nanocomposites, and when assembled with biopolymers and other green polymers, they gave rise to functional bio nanocomposites. Recent findings in fabrication of drug delivery vehicles and tissue engineering constructs are highlighted. Amazing examples of interfacing biological entities (cells, tissues, DNA) with nanoclay particles are presented here as well.

In this book you will find contributions about biomedical composites, including nanosafety characterization, described in Chapters 12–14. Overall, it appears that nanoclays are safe and biocompatible nanomaterials. Notable advancements in functional substance-loaded halloysite for fibers (Chapter 1) and rubbers (Chapters 3 and 11), flame retardancy (Chapter 9), ultrafiltration and proton exchange membranes (Chapter 10), and immobilized biomacromolecules (Chapter 12) have been collected. This book also includes advancements in attapulgite–rubber composites for engineering applications (Chapter 2) and wraps up with rectorite composites for enhanced permeability (Chapter 3). The book provides one with the most up to date information on a field that is advancing at a blazing pace, with promising new materials constantly being discovered that will certainly make their way into significant applications. I dare to predict that in five to ten years we shall discover even more fascinating properties and technological applications of nanoclay materials, especially in the area of biology which we just begin to uncover.

Nicholas A. Kotov
University of Michigan
http://www.umkotov.com/profkotov/

Preface

Clay–polymer materials are important nano–micro composites with enormous possibilities for industrial scale up. Demand is rising for these materials on account of their excellent characteristics, including toughness, stiffness, dimensional stability, thermal resistance, chemical resistance, enhanced barrier properties and flexibility at high temperatures, and is expected to boost growth over the coming years. Nanoclay is a key factor in improving the material properties of common plastics such as polyolefins, engineering polymers and rubbers. Loading polymers with clay particles essentially increases the material's strength as well as the plasticity and elongation for elastomers and rubbers; however, only recently have such composites been prepared with control of the material's organization at the nanoscale. The design of organized clay–polymeric composites demands well defined clay nanoparticles. New methods of clay purification and techniques in surface modification have been developed to produce functional polymer–clay nanocomposites. The effectiveness of nanoclays in polymeric materials depends not only on the clay composition but also on the clay's shape and size. Clay nanoparticles have been prepared as sheets a hundred nanometers wide (such as kaolin, montmorillonite, bentonite and rectorite) or elongated one to two micrometer tubular and fibrous clays such as halloysite, imogolite and sepiolite.

Clay nanotubes may be loaded with anticorrosion, antioxidant, flame-retardant or antibacterial agents, and other chemically active compounds. Controlled release of chemical agents from nanotubes embedded in a polymeric matrix provides new long lasting properties for material composites, often called "smart" composites, with self-healing properties. The release time of chemical agents may be adjusted for weeks, months and years providing sustained functionality. Fibrous clays are particularly interesting for applications with biopolymers and other "green" polymers. These long

RSC Smart Materials No. 22
Functional Polymer Composites with Nanoclays
Edited by Yuri Lvov, Baochun Guo and Rawil F. Fakhrullin

Published by the Royal Society of Chemistry, www.rsc.org

needle-like structures imbue composites with greatly enhanced mechanical properties and can be functionalized similarly to tubular structures. Fibrous and tubular clays, in particular, are attracting particular interest in compounding with drugs, proteins and nucleic acids. Traditional applications of clay in the cosmetics industry can be extended to more efficient pharmacological uses, including controllable drug release and intracellular delivery of polymer–nanoclay drug formulations.

This book fills a gap in the description of nanoclay research by analyzing clay structures at the nano, micro and macro scales rather than the earlier approaches which were primarily concerned with the bulk properties of the clay. Clay is a natural compound and in most cases it is a biocompatible material, which means that clay materials pose little to no risk to the environment even at mass industrial applications. This 15-chapter book combines the efforts of US, Chinese and European authors and emphasizes eco-friendly "green" nanomaterials with sustained functionality based on natural clay composites. Special attention is devoted to nanosafety aspects which provide an optimistic prognosis for the mass application of nanoclay in household products, which has a historical background starting from porcelain and pottery production traditions.

Yuri Lvov, Baochun Guo and Rawil F. Fakhrullin

Contents

RSC Smart Materials No. 22
Functional Polymer Composites with Nanoclays
Edited by Yuri Lvov, Baochun Guo and Rawil F. Fakhrullin

Published by the Royal Society of Chemistry, www.rsc.org

CHAPTER 1

Functional Nanocomposites Based on Fibrous Clays

EDUARDO RUIZ-HITZKY*[a], MARGARITA DARDER[a], ANA C. S. ALCÂNTARA[b], BERND WICKLEIN[a] AND PILAR ARANDA[a]

[a]Instituto de Ciencia de Materiales de Madrid, CSIC, c/ Sor Juana Inés de la Cruz 3, 28049 Madrid, Spain; [b]Universidade Federal do Maranhão (UFMA), Departamento de Química (DEQUI), São Luís-MA, Brazil
*E-mail: eduardo@icmm.csic.es

1.1 Introduction

Since the beginning of the *polymer age*, silicates – like clays and other finely particulated solids such as silica, calcium carbonate and carbon black – have been incorporated as fillers at the micrometer dimension into plastics and elastomers with the aim of improving the mechanical and rheological properties of these polymers. Kaolinite was the clay mineral initially most widely used as a silicate filler of diverse polymeric matrices.[1] More recently, swelling clays such as smectites have received great interest due to their ability to exfoliate, giving rise to elemental silicate platelets of 1 nm thickness, which represent a way to develop fillers at the nanometer dimension (*nanofillers*). In this context, the concept of clay delamination encompassed by its high dispersion in polymers, introduced by Fukushima and other researchers at the Toyota Central Laboratory almost three decades ago,[2,3] represents a

RSC Smart Materials No. 22
Functional Polymer Composites with Nanoclays
Edited by Yuri Lvov, Baochun Guo and Rawil F. Fakhrullin

Published by the Royal Society of Chemistry, www.rsc.org

revolutionary idea, not only for the use of clays as reinforcing charges but also to introduce functionality in the resulting materials.

Very quickly, polymer–clay nanocomposites became a popular topic with a rapid increase of publications and registered patents and, in the last 20 years, about 13 000 articles and 200 patents have been published or registered, according to the ISI Web of Science data. Practically all these works make reference to polymer–clay nanocomposites belonging to the smectite layered clays family, *i.e.* 2 : 1 charged phyllosilicates, such as montmorillonite, hectorite and saponite natural clay minerals as well as analogous synthetic hectorites and fluoro-hectorites (*e.g.* LAPONITE® and the so-called "synthetic mica").[4] The smectite layers exhibit a high aspect ratio as each one is approximately 1 nm thick, while the diameter may rise to several microns or larger, as is the case with vermiculites. So, polymer–clay nanocomposites based on smectites have been extensively studied from basic aspects to applications, and a significant number of reviews have been published on this topic (see, for instance, ref. 5–11). Compared to smectites, other types of clay minerals, such as kaolinite, halloysite, imogolite, sepiolite and palygorskite, have barely been studied as nanofillers of polymers. However, as we report below in this chapter, the structural and textural characteristics of the fibrous clays could be of great benefit for the properties and applications of the novel polymer–clay nanocomposites derived from them.

The sepiolite and palygorskite fibrous clays also appear to be attractive nanofillers to reinforce polymer matrices.[12–19] They do not exhibit intercalation capacity, but these two natural silicates offer interesting characteristics, such as microporosity and large specific surface area. Interestingly, the presence of hydroxyls (silanol groups, Si–OH) at their external surface allows easy functionalization by controlled modification based on chemical reactions, *e.g.* with coupling agents or by anchoring of nanoparticles (NPs).[16,20] This behavior allows the preparation of a wide variety of polymer–clay nanocomposites provided with diverse functionality.

Sepiolite is a natural hydrated magnesium silicate with fibrous morphology, displaying a crystal structure consisting of talc-like ribbons arranged parallel to the fiber direction (*c*-axis) with $Si_{12}O_{30}(OH)_4(H_2O)_{24}\cdot 8H_2O$ as the ideal formula[21,22] (Figure 1.1A). To a variable extent, isomorphous substitution of magnesium at the octahedral layer by trivalent metals, mainly Al(III), provokes a charge deficiency in the structure that is compensated by extra-framework cations. This is the origin of the cation-exchange capacity (CEC) of sepiolite samples, which has been generally established in the 10–20 mEq 100 g^{-1} range, depending on the origin of the clay and always being about 4–5 times inferior to typical CEC values found in smectite clays. As often occurs in clays formed in lacustrine sediments, certain sepiolite samples show that hydroxyls located at the octahedral layers (mainly Mg–OH) can be partially substituted by fluorine.[23] One of the most interesting features of this silicate is the existence of microporosity ascribed, as occurring in zeolites, to their backbone structural arrangement – in this case with discontinuity of the phyllosilicate layers. This characteristic results in

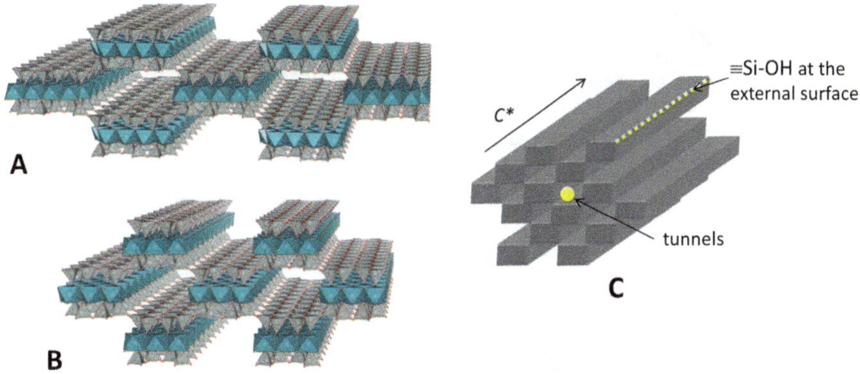

Figure 1.1 Schematic representation of the crystal structure (cross-section) of sepiolite (A) and palygorskite (B) fibrous clays, as well as a model of blocks arrangement along the *c** axis of the fiber (C).

an alternating distribution of structural blocks, each one composed of two tetrahedral silica sheets and a central sheet of magnesium hydroxide, which determine the presence of structural cavities (*tunnels*) that grow along the *c*-axis direction (Figure 1.1A). The cross-section of the sepiolite tunnels is 1.06×0.37 nm^2, which determines a structural microporosity typically superior to 0.3 cm^3 g^{-1} as determined from nitrogen adsorption isotherms.[24] Palygorskite (Figure 1.1B) is a related silicate with a higher content of aluminum and shorter dimensions of the structural blocks than sepiolite, with tunnel dimensions of 0.64×0.37 nm^2. The irregular arrangement of blocks and tunnels gives rise to fibers, as represented for instance in Figure 1.1C, and determines an elevated surface area in both silicates. Sepiolite and palygorskite exhibit BET (Brunauer–Emmett–Teller theory) specific surface area values in the order of 300 and 200 m^2 g^{-1}, respectively.[24]

The roughness of fibers at the nanometric scale in both silicates favors the interaction with their surroundings, which makes these materials excellent adsorbents for many applications in diverse sectors such as industry, agronomy and environmental remediation. The peculiar morphology and textural characteristics of sepiolite and palygorskite make these silicates also appropriate as fillers in polymer matrices. The fiber length of sepiolite and palygorskite clays depends on their origin: for instance, sepiolite from Taxus basin deposits in Spain are in the micrometer dimensions, whereas fibers from Finland and China can be much longer. The high aspect ratio[18,25] and the mechanical properties of sepiolite fibers[26] make these silicates useful as fillers for the reinforcement of polymers. The elastic properties of individual sepiolite fibers have been measured by atomic force microscopy (AFM) (Figure 1.2), obtaining values of the modulus in the order of 10 GPa in bending mode.[26] The abovementioned characteristics indicate that sepiolite and palygorskite could act as efficient fillers of polymers to improve their structural properties.

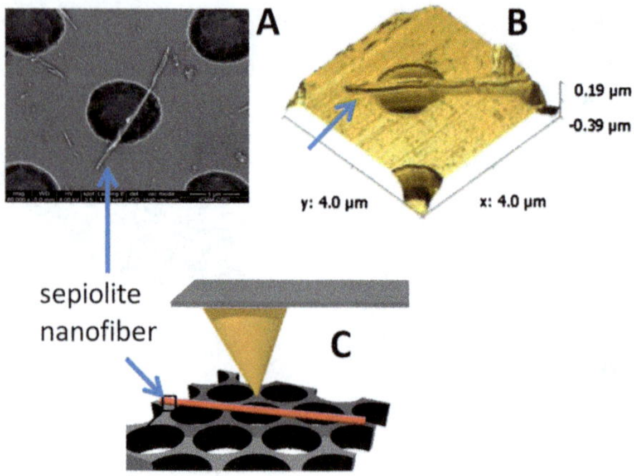

Figure 1.2 Determination of the elastic properties of a sepiolite nanofiber. (A) FE-SEM and (B) AFM images of sepiolite nanofiber prepared for measurements; (C) disposition of the AFM cantilever for deflections measurements.[26,29] Adapted from ref. 26 with permission from The Royal Society of Chemistry.

Fibrous clays are present in Nature as aggregates of the fibers, often forming interwoven bundles as observed by electronic microscopy techniques.[27] Disaggregation can be conducted under mechanical or sonomechanical treatment of the fiber aggregates in water dispersion, which after drying, results in xerogels that in turn produce highly viscous aqueous dispersions useful for many applications.[28] These processes only produce a small proportion of individualized fibers, and an entire defibrillation of sepiolite and palygorskite appears to be extremely difficult. However, it is evident that high disaggregation degrees can facilitate the fibrous clay dispersions into polymer matrices, leading to the corresponding clay nanocomposites.

The aim of this chapter is to show how sepiolite and palygorskite fibrous clays can be incorporated as efficient fillers into polymer matrices, not only to improve their mechanical properties but also to provide functionality to the resulting polymer-clay nanocomposites. To achieve this objective, diverse approaches may be needed as preliminary procedures, including the disaggregation of the bundles of fibers by physical methods. In some cases, as occurs in other types of clays used as nanofillers, it could also be necessary to modify the nature of the silicate surface in view of making it compatible with the polymer.

1.2 Modifications of Fibrous Clays for Use as Nanofillers

As indicated above, raw fibrous clays are present as aggregates of fibers or laths that form interwoven bundles, leading to rigid particles. Their disaggregation, based, for instance, on micronization processes conducted at the industrial level, represents an essential route for applications as fillers and rheological additives for water-based systems.[27] For many applications, such as those based on rheological properties, it is crucial that sepiolite or palygorskite are separated into individual fibers. This is the case for important uses such as drilling fluids, construction, asphalt and bitumen, paints and coatings, liquid animal feeds, mortars, fluid fertilizers or grease thickener.[27] The micronized fibrous clays, *e.g.* Pangel® commercialized by TOLSA S.A., are also useful as filler for rubber and provide reinforcing characteristics with polar polymers. This type of material is produced by a patented milling process of the fibrous clays using humidified samples, leading to defibrillation and producing the commercially called *sepiolite of rheological grade*, capable of producing very viscous suspensions at relatively low concentrations compared with other clays.[30] More recently, a new patented invention claims the replacement of the grinding step and wet disaggregation by treatments based on ultrasound irradiation.[28]

The high viscosity developed by sepiolite samples conveniently disaggregated in water can be exploited as a so-called *suspension capacity enhancer*[20] that can be applied in diverse systems. As an illustrative example, multiwalled carbon nanotubes (MWCNTs) can be maintained in aqueous suspension by assembly with sepiolite under ultrasound irradiation.[31,32] These carbon nanotube (CNT) dispersions are extraordinarily stable, avoiding settling over very long periods of time, greater than a year (Figure 1.3). The resulting sepiolite–MWCNT materials can be easily isolated from the dispersions giving rise to the so-called *hybrid buckypapers* – similar to self-supported films of MWCNTs alone.[31] These materials exhibit electrical conductivity as well as reinforcing ability when incorporated as fillers in polymers such as polyvinyl alcohol (PVA).[31] In this way, functionalization of sepiolite introduced by means of physical processes represents an option for the further production of functional polymer–clay nanocomposites, in this case related to conducting nanocomposites.

As the external surface of fibrous clays is covered by silanol groups (≡Si–OH), a useful approach to introduce chemical modifications is based on their reaction with coupling agents such as organosilanes and other reagents such as epoxides and isocyanates.[33] Organosilane coupling agents containing Si–X groups (X = OR, Cl) can react with the silanol groups; the organic part remains grafted to the mineral surface through very stable siloxane bridges (≡Si–O–Si≡).[34] In agreement with thermogravimetry (TG) and Differential Thermal Analysis (DTA) data, the grafted groups are very stable, being eliminated by heating at temperatures above 400 °C. In this way, the characteristic hydrophilic surface becomes organophilic and the fibrous clays can then be easily dispersed in low-polar polymers.[17] Sepiolite has been functionalized by treatment with silanes containing unsaturated or thiol groups, such as vinyl and methacryloxy or 3-propylmercapto, respectively, giving organic derivatives of sepiolite capable of further copolymerization reactions.[17] Surface modifications by treatment of sepiolite with alkyl and functional silanes in the form of aqueous gels have been recently reported.[35] Other reagents used to

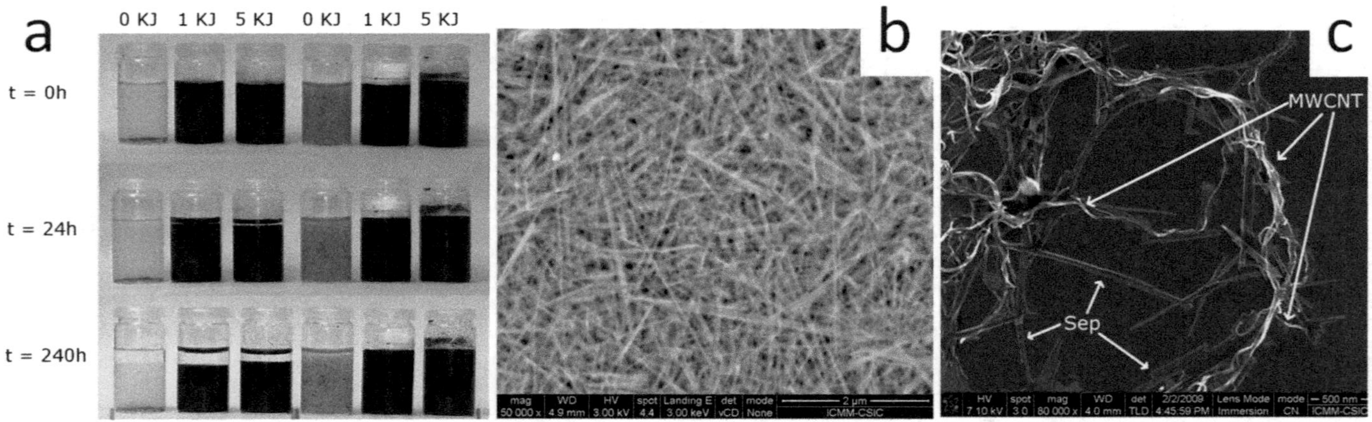

Figure 1.3 Simultaneous dispersion of sepiolite and MWCNTs in water using ultrasounds. (a) Sep–MWCNTs dispersion stability as function of time after sonication and amount of energy applied to the system; (b) FE-SEM and (c) TEM images of sepiolite MWCNT 5 : 1 dry residue (*hybrid buckypaper*) after filtration. (Adapted from *Carbon*, **72**, F. M. Fernandez and E. Ruiz-Kitzky, Assembling nanotubes and nanofibres: Cooperativeness in sepiolite-carbon nanotube materials, 296–303, Copyright 2014 with permission from Elsevier.[31])

functionalize sepiolite by grafting reactions contain epoxide and isocyanate groups, which react with silanol groups, remaining covalently attached to the silicate through ≡Si–O–C≡ bonds and showing lower stability than siloxane bridges.[33] However, the most usual approach to modify the external surface of fibrous clays to reduce their elevated polarity is based on treatments with alkylammonium salts, leading to fillers showing good compatibility with low-polar polymer matrices.

Organoclays derived from sepiolite and palygorskite can be easily prepared by treatment with neutral or cationic surfactants in aqueous solution. In this last case, treatment with long-chain alkylammonium salts[36,37] following an ion-exchange process is carried out in a similar way to that in montmorillonite and other smectite clay minerals. These organically modified fibrous clays are attaining great importance because they are commercialized not only for applications as nanofillers of polymers, like in epoxy systems and plastisols, but also in solvent-based paints, greases, solvent asphalt coatings, inks, foundry washes and adhesives.[20]

New possibilities to modify sepiolite and palygorskite in view of their use as fillers for diverse polymer matrices are still open and novel developments in this sense may find applications in the future. One of them consists of the encapsulation of molecular species into the structural cavities of nanometric dimensions grown along the silicate fibers (*tunnels*), and the other one is based on the assembly of diverse NPs to the external surface of those clays. The presence of tunnels allows the access of small molecular species into the interior of the silicate by replacing the *zeolitic water* that is usually filling these structural cavities and is reversibly adsorbed and desorbed by heating or by vacuum exposure.[24] Organic dyes such as methylene blue can penetrate at least partially into the tunnels of sepiolite and palygorskite.[24,38] In fact, several centuries ago Maya people in Mesoamerica developed a technique to encapsulate indigo into palygorskite clay, giving rise to a pigment known as Maya Blue.[39] It was used during the pre-Spanish period and later-on; its use was continued in Spain till the 17th century and until the 19th century in Cuba, where it was known as Havana Blue.[40,41] This clay-dye hybrid material shows a remarkable stability against weathering and microbiological activity, also being resistant to thermal treatments and solvent extractions, attributed to the encapsulation of the dye molecules inside the silicate tunnels.[42] Due to this quality, the existence of a strong interaction between the dye and the host silicate could be admitted, giving rise to the high stability of the Maya Blue pigment.[41–44] Maya Blue analogs prepared by adsorption of molecular dyes such as methylene blue (MB) and methyl red (MR) by both sepiolite and palygorskite have been used for coloring polymer matrices, opening the way to introduce stable colors in polymer–clay nanocomposites.[45] More recently, Ouellet-Plamondon *et al.*[46] reported the functionalization of inorganic polymers (geopolymers) using sepiolite-dye hybrids inspired by Maya Blue. In this case, it is also confirmed that the encapsulation of MB and MR dyes into sepiolite leads to the stability of the pigment in the geopolymer matrix despite the chemical aggressivity of that system in showing strong alkalinity. The protection of the colors towards light and external reagents such as hydrogen peroxide and acids has been clearly shown in the resulting materials.[46]

As indicated above, another possibility for introducing functionality in fibrous clays used as polymer fillers is based on the immobilization of diverse types of NPs that, in some cases, can be anchored on the silicate surface by interaction with the external silanol groups.[16] This method could represent an excellent opportunity for the introduction of specific characteristics in fibrous clays-based nanocomposites such as antimicrobial activity, for instance by incorporation of Ag NPs,[16] superparamagnetic properties, for instance by introduction of magnetite–maghemite NPs,[47,48] or photoactivity by assembly of TiO_2 NPs.[49,50]

1.3 Functional Polymer–Fibrous Clay Nanocomposites

Research on polymer–clay nanocomposites is a discipline with a continuous increase in the number of publications since they were reported for the first time around 25 years ago.[2] For a long time, the majority of the publications were related to smectite-based nanocomposites, however in recent years, the use of fibrous clays, sepiolite and palygorskite, has attracted increasing interest. Since 2014 until September 2015 more than 150 publications have been reported in the Web of Science, searched using "(sepiolite* or palygorskite* or attapulgite*) and nanocomposite*" for a total number of around 670 entries.[51] The increasing interest in this type of nanocomposite may be related to the potentiality that the incorporation of fibrous clays offers in comparison to layered clays from the point of view of functionality. Actually, the most part of the interest in polymer–smectite-based nanocomposites has been directed to the improvement of mechanical properties of polymers. Additionally, the presence of delaminated clay nanosheets results of interest in the improvement of other properties such as the reduction of gas diffusion, as they introduce more tortuous paths which work as a barrier for gases. Hence, for a long time, fibrous clays have scarcely been employed for those purposes, but they may prove interesting for introducing other types of properties and functionalities.

In this section, diverse examples will be introduced in order to illustrate different synthetic polymer–fibrous clay nanocomposites with special emphasis on their implications for developing functional materials. In this way, the use of fibrous clays as nanofillers of thermoplastics and other conventional polymers has been explored as their high aspect ratio and specific surface area may prove favorable for the improvement of certain properties when used instead of layered clays. In this case, the application of specific strategies to overcome the high hydrophilicity of these clay minerals must be considered to promote adequate interfaces with conventional polymers such as polyethylene (PE) or polypropylene (PP).[16–18] In some cases the clay is modified just to render the interface organophilic and hence, compatible, but others preferred to use approaches involving the assembly of monomeric species that could be further used in copolymerization reactions. As mentioned above, more recent advances try to incorporate additional functionalities, for instance by combining two or more types of polymers, addition of CNTs or other NPs, for instance magnetite, that may lead to new applications. Anyway, from a practical point of view, advances in this field may consider that the synthesis of nanocomposites based on those polymers should be integrated in processes that are already well established for large scale production.

1.3.1 Functional Nanocomposites Based on Conventional Polymers

As in the case of nanocomposites based on layered silicates and highly hydrophobic thermoplastics such as PE, PP, polybutylene (PB), polystyrene (PS), polyethylene terephthalate (PET), polyvinyl chloride (PVC), polymethylmethacrylate (PMMA) or polyamides (*e.g.* Nylon®, polyamide 6 (PA6),...), the use of fibrous clays also requires their modification to procure compatibility between the components. As mentioned above, the specific reactivity of the external surface of palygorskite and sepiolite fibrous clays facilitates the incorporation of organophilicity, as well as other particular functionalities, to improve the interactions at the polymer–clay interface region, which in general drives to enhanced thermal and mechanical properties.[16,17] Moreover, such reactivity is currently also exploited to achieve the integration of other NPs or functionalities, making the development of new functional plastics possible.

Polyamides such as PA6 have been frequently reinforced with sepiolite or palygorskite directly[52] or after modification with organosilanes.[53] The fibrous shape of these nanofillers may contribute to enhancing mechanical properties during their processing. For instance, by controlling experimental conditions during melt spinning of fibers of PA6 reinforced with palygorskite modified with *N-b*-aminoethyl-caminopropyltrimethoxy silane it is possible to favor the orientation of the nanofillers in the drawing direction, resulting in improved tensile properties.[54] The presence of the fibrous clay enhanced the effect of fire retardant agents, *e.g.* the aluminum salt of diethylphosphinic acid (AlPi), as well as the thermal stability and char yield of PA66 nanocomposites incorporating sepiolite.[55] In this case, the presence of sepiolite shows an effective role favoring the formation of a compact char layer during burning (Figure 1.4). PA6–fibrous clay nanocomposites have also been prepared by *in situ* polymerization in the presence of the clay modified with appropriate organosilanes.[56] This methodology can be used to prepare, in two steps, nanocomposites containing two types of polymers, for instance PA6 and poly(ethylene glycol) (PEG) reinforced with palygorskite. Thus, PA6–palygorskite was firstly *in situ* polymerized and then combined with PEG to produce nanocomposites of PA6–PEG block-copolymers with enhanced

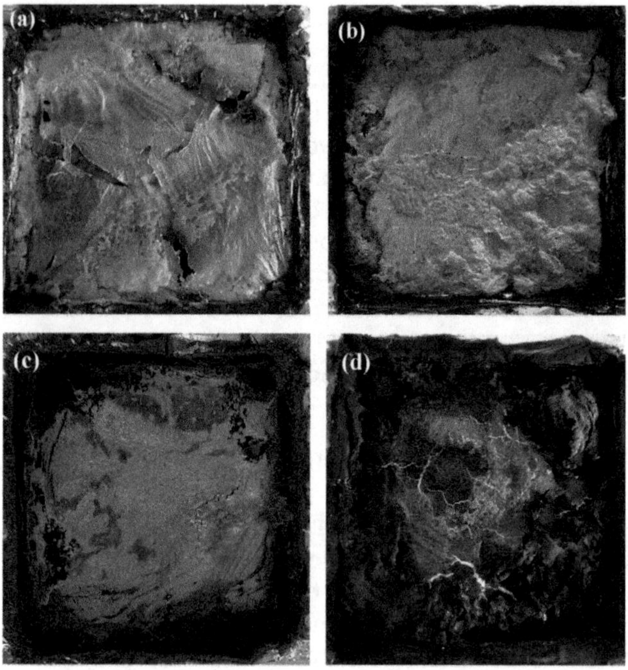

Figure 1.4 Photographs of (a) pure PA66, and various PA66/sepiolite nanocomposites containing (b) A66/0.3 wt% sepiolite, (c) PA66/14 wt% AlPi and (d) PA66/9.7 wt% AlPi/0.3 wt% sepiolite after cone calorimeter tests. (Reprinted from *Polymer Degradation and Stability*, **117**, Z. Zhan *et al.*, Synergistic effects of sepiolite on the flame retardant properties and thermal degradation behaviors of polyamide 66–aluminum diethylphosphinate composites, Copyright (2015), with permission from Elsevier.[55])

heat-resistant and flame retardant properties by conveniently adjusting the ratio of the components.[57]

In the case of polyolefins, such as PE, the presence of sepiolite or palygorskite contributes to reducing thermo-oxidation processes and thus improves the thermostability and mechanical properties.[58] In general, the processing methodology may affect the final characteristics of the resulting nanocomposites. Dynamic packing injection has been proposed over conventional injection molding for favoring the dispersion of the nanofiller in the preparation of PE–palygorskite nanocomposites.[59] The way in which the clay is modified also affects the final properties of the nanocomposites. Thus, for instance, modification of sepiolite with vinyl triethoxy silane before (*ex situ*) or during (*in situ*) heat mixing produces, in the latter case, nanocomposites with improved thermal stability and tensile strength and modulus but reduced elongation at break.[60] Specific catalysts have also been supported on the clay to act during the polymerization in the presence of the nanofiller.[61] When the co-catalyst methylaluminoxane (MAO) is directly grafted on sepiolite, the process maximizes the interaction between the matrix and the nanofiller, which leads to PE–sepiolite nanocomposites that show better performance than the ones obtained by melt intercalation.[62] In the same way, it is possible to graft other reactive species to the fibrous clays, such as octadecyl acrylate, which can react, for instance, with atactic PP producing nanocomposites with improved mechanical properties.[63] Palygorskite modified by grafting reaction with aminopropyltrimethoxysilane (APTMS) and treated with ethylene octene elastomer-grafted maleic anhydride (POE-*g*-MAH) was used to prepare PP nanocomposites in the presence of metallocene linear low density PE (*m*-LLDPE). Interestingly, the resulting PP–*m*-LLDPE nanocomposites present enhanced ultraviolet (UV) photoresistance with little change in their mechanical properties.[64] The incorporation of flame retardant additives (*e.g.* magnesium hydroxide) and vinyltriethoxysilane as a crosslinking agent in PE–sepiolite nanocomposites results not only in improved flame retardancy but also thermal stability and mechanical properties of the resulting systems.[65,66] Thus, PP–sepiolite nanocomposites prepared from organosepiolites containing cethyltrimethylammonium ions and zinc borate[67] or from sepiolite and MWCNTs,[68] present improved fire retardancy properties ascribed to the superior char formation. The filler may introduce other functionalities to the nanocomposites, thus PE–sepiolite and PS–sepiolite nanocomposites showing plasmonic properties have been prepared by previous assembly of Au and Ag NPs to sepiolite.[69] The presence of FeCo NPs associated to a sepiolite that is treated with methyltrimethoxysilane results in a functional nanofiller for PS-based nanocomposites that show magneto-optical Faraday activity.[70] Similar properties can be achieved when using sepiolite to which γ-Fe_2O_3 NPs of around 10–30 nm were previously associated.[71] Magnetic PS–palygorskite nanocomposites have been prepared by *in situ* polymerization of styrene in the presence of a palygorskite, to which magnetite NPs were previously assembled to avoid their aggregation.[72]

Sepiolite has also been employed as nanofiller in the production of ethylene-vinyl acetate (EVA) with improved mechanical and thermal properties, including flame retardancy, due to the presence of the clay.[73] The use of sepiolite modified by APTMS in EVA allows the preparation of reinforced plastics that show interesting properties as membranes for gas separation – with an increase in the permeability of O_2, CO_2, CH_4, and N_2 gases with just the addition of 3 wt% of sepiolite and the CO_2/CH_4 and CO_2/N_2 selectivity also being improved.[74] Sepiolite modified with 3-aminopropyltriethoxysilane (APTES) has recently been explored as a nanofiller of polynorbornene matrices, resulting in hybrid nanocomposite films with enhanced mechanical and oxygen barrier properties, which may be of interest in packaging applications.[75]

Fibrous clay has also been employed to produce thermoplastic PMMA-based nanocomposites. In general, a previous modification of the clay surface is necessary, for instance with cationic surfactants[76] or isocyanates,[77] to improve their compatibility. In some cases, the modifier may act as a crosslinking agent during the melt-compounding process.[76] Most frequently PMMA is prepared by *in situ* polymerization using different approaches.[78,79] The clay modification by grafting specific groups, for instance by treatment with mercaptopropyltrimethoxysilane (MPTMS), favors their ulterior coupling with *N,N*-dimethylaminoethyl methacrylate (DMAEMA) inducing copolymerization *via* Ce(IV) induced redox polymerization with reactive thiol groups.[80] In the same way, Cu(0)-mediated radical polymerization of DMAEMA has been coupled to a palygorskite containing 2-bromoisobutyryl groups.[81] Interestingly, the resulting poly(2-(diethylamino)ethyl methacrylate) (PDEAEMA)–palygorskite nanocomposite particles can be used as recyclable Pickering emulsifiers responsive to pH (Figure 1.5). Moreover, the possibility of using methacrylate monomers provided with two sites of polymerization (*e.g.* ethylene glycol dimethacrylate) allows for the incorporation of additional functionalities to the final nanocomposite. For instance, their combination with organoclays produces nanocomposites with interesting properties for oil adsorption.[82] In other cases, copolymerization of functional monomers such as 2-(2-methoxyethoxy) ethyl methacrylate, oligo (ethylene glycol) methacrylate and acrylic acid in the presence of palygorskite results in the formation of stimuli-responsive nanocomposite hydrogels.[83] Further functionalization can be reached when magnetite NPs are previously assembled to palygorskite, in this case rendering magnetic hydrogels with improved mechanical properties and sensitivity to pH and temperature changes.[84,85]

Another relevant group of polymer–clay nanocomposites involves epoxy, polyurethane (PU) and other thermosetting resins of importance in construction and aviation, amongst other areas of application. In those nanocomposites, the clay is incorporated mainly to procure mechanical properties.[86] The use of fibrous clays as nanofillers plays a double role as,

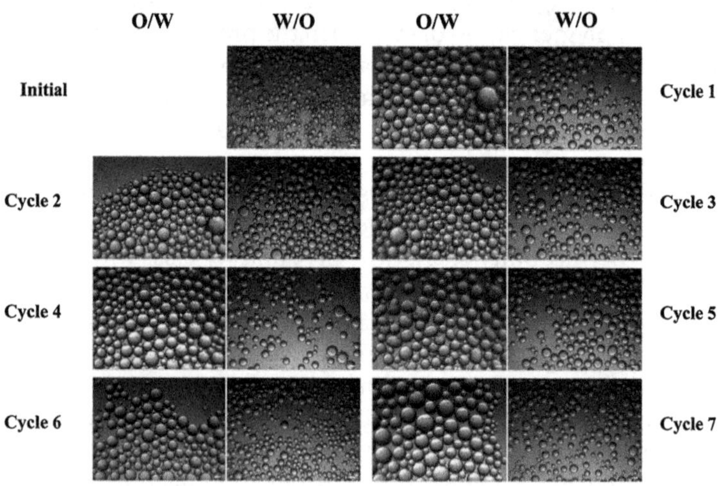

Figure 1.5 Optical micrographs of PDEAEMA–palygorskite nanocomposite emulsions used as recyclable Pickering emulsifiers tested in cycles switched by reversing pH. (Reproduced from ref. 81 with permission from The Royal Society of Chemistry.)

besides improvement of mechanical and thermal properties, they may cause crosslinking effects that affect the mobility of the polymer chains.[16] Thus, for instance, the incorporation of commercial organosepiolite in glassy bisphenol A-based epoxy resin has only a slight effect on the thermal stability of the polymer but enhances its storage modulus in flexion.[87] In the same way, epoxy resins filled with palygorskite, either pristine or modified with cetyltrimethylammonium ions and APTES, present a better distribution of the nanofiller as well as stronger interfacial adhesion when using the modified clay.[88] The presence of the clay improves the thermal and mechanical properties of the resulting nanocomposites, although in the case of the nanofiller that incorporates grafted amino groups, the degradation of these properties is strongly affected by the loading, where a maximum effect is reached at around 2 wt%. When the nanofiller is modified with more specific agents, for instance glycidyl silane, it is possible to get better toughness although the concentration of the nanofiller has to be kept below certain concentrations to avoid clay agglomeration that may compromise the final mechanical behavior.[89] Clearly the rheology of sepiolite–epoxy systems is affected by the type of modification introduced to the clay to achieve a better compatibility between the two components, and in general, only physical gels are formed when adding modified clays.[90] The use of modifiers containing epoxy functionalities, for instance prepared from surfactants of natural origin, procures sites for reaction with the epoxy monomers favoring compatibility between the two phases, which results in enhanced properties in the final nanocomposites.[91] Alternative approaches imply the modification of palygorskite with acrylic resins and its further compatibilization with the epoxy resin to reach more efficient interactions that result in remarkable enhancement in the impact strength, Young's modulus and the tensile strength relative to the neat epoxy resin.[92] In some cases, organosepiolites have been modified with SiO_2 NPs revealing a more effective adhesion between the two components, probably due to the existence of stronger interactions between the nanofiller and the polymer in the presence of the interfacial agent.[93] Moreover, the use of a magnetic sepiolite, prepared by chemical coupling of aminosilane-functionalized sepiolite with epoxysilane-functionalized magnetite, allows for the preparation of epoxy resin systems with a reinforcing effect ascribed to the alignment of the magnetic nanofiller in the presence of a magnetic field.[94] Most sophisticated systems involve the preparation of epoxy-based nanocomposites incorporating palygorskite and graphite oxide NPs under ultrasound irradiation, in which the presence of both types of NPs results in synergies with improved properties compared to nanocomposites containing only one type of NP.[95]

In the same way, most research related to PU–fibrous clay nanocomposites is focused on the improvement of thermal properties due to the necessity of fulfilling the regulations for applications in the construction and building sector.[96–100] There is recent interest in exploring the use of the so-called water-borne polyurethane (WPU); a non-toxic, nonflammable and non-pollutant polymer. However, it is insoluble and shows lower thermal stability

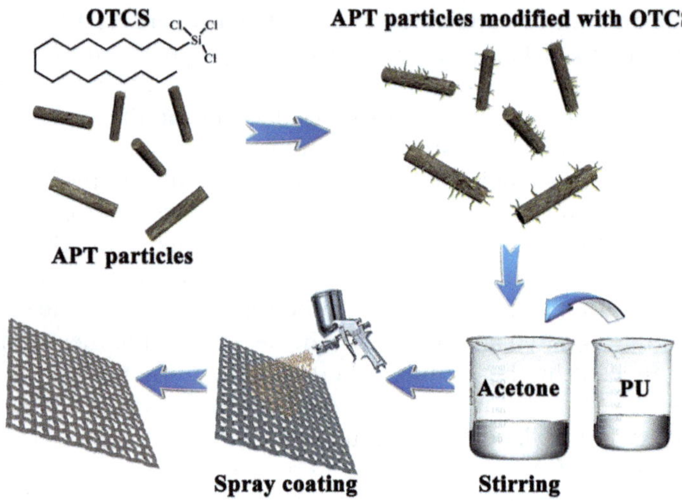

Figure 1.6 Scheme of the preparation method of superhydrophobic metal meshes modified with WPU–palygorskite nanocomposite coatings for removal of oils from water. (Reproduced from ref. 103 with permission from The Royal Society of Chemistry.)

and mechanical properties than other PU. In this way, the incorporation of nanofillers such as palygorskite to this system has been explored with the view of producing nanocomposites with improved mechanical properties.[101] The incorporation of other types of NPs, such as carbon black, results in nanocomposites that may also show electrical conductivity.[102] WPU combined with palygorskite modified with silanes providing organophilic groups (*e.g.* octadecyltrichlorosilane, OTCS) has been used to modify stainless steel meshes (Figure 1.6) and provide them with superhydrophobic and superoleophilic properties.[103] Interestingly, they can be used to separate diverse oil–water mixtures (*e.g.* kerosene, chloroform and petroleum) with efficiencies as high as for the kerosene–water system. The application of such coatings to Cu-based meshes results in superoleophobic systems with efficiencies over 99.0% in the removal of kerosene and showing stable recyclability after 50 separation cycles.[104] The coating of the mesh may be prepared from other palygorskite-based nanocomposites, for instance epoxy resins, allowing for the separation of diverse oil–water mixtures, containing for instance hexadecane, lubrication oil or paroline, that can be filtered using the coated mesh with a separation efficiency above 98%.[105]

Another relevant group of functional fibrous clay nanocomposites based on conventional polymers are based on certain hydrophilic polymers such as PVA, polyacrylic acid (PAA) or polyacrylamide (PAAm) related polymers. Since these polymers, as well as their related monomers, are hydrophilic, they can be easily associated with sepiolite and palygorskite by interaction with the silanol groups located at the external surface of the clay. In this way, the simple mixture of the sepiolite with PVA results in the formation of the nanocomposite.[106] Aqueous suspensions of them alone or incorporating other biocompatible polymers, such as PAA, can be directly submitted to freeze-drying to produce macroporous nanocomposites that can be explored as scaffolds in tissue regeneration applications.[107] However, the majority of the research related to nanocomposites generated by polymerization (or copolymerization) of acrylic acid and/or acrylamide and other related monomers incorporating sepiolite or palygorskite intends to develop superadsorbent materials. The strong hydrophilic character of fibrous clays favors their dispersion in hydrogels and imparts greater consistency to the generated systems, which results in improved water uptake properties, relevant in agriculture and horticulture, sanitary goods, drug-delivery systems or water remediation. Besides applications as water superabsorbents PAAm-based nanocomposites can be used in the removal of pollutants in water such as Pb(II) and Cu(II)[108,109] or dyes (*e.g.* methylene blue).[110] The preparation of these nanocomposites usually employs crosslinking agents (*e.g.* *N*,*N*′-methylenebisacrylamide) that in the presence of an initiator (*e.g.* ammonium persulfate) produces the growth of PAA, PAAm or PAA–PAAm copolymers that become bonded to the silanol groups at the external surface of sepiolite and palygorskite.[111,112] The final water absorbency properties of the developed nanocomposites are strongly dependent on the amount of clay, cross-linker content, initiator dosage, nature of monomer or ratio of monomers in AA–AAm copolymer systems amongst other parameters of synthesis.[112–114] The use of modified clays has also been tested, for instance palygorskites treated with APTMS and methacryloxypropyltrimethoxysilane (MAPTMS) were simultaneously incorporated to AA, where the first clay-hybrid (APTMS–palygorskite) acted as initiator and the latter MAPTMS–palygorskite hybrid acted as crosslinker in the polymerization reaction.[115] Microgels are obtained when the polymerization of organopalygorskites, *e.g.* modified with APTMS, occurs in the presence of an emulsifier and paraffin.[108] The presence of vinyl groups in certain organopalygorskites can be exploited to copolymerize with 2-acrylamido-2-methylpropane-sulfonic acid, giving rise to the formation of a nanocomposite network in which acrylamide can be further polymerized to produce a double network hydrogel with improved mechanical properties.[116]

1.3.2 Polymer–Clay Nanocomposites with Electrical Conductivity

The development of polymer–clay nanocomposites equipped with electrical and ionic conductivity was probably one of the first research fields regarding the incorporation of functionality with implications in the development of electrochemical and electroanalytical applications.[117] Although smectites have mainly been used in the preparation of such nanocomposites, diverse examples of systems based on sepiolite and palygorskite fibrous clays have emerged lately. In the case of smectites, the presence of conducting polymers between the clay layers improves their stability and often determines anisotropy in the observed electrical behavior.[117,118] Thus, this section will focus on the description of the main features and the latest advances related to polymer–fibrous clay nanocomposites exhibiting ionic and/or electrical conductivity.

Polyaniline (PANI), polypyrrole (PPy) or poly(3,4-ethylenedioxythiophene) (PEDOT) are typical electron conducting polymers exhibiting characteristic semiconductor behavior. Although the majority of them are practically insoluble in most common solvents, they can be easily prepared from their corresponding unsaturated monomers (*e.g.* aniline, pyrrole, thiophene, isoprene...), which has been exploited to produce diverse examples of conducting polymer–fibrous clay nanocomposites.[119,120] The small size of these monomers may result in their incorporation into the structural tunnels of the clay, which may then act as templating pores for the growth of single chains of the corresponding conducting polymer. Thus, the formation of PPy–sepiolite nanocomposites by *in situ* polymerization of pyrrole has been reported, which after doping with I_2 vapors, introduces conducting properties to the nanocomposites. This conductivity has been explained by considering polaron and bipolaron mechanisms on the basis of ultraviolet-visible (UV-vis) and *electron paramagnetic resonance* (EPR) evidence.[121] In the case of palygorskite, PPy polymerization in the interior of the tunnels does not seem as evident but PPy–palygorskite nanocomposites containing rhodamine B (RhB) have been reported.[122] The resulting nanocomposites show higher conductivity values and better electrochemical behavior when the RhB dye is incorporated, which has been interpreted considering the existence of π–π interactions and stacking between RhB molecules and the pyrrole rings.[122] This effect should be exploited and so the incorporation of chromophores in other PPy–fibrous clay nanocomposites could lead to the development of, for instance, novel photoconductor materials. PPy can also be *in situ* polymerized at the external surface of sepiolite fibers modified with alkylammonium-based surfactants (*e.g.* CTAB), giving rise to core–shell structures in which PPy encapsulates the clay fibers.[123] Interestingly, these nanocomposites exhibit sufficient conductivity, as well as good electrical double-layer capacitance and specific capacitance, to be explored as an alternative to pure PPy electrodes for supercapacitor devices.[123] Organopalygorskites modified with hexadecylpyridinium ions have also been used to produce PPy–palygorskite nanocomposites in which the conducting polymer encapsulates the clay fibers (Figure 1.7). These nanocomposites exhibit excellent electrical conductivity values (around 85 S cm^{-1}) as well as a higher thermal degradation temperature (>240 °C) after doping with HCl, which could be of interest in view of enlarging the field of applications of this type of nanocomposite to anti-corrosion coatings, shielding of electromagnetic interferences and other areas.[124]

PPy–fibrous clay nanocomposites have also been evaluated as adsorbents in the removal of anionic pollutants, such as chromate ions, from industrial wastewater.[125,126] This property seems to be ascribed to the ion-exchange ability of the formed polymer, which after doping, results in the presence of PPy^+X^- sites at the external surface of the clay. Alternative approaches

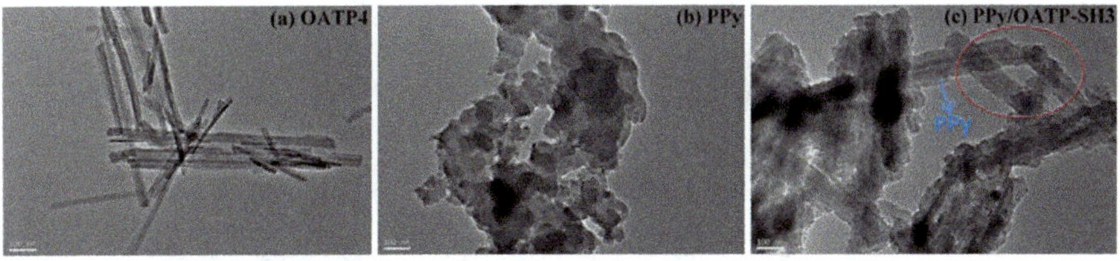

Figure 1.7 TEM images of (a) OATP4 (organosepiolite modified with hexadecylpyridinium ions), (b) PPy and (c) PPy–OATP-SH3 material corresponding to an organopalygorskite–PPy nanocomposite prepared from a 1/0.8 $m_{organosepiolite}/m_{Py}$ mass ratio. (Reproduced from ref. 124 with permission from John Wiley and Sons. Copyright © 2014 Wiley Periodicals, Inc.)

intend to profit from the possibility of different associated active species on the fibrous clay that are later assembled to *in situ* electropolymerized PPy to produce drug delivery systems. In this way, PPy nanocomposites incorporating aspirin-loaded palygorskite were tested to produce a controlled aspirin release by using an external electrical stimulus.[127]

PANI–fibrous clay nanocomposites have been prepared using diverse experimental approaches with the polymer assembled only on the external surface of palygorskite and sepiolite.[128,129] Here again, the role of the clay interface influences the growth and characteristics of the formed polymer. Thus, palygorskite submitted to an acid treatment to generate fresh silanol groups before modification with APTES notably improves the assembly of PANI, favored by the presence of the aminopropyl functions.[128] The electrical conductivity of these palygorskite–PANI nanocomposites may reach values of around 2.2 S cm^{-1}, which is higher than that of the pure PANI prepared under analogous experimental conditions. This result points to a favorable effect of the clay interface that favors the growth of the polymer in a more favorable conformation leading to PANI chains with a greater conjugation length, which results in enhanced conductivity of the nanocomposite.[128] Polymerization of PANI in the presence of sepiolite slurry results in nanocomposites in which PANI is adhered to the sepiolite surface, giving rise to interesting electrorheological properties in systems containing up to 50% of conducting polymer.[129] This type of conducting PANI–sepiolite nanocomposite has been explored due to its potential interest as an electrorheological system.[130] The electrorheological behavior of pure polyaniline compared to that of sepiolite coated with polyaniline has been ascribed to a higher aspect ratio of sepiolite fibers,[131] which can be used as fillers to reinforce, for instance, as an alternative to the incorporation of magnetite-clay into epoxy resins with the aim of producing a specific orientation of the fibers before polymerization.[94] Poly(2,3-dimethylaniline) has been grown in the presence of palygorskite previously modified with APTES to produce nanocomposites that have been postulated as self-healing coatings with improved corrosion resistance.[132] In the same way, palygorskite modified with diverse types of cationic surfactants has been explored as a nanofiller of PANI–PVA based nanocomposite

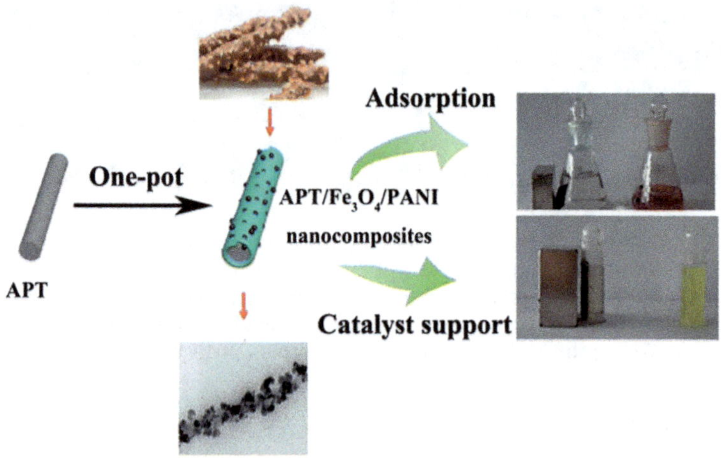

Figure 1.8 Scheme showing the approach for the preparation of the superparamagnetic PANI–Fe_3O_4–palygorskite nanocomposites from palygorskite (APT) fibers that are covered with Fe_3O_4 nanoparticles (APT–Fe_3O_4–PANI), whose aspect observed by TEM is comparable to a peanut chocolate bar structure. Pictures at the right show their applicability as an adsorbent and a catalyst. (Reproduced from ref. 134 with permission from The Royal Society of Chemistry.)

films. It was observed that the type of clay modifier strongly influences the conducting properties of the resulting materials.[133] Multifunctional nanocomposites have been prepared in a one-pot process using Fe(III) salts that act as a precursor to Fe_3O_4 NP formation and oxidant for aniline polymerization in the presence of attapulgite.[134] The morphology (Figure 1.8) and magnetic properties of the nanocomposites can be tuned by controlling the molar ratio of aniline to Fe(III). These superparamagnetic PANI–Fe_3O_4–palygorskite nanocomposites can be tested as adsorbents for the removal of dyes and the enrichment of Au(III) from the solution, with the possibility of being easily removed from the solution with the help of a magnet (Figure 1.8). Interestingly, these nanocomposites also exhibit catalytic activity toward the catalytic reduction of 4-nitrophenol in the presence of $NaBH_4$.[134]

Other polymers of great interest in electrochemical applications are those related to polyelectrolytes such as poly(ethylene oxide) (PEO) and nanocomposites based on layered clays that have been studied for a long time in view of developing ion-conductors for application as solid electrolytes of solid state batteries.[117,118,135–138] In those cases, the 2D structure of the inorganic host acted as the counter-ion, where the interlayer cations associated to the intercalated polymer are the only mobile species, *i.e.*, the transport number of anions is practically zero ($t- \approx 0$) in the resulting nanocomposites.[117,139] PEO has also been assembled to sepiolite, either from solution in acetonitrile or from melt by MW (microwave) irradiation, giving rise to nanocomposites in which a partial penetration of the polymer chains takes place.[24,117]

More recently, diverse studies have been reported focusing on aspects such as the control and modulation of the silicate interphase in order to understand diverse physical characteristics of PEO–sepiolite nanocomposites.[140] Sepiolite modified with various types of PEG and other additives such as vitamin E, tocopherol or polyethylene glycol succinate (TPGS) has been tested as a mechanical reinforcer of PEO in the preparation of ethylene carbonate (EC) containing PEG and PEO blends, together with lithium triflate (LiTf) salt, in the preparation of liquid and solid nanocomposite polymer electrolytes.[141] From this study, it is established that TPGS–sepiolite disperses better in EC–PEG mixtures, being completely stable for weeks or even longer periods of time, in contrast to neat sepiolite or PEG–sepiolite which sediment progressively under the same conditions. Sepiolite-based nanocomposites prepared with an EC : PEO weight ratio of around 1 present ionic conductivity close to 10^{-3} S cm^{-1} at 30 °C, which is in the order of that reported for liquid electrolytes.[142] Moreover, ionic conductivity on these nanocomposites varies in the range of 10^{-6} to 10^{-4} S cm^{-1} at 30 °C with increasing concentration of EC and LiTf, being close to that of liquid polyelectrolytes, while they show solid-like mechanical properties above the melting point of PEO.[143] This combination of properties appears very promising in view of developing new clay-based solid polymer electrolytes.

Nanocomposites based on fibrous clay showing proton conductivity have also been explored as conducting membranes for fuel-cell applications. For this purpose, Nafion® membranes containing neat sepiolite or sepiolite modified with sulfonic groups have been tested and showed enhanced properties when compared to the pure polymer, especially when working at low relative humidity and high temperature.[144] Conducting membranes based on sulfonated poly(ether sulfone) containing palygorskite and phosphotungstic acid have also been evaluated for application in direct methanol fuel cells (DMFCs).[145] Membranes containing 10% of palygorskite show very attractive properties, reducing methanol permeability by up to 25% that of Nafion® membranes, with proton conductivity values of around 3×10^{-2} S cm^{-1}, although a drop of the tensile strength has been observed. Anyway, the study confirms that mechanical properties still remain adequate for palygorskite loadings up to 15%.[145] Palygorskite modified with quaternary ammonium has been assembled to bromide polyphenyl ether (PPO) to produce nanocomposite membranes for application in alkaline fuel cells. These membranes exhibit anion-exchange properties (1.07 meq g^{-1}), good thermal and chemical stability as well as excellent mechanical properties in the hydrated state, with a conductivity ranging from 1.6×10^{-2} S cm^{-1} at 20 °C to 2.1×10^{-2} S cm^{-1} at 80 °C.[146]

Certain polymer–fibrous clay nanocomposites have been explored as a precursor of conducting materials by controlled thermal transformation of the involved polymer. For instance, polyacrylonitrile (PAN) produced inside the sepiolite tunnels by *in situ* polymerization of acrylonitrile monomers can be further transformed into a conducting carbon by treatment at temperatures of around 700 °C in an inert atmosphere.[147,148]

The produced carbon–sepiolite nanocomposites show electrical conductivity that has been attributed to the presence of graphene-like species that remain strongly associated with the silicate substrate, similar to that occurring to PAN intercalated into montmorillonite.[149] The formed carbonaceous material can be removed from the nanocomposite by its successive treatment with HCl and HF acids, although the carbon–sepiolite nanocomposites can be directly employed as an electrode in secondary battery electrodes and supercapacitor electrochemical devices.[117,150] Interestingly, the carbon recovered from sepiolite-based nanocomposites, compared to that templated on smectites, shows better properties for being used as an electrode in rechargeable Li-batteries.[151]

1.4 Functional Biopolymer–Fibrous Clay Nanocomposites

Fillers are introduced in a continuous matrix with the aim of improving a given property of a material.[152] Relevant examples of this behavior are the improvement of mechanical properties, the increase of thermal stability or the flame retardancy, and the enhancement of the gas–vapor barrier properties of biopolymers. Such properties, together with the biodegradable and biocompatible character of biopolymer compounds, make these bionanocomposites based on sepiolite and palygorskite susceptible materials for diverse applications, *e.g.* drug delivery, environmental remediation, bioplastics, packaging and other food contact surfaces.

1.4.1 Polylactic Acid and Other Biodegradable Polyesters

Due to their biodegradable character when composted industrially, biopolyesters such as poly(lactic acid) (PLA) or polycaprolactone (PCL) are ecological alternatives to conventional polymers in the production of bioplastic materials. Nevertheless, the use of these compounds as main raw materials for bioplastic is still limited, mainly due to their brittleness, slow crystallization rate and high gas permeability.[153] In order to overcome these problems, clay minerals, including sepiolite and palygorskite clays, have been employed as a nanofiller in PLA and PCL matrices. Within this perspective, PLA and PCL films loaded with sepiolite clay showed a thermal–mechanical improvement, which is related to the good compatibility with biodegradable polyester matrices.[154] The control of the biodegradation rate of both PLA and PCL films reinforced with sepiolite was also investigated, and the results obtained indicated a significant level of degradation in compost at 40 °C, together with evidence of erosion after 35 days.[14,154] On the other hand, hydrolytic degradation studies suggest that the presence of sepiolite favors a delay in degradation of the PLA matrix at 37 °C, likely due to the increase of the PLA crystallization effect induced by the clay mineral.[15] Based on this fact, Russo and co-authors[155] have studied the influence of sepiolite on the crystallinity of the PLA, and according to the authors, the presence of sepiolite not only could modify the crystallinity of the PLA, but also influence both gas barrier and mechanical properties of the resulting PLA–sepiolite films.

Bionanocomposites based on PLA and palygorskite prepared *via* melt blending were studied by Jiang and co-authors for application in agricultural packaging.[156] In addition to significant enhancements in tensile properties, which are likely occasioned by the good integration of palygoskite fibrils in the PLA matrix as revealed by scanning electron microscopy (SEM) and transmission electron microscopy (TEM) images, spectroscopy measurements such as Fourier transform infrared spectroscopy (FTIR) and X-ray photoelectron spectroscopy (XPS) indicated interfacial interactions through hydrogen bonds between palygorskite and PLA. Alternatively, previous surface modification of fibrous clays by silane compounds was the strategy employed by Moazeni *et al.*[157] Although the addition of modified sepiolite into PLA resulted in a reduction in the elongation at break, noteworthy increases of tensile modulus and oxygen and carbon dioxide barrier properties were evidenced, showing in both cases the best results for the PLA film that incorporates 1.5 wt% silane treated sepiolite. According to the authors, this behavior is due to better dispersion and interaction of the sepiolite nanofiller within the PLA matrix, which are evidenced by SEM and spectroscopy analyses.

The capacity of sepiolite and palygorskite to act as a nucleating agent in PLA or in other biodegradable polyesters was recently explored by several authors.[158–160] In the case of PLA, the crystallinity, nucleation density and the *G* constant for a specific crystallization temperature, were significantly enhanced in the presence of 1.0 wt% sepiolite, whilst a reduction of crystallization time from 17.7 to 2.1 min at the optimum crystallization temperature (110 °C) was evidenced. On the other hand, FTIR and WAXD results revealed that the crystal structure of PLA–sepiolite samples adopted a common α-form, which indicates that the introduction of sepiolite in the PLA matrix did not induce the formation of a new crystal structure.[160] Alternatively, biodegradable materials prepared from a combination of organo-palygorskite and poly(3-hydroxybutyrate-*co*-3-hydroxyvalerate) (PHBV) can also be of great interest in the plastics industry. Recent studies[158] demonstrated that PHBV films loaded with organophilic palygorskite showed a decrease of the thermal stability due to the presence of quaternary ammonium compounds used as modifier agents in the clay. Importantly, however, an improvement of the crystallization ability of PHBV was evidenced, which can be attributed to a nucleation effect of the clay in the polymer matrix during the non-isothermal crystallization process. Palygorskite clay was also used as a nucleating agent and filler in poly(butylene succinate) (PBS) using 1,2-octanediol as comonomer to synthesize branched PBS copolymers.[159] Besides the improved thermal stability and the crystallization temperature of the PBS matrix, which was promoted by the presence of palygorskite fibrils, the role of the filler agent was also evidenced, where clay-reinforced branched PBS containing 3 wt% of silicate in PBS matrix resulted in break elongation values of 550%. All these salient features are due to homogeneous dispersion of palygorskite fibrils in the PBS matrix, which results from strong hydrogen bonding interactions established between the components.

1.4.2 Polysaccharides

Polysaccharides are composed of sugar units joined by glycosidic linkages. These polymeric carbohydrates, the most abundant polymers in Nature, are emerging as a very attractive alternative to replace synthetic polymers in the development of nanocomposite materials. Thus, the resulting bionanocomposites show biocompatibility and biodegradability, together with low cost and positive ecological impact.

1.4.2.1 Bioplastics

Due to abundance, biodegradable and biocompatible properties, hierarchical structure and good film-forming ability, the polysaccharides are attractive matrices for the development of "green" nanocomposites. Although the layered clays such as montmorillonite are inorganic solids traditionally employed as filler in polymeric matrices, the use of fibrous clays (sepiolite and palygorskite) as promising reinforcing agents of biopolymers in the development of bionanocomposite materials has emerged in recent years. In this sense, it is worth mentioning the pioneering work reported by Lynch and co-authors,[161] who studied the effect of palygorskite fibrils in the degradation of a cellulose dextrin matrix in the presence of soil suspensions containing microorganisms. Thenceforward, studies on the properties and applications resulting from the assembling of both types of natural components, microfibrous clays and biopolymers, began to explore the development of new ecological nanocomposites provided with structural and/or functional properties.

Although it is well known that cellulose is the most abundant renewable and biodegradable natural polymer on Earth, the investigation focusing on cellulose films reinforced with fibrous clays is still recent. In contrast to the major polysaccharides, cellulose shows low solubility in water or other common solvents as a result of its strong inter- and intra-molecular hydrogen bonds, which causes a certain limitation in the preparation of cellulose bionanocomposites based on fibrous clay minerals. To overcome this problem, new synthetic strategies were employed, such as for example, the use of ionic liquids as solvent.[162,163] Ionic liquids are considered "green" solvents and have attracted a great deal of scientific attention, mainly due to their unique physicochemical properties such as chemical and thermal stability, non-flammability, immeasurably low vapor pressure and recyclability.[163] Thereby, using a simple, cost effective and "green" method, novel regenerated cellulose–sepiolite bionanocomposite films have been prepared using the environmentally friendly ionic liquid 1-butyl-3-methylimidazolium chloride (BMIMCl) as main solvent.[164] The results obtained from physicochemical studies showed that interactions between sepiolite and regenerated cellulose were established, where images from field-emission scanning electron microscopy (FE-SEM) and TEM revealed that the sepiolite fibers are uniformly dispersed in the cellulose matrix with contact between them. Mechanical properties and thermal stability of the bionanocomposite films were significantly improved when compared to pure cellulose films. In the case of the mechanical properties, the Young's modulus of bionanocomposite films with 8 wt% sepiolite increased by 166% compared to pristine cellulose films. Interestingly, in contrast to other biopolymer films that incorporate fibrous clays as filler, cellulose–sepiolite films showed an improvement of the ductility when sepiolite was added, where the incorporation of 6 wt% sepiolite resulted in an increase of 35% elongation at break, maintained nearly constant for higher sepiolite loadings. The incorporation of sepiolite in the regenerated cellulose matrix was also reflected in the water resistance and gas barrier properties. Thus, O_2 permeability values decreased gradually with the increasing sepiolite content, reaching values of about 56% of the initial value by addition of 8 wt% sepiolite. All the properties shown by cellulose–sepiolite bionanocomposite films make these materials promising for application in the areas of biomaterials, membranes and food packaging.

The affinity of sepiolite and palygorskite towards cellulose derivatives, such as carboxymethylcellulose (CMC) and hydroxypropylmethylcellulose (HPMC), in the development of bionanocomposite films for packaging applications was also evaluated. On this occasion, self-standing films loaded with different amounts of fibrous clays not only exhibited enhanced mechanical properties and water resistance, but also significant barrier properties against UV light (Figure 1.9) and reduced water absorption capacity, which make these bionanocomposites very attractive as bioplastics in the food packaging sector.[19] These properties are a consequence of the strong interaction by hydrogen bonding between the hydroxyl groups of the biopolymers and the silanol groups located on the surface of the fibrous silicates. Xylan-type

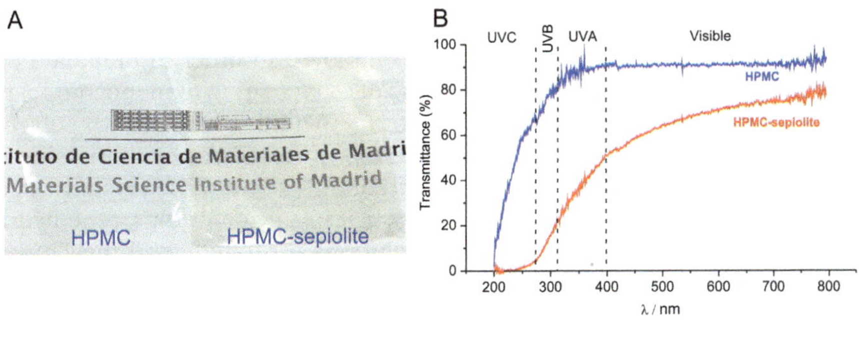

Figure 1.9 (A) Image of films of pristine HPMC and HPMC bionanocomposite loaded with 17% wt sepiolite. (B) Transmittance measurements in the 200–800 nm wavelength range of the HPMC and bionanocomposite films shown in (A).

hemicelluloses, such as arabinoxylan, are also interesting candidates in the preparation of bionanocomposites based on sepiolite.[165] In addition to a significant increase in the Young's modulus and tensile strength exhibited by arabinoxylan–sepiolite films compared to those previously reported for xylan–layered clay systems,[166] the drawback of the high water uptake typical of hemicellulose matrices was overcome by the introduction of sepiolite clay in the biopolymer film. Thus, materials become more resistant to water and are promising for use in the food sector.

Besides cellulose, interesting structural and functional properties in bionanocomposites based on fibrous clays were also observed with other abundant polysaccharides like chitosan. In this case, Darder and co-authors observed that the assembly of positively charged chitosan to sepiolite produced a clear reinforcing effect in the polymer film, which derived mainly from the good distribution of sepiolite fibrils in the chitosan matrix, originating from electrostatic interactions between the charged amine groups of the biopolymer and silanol groups of the silicate as well as from hydrogen bonding interactions.[167] In addition to excellent mechanical properties, these chitosan–sepiolite bionanocomposite films showed interesting results as membranes for N_2 separation, being an attractive material in membrane processes.[168]

Since the main goal in the use of fibrous silicates as nanofiller materials in polymer matrices is related to providing improved mechanical properties, this reinforcement strategy was employed by different authors in order to enhance the elastic modulus of polymer blends. Thus, a series of reinforced nanocomposite films prepared by a solution casting method using sepiolite[169] and palygorskite[170] as filler in chitosan–PVA blends was prepared. In these studies, the authors indicated that, due to good compatibility occasioned by the interactions through hydrogen bonds between hydroxyl groups of the polymer matrix and the silicate, the incorporation of fibrous clays can enhance the mechanical and barrier properties and water resistance of bionanocomposites compared to pristine blend films, favoring their application as a bioplastic in the food sector. Although the incorporation of sepiolite or palygorskite in chitosan–PVA matrices seems to be well studied, assorted synthesis strategies have recently been employed in order to modulate the crystallization and mechanical behaviors of the resulting bionanocomposite film. Within this context, once the chitosan–PVA films loaded with palygorskite were prepared, Lu *et al.*[171] adopted a post treatment by immersion of these films in a solution of NaOH and ethanol to control the crystallization degree of the polymer chains. In this work, it was evidenced that the hybrid film treated with NaOH and ethanol solutions exhibited good mechanical performance, likely due to higher crystallinity of the NaOH–ethanol treated films. On the one hand, this effect can be attributed to a decrease of the solubility of chitosan by a consequent neutralization effect of the acetic acid used in the solubilization of chitosan by NaOH solution. Otherwise, the films treated by ethanol showed different crystallinities, suggesting that the mechanical properties of these bionanocomposite films could be controlled by altering the dehydration process occasioned by the treatment with ethanol.

In recent years, Avérous and co-authors[12,172] developed different synthetic approaches in order to obtain bionanocomposite films based on thermoplastic starch and fibrous clays. One of the strategies employed in this case was the surface modification of sepiolite by a cationic starch, achieving bio-organosepiolites which were used as filler in starch matrices. The compatibility provided by the assembly of cationic starch to sepiolite produced a considerable improvement of the mechanical properties in the starch bionanocomposite films, increasing the Young's modulus by a factor of up to 2.5 as compared to those analogous materials based on montmorillonite.[12] Surface modification of sepiolite was also the strategy employed by Samper-Madrigal *et al.*[173] in studies addressing the combination of hydrophilic thermoplastic starch and hydrophobic "green" PE obtained from bioethanol derived from sugarcane. In order to compatibilize the system, pristine sepiolite was modified with hydrophobic groups from the silane (PTMS) resulting in a filler with dual functionality, where modified sepiolite is able to act as a bridge between hydrophobic PE chains and hydrophilic plasticized starch chains. The usefulness of silane-treated sepiolite as a filler in starch–PE blends was corroborated by a significant increase in the elongation at break of films from 90% (uncompatibilized starch–PE blend) up to values of 260% in the case of starch–PE films reinforced with 3 and 5 wt% of modified sepiolite.

Recently, a study focused on the incorporation of nano-sized sepiolite clay into thermoplastic starch–poly(butyrate adipate terephthalate) (PBAT) blends. According to the authors,[172] the addition of only 3 wt% sepiolite resulted in an important enhancement of the mechanical properties of starch–PBAT films, reaching up to 300% and 150% increase in the Young's modulus and tensile strength, respectively. These results confirm that fibrous clays can act as a filler, representing an interesting route to produce more resistant polysaccharide-based films.

This type of bionanocomposite involving polysaccharides and fibrous clays can also be processed as cellular solids[174] aimed at applications in insulation and cushioning. The available techniques for preparation of porous materials include phase separation/emulsion freeze-drying, solvent casting and particle leaching, gas foaming or fiber knitting, freeze-drying and supercritical drying. Several bionanocomposites based on smectites were prepared by supercritical drying with CO_2[175] or by foaming in a heated press in the presence of a blowing agent,[176] and these technologies could be analogously extended to the development of fibrous clay-based foams. However, the most commonly employed technique for processing polysaccharide-based materials as cellular solids is freeze-drying, due to the use of water as solvent in the preparation of these bionanocomposites.[177] This technique involves an ice-templating process in which the solid fraction in the aqueous suspension is pushed away towards the voids between adjacent crystals as the ice crystals

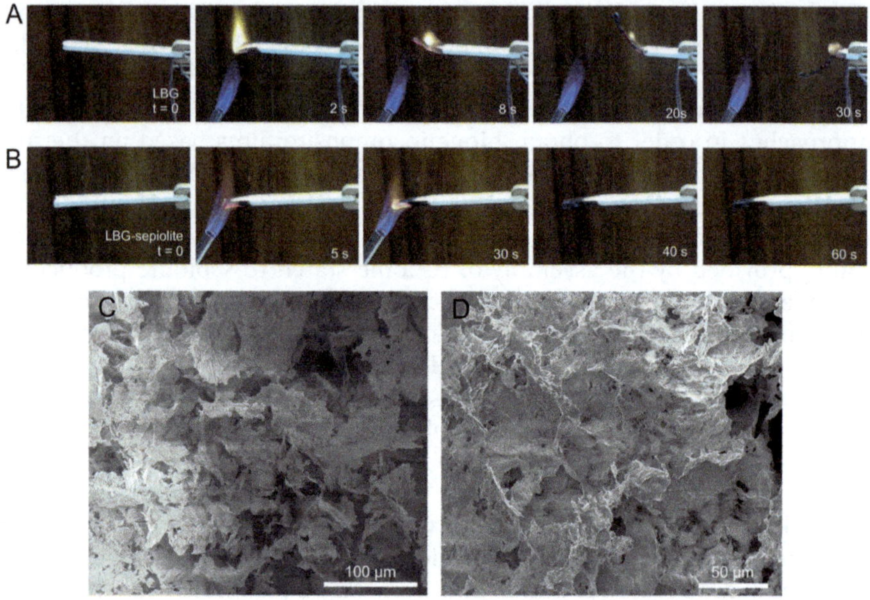

Figure 1.10 Fire resistance tests of the (A) locust bean gum (LBG) and (B) LBG–sepiolite foams with a 50% (w/w) silicate content. FE-SEM images of this LBG-sepiolite foam (C) before and (D) after the fire resistance test.

grow. Finally, sublimation of the ice under vacuum leads to the generation of the porous structure. As a result of this process, low density materials exhibiting macroporosity are produced, and diverse examples of polysaccharide–fibrous clay-based materials have been reported in recent years.[178] For instance, cellular materials involving sepiolite or palygorskite and a wide variety of polysaccharides (alginate, starch, carrageenan, locust bean gum or xanthan gum) showed macropores with an average diameter around 100 μm and apparent density values below 0.20 g cm^{-3}. However, they offered good mechanical resistance and those samples with a sepiolite content around 50–75 wt% reached values of compression modulus of about 40 MPa.[179] The presence of the fibrous clays can help to increase the fire retardancy of the resulting bionanocomposite foams (Figure 1.10), and it was reported that some polysaccharide–sepiolite materials with a clay content higher than 25% avoid the propagation of flames, thus behaving as auto-extinguishable foams that could replace polyurethane and analogue foams as building materials.[180] Unidirectional freezing leading to well-aligned microchannels in the freezing direction was applied in a recent work in order to obtain anisotropic foams with thermal insulation and fire retardancy properties from a mixture of nanocellulose fibers, graphene oxide and sepiolite.[181] The resulting ultra-lightweight materials showed a lower thermal conductivity in the radial direction than conventionally used insulation materials like polyurethane, good mechanical properties and moisture resistance, as well as fire retardant behavior, avoiding the self-propagation of the flame in a vertical burning test.

1.4.2.2 Environmental Remediation

Another application of polysaccharide-based bionanocomposites is their use as low cost biosorbents in environmental remediation.[182] Several examples of polysaccharide–fibrous clay materials applied for the removal of pollutants can be found in the literature. For this purpose, a wide variety of charged and neutral polysaccharides have been evaluated. The recently discovered polysaccharide sacran, an anionic megamolecule extracted from the cyanobacterium *Aphanothece sacrum*, shows a great affinity towards heavy metal and rare earth ions.[183] Such a property was exploited for the development of biosorbent materials with increased stability in water and the ability to retain neodymium ions selectively.[184] For this purpose, the assembly of sacran with sepiolite fibers was carried out in the optimal conditions that lead to the organization of sacran chains as liquid crystals within the bionanocomposite, which seems to favor the adsorption of neodymium in comparison to other lanthanide ions (Figure 1.11).

Bionanocomposites based on the assembly of sepiolite or palygorskite with alginate, chitosan and starch were tested in the uptake of copper and lead from aqueous solution.[185] All these materials showed better mechanical properties and enhanced stability in water than the pristine polymers, and the best results were obtained for alginate-based systems in the case of lead uptake and for chitosan-based materials in the removal of copper ions. Chitosan–palygorskite bionanocomposites were also applied in the removal of Fe(III) and Cr(III) cations by a complexation mechanism, showing a higher sorption capacity than the individual components.[186] The removal of uranium from aqueous solutions is also of great interest.

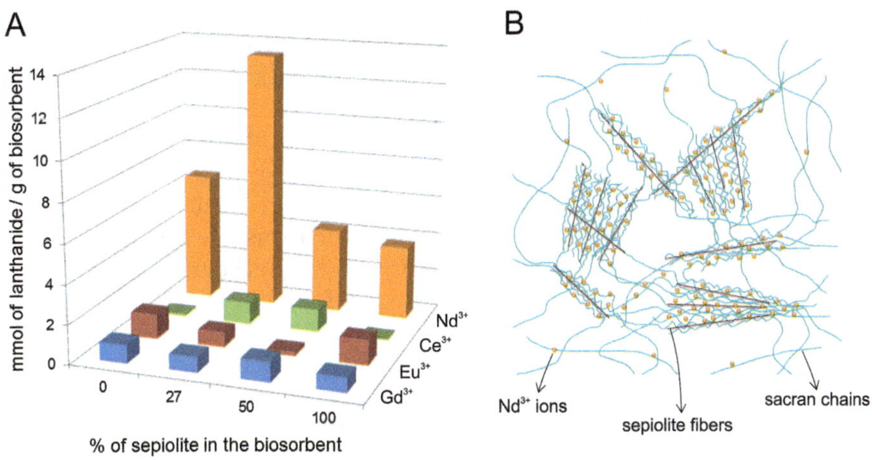

Figure 1.11 (A) Sorption capacity of diverse lanthanide(III) ions by sacran–sepiolite bionanocomposites with different sepiolite content. (B) Schematic representation of the possible organization of sacran chains as crystalline domains favored by the silicate fibers and leading to enhanced Nd^{3+} adsorption.

The same chitosan–palygorskite biosorbent[187] retained uranium ions based on the interaction of U(VI) cations with the –OH and $-NH_2$ groups of chitosan. The optimal adsorption takes place at pH 5.5, when the competing effects with other charged species at extreme pH values are minimized. In other examples, the ion-imprinting concept was used in the production of biosorbents for selective removal of heavy metal ions. Thus, chitosan, previously mixed with a solution of a Co(II) salt, was combined with γ-glycidoxypropyltrimethoxysilane and palygorskite in order to produce an ion-imprinted material with selectivity towards Co(II) ions.[188] In a recent work, a similar Cu(II)-imprinted material was reported, based on chitosan, palygorskite and ethylene glycol dimethacrylate (EGDMA),[189] which showed a selective adsorption of copper ions.

The association of chitosan to fibrous clays takes place mainly by hydrogen bonding interaction between the hydroxyl groups on the biopolymer and the silanol groups located at the external surface of the silicate fibers, and in a minor extent through electrostatic interaction due to the protonated amino groups of chitosan.[167] Due to the low cation-exchange capacity of sepiolite and palygorskite (<25 mEq 100 g^{-1})[20] to compensate the protonated amino groups of chitosan, the resulting bionanocomposites usually show anionic-exchange ability due to the excess of surface positive charges, thus being useful for the removal of diverse anionic pollutants. For instance, a chitosan–palygorskite biosorbent was successful in the removal of tannic acid from drinking water by electrostatic interaction, together with hydrogen bonding and van der Waals interactions established with the uncharged parts of the polysaccharide.[190] A similar system was prepared for the same purpose, but in this case an emulsified chitosan–palygorskite mixture was processed as microspheres using formaldehyde and glutaraldehyde as crosslinking agents.[191] Other chitosan-based bionanocomposites were applied in the removal of anionic dyes like reactive yellow 3RS, such as for instance, the material prepared with sepiolite fibers modified with APTES and chitosan, which were crosslinked with glutaraldehyde.[192]

In a different type of application, the affinity of chitosan–sepiolite and chitosan–palygorskite materials towards anionic dyes like Acid Yellow 25 or chrysophenine G, among others, was exploited to create a photoprotective matrix for *Beauveria bassiana* conidia, which was used as a microbial biocontrol agent.[193] The biocompatibility of polysaccharide–fibrous clay materials allows their application as a support for living species, such as for instance, microalgal cells that were able to colonize chitosan–sepiolite bionanocomposites processed as foams.[194] Similarly, a recent work described the immobilization of *Pseudomonas* sp. DG17 in sodium alginate–palygorskite materials,[195] and the resulting system was successfully applied in petroleum hydrocarbon biodegradation.

Polysaccharide–fibrous clay bionanocomposites can also be applied to treatment of wastewater effluents by means of coagulation–flocculation processes. A recent example of such applications is the use of chitosan–sepiolite materials for clarification of winery wastewater by neutralization of the colloids with the highly charged bionanocomposites, as well as of olive mill wastewater by van der Waals or OH interactions between specific organic colloids and the glucosamine units of the chitosan in low charge nanocomposites.[196] Similarly, chitosan–sepiolite bionanocomposites were used as flocculants to remove harmful microalgal cells from the Taihu Lake, showing a good removal efficiency with a reduced competing effect from humic acids.[197]

Several multifunctional materials involving polysaccharides and previously modified fibrous clays have been reported. For instance, palygorskite fibers can be modified with superparamagnetic NPs and then coated by chitosan and cysteine modified β-cyclodextrin deposited by means of a layer-by-layer process.[198] The resulting multifunctional bionanocomposite was applied to the recovery of precious metal ions, showing a preferential adsorption of Ag^{+} and Pd^{2+} over Pt^{4+}. Wang *et al.*[199] reported the preparation of a bionanocomposite material involving Au NPs assembled by electrostatic interaction to a chitosan layer that covers Fe_3O_4-modified palygorskite. The catalytic activity of this material was exploited in the decoloration of Congo red solutions. In both examples, the superparamagnetic properties of the materials are helpful to facilitate the recovery of the bionanocomposites by means of a magnet.

Another type of multifunctional bionanocomposite involves the combination of the polysaccharides and the fibrous clays with PAA or PAAm. This additional polymer confers superabsorbent properties to the final system that can be of benefit for a wide variety of applications. Amongst them, some of these bionanocomposite hydrogels are devoted to agriculture and horticulture, for instance for the controlled release of nutrients. Such fertilizer granules were prepared by coating a mixture of alginate, palygorskite, urea and KH_2PO_4 with an alginate-*g*-PAA-*co*-PAAm–humic acid hydrogel.[200] In another recent example, the carboxymethyl chitosan-*g*-PAA–attapulgite composite was used as an outer coating of a complex system entrapping co-granulated ammonium zinc phosphate and urea as fertilizer core.[201] In these examples, the goal is the reduction of environmental pollution due to the excessive use of fertilizers, in a similar way to other bionanocomposites based on alginate and palygorskite and processed as microbeads that were used for the controlled release of the organophosphate insecticide chlorpyrifos in order to minimize the adverse environmental impact.[202]

Another interesting application of this type of hydrogel system is related to their use as sorbents in environmental remediation, with special interest in the removal of heavy metal ions. For instance, CMC-*g*-PAA–palygorskite systems were applied as adsorbents of Pb(II),[203] chitosan-*g*-PAA–palygorskite was useful in the removal of Hg(II)[204] and, more recently, alginate-*g*-poly(acrylic acid-*co*-2-acrylamido-2-methyl-1-propane sulfonic acid)–attapulgite was successfully applied in the uptake of Pb(II), Cd(II), Cu(II) and Zn(II),[205] while the hydroxypropyl cellulose-*g*-PAA–palygorskite hydrogel was an effective sorbent for the recovery of rare earth elements like La(III) and Ce(III).[206] There is also great interest in the removal of ammonium nitrogen (NH_4^{+}–N), which is a primary

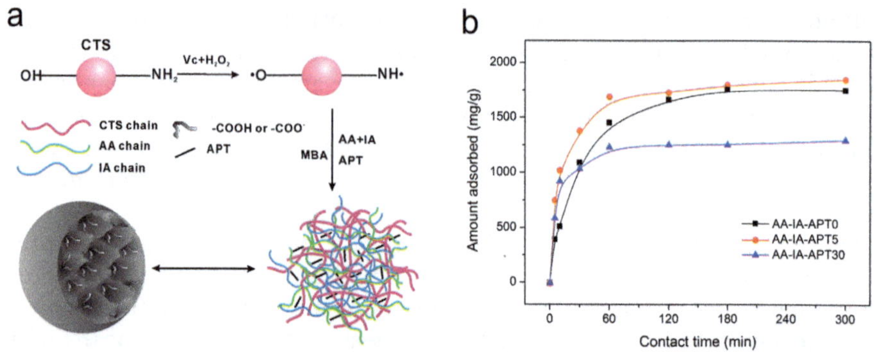

Figure 1.12 (a) Schematized preparation of the AA–IA–APT hydrogel beads. (b) Adsorption of malachite green as a function of time with AA–IA–APT hydrogel containing 0, 5% or 30% of palygorskite (Reprinted from *Chemical Engineering Journal*, **257**, Y. Zheng *et al.*, Highly efficient and selective adsorption of malachite green onto granular composite hydrogel, 66–73, Copyright (2014), with permission from Elsevier.[210])

indicator of water quality[207] and also of dyes like crystal violet, which was effectively sorbed on a κ-carrageenan-*g*-PAAm–sepiolite,[208] as well as methylene blue[209] or malachite green,[210] which were removed making use of materials involving chitosan-*g*-PAA and palygorskite fibers processed as beads (Figure 1.12a). The presence of palygorskite can help to enhance the adsorption kinetics, as shown in the removal of malachite green (Figure 1.12b).

In some applications, polysaccharide–fibrous clay bionanocomposites have been used as intermediates in the preparation of adsorbent carbonaceous materials. Several recent examples report the use of chitosan[211,212] and cellulose[213,214] as low cost raw materials in combination with palygorskite. For instance, chitosan was transformed into carbon by calcination at temperatures above 280 °C,[211] which gave rise to larger pores and increased the adsorption sites, being successfully applied in the bleaching of palm oil. Alternatively, chitosan[212] or cellulose[213,214] can be submitted to a hydrothermal carbonization process at lower temperatures, below 250 °C, resulting in carbon–silicate materials that were effective in the removal of methylene blue or phenol from aqueous solution. Similarly, caramel–sepiolite materials derived from the disaccharide sucrose were transformed at around 800 °C into conducting carbon–sepiolite nanocomposites, useful as electrode materials after functionalization with appropriate organoalkoxysilanes.[215] The covalent grafting of thiol groups results in electrochemical sensors that could be applied in the amperometric determination of heavy metal ions in aqueous solution.[216]

1.4.2.3 Biomedical Applications

Nanoparticulated materials have been receiving enormous interest for biomedical applications such as drug delivery due to the high loading capacities of pharmacological molecules related with their large specific surface area and the possibility to direct these particles to the required site.[217–219] The combination of nanomaterials such as clays with biopolymers (*e.g.* polysaccharides) offers several advantages: (i) control over the drug release profile, (ii) control over carrier assembly and its microstructure and (iii) improved uptake.[220,221]

Recently, Wu *et al.*[222] prepared emulsion crosslinked microbeads of chitosan and organo-palygorskite. The clay was modified with the amphoteric surfactant hexadecyl betaine (BS-16) to improve the compatibility and affinity with the chitosan matrix. Furthermore, the incorporated clay increased the encapsulation of diclofenac sodium (DS) and slowed down the drug release kinetics. A similar delivery system was reported by Wang *et al.*,[223] who prepared hydrogel microbeads composed of chitosan-*g*-poly(acrylic acid)–attapulgite–sodium alginate (CTS-*g*-PAA–APT–SA) suitable for pH-sensitive delivery of DS. The model drug was loaded into CTS-*g*-PAA–APT microparticles and, subsequently, the microparticles were confined in Ca^{2+}-crosslinked sodium alginate hydrogel beads. This configuration prolongs the drug diffusion paths and, hence, retards the drug release. Importantly, the drug can be released by changing the pH of the medium. This ensures that the release of DS in gastric fluid (acidic pH) is avoided, but allows for its complete and controlled delivery in intestinal fluid (neutral pH). In acidic conditions, hydrogen bonding between the polyelectrolytes and the clay results in non-swollen microbeads, which greatly limits the migration of DS. Contrarily, at a pH above 5.5 both the clay and sodium alginate are negatively charged, which results in electrostatic repulsion, in extensive swelling and, consequently, in pronounced release of DS. In another work, pH-sensitive microbeads were prepared by spray-drying of glutaraldehyde crosslinked chitosan–attapulgite gels.[224] The microbeads allowed the controlled release of DS in the simulated intestinal fluid. The presence of attapulgite has also been observed to favor the production of small and uniform microbeads of narrow size distribution around 5 μm. The same authors have also proposed composite hydrogel beads composed of carboxymethyl cellulose-*g*-poly(acrylic acid)–attapulgite–sodium alginate (CMC-*g*-PAA/APT/SA) as a pH-sensitive release system for DS.[225] The bionanocomposite was mixed with the model drug and dropped into Ca^{2+} crosslinking solution to obtain the desired nanocomposite beads. A clay content of up to 20 wt% ensured reduced swellability, and thus, prevented undesired burst release of the drug. The cumulative release ratio of DS from the hydrogel beads in simulated gastric fluid was only 3.7% within 3 h, but in simulated intestinal fluid was about 50% for 3 h and up to 90% after 24 h.

Needle-free, nasal vaccine administration[226] is another field where clay–polysaccharide materials can make an important contribution as adjuvants. An interesting aspect of using microfibrous clays such as sepiolite as vaccine adjuvants is connected with their needle-like morphology that is thought to enhance mucosal immune responses provoked by irritation of the nasal mucous through the fibers.[227] This can help to improve the efficacy of vaccination and thus reduce the dose necessary for immunization. However, the bare surface of inorganic carriers can cause adsorption-induced alterations of the protein structure of antigens provoked by strong electrostatic interactions. This can result in diminished immunogenicity of the antigen, as observed for pristine sepiolite[227] and sometimes even in the case of standard aluminum containing adjuvants.[228,229] Therefore, sepiolite was modified with the biopolymer xanthan gum to mimic the surface of nasal mucous, the natural adsorption site of the influenza virus during entry into an organism.[227] Mimicking this environment in terms of surface charge, water retention and functional sugar groups is considered favorable for the interaction between carrier and virion and thus determining the preservation of immunogenic activity of the antigen. Influenza virus particles of the strain PR/8(H1N1) were immobilized on sepiolite–xanthan and elicited high levels of seroprotection in a balb/c mouse model.[227]

1.4.3 Proteins

Proteins are biopolymers composed of amino acid units that are linked by peptide bonds to form polypeptide chains. They can be obtained from plants, animals or bacteria and can be classified as fibrous, globular or membrane proteins depending on their shape and solubility. Proteins have been used in the preparation of nanocomposite materials, although in a lesser extent than polysaccharides. Common non-food applications of proteins include adhesives, glues, paper coatings, paints, textile fibers, as well as molded plastic items.

1.4.3.1 Bioplastics

Among the diverse bionanocomposites, maybe the most interesting are those derived from proteins. These biomacromolecules have become very attractive due the variety of functional groups present on their peptide chain, which may result in interaction points with inorganic solids and thus provide the resulting bionanocomposite materials with diversified structural and functional properties.

Analogously to carbohydrates, various routes and strategies of synthesis were explored using fibrous clays as filler in order to improve the water resistance and mechanical properties of protein-based films. In this sense, films based on gelatin, a structural protein, were reinforced with sepiolite clay, varying the sepiolite content between 0.05 and 0.50 mass fraction.[13] The most salient feature of these gelatin–sepiolite films was the substantial improvement of the elastic modulus, which showed an increase of up to 250% at 50 wt% clay loading compared to unmodified gelatin films. This is a consequence of interactions established between the sepiolite and the available functional groups of gelatin.[13,230] More recently, Fernandes *et al.*[230] evaluated the mechanical properties of gelatin films, as function of the crystallinity of sepiolite. Interestingly, the changes in the nature of the silicate induced by thermal treatments resulted in different values of the Young's modulus, where elastic modulus increased up to 220% at 20% clay loading using raw sepiolite, while no improvement was found at 20% clay loading using the protoenstatite form. Bionanocomposite films resulting from the assembly of type I collagen, another structural protein, to palygorskite were evaluated by Su and co-authors.[231] In addition to enhanced mechanical properties in comparison with neat collagen films, it was evidenced that the increase of palygorskite content in collagen–palygorskite films contributes to better thermal behavior. In this case, although spectroscopic studies point out that the presence of palygorskite in a collagen matrix causes a contraction and aggregation of the collagen chains, the triple-helix backbone of the protein is preserved throughout the synthesis process.

The role of sepiolite fibers in the release of bioactive compounds from gelatin–egg white films containing clove essential oil was investigated by Giménez *et al.*[232] Although an improvement of the Young's modulus from the introduction of the clay mineral was evidenced, the presence of clove essential oil in the nanocomposite films promoted a slight decrease in the water vapor permeability (WVP), this latter compound acting as a plasticizer of the system. Furthermore, the release of protein components and eugenol from the matrix was promoted by the presence of sepiolite in gelatin–egg white films. This contributed to the antioxidant and antimicrobial activity of the films, which makes them very attractive as bioactive packaging for the food sector.

Another interesting protein for the production of reinforced films is zein, a storage protein extracted from corn. Due to non-polar amino acid residues, zein shows a relatively strong hydrophobic character, which makes it a very attractive biopolymer in the preparation of moisture resistant materials. In this context, Alcântara *et al.*[19,233–235] have reported the assembly of zein to sepiolite and palygorskite giving rise to new bio-organohybrids that can be used as a "green" alternative for the development of clay hybrids for food packaging. The incorporation of zein–fibrous clays as a biofiller in alginate films produced a reinforcing effect, helping to overcome the drawback related with the low water resistance of the alginate matrix. Moreover, these reinforced alginate films showed improved water vapor barrier properties and gas permeation toward CO_2, N_2, He and O_2, even at high humidity conditions.[19] It was observed that oxygen permeability and water uptake are significantly influenced by the total zein content in the alginate matrix, reducing the values of both properties, which is an advantageous feature that would allow the application of these materials in the food packaging sector (Figure 1.13).

Pristine palygorskite clay was also used as a nanofiller in wheat gluten, another storage protein. In this study, carried out by Yuan *et al.*,[236] the structure and biodegradable properties of transparent thermos-pressed films were studied. Thanks to the good dispersion of palygorskite particles into the wheat gluten matrix, a significant enhancement of the mechanical and viscoelastic properties was observed, where the best results were found for those bionanocomposite films loaded with 7 wt% of clay. Furthermore, an increase in the biodegradability due the presence of palygorskite was evidenced, leading to the disaggregation and dwindling of wheat gluten–palygorskite films after 15 days of burial.

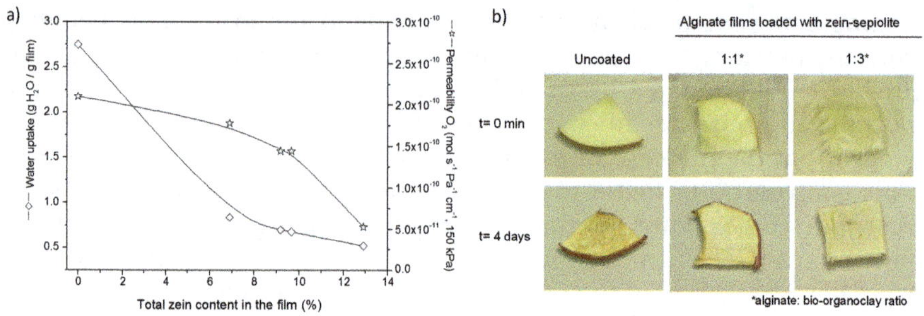

Figure 1.13 (a) Evolution of water uptake values and oxygen permeability as a function of zein content in alginate films and (b) evaluation of the protective role of different alginate films loaded with zein–sepiolite against oxidation of apple slices.

Recently, several works reporting the use of soybean protein as organic counterpart in the preparation of hybrid films have emerged.[237–239] In most cases, this class of protein has been associated with unmodified and silylated palygorskite clay resulting in homogeneous nanocomposite films. For instance, Wang and co-authors[238,239] investigated the structure and loading effects of raw and silylated palygorskite on the properties of soy polyol-derived polyurethane nanocomposites. In addition to an increase of the glass transition temperature, the mechanical properties were strongly influenced by the presence of palygorskite fibrils, where at 12 wt% pristine palygorskite and silylated palygorskite loadings, a 223% and 100% increase in tensile strength and 455% and 182% increase in Young's modulus were obtained, respectively. These new nanocomposites represent a promising alternative in preparing environmentally friendly materials with excellent structural properties for practical applications.

1.4.3.2 *Biomedical Applications*

Together with smectites, kaolinite, talc and other phyllosilicates, fibrous clays are also widely used as excipients[240] and even as active ingredients[241] in both the pharmaceutical and the cosmetic industries. As excipients, sepiolite and palygorskite are used for diverse purposes as thickening and anticaking agent, flavor corrector or carrier–releaser of active principles, while for their use as active ingredients they can be orally administered as antacid, gastrointestinal protector or anti-diarrheal drugs. Another application of palygorskite in the health field is related to its topical use as wound dressings due to its anti-inflammatory action and its contribution to providing better rehydration of the damaged area.[242] Recently, a detailed study on the anti-inflammatory properties of both sepiolite and palygorskite has been carried out by Cervini-Silva *et al.*[243] These authors have determined that clay hydration due to structural water in these fibrous clays together with the presence of external silanol groups contribute to edema inhibition and migration of neutrophils. They have also confirmed the non-cytotoxic effect of these fibrous clays when the cells are exposed to low concentrations of the silicates.

Bionanocomposites involving fibrous clays and proteins have been reported for biomedical applications, with special emphasis on regenerative medicine. The first works reporting the biocompatibility of collagen–sepiolite complexes and their potential use in bone tissue repair appeared about three decades ago.[244,245] Initially, *in vitro* experiments confirmed the suitability of these bionanocomposites to favor the adhesion and proliferation of fibroblasts, and some years later *in vivo* experiments were carried out with the implantation of the collagen–sepiolite materials in rat calvaria defects.[246] These studies confirmed the non-cytotoxicity of these bionanocomposites and their ability to promote bone regeneration in the damaged area, which was accompanied by a complete resorption of the implanted material. The rate of resorption of the collagen–sepiolite implants was tuned by means of crosslinking with 1% glutaraldehyde, which contributed to extend the resorption for several months after subcutaneous implantation.[247] Recent studies have been focused on the assembly between collagen or its derivative gelatin and fibrous clays,[230,231] as well as on the development of macroporous bionanocomposite scaffolds[248] that could be applied in reparative medicine. In this last example, gelatin–sepiolite bionanocomposites were freeze-dried in order to produce macroporous foams, showing a porosity around 98% and a maximum Young's modulus of 6 MPa.[248]

Another interesting application of bionanocomposites related to health is drug delivery, benefiting from the swelling properties, bioadhesion or possibility of cell uptake of these systems, which allow the controlled release of a constant dose of the entrapped drug.[220] Similarly to those reported systems involving layered silicates, bionanocomposites based on proteins and fibrous clay minerals can be used for drug delivery. A recent work reports the use of silk fibroin–palygorskite materials for controlled release of diclofenac.[249] In this case, a core–shell structure was formed by coating the drug-loaded palygorskite with silk fibroin produced by spiders, the larvae of *Bombyx mori* and other insects, and the resulting system showed a reduction in the release rate of the entrapped diclofenac by about 50% with respect to the pristine clay. Other examples are related to the use of bionanocomposites involving antibiotics for wound-dressing applications. For instance, the controlled release of neomycin from sepiolite-based bionanocomposites prepared with the corn protein zein was recently studied.[250] In this case, zein was combined with CMC in order to increase the water resistance of the polysaccharide and to reduce the swelling degree of the resulting biopolymer blend. All the prepared systems involving neomycin showed antibacterial properties, but the release rate of this antibiotic was reduced when it was previously loaded on sepiolite, in comparison to systems with neomycin directly added to the biopolymer blend.

1.4.3.3 Biocatalytic Applications

Enzymes are globular proteins that act as catalysts in complex chemical reactions, and for many applications in biosensors and bioreactors they are previously immobilized on a solid support, including fibrous clays. These silicates can provide a protective support for the biological molecule, preventing its denaturation and loss of catalytic activity, and also enhancing the thermal and storage stabilities of the enzymes, as determined in catalase supported on sepiolite[251] or alcohol dehydrogenase on palygorskite.[252] In addition, the solid support allows the recovery of the enzyme, which can be reused in consecutive cycles. Similarly, invertase immobilized on sepiolite showed higher thermal stability and increased resistance to leaching, which was exploited for its use in a packed bed reactor.[253]

In many cases, the enzyme is directly adsorbed on the fibrous clays. For instance, *Rhizomucor miehei* and *Candida cylindracea* lipases were immobilized on sepiolite and palygorskite by ionic adsorption through the positively charged groups of the proteins and the silanol groups of the silicate surface,[254] and the resulting material yielded higher rates of hydrolysis of ethyl carboxylates than analogous systems involving layered silicates. In contrast, sometimes the enzymatic activity can be partially lost due to the immobilization process. This was the case for a pig pancreatic lipase adsorbed on sepiolite, which showed a decrease in the enzymatic activity up to 42% in comparison to the free enzyme.[255] This fact is most likely due to a partial deactivation of the active sites as a consequence of adsorption or to a steric effect of the supported lipase in the reaction. In spite of this reduction, the supported lipase was applied in the production of biofuel from sunflower oil, showing high stability and easy recyclability.[255] A previous modification of the surface of the fibrous clays can be carried out before enzyme immobilization in order to avoid the decrease in the catalytic activity. For instance, sepiolite was modified with phospholipids forming a biomimetic supported membrane for immobilization of urease and cholesterol oxidase,[256] as will be detailed in the next section. The fibrous clays can also be functionalized in order to facilitate the covalent immobilization of the enzymes and enlarge their stability during reuse, as shown in a recent work by Luna *et al.*[257] In this example, sepiolite was coated with an external $AlPO_4$ layer to which *Rhizopus oryzae* lipase was covalently grafted through phosphamide bonds. The immobilized lipase was applied in the production of biofuel by ethanolysis of sunflower oil, minimizing the glycerol waste production and showing a significant stability together with the possibility of reusing it more than five times. Similarly, acid activated palygorskite was modified with APTES in order to facilitate the covalent immobilization of lipase from *Burkholderia cepacia*, using glutaraldehyde as crosslinker.[258] This system was used to obtain biodiesel from jatropha oil, showing a higher stability of the immobilized lipase even after 15 cycles of reuse.

The fibrous clays used as a support for enzymes can also be modified with magnetic particles in order to facilitate the recovery of the enzymatic system after the catalytic cycle. In this way, palygorskite was modified with γ-Fe_2O_3 assembled to its external surface, and then *Candida* sp. 99–125 lipase was adsorbed on the functionalized silicate.[259] The immobilized lipase showed a good performance in the hydrolysis of olive oil and higher stability to changes in pH and temperature compared with free lipase. This high stability allowed its repeated use in at least eight consecutive cycles, profiting from easy recovery from the reaction medium by means of a magnet.

Immobilization of enzymes on solid supports is also of great interest for biosensing applications. In certain cases, the fibrous clay is used to develop surface-modified biosensors, in which the silicate is deposited as a film that entraps the enzyme, preserving its catalytic activity. This is the case for palygorskite used to immobilize tyrosinase,[260] leading to the construction of a biosensor that was used in the electrochemical determination of phenol. Similarly, glucose oxidase was immobilized with palygorksite on the electrode surface, and the resulting biosensor was used to determine glucose in blood and urine samples.[261] In other cases, bulk-modified electrodes are developed. A recent work describes the immobilization of peroxidase extracted from a fruit native to South America on sepiolite,[262] and this system is subsequently mixed with graphite powder, MWCNTs, mineral oil and nafion perfluorinated resin (Figure 1.14). The resulting composite can act

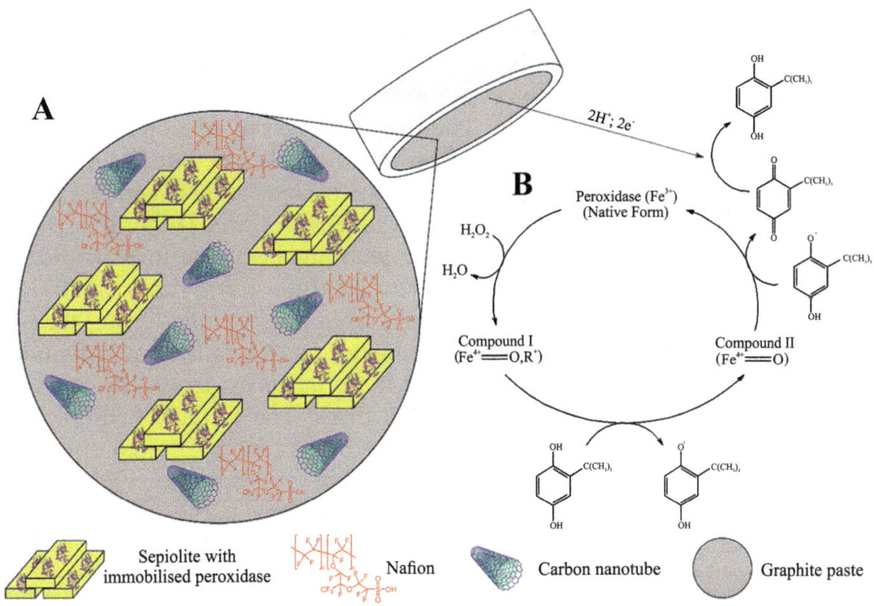

Figure 1.14 (A) Peroxidase immobilized on sepiolite and dispersed within the graphite paste together with carbon nanotubes and nafion. (B) Electrocatalytic determination of *tert*-butylhydroquinone (TBHQ) with the peroxidase-based biosensor. (Reproduced from ref. 262 with permission from The Royal Society of Chemistry.)

simultaneously as the sensing and conducting phase of the biosensor, which was applied in the determination of *tert*-butylhydroquinone, used as an antioxidant in foods.

1.4.4 Lipids

Naturally occurring surfactants such as fatty acids, phospholipids, lipopeptides, *etc.*[263] are versatile and environmentally benign alternatives to synthetic surfactants like quaternary alkyl ammonium salts[8,264] for the organophilization of clay minerals. Organoclays find important application as thickeners or nanofillers in polymers and resins.[8,265] However, an important drawback of organoclays is connected with the toxicity of many quaternary alkylammonium components and hence makes their extension into biological fields difficult.[266] Therefore, increased attention has been directed toward biosurfactants as biocompatible and eco-friendly alternatives.[263,267] Moreover, these molecules can form solid-supported bilayers with certain biomimetic features, which can be prepared *via* self-assembly routes[268] and find potential applications in biomedical or other biotechnological fields.[269–271]

The first example of a fibrous clay–lipid hybrid was reported by Wicklein *et al.* who adsorbed phosphatidylcholine (PC) on sepiolite.[272,273] Lipid deposition was carried out either by liposome adsorption from aqueous media or by molecular adsorption from organic media.[229,273] In both cases, the adsorption isotherms show the characteristic shape of a mono- and bilayer formation as a function of the equilibrium PC concentration. This study has also shown that liposomal deposition is more effective for the formation of a lipid membrane in this specific system. The reason is that in aqueous media PC adsorbs as aggregates (*i.e.* bilayered liposomes) while in ethanol single PC molecules adsorb on the clay. The molecular interaction between the lipid headgroup and the clay surface was investigated by infrared (IR) and nuclear magnetic resonance (NMR) spectroscopy techniques. These investigations strongly indicated hydrogen bonding between the ester and phosphatidyl groups of the lipid and the silanol groups located on the sepiolite surface.[272] The adsorption of phospholipids on sepiolite also has a significant influence on the surface wettability of the resultant material, as evidenced by contact angle measurements on cast sepiolite–PC films. The initially hydrophilic sepiolite becomes more hydrophobic as the first monolayer is completed, which corroborates the presence of hydrocarbon chains at the solid–liquid interface. With subsequent deposition of lipid molecules the contact angle gradually decreases again as hydrophilic lipid headgroups assemble at the external surface of the lipid bilayer.

Self-assembly of biosurfactants can be used to prepare mixed lipid membranes that offer diverse functionalities, which can be of interest in different technological applications.[274] These hybrid layers are composed of individual membrane leaflets of different molecules.[275–277] Such a hybrid interface was assembled on sepiolite, where the inner leaflet was comprised of PC and the outer leaflet of *n*-octyl-β-D-galactoside (OGal). The mixed membrane resulted from subsequent monolayer deposition from the corresponding aqueous biosurfactant solutions.[256] It has been assumed that the PC monolayer serves as nucleation site for the growth of the OGal outer membrane leaflet by hydrophobic interaction between the hydrocarbon chains.[278] This assumption is also supported by Persson and co-workers who investigated adsorption of *n*-dodecyl-β-D-maltopyranoside on hydrophobized surfaces based on silane and concluded that insertion of alkyl chains of the sugar-based surfactant into the hydrophobic silane layer acts as an anchoring mechanism.[279] The adsorption of OGal molecules causes an inversion of the surface hydrophilicity of the resultant material, as evidenced from water adsorption isotherms.[256] Nonionic sugar-based surfactants (SBSs) are a type of biosurfactant which are particularly interesting for incorporation into these hybrid layers, *e.g.* for their antifouling properties[280,281] or protein stabilization ability.[282–284]

Thanks to their biocompatibility, clay–lipid hybrids are interesting candidates for *in vivo* applications like enterosorbents for mycotoxins. Infestation of crops and foodstuffs by fungi-produced toxic compounds such as mycotoxins has been a severe threat for millennia, with occasional catastrophic consequences for the lives and health of both animals and humans.[285] A convenient and efficient method for elimination of these compounds from animal alimentation is their *in vivo* adsorption and sequestration (*i.e.* enterosorption) by food additives like clay and soil minerals.[286] As many of these mycotoxins are rather hydrophobic (*e.g.* alflatoxins) the use of organophilic sorbents would increase their retention. In fact, bio-organoclays like sepiolite–lipid hybrids demonstrated higher *in vitro* sequestration efficiencies for aflatoxin B1 compared to the neat clay analogue[272] and could thus be a promising candidate material for *in vivo* applications.

Another *in vivo* application of these hybrids is their employment as vaccine adjuvants where they can act as an antigen carrier. Recently, the bilayer lipid membrane supported on sepiolite fibers was reported to serve as a promising immobilization site for influenza A antigens (*i.e.* whole virus particles, hemagglutinin protein).[229] The rationale was to limit the strong support–antigen interactions that are frequently observed for inorganic vaccine adjuvants and which are prone to compromise the immunogenicity of the vaccine.[228] Functional studies in mice revealed that, in contrast to a standard aluminum hydroxide adjuvant, influenza vaccines based on sepiolite–lipid induced high titers of specific antibodies with a Th1 profile that is often associated with efficient clearance of viral infections.[287]

The biomimetic character of the clay-supported lipid interface is also beneficial for the immobilization of certain proteins, especially membrane-bound enzymes. The catalytic activity of such enzymes is very sensitive to the immobilization site and can be easily compromised by protein structure deformations. It could be shown in electrochemical assays that, for instance, cholesterol oxidase (COx) maintained its catalytic activity when supported on sepiolite, offering a bilayer lipid membrane.[256] Contrarily, after immobilization on sepiolite hybrids displaying cetyltrimethylammonium or hybrid lipid–octyl-galactoside layers, the activity of COx was greatly reduced, underlining the importance of biomimetic interfaces like the bilayer lipid membrane for good stabilization of such enzymes. This offers possibilities for preparing, for instance, selective biocatalysts[288] and sensitive biosensors[256] employing immobilized enzymes on sepiolite–lipid hybrids as active species.

1.4.5 Nucleic Acids

Fibrous silicates have been proposed as non-viral vectors for gene transfection. The spontaneous assembly of short chain DNA molecules on sepiolite was recently reported. In these bionanocomposites, the silicate can provide a protective environment limiting the DNA degradation by DNase,[289] as observed for other clays of the smectite family.[290] The assembly of the negatively charged nucleic acids on fibrous silicates is most likely due to hydrogen bonding interactions with the silanol groups in their structure. In order to evaluate the possible application of DNA–sepiolite materials for gene transfection, they were mixed with bacteria and deposited on agar plates. According to the reported Yoshida effect,[291] nano-sized acicular materials, stimulated by sliding friction on the surface of the hydrogel, are able to form penetration intermediates that can incorporate exogenous DNA in the recipient bacteria, and therefore, the nucleic acid can be expressed therein.[292] Similarly, sepiolite fibers with a diameter at the nanometer scale, obtained from Spanish deposits, can be used as a carrier for the transfer of exogenous DNA into *Escherichia coli* and other bacterial cells.[293] The use of this type of sepiolite with a diameter around 50 nm is due to its apparent lack of carcinogenic potential and lower biological activity compared to sepiolite from other origins.[294] Further studies were focused on the optimization of the Yoshida effect, making use of sepiolite fibers to transfect *Escherichia coli*, *Yersinia enterocolitica* and *Acinetobacter baumannii* into bacterial cells.[295] Later on, it was demonstrated that the use of sonicated sepiolite increases the bacterial transformation efficiency from 20 to 30-fold compared to methods based on the Yoshida effect.[296] Interestingly, the use of sepiolite–DNA systems has been recently extended to gene transfection to eukaryotic cells.[296]

1.5 Conclusions

Conventional polymer–clay nanocomposites are based on the delamination ability of phyllosilicates like smectites, giving rise to highly dispersed elemental silicate layers in the polymer matrix that introduce significant improvement in key properties for advanced applications. In the case of sepiolite and palygorskite fibrous clays, which are not exfoliable, it is necessary to use fibrillated silicates produced by physical processes that may help to enable homogeneous distribution in the polymer matrix. This is especially useful when hydrophilic polymers are used, as for instance, in the case of bionanocomposites based on polysaccharides, proteins and nucleic acids.

The functionality of the resulting polymer–clay nanocomposites can be provided not only by the nature of the polymer matrix but also by the functions deliberately introduced in the starting silicates by chemical reactions or by NP assembly, always based on interactions with silanol groups located at the external surface of the clays. In this way, the versatility of functions that can be introduced in the nanocomposites derived from sepiolite and palygorskite can be useful for a wide variety of applications. So, it can be expected that not only typical applications, such as reinforcing nanofillers for rubbers and thermoplastics, but also novel uses emerge based on the development of functional composites with magnetic, optical and conducting properties, as well as the development of active phases for sensor devices, membranes and bioplastics for food packaging, drug delivery systems, adjuvant of vaccines and gene vectors in non-viral transfection processes. A great future can therefore be estimated for this type of advanced material, which still require important progress in the basic and applied research on this topic.

Acknowledgements

This work was partially supported by the CICYT, Spain (project MAT2012-31759) and the EU COST Action MP1202.

References

1. *Particulate fillers for polymers, Rapra Rev. Rep. 12(9), Report 141*, ed. R. N. Rothon, Rapra Technology Ltd., 2001, vol. 12.
2. Y. Fukushima and S. Inagaki, *J. Inclusion Phenom.*, 1987, **5**, 473–482.
3. Y. Fukushima, A. Okada, M. Kawasumi, T. Kurauchi and O. Kamigaito, *Clay Miner.*, 1988, **23**, 27–34.
4. *Handbook of Clay Science*, ed. F. Bergaya and G. Lagaly, Elsevier Science Ltd, Oxford, 2nd edn, 2013.
5. *Polymer-Clay Nanocomposites*, ed. T. J. Pinnavaia and G. Beall, John Wiley & Sons, New York, 2000.
6. M. Alexandre and P. Dubois, *Mater. Sci. Eng., R*, 2000, **28**, 1–63.
7. S. S. Ray and M. Okamoto, *Prog. Polym. Sci.*, 2003, **28**, 1539–1641.
8. E. Ruiz-Hitzky and A. Van Meerbeek, in *Handbook of clay science. Development in clay science*, ed. F. Bergaya, B. K. G. Theng and G. Lagaly, Elsevier, Amsterdam, 2006, pp. 583–621.
9. D. R. Paul and L. M. Robeson, *Polymer*, 2008, **49**, 3187–3204.
10. S. Pavlidou and C. D. Papaspyrides, *Prog. Polym. Sci.*, 2008, **33**, 1119–1198.
11. J.-F. Lambert and F. Bergaya, in *Handbook of Clay Science*, ed. F. Bergaya and G. Lagaly, Elsevier Science Ltd, Oxford, 2nd edn, 2013, pp. 679–706.
12. F. Chivrac, E. Pollet, M. Schmutz and L. Averous, *Carbohydr. Polym.*, 2010, **80**, 145–153.
13. F. M. Fernandes, A. I. Ruiz, M. Darder, P. Aranda and E. Ruiz-Hitzky, *J. Nanosci. Nanotechnol.*, 2009, **9**, 221–229.

14. K. Fukushima, D. Tabuani, C. Abbate, M. Arena and L. Ferreri, *Polym. Degrad. Stab.*, 2010, **95**, 2049–2056.
15. K. Fukushima, D. Tabuani, M. Dottori, I. Armentano, J. M. Kenny and G. Camino, *Polym. Degrad. Stab.*, 2011, **96**, 2120–2129.
16. E. Ruiz-Hitzky, P. Aranda, A. Alvarez, J. Santarén and A. Esteban-Cubillo, in *Developments in Palygorskite-Sepiolite Research. A New Outlook on these Nanomaterials*, ed. E. Galán and A. Singer, Elsevier B.V., Oxford, UK, 2011, pp. 393–452.
17. E. Ruiz-Hitzky, P. Aranda, M. Darder and F. M. Fernandes, in *Handbook of Clay Science. Part A: Fundamentals*, ed. F. Bergaya and G. Lagaly, Elsevier, Amsterdam & Oxford, 2nd edn, 2013, pp. 721–741.
18. E. Ruiz-Hitzky, M. Darder, A. S. Alcântara, B. Wicklein and P. Aranda, in *Organic-Inorganic Hybrid Nanomaterials*, ed. S. Kalia and Y. Haldorai, Springer International Publishing, 2015, vol. 267, pp. 39–86.
19. A. C. S. Alcântara, M. Darder, P. Aranda, A. Ayral and E. Ruiz-Hitzky, *J. Appl. Polym. Sci.*, 2016, **133**, 42362.
20. A. Álvarez, J. Santarén, A. Esteban-Cubillo and P. Aparicio, in *Developments in Palygorskite-Sepiolite Research. A New Outlook on these Nanomaterials*, ed. E. Galán and A. Singer, Elsevier B.V., Oxford, UK, 2011, pp. 281–298.
21. K. Brauner and A. Pressinger, *Mineral. Petrol.*, 1956, **6**, 120–140.
22. W. F. Bradley, *Am. Mineral.*, 1940, **25**, 405–410.
23. J. Santarén, J. Sanz and E. Ruiz-Hitzky, *Clay Miner.*, 1990, **38**, 63–68.
24. E. Ruiz-Hitzky, *J. Mater. Chem.*, 2001, **11**, 86–91.
25. E. Ruiz-Hitzky, M. Darder, F. M. Fernandes, B. Wicklein, A. C. S. Alcantara and P. Aranda, *Prog. Polym. Sci.*, 2013, **38**, 1392–1414.
26. F. M. Fernandes, L. Vazquez, E. Ruiz-Hitzky, A. Carnicero and M. Castro, *RSC Adv.*, 2014, **4**, 11225–11231.
27. *Developments in Palygorskite-Sepiolite Research. A new Outlook on these Nanomaterials*, ed. E. Galán and A. Singer, Elsevier B.V., Oxford, UK, 2011.
28. C. Ruiz-García, P. Aranda, F. M. Fernandes and E. Ruiz-Hitzky, Spanish Pat. P201330615, 2013.
29. F. M. Fernandes, PhD thesis, Autonomous University of Madrid, 2011.
30. A. Alvarez Berenguer, A. Perez Castells, J. J. Aragon Martinez and M. A. Esteban Aldezabal, Spanish Pat. EP 0170299 B1, 1990.
31. F. M. Fernandes and E. Ruiz-Hitzky, *Carbon*, 2014, **72**, 296–303.
32. E. Ruiz-Hitzky and F. M. Fernandes, WO2011070208, 2011.
33. E. Ruiz-Hitzky, in *Functional Hybrid Materials*, ed. P. Gómez-Romero and C. Sanchez, Wiley-VCH, Weinheim, 2004, pp. 15–49.
34. E. Ruiz-Hitzky and J. J. Fripiat, *Clays Clay Miner.*, 1976, **24**, 25–30.
35. N. García, J. Guzmán, E. Benito, A. Esteban-Cubillo, E. Aguilar, J. Santarén and P. Tiemblo, *Langmuir*, 2011, **27**, 3952–3959.
36. A. Alvarez, J. Santaren, R. Perez-Castells, B. Casal, E. Ruiz-Hitzky and P. Levitz, in *Proceedings of the International Clay Conference Denver, 1985*, ed. L. G. Schultz, H. van Olphen and F. A. Mumpton, The Clay Minerals Society, Bloomington, IN, USA, 1987, pp. 370–374.

37. Z. H. Li, C. A. Willms and K. Kniola, *Clays Clay Miner.*, 2003, **51**, 445–451.
38. A. J. Aznar, B. Casal, E. Ruiz-Hitzky, I. Lopez-Arbeloa, F. Lopez-Arbeloa, J. Santaren and A. Alvarez, *Clay Miner.*, 1992, **27**, 101–108.
39. M. Sánchez del Río, A. Doménech, M. T. Doménech-Carbó, M. L. Vázquez de Agredos Pascual, M. Suárez and E. García-Romero, in *Developments in Palygorskite-Sepiolite Research. A New Outlook on these Nanomaterials*, ed. E. Galán and A. Singer, Elsevier B.V., Oxford, UK, 2011, pp. 453–481.
40. P. Tagle, A. Alberto, G. Infante de los Reyes, H. Paschinger and H. Richard, Chemical physical characterization of a blue pigment from colonial Havana, ICCROM, 1989.
41. B. Hubbard, W. X. Kuang, A. Moser, G. A. Facey and C. Detellier, *Clays Clay Miner.*, 2003, **51**, 318–326.
42. R. Kleber, L. Masschelein-Kleiner and J. Thissen, *Stud. Conserv.*, 1967, **12**, 41–56.
43. H. van Olphen, *Science*, 1966, **154**, 645–646.
44. M. Sanchez del Rio, E. Boccaleri, M. Milanesio, G. Croce, W. van Beek, C. Tsiantos, G. D. Chyssikos, V. Gionis, G. H. Kacandes, M. Suarez and E. Garcia-Romero, *J. Mater. Sci.*, 2009, **44**, 5524–5536.
45. N. Volle, L. Challier, A. Burr, F. Giulieri, S. Pagnotta and A. M. Chaze, *Compos. Sci. Technol.*, 2011, **71**, 1685–1691.
46. C. Ouellet-Plamondon, P. Aranda, A. Favier, G. Habert, H. Van Damme and E. Ruiz-Hitzky, *RSC Adv.*, 2015, **5**, 98834–98841.
47. Y. González-Alfaro, P. Aranda, F. M. Fernandes, B. Wicklein, M. Darder and E. Ruiz-Hitzky, *Adv. Mater.*, 2011, **23**, 5224–5228.
48. M. Darder, Y. Gonzalez-Alfaro, P. Aranda and E. Ruiz-Hitzky, *RSC Adv.*, 2014, **4**, 35415–35421.
49. P. Aranda, R. Kun, M. A. Martin-Luengo, S. Letaief, I. Dekany and E. Ruiz-Hitzky, *Chem. Mater.*, 2008, **20**, 84–91.
50. A. Alvarez Berenguer, A. M. Casado Barrasa, A. Esteban Cubillo, J. Gravalos Moreno, A. J. Sanchez Rojo, J. Santaren Rome and J. Vera Agullo, WO 2013020972 A2, 2013.
51. Web of Science, www.webofknowledge.com/ accessed on October 21st, 2015.
52. S. B. Xie, S. M. Zhang, F. S. Wang, M. S. Yang, R. Seguela and J. M. Lefebvre, *Compos. Sci. Technol.*, 2007, **67**, 2334–2341.
53. D. García-López, J. F. Fernández, J. C. Merino and J. M. Pastor, *Composites, Part B*, 2013, **45**, 459–465.
54. F. C. Tsai, P. Li, Z. W. Liu, G. Feng, P. Zhu, C. K. Wang, K. N. Chen, C. Y. Huang and J. T. Yeh, *J. Appl. Polym. Sci.*, 2012, **126**, 1906–1916.
55. Z. Zhan, M. Xu and B. Li, *Polym. Degrad. Stab.*, 2015, **117**, 66–74.
56. L. Shen, Y. J. Lin, Q. G. Du, W. Zhong and Y. L. Yang, *Polymer*, 2005, **46**, 5758–5766.
57. E. Bilotti, E. Duquesne, H. Deng, R. Zhang, F. Quero, S. N. Georgiades, H. R. Fischer, P. Dubois and T. Peijs, *Eur. Polym. J.*, 2014, **56**, 131–139.
58. N. García, M. Hoyos, J. Guzmán and P. Tiemblo, *Polym. Degrad. Stab.*, 2009, **94**, 39–48.

59. J. Gao, Q. Zhang, K. Wang, Q. Fu, Y. Chen, H. Chen, H. Huang and J. M. Rego, *Composites, Part A*, 2012, **43**, 562–569.
60. M. Shafiq, T. Yasin and S. Saeed, *J. Appl. Polym. Sci.*, 2012, **123**, 1718–1723.
61. A. Carrero, R. van Grieken, I. Suarez and B. Paredes, *J. Appl. Polym. Sci.*, 2012, **126**, 987–997.
62. K. Núñez, R. Gallego, J. M. Pastor and J. C. Merino, *Appl. Clay Sci.*, 2014, **101**, 73–81.
63. J. Chen, J. Chen, S. Zhu, Y. Cao and H. Li, *J. Appl. Polym. Sci.*, 2011, **121**, 899–908.
64. B. Panda, S. Mohanty and S. Nayak, *J. Mater. Eng. Perform.*, 2014, **23**, 3229–3244.
65. R. Gul, A. Islam, T. Yasin and S. Mir, *J. Appl. Polym. Sci.*, 2011, **121**, 2772–2777.
66. M. Shafiq and T. Yasin, *Radiat. Phys. Chem.*, 2012, **81**, 52–56.
67. M. He, W. C. Cao, L. J. Wang and C. A. Wilkie, *Polym. Adv. Technol.*, 2013, **24**, 1081–1088.
68. T. D. Hapuarachchi, T. Peijs and E. Bilotti, *Polym. Adv. Technol.*, 2013, **24**, 331–338.
69. P. Tiemblo, E. Benito, N. Garcia, A. Esteban-Cubillo, R. Pina-Zapardiel and C. Pecharroman, *RSC Adv.*, 2012, **2**, 915–919.
70. L. Fernández-García, C. Pecharromán, A. Esteban-Cubillo, P. Tiemblo, N. García and J. L. Menéndez, *J. Nanopart. Res.*, 2013, **15**, 1–6.
71. J. L. Menendez, L. Fernandez-Garcia, C. Pecharroman, I. Montero, A. Esteban-Cubillo, P. Tiemblo and N. Garcia, *Opt. Mater. Express*, 2015, **5**, 1927–1933.
72. W. Zhong, P. Liu and A. Wang, *Mater. Lett.*, 2012, **85**, 11–13.
73. H. C. Bidsorkhi, M. Soheilmoghaddam, R. H. Pour, H. Adelnia and Z. Mohamad, *Polym. Test.*, 2014, **37**, 117–122.
74. H. Adelnia, H. C. Bidsorkhi, A. F. Ismail and T. Matsuura, *Sep. Purif. Technol.*, 2015, **146**, 351–357.
75. H. J. Cho, S. Park and C.-S. Ha, *Polym. Int.*, 2015, **64**, 96–104.
76. N. Huang, Z. Chen, H. Liu and J. Wang, *J. Macromol. Sci., Part B*, 2012, **52**, 521–529.
77. F. Chen, D. Lou, J. Yang and M. Zhong, *Polym. Adv. Technol.*, 2011, **22**, 1912–1918.
78. Y. Liu, P. Liu and Z. Su, *Polym. Int.*, 2008, **57**, 306–310.
79. H. Zhang, C. Li, L. Zang, J. Luo and J. Guo, *J. Macromol. Sci., Part A: Pure Appl. Chem.*, 2012, **49**, 154–159.
80. H. Xu, H. Yang, S. Xue, J. Pan, Q. Ni and F. Gong, *J. Appl. Polym. Sci.*, 2015, **132**, 42762.
81. J. Lu, W. Zhou, J. Chen, Y. Jin, K. B. Walters and S. Ding, *RSC Adv.*, 2015, **5**, 9416–9424.
82. J. Wang, Q. Wang, Y. Zheng and A. Wang, *Polym. Compos.*, 2013, **34**, 274–281.
83. Y. Wang and D. Chen, *J. Colloid Interface Sci.*, 2012, **372**, 245–251.
84. Y. Wang, A. Dong, Z. Yuan and D. Chen, *Colloids Surf., A*, 2012, **415**, 68–76.

85. Z. Yuan, Y. Wang and D. Chen, *J. Mater. Sci.*, 2014, **49**, 3287–3296.
86. L. Zhao, P. Liu, G. Liang, A. Gu, L. Yuan and Q. Guan, *Appl. Surf. Sci.*, 2014, **288**, 435–443.
87. A. Nohales, L. Solar, I. Porcar, C. I. Vallo and C. M. Gomez, *Eur. Polym. J.*, 2006, **42**, 3093–3101.
88. R. Wang, Z. Li, Y. Wang, W. Liu, L. Deng, W. Jiao and F. Yang, *Polym. Compos.*, 2013, **34**, 22–31.
89. D. Foix, M. T. Rodríguez, F. Ferrando, X. Ramis and A. Serra, *Prog. Org. Coat.*, 2012, **75**, 364–372.
90. A. Nohales, D. López, M. Culebras and C. M. Gómez, *Polym. Int.*, 2013, **62**, 397–405.
91. P. Verge, T. Fouquet, C. Barrère, V. Toniazzo, D. Ruch and J. A. S. Bomfim, *Compos. Sci. Technol.*, 2013, **79**, 126–132.
92. X. Niu, L. Huo, C. Cai, J. Guo and H. Zhou, *Ind. Eng. Chem. Res.*, 2014, **53**, 16359–16365.
93. A. Gomez-Aviles, P. Aranda, F. M. Fernandes, C. Belver and E. Ruiz-Hitzky, *J. Nanosci. Nanotechnol.*, 2013, **13**, 2897–2907.
94. J. A. Marins, A. Mija, J.-M. Pin, F. Giulieri, B. G. Soares, N. Sbirrazzuoli, P. Lançon and G. Bossis, *Compos. Sci. Technol.*, 2015, **112**, 34–41.
95. R. Wang, Z. Li, W. Liu, W. Jiao, L. Hao and F. Yang, *Compos. Sci. Technol.*, 2013, **87**, 29–35.
96. H. Chen, M. Zheng, H. Sun and Q. Jia, *Mater. Sci. Eng., A*, 2007, **445**, 725–730.
97. C.-H. Wang, M. L. Auad, N. E. Marcovich and S. Nutt, *J. Appl. Polym. Sci.*, 2008, **109**, 2562–2570.
98. Z. Lei, Q. Yang, S. Wu and X. Song, *J. Appl. Polym. Sci.*, 2009, **111**, 3150–3162.
99. B. Xu, W. M. Huang, Y. T. Pei, Z. G. Chen, A. Kraft, R. Reuben, J. T. M. De Hosson and Y. Q. Fu, *Eur. Polym. J.*, 2009, **45**, 1904–1911.
100. H. Chen, H. Lu, Y. Zhou, M. Zheng, C. Ke and D. Zeng, *Polym. Degrad. Stab.*, 2012, **97**, 242–247.
101. L. Peng, L. Zhou, Y. Li, F. Pan and S. Zhang, *Compos. Sci. Technol.*, 2011, **71**, 1280–1285.
102. Y. Bao, Q. Li, P. Xue, J. Wang and C. Wu, *Colloid Polym. Sci.*, 2012, **290**, 1527–1536.
103. J. Li, L. Yan, H. Li, J. Li, F. Zha and Z. Lei, *RSC Adv.*, 2015, **5**, 53802–53808.
104. J. Li, L. Yan, H. Li, W. Li, F. Zha and Z. Lei, *J. Mater. Chem. A*, 2015, **3**, 14696–14702.
105. J. Yang, Y. Tang, J. Xu, B. Chen, H. Tang and C. Li, *Surf. Coat. Technol.*, 2015, **272**, 285–290.
106. M. Alkan and R. Benlikaya, *J. Appl. Polym. Sci.*, 2009, **112**, 3764–3774.
107. D. Killeen, M. Frydrych and B. Chen, *Mater. Sci. Eng., C: Biomimetic Supramol. Syst.*, 2012, **32**, 749–757.
108. P. Liu, L. Jiang, L. Zhu and A. Wang, *React. Funct. Polym.*, 2014, **74**, 72–80.
109. P. Liu, L. Jiang, L. Zhu and A. Wang, *Ind. Eng. Chem. Res.*, 2014, **53**, 4277–4285.

110. S. Ekici, Y. Isikver and D. Saraydin, *Polym. Bull.*, 2006, **57**, 231–241.
111. A. Li, A. Q. Wang and J. M. Chen, *J. Appl. Polym. Sci.*, 2004, **92**, 1596–1603.
112. F. Q. Zhang, Z. J. Guo, H. Gao, Y. C. Li, L. Ren, L. Shi and L. X. Wang, *Polym. Bull.*, 2005, **55**, 419–428.
113. A. Li, A. Wang and J. Chen, *J. Appl. Polym. Sci.*, 2004, **94**, 1869–1876.
114. J. Chen, S. Ding, Y. Jin and J. Wu, *J. Appl. Polym. Sci.*, 2013, **128**, 1779–1784.
115. L. Zhu, P. Liu and A. Wang, *Ind. Eng. Chem. Res.*, 2014, **53**, 2067–2071.
116. G. Gao, G. Du, Y. Cheng and J. Fu, *J. Mat. Chem. B*, 2014, **2**, 1539–1548.
117. P. Aranda, M. Darder, R. Fernandez-Saavedra, M. Lopez-Blanco and E. Ruiz-Hitzky, *Thin Solid Films*, 2006, **495**, 104–112.
118. P. Aranda and E. Ruiz-Hitzky, *Chem. Mater.*, 1992, **4**, 1395–1403.
119. S. Inagaki, Y. Fukushima, H. Doi and O. Kamigaito, *Clay Miner.*, 1990, **25**, 99–105.
120. S. Inagaki, Y. Fukushima and M. Miyata, *Res. Chem. Intermed.*, 1995, **21**, 167–180.
121. Y. Kitayama, H. Katoh, T. Kodama and J. Abe, *Appl. Surf. Sci.*, 1997, **121**, 331–334.
122. Y. Wang, P. Liu, C. Yang, B. Mu and A. Wang, *Electrochim. Acta*, 2013, **89**, 422–428.
123. Y. Chang, Z. Liu, Z. Fu, C. Wang, Y. Dai, R. Peng and X. Hu, *Ind. Eng. Chem. Res.*, 2014, **53**, 38–47.
124. H. Feng, B. Wang, N. Wang, J. Qiu, L. Tan and N. Chen, *J. Appl. Polym. Sci.*, 2015, **132**, 41407.
125. C. Yao, Y. Xu, Y. Kong, W. Liu, W. Wang, Z. Wang, Y. Wang and J. Ji, *Appl. Clay Sci.*, 2012, **67–68**, 32–35.
126. J. Chen, X. Hong, Q. Xie, D. Li and Q. Zhang, *J. Chem. Eng. Data*, 2014, **59**, 2275–2282.
127. Y. Kong, H. Ge, J. Xiong, S. Zuo, Y. Wei, C. Yao and L. Deng, *Appl. Clay Sci.*, 2014, **99**, 119–124.
128. L. Shao, J. Qiu, L. Lei and X. Wu, *Synth. Met.*, 2012, **162**, 2322–2328.
129. J. A. Marins, F. Giulieri, B. G. Soares and G. Bossis, *Synth. Met.*, 2013, **185**, 9–16.
130. D. S. Jang and H. J. Choi, *Colloids Surf., A*, 2015, **469**, 20–28.
131. G. Bossis, J. A. Marins, P. Kuzhir, O. Volkova and A. Zubarev, *J. Intell. Mater. Syst. Struct.*, 2015, **26**, 1871–1879.
132. H. Hu, M. Gan, L. Ma, Z. Li, Y. Li, C. Ge, Y. Tu, L. Yu, H. Huang and F. Yang, *Compos. Sci. Technol.*, 2014, **104**, 9–17.
133. M. Oliveira, L. de Miranda Santos, M. de Gois da Silva, H. da Cunha, E. da Silva Filho and C. da Silva Leite, *J. Therm. Anal. Calorim.*, 2015, **119**, 37–46.
134. B. Mu and A. Wang, *J. Mater. Chem. A*, 2015, **3**, 281–289.
135. E. Ruiz-Hitzky and P. Aranda, *Adv. Mater.*, 1990, **2**, 545–547.
136. E. Ruiz-Hitzky, P. Aranda, B. Casal and J. C. Galvan, *Adv. Mater.*, 1995, **7**, 180–184.
137. E. Ruiz-Hitzky, *Adv. Mater.*, 1993, **5**, 334–340.

138. R. A. Vaia, S. Vasudevan, W. Krawiec, L. G. Scanlon and E. P. Giannelis, *Adv. Mater.*, 1995, **7**, 154–156.
139. P. Aranda, Y. Mosqueda, E. Perez-Cappe and E. Ruiz-Hitzky, *J. Polym. Sci., Part B: Polym. Phys.*, 2003, **41**, 3249–3263.
140. A. Mejia, N. Garcia, J. Guzman and P. Tiemblo, *Eur. Polym. J.*, 2013, **49**, 118–129.
141. A. Mejía, N. García, J. Guzmán and P. Tiemblo, *Appl. Clay Sci.*, 2014, **95**, 265–274.
142. A. Mejía, N. García, J. Guzmán and P. Tiemblo, *Solid State Ionics*, 2014, **261**, 74–80.
143. A. Mejía, N. García, J. Guzmán and P. Tiemblo, *Electrochim. Acta*, 2014, **137**, 526–534.
144. C. Beauger, G. Lainé, A. Burr, A. Taguet, B. Otazaghine and A. Rigacci, *J. Membr. Sci.*, 2013, **430**, 167–179.
145. S. Wen, C. Gong, Y.-C. Shu, F.-C. Tsai and J.-T. Yeh, *J. Appl. Polym. Sci.*, 2012, **123**, 646–656.
146. J. Luo, C. Liu, Y. Song, P. Cui and Y. Zhang, *Int. J. Polym. Mater. Polym. Biomater.*, 2015, **64**, 831–837.
147. R. Fernández-Saavedra, P. Aranda and E. Ruiz-Hitzky, *Adv. Funct. Mater.*, 2004, **14**, 77–82.
148. R. Fernandez-Saavedra, P. Aranda, K. A. Carrado, G. Sandi, S. Seifert and E. Ruiz-Hitzky, *Curr. Nanosci.*, 2009, **5**, 506–513.
149. T. Kyotani, N. Sonobe and A. Tomita, *Nature*, 1988, **331**, 331–333.
150. R. Fernández-Saavedra, M. Darder, A. Gómez-Avilés, P. Aranda and E. Ruiz-Hitzky, *J. Nanosci. Nanotechnol.*, 2008, **8**, 1741–1750.
151. G. Sandi, K. A. Carrado, R. E. Winans, C. S. Johnson and R. Csencsits, *J. Electrochem. Soc.*, 1999, **146**, 3644–3648.
152. P. C. LeBaron, Z. Wang and T. J. Pinnavaia, *Appl. Clay Sci.*, 1999, **15**, 11–29.
153. V. Ojijo and S. S. Ray, *Prog. Polym. Sci.*, 2013, **38**, 1543–1589.
154. K. Fukushima, D. Tabuani and G. Camino, *Mater. Sci. Eng., C: Biomimetic Supramol. Syst.*, 2009, **29**, 1433–1441.
155. P. Russo, S. Cammarano, E. Bilotti, T. Peijs, P. Cerruti and D. Acierno, *J. Appl. Polym. Sci.*, 2014, **131**, n/a-n/a.
156. Y. Jiang, S. Han, S. Zhang, J. Li, G. Huang, Y. Bi and Q. Chai, *Polym. Compos.*, 2014, **35**, 468–476.
157. N. Moazeni, Z. Mohamad and N. Dehbari, *J. Appl. Polym. Sci.*, 2015, **132**, n/a-n/a.
158. R. M. d. S. M. Thiré, L. C. Arruda and L. S. Barreto, *Mater. Res.*, 2011, **14**, 340–344.
159. Z. Qi, H. Ye, J. Xu, J. Peng, J. Chen and B. Guo, *Colloids Surf., A*, 2013, **436**, 26–33.
160. J. Wu, X. X. Zou, B. Jing and W. L. Dai, *Polym. Eng. Sci.*, 2015, **55**, 1104–1112.
161. D. L. Lynch, L. M. Wright and L. J. Cotnoir, *Nature*, 1957, **179**, 1131.
162. B. Kosan, C. Michels and F. Meister, *Cellulose*, 2008, **15**, 59–66.

163. W. Lan, C.-F. Liu, F.-X. Yue, R.-C. Sun and J. F. Kennedy, *Carbohydr. Polym.*, 2011, **86**, 672–677.
164. M. Soheilmoghaddam, M. U. Wahit, A. A. Yussuf, M. A. Al-Saleh and W. T. Whye, *Polym. Test.*, 2014, **33**, 121–130.
165. Z. Sárossy, T. O. J. Blomfeldt, M. S. Hedenqvist, C. B. Koch, S. S. Ray and D. Plackett, *ACS Appl. Mater. Interfaces*, 2012, **4**, 3378–3386.
166. C. H. Ünlü, E. Günister and O. Atıcı, *Carbohydr. Polym.*, 2009, **76**, 585–592.
167. M. Darder, M. Lopez-Blanco, P. Aranda, A. J. Aznar, J. Bravo and E. Ruiz-Hitzky, *Chem. Mater.*, 2006, **18**, 1602–1610.
168. P. Martínez-Frías, Industrial Engineering Degree Dissertation, Autonomous University of Madrid, 2008.
169. D. Huang, B. Mu and A. Wang, *Mater. Lett.*, 2012, **86**, 69–72.
170. D. Huang, W. Wang, J. Xu and A. Wang, *Chem. Eng. J.*, 2012, **210**, 166–172.
171. H. Lu, W. Wang and A. Wang, *RSC Adv.*, 2015, **5**, 17775–17781.
172. J. B. Olivato, J. Marini, E. Pollet, F. Yamashita, M. V. E. Grossmann and L. Avérous, *Carbohydr. Polym.*, 2015, **118**, 250–256.
173. M. D. Samper-Madrigal, O. Fenollar, F. Dominici, R. Balart and J. M. Kenny, *J. Mater. Sci.*, 2015, **50**, 863–872.
174. L. J. Gibson and M. F. Ashby, *Cellular solids: structure and properties*, Cambridge University Press, Cambridge, UK, 2nd edn, 1997.
175. M. Okamoto, in *Bio-inorganic hybrid nanomaterials. Strategies, syntheses, characterization and applications*, ed. E. Ruiz-Hitzky, K. Ariga and Y. M. Lvov, Wiley-VCH, Weinheim, 2008, pp. 271–312.
176. M. Chen, B. Q. Chen and J. R. G. Evans, *Nanotechnology*, 2005, **16**, 2334–2337.
177. S. Deville, E. Saiz, R. K. Nalla and A. P. Tomsia, *Science*, 2006, **311**, 515–518.
178. M. Darder, P. Aranda, M. L. Ferrer, M. C. Gutiérrez, F. del Monte and E. Ruiz-Hitzky, *Adv. Mater.*, 2011, **23**, 5262–5267.
179. E. Ruiz-Hitzky, P. Aranda, M. Darder, F. M. Fernandes and C. R. S. Matos, Spanish Pat. 2342871, 2010.
180. M. Darder, C. R. S. Matos, P. Aranda and E. Ruiz-Hitzky, *SEA-CSSJ-CMS Trilateral Meeting on Clays*, Sevilla, Spain, 2010.
181. B. Wicklein, A. Kocjan, G. Salazar-Alvarez, F. Carosio, G. Camino, M. Antonietti and L. Bergström, *Nat. Nano*, 2015, **10**, 277–283.
182. V. K. Gupta and Suhas, *J. Environ. Manage.*, 2009, **90**, 2313–2342.
183. M. K. Okajima, S. Miyazato and T. Kaneko, *Langmuir*, 2009, **25**, 8526–8531.
184. A. C. S. Alcantara, M. Darder, P. Aranda, S. Tateyama, M. K. Okajima, T. Kaneko, M. Ogawa and E. Ruiz-Hitzky, *J. Mater. Chem. A*, 2014, **2**, 1391–1399.
185. A. C. S. Alcantara, M. Darder, P. Aranda and E. Ruiz-Hitzky, *Appl. Clay Sci.*, 2014, **96**, 2–8.
186. X. Zou, J. Pan, H. Ou, X. Wang, W. Guan, C. Li, Y. Yan and Y. Duan, *Chem. Eng. J.*, 2011, **167**, 112–121.

187. C. Pang, Y. Liu, X. Cao, R. Hua, C. Wang and C. Li, *J. Radioanal. Nucl. Chem.*, 2010, **286**, 185–193.
188. C. Li, J. Pan, X. Zou, J. Gao, J. Xie and Y. Yongsheng, *Int. J. Environ. Anal. Chem.*, 2011, **91**, 1035–1049.
189. Y. Y. Shi, Q. H. Zhang, L. D. Feng, Q. P. Xiong and J. Chen, *Korean J. Chem. Eng.*, 2014, **31**, 821–827.
190. Y. Deng, L. Wang, X. Hu, B. Liu, Z. Wei, S. Yang and C. Sun, *Chem. Eng. J.*, 2012, **181**, 300–306.
191. J. Wu and J. Chen, *J. Appl. Polym. Sci.*, 2013, **127**, 1765–1771.
192. Y. Peng, D. Chen, J. Ji, Y. Kong, H. Wan and C. Yao, *Appl. Clay Sci.*, 2013, **74**, 81–86.
193. E. Cohen and T. Joseph, *Appl. Clay Sci.*, 2009, **42**, 569–574.
194. E. Ruiz-Hitzky, M. Darder, P. Aranda and K. Ariga, *Adv. Mater.*, 2010, **22**, 323–336.
195. H. Q. Wang, F. Hua, Y. C. Zhao, Y. Li and X. Wang, *Biotechnol. Biotechnol. Equip.*, 2014, **28**, 834–842.
196. G. Rytwo, R. Lavi, Y. Rytwo, H. Monchase, S. Dultz and T. N. Koenig, *Sci. Total Environ.*, 2013, **442**, 134–142.
197. G. Pan, H. Zou, H. Chen and X. Z. Yuan, *Environ. Pollut.*, 2006, **141**, 206–212.
198. B. Mu, Y. Kang and A. Wang, *J. Mater. Chem. A*, 2013, **1**, 4804–4811.
199. W. Wang, F. Wang, Y. Kang and A. Wang, *RSC Adv.*, 2013, **3**, 11515–11520.
200. B. Ni, M. Liu, S. Lue, L. Xie and Y. Wang, *J. Agric. Food Chem.*, 2010, **58**, 12373–12378.
201. X. Wang, S. Lu, C. Gao, X. Xu, Y. Wei, X. Bai, C. Feng, N. Gao, M. Liu and L. Wu, *RSC Adv.*, 2014, **4**, 18382–18390.
202. X. Zhou, Q. Liu, G. Ying and Y. Cui, in *Advanced Materials and Processes Ii*, Pts 1–3, ed. H. B. Ji, Y. Chen and S. Z. Chen, 2012, vol. 557–559, pp. 1528–1532.
203. Y. Liu, W. B. Wang and A. G. Wang, *Desalination*, 2010, **259**, 258–264.
204. X. Wang and A. Wang, *Sep. Sci. Technol.*, 2010, **45**, 2086–2094.
205. L. H. Zhu, L. L. Zhang, Y. J. Tang and X. Y. Kou, *Polym.-Plast. Technol. Eng.*, 2014, **53**, 74–79.
206. Y. Zhu, Y. Zheng and A. Wang, *Int. J. Biol. Macromol.*, 2015, **72**, 410–420.
207. Y. Zheng, J. Zhang and A. Wang, *Chem. Eng. J.*, 2009, **155**, 215–222.
208. G. R. Mahdavinia and A. Asgari, *Polym. Bull.*, 2013, **70**, 2451–2470.
209. L. Wang, J. Zhang and A. Wang, *Desalination*, 2011, **266**, 33–39.
210. Y. Zheng, Y. Zhu and A. Wang, *Chem. Eng. J.*, 2014, **257**, 66–73.
211. G. Tian, W. Wang, B. Mu, Y. Kang and A. Wang, *J. Taiwan Inst. Chem. Eng.*, 2015, **50**, 252–258.
212. Q. Zhou, Q. Gao, W. Luo, C. Yan, Z. Ji and P. Duan, *Colloids Surf., A*, 2015, **470**, 248–257.
213. X. Wu, P. Gao, X. Zhang, G. Jin, Y. Xu and Y. Wu, *Appl. Clay Sci.*, 2014, **95**, 60–66.
214. X.-p. Wu, Y.-q. Xu, X.-l. Zhang, Y.-c. Wu and P. Gao, *New Carbon Mater.*, 2015, **30**, 71–78.

215. A. Gómez-Avilés, M. Darder, P. Aranda and E. Ruiz-Hitzky, *Appl. Clay Sci.*, 2010, **47**, 203–211.
216. P. Aranda, M. Darder, A. Gómez-Avilés, E. Zatile and E. Ruiz-Hitzky, *2010 SEA-CSSJ-CMS Trilateral Meeting on Clays General Meeting*, Seville, 2010.
217. F. M. Kievit and M. Zhang, *Adv. Mater.*, 2011, **23**, H217–H247.
218. T. L. Doane and C. Burda, *Chem. Soc. Rev.*, 2012, **41**, 2885–2911.
219. S. Venkataraman, J. L. Hedrick, Z. Y. Ong, C. Yang, P. L. R. Ee, P. T. Hammond and Y. Y. Yang, *Adv. Drug Delivery Rev.*, 2011, **63**, 1228–1246.
220. C. Viseras, C. Aguzzi, P. Cerezo and M. C. Bedmar, *Mater. Sci. Technol.*, 2008, **24**, 1020–1026.
221. I. Salcedo, C. Aguzzi, G. Sandri, M. C. Bonferoni, M. Mori, P. Cerezo, R. Sanchez, C. Viseras and C. Caramella, *Appl. Clay Sci.*, 2012, **55**, 131–137.
222. J. Wu, S. Ding, J. Chen, S. Zhou and H. Ding, *Int. J. Biol. Macromol.*, 2014, **68**, 107–112.
223. Q. Wang, J. P. Zhang and A. Q. Wang, *Carbohydr. Polym.*, 2009, **78**, 731–737.
224. Q. Wang, J. Wu, W. Wang and A. Wang, *J. Biomater. Nanobiotechnol.*, 2011, **2**, 250–257.
225. Q. Wang, W. Wang, J. Wu and A. Wang, *J. Appl. Polym. Sci.*, 2012, **124**, 4424–4432.
226. J.-P. Amorij, W. L. J. Hinrichs, H. W. Frijlink, J. C. Wilschut and A. Huckriede, *Lancet Infect. Dis.*, 2010, **10**, 699–711.
227. E. Ruiz-Hitzky, M. Darder, P. Aranda, M. Á. Martin del Burgo and G. del Real, *Adv. Mater.*, 2009, **21**, 4167–4171.
228. T. Clapp, P. Siebert, D. Chen and L. Jones Braun, *J. Pharm. Sci.*, 2011, **100**, 388–401.
229. B. Wicklein, M. Á. Martín del Burgo, M. Yuste, M. Darder, C. Escrig Llavata, P. Aranda, J. Ortín, G. del Real and E. Ruiz-Hitzky, *Eur. J. Inorg. Chem.*, 2012, **2012**, 5186–5191.
230. F. M. Fernandes, I. Manjubala and E. Ruiz-Hitzky, *Phys. Chem. Chem. Phys.*, 2011, **13**, 4901–4910.
231. D. Su, C. Wang, S. Cai, C. Mu, D. Li and W. Lin, *Appl. Clay Sci.*, 2012, **62–63**, 41–46.
232. B. Gimenez, M. C. Gomez-Guillen, M. E. Lopez-Caballero, J. Gomez-Estaca and P. Montero, *Food Hydrocolloids*, 2012, **27**, 475–486.
233. A. C. S. Alcantara, P. Aranda, M. Darder and E. Ruiz-Hitzky, *Abstr. Pap., Am. Chem. Soc.*, 2011, **241**, 114–115.
234. A. C. S. Alcântara, M. Darder, P. Aranda and E. Ruiz Hitzky, *Eur. J. Inorg. Chem.*, 2012, **2012**, 5216–5224.
235. E. Ruiz Hitzky, P. Aranda, M. Darder and A. C. S. Alcântara, Spanish Pat. 2353100, 2010.
236. Q. Yuan, W. Lu and Y. Pan, *Polym. Degrad. Stab.*, 2010, **95**, 1581–1587.
237. H. Li, C. Li, Q. Gao, S. Zhang and J. Li, *Ind. Crops Prod.*, 2014, **59**, 35–40.
238. C. Wang, L. Ding, Q. Wu, F. Liu, J. Wei, R. Lu, H. Xie and R. Cheng, *Ind. Crops Prod.*, 2014, **57**, 29–34.
239. C. Wang, Q. Wu, F. Liu, J. An, R. Lu, H. Xie and R. Cheng, *Appl. Clay Sci.*, 2014, **101**, 246–252.

240. M. I. Carretero and M. Pozo, *Appl. Clay Sci.*, 2009, **46**, 73–80.
241. M. I. Carretero and M. Pozo, *Appl. Clay Sci.*, 2010, **47**, 171–181.
242. M. L. D. da Silva, A. C. Fortes, A. D. Tome, E. C. da Silva, R. M. de Freitas, J. L. Soares-Sobrinho, C. M. D. Leite and M. F. D. Soares, *Braz. J. Pharm. Sci.*, 2013, **49**, 729–736.
243. J. Cervini-Silva, C. Antonio Nieto, M. T. Ramirez-Apan, V. Gomez-Vidales, E. Palacios, A. Montoya and E. R. de Jesus, *Colloids Surf., B*, 2015, **129**, 1–6.
244. R. Perez-Castells, A. Alvarez, J. Gavilanes, M. A. Lizarbe, A. Martinez Del Pozo, N. Olmo and J. Santaren, in *Proceedings of the International Clay Conference Denver, 1985*, ed. L. G. Schultz, H. van Olphen and F. A. Mumpton, The Clay Minerals Society, Bloomington, IN, USA, 1987, pp. 359–362.
245. M. A. Lizarbe, N. Olmo and J. G. Gavilanes, *J. Biomed. Mater. Res.*, 1987, **21**, 137–144.
246. J. I. Herrera, N. Olmo, J. Turnay, A. Sicilia, A. Bascones, J. G. Gavilanes and M. A. Lizarbe, *Biomaterials*, 1995, **16**, 625–631.
247. N. Olmo, J. Turnay, J. I. Herrera, J. G. Gavilanes and M. A. Lizarbe, *J. Biomed. Mater. Res.*, 1996, **30**, 77–84.
248. M. Frydrych, C. Wan, R. Stengler, K. U. O'Kelly and B. Chen, *J. Mater. Chem.*, 2011, **21**, 9103–9111.
249. W.-Z. Li, G.-F. Li and J.-L. Wang, *Int. J. Polym. Mater. Polym. Biomater.*, 2013, **62**, 358–361.
250. E. P. Rebitski, Master, Universidade Federal do Rio Grande do Norte, RN, Brazil, 2015.
251. S. Cengiz, L. Çavaş and K. Yurdakoç, *Appl. Clay Sci.*, 2012, **65–66**, 114–120.
252. Q. Zhao, Y. Hou, G.-H. Gong, M.-A. Yu, L. Jiang and F. Liao, *Appl. Biochem. Biotechnol.*, 2010, **160**, 2287–2299.
253. R. M. Prodanovic, M. B. Simic and Z. M. Vujcic, *J. Serb. Chem. Soc.*, 2003, **68**, 819–824.
254. I. E. de Fuentes, C. A. Viseras, D. Ubiali, M. Terreni and A. R. Alcántara, *J. Mol. Catal. B: Enzym.*, 2001, **11**, 657–663.
255. V. Caballero, F. M. Bautista, J. M. Campelo, D. Luna, J. M. Marinas, A. A. Romero, J. M. Hidalgo, R. Luque, A. Macario and G. Giordano, *Process Biochem.*, 2009, **44**, 334–342.
256. B. Wicklein, M. Darder, P. Aranda and E. Ruiz-Hitzky, *ACS Appl. Mater. Interfaces*, 2011, **3**, 4339–4348.
257. C. Luna, C. Verdugo, E. D. Sancho, D. Luna, J. Calero, A. Posadillo, F. M. Bautista and A. A. Romero, *Molecules*, 2014, **19**, 11419–11439.
258. Q. H. You, X. L. Yin, Y. P. Zhao and Y. Zhang, *Bioresour. Technol.*, 2013, **148**, 202–207.
259. Y. Li, J. Hu and P. Han, *Chin. J. Chem. Eng.*, 2015, **23**, 822–826.
260. J. Chen and Y. Jin, *Microchim. Acta*, 2010, **169**, 249–254.
261. J. Xu, W. Han, Q. Yin, J. Song and H. Zhong, *Chin. J. Chem.*, 2009, **27**, 2197–2202.
262. T. R. de Oliveira, G. F. Grawe, S. K. Moccelini, A. J. Terezo and M. Castilho, *Analyst*, 2014, **139**, 2214–2220.

263. M. Pacwa-Plociniczak, G. A. Plaza, Z. Piotrowska-Seget and S. S. Cameotra, *Int. J. Mol. Sci.*, 2011, **12**, 633–654.
264. G. Lagaly, *Solid State Ionics*, 1986, **22**, 43–51.
265. E. Ruiz-Hitzky, P. Aranda, M. Darder and G. Rytwo, *J. Mater. Chem.*, 2010, **20**, 9306–9321.
266. C. Abbate, M. Arena, A. Baglieri and M. Gennari, *J. Hazard. Mater.*, 2009, **168**, 466–472.
267. S. Lang, *Curr. Opin. Colloid Interface Sci.*, 2002, **7**, 12–20.
268. K. Ariga, J. P. Hill, M. V. Lee, A. Vinu, R. Charvet and S. Acharya, *Science and Technology of Advanced Materials*, 2008, vol. 9, pp. 1–96.
269. T. Sun and G. Qing, *Adv. Mater.*, 2011, **23**, H57–H77.
270. X. Zhang, N. Zhao, S. Liang, X. Lu, X. Li, Q. Xie, X. Zhang and J. Xu, *Adv. Mater.*, 2008, **20**, 2938–2946.
271. K. Mark, J. Park, S. Bauer and P. Schmuki, *Cell Tissue Res.*, 2010, **339**, 131–153.
272. B. Wicklein, M. Darder, P. Aranda and E. Ruiz-Hitzky, *Langmuir*, 2010, **26**, 5217–5225.
273. B. Wicklein, Thesis, Autonomous University of Madrid, 2011.
274. A. L. Plant, *Langmuir*, 1999, **15**, 5128–5135.
275. J. B. Hubbard, V. Silin and A. L. Plant, *Biophys. Chem.*, 1998, **75**, 163–176.
276. A. Hosseini, C. J. Barile, A. Devadoss, T. A. Eberspacher, R. A. Decreau and J. P. Collman, *J. Am. Chem. Soc.*, 2011, **133**, 11100–11102.
277. H. Xie, K. Jiang and W. Zhan, *Phys. Chem. Chem. Phys.*, 2011, **13**, 17712–17721.
278. Q. Zhou and P. Somasundaran, *J. Colloid Interface Sci.*, 2009, **331**, 288–294.
279. C. M. Persson, P. M. Claesson and K. Lunkenheimer, *J. Colloid Interface Sci.*, 2002, **251**, 182–192.
280. M. Hederos, P. Konradsson and B. Liedberg, *Langmuir*, 2005, **21**, 2971–2980.
281. T. Fyrner, H.-H. Lee, A. Mangone, T. Ekblad, M. E. Pettitt, M. E. Callow, J. A. Callow, S. L. Conlan, R. Mutton, A. S. Clare, P. Konradsson, B. Liedberg and T. Ederth, *Langmuir*, 2011, **27**, 15034–15047.
282. G. W. Stubbs, H. G. Smith and B. J. Litman, *Biochim. Biophys. Acta*, 1976, **426**, 46–56.
283. G. G. Privé, *Methods (Amsterdam, Neth.)*, 2007, **41**, 388–397.
284. D. Mukherjee, M. May and B. Khomami, *J. Colloid Interface Sci.*, 2011, **358**, 477–484.
285. L. A. Goldblatt, *J. Am. Oil Chem. Soc.*, 1977, **54**, A302–A309.
286. W. F. Jaynes, R. E. Zartman and W. H. Hudnall, *Appl. Clay Sci.*, 2007, **36**, 197–205.
287. T. M. Moran, H. Park, A. Fernandez-Sesma and J. L. Schulman, *J. Infect. Dis.*, 1999, **180**, 579–585.
288. B. Wicklein, P. Aranda, E. Ruiz-Hitzky and M. Darder, *J. Mater. Chem. B*, 2013, **1**, 2911–2920.

289. E. Ruiz-Hitzky, M. Darder, B. Wicklein, F. M. Fernandes, F. A. Castro-Smirnov, M. A. M. del Burgo, G. del Real and P. Aranda, in *Nanosystems in Engineering and Medicine*, ed. S. H. Choi, J. H. Choy, U. Lee and V. K. Varadan, Spie-Int Soc Optical Engineering, Bellingham, 2012, vol. 8548.
290. E. Paget, L. J. Monrozier and P. Simonet, *FEMS Microbiol. Lett.*, 1992, **97**, 31–39.
291. N. Yoshida, *Recent Pat. Biotechnol.*, 2007, **1**, 194–201.
292. N. Yoshida and K. Ide, *Appl. Microbiol. Biotechnol.*, 2008, **80**, 813–821.
293. N. Yoshida and M. Sato, *Appl. Microbiol. Biotechnol.*, 2009, **83**, 791–798.
294. B. Bellmann, H. Muhle and H. Ernst, *Environ. Health Perspect.*, 1997, **105**, 1049–1052.
295. G. Wilharm, D. Lepka, F. Faber, J. Hofmann, T. Kerrinnes and E. Skiebe, *J. Microbiol. Methods*, 2010, **80**, 215–216.
296. F. A. Castro-Smirnov, PhD Thesis, Université Paris-Sud, 2014.

CHAPTER 2

Fibrillar Attapulgite–Rubber Nanocomposites

NANYING NING[a,b], GUANGYU CHU[a], YUXING FENG[a,b], XIUCHUN ZHEN[a], MING TIAN*[a,b] AND TUNG W. CHAN[c]

[a]State Key Lab of Organic–Inorganic Composites, Beijing University of Chemical Technology, Beijing 100029, China; [b]Key Laboratory of Carbon Fiber and Functional Polymers, Ministry of Education, Beijing University of Chemical Technology, Beijing 100029, China; [c]Department of Materials Science and Engineering, Virginia Polytechnic Institute and State University, Blacksburg, VA 24061, USA
*E-mail: tianm@mail.buct.edu.cn

2.1 Introduction

The introduction of reinforcing agents is an effective, widely applicable, and important way to enhance the strength, modulus, and heat resistance of polymers. Without carbon black and silica, the rubber industry and the tire industry would not be what they are today. In the plastics industry, the reinforcing agents with relatively large shape coefficients, such as glass fiber, talc, mica, and wollastonite, find many applications. The most commonly used and effective reinforcing agent for plastics is glass fiber. The addition of 10–30% of glass fibers can greatly improve the strength and stiffness of the polymer matrix. However, the use of glass fibers as a reinforcing agent has some disadvantages, such as the poor liquidity and serious wear and tear

RSC Smart Materials No. 22
Functional Polymer Composites with Nanoclays
Edited by Yuri Lvov, Baochun Guo and Rawil F. Fakhrullin

Published by the Royal Society of Chemistry, www.rsc.org

on equipment. In addition, it is difficult for glass-fiber-reinforced polymers to pass through intricate flow paths because of the high viscosity, and thus some high-performance precision devices cannot be fabricated from these reinforced polymers. Wollastonite, with a needle-like structure, and talc and mica, with lamellar structures, show a moderate reinforcing effect on polymers because of their lower aspect ratios. In addition, carbon fibers, metal fibers, and nylon fibers are used only for special needs because of their high costs. Whiskers have also failed to become a new generation of polymer reinforcing agents because they are expensive (such as SiC whiskers) or are easily broken during processing (such as $CaSO_4$ whiskers).

On the other hand, carbon black, nano-silica, and short organic fibers[1,2] can efficiently reinforce rubbers, but the production of carbon black depends on petroleum resources. Silica also has great potential to reinforce rubber. The disadvantage of silica is the high price and the lower reinforcing effect than that of carbon black. Thus, silica needs an expensive silane coupling agent as an auxiliary. Short fiber is a good reinforcing filler for rubber, because it gives rubbers a high modulus at low elongations. However, the diameter of the organic fibers used in composites is large (the same as that of glass fibers), and so is the aspect ratio, resulting in a high viscosity of the composites.[2] Therefore, the development of new low-cost nanoparticles and nanofiber fillers is required for reinforcing rubbers.

Layered silicates, with unique physical and chemical properties, can effectively improve the mechanical properties, the gas barrier property, and the flame retardancy of polymers and thus have attracted much attention.[3–9] The nanocrystalline layers of a layered silicate are the basic dispersion unit, with a thickness of about 1 nm and a width of 100 to 1000 nm. A disadvantage of layered silicate is the high binding force of the nanocrystalline layers. Thus, the exfoliated layered silicate in polymers can only be obtained by using special methods (such as *in situ* intercalation polymerization). Today, very few layered silicate–polymer composites, such as clay–nylon nanocomposites, have been commercialized.

Fibrillar silicate (FS) is another kind of natural clay mineral. There are several types of fibrillar silicate minerals found in nature. The most abundant type is known as attapulgite or palygorskite (AT), which is widely found in the United States and China. The main mineral of attapulgus clay, commonly called attapugite, is a crystalline hydrated magnesium aluminum silicate with a unique chain structure that gives it a nanofibrillar morphology. The high surface area, thermal resistance, and fibrous morphology of AT provide much potential for diverse applications. The primary structural units of FS are silicate single crystals that are 100–3000 nm in length and 10–25 nm in diameter. These single crystals stack into microparticles.[10,11] The FS single crystals possess a high degree of structural perfection and superior mechanical properties. The strength and modulus of the single crystals are over 50 and 500 GPa, respectively, which are at least five times higher than those of conventional fibers. Unlike layered silicates such as montmorillonite, FSs are much easier to separate into single nanocrystals and to uniformly distribute in polymer matrices[12] because the spacing among the aggregated crystals in FSs is much larger than layered silicates. As a result, the interaction (van der Waals force) between the single crystals in FSs is considerably weaker than that between the layers in layered silicates. A FS can be uniformly distributed in a polymer matrix by a simple extrusion process, even without the chemical substitution of metal ions with surfactants such as tertiary amines. Additionally, there are abundant Si–OH groups on the surface of FS single crystals, and these groups can react with silane coupling agents such as (3-glycidyloxypropyl) trimethoxysilane (GOPTMS).[13] Because of its particular structure and properties, AT has received considerable attention and is often used as an adsorbent, thickening agent, adhesive agent, decontaminant, and drier.[14,15] In recent years, the application of AT in highly profitable fields has become more and more attractive.[16,17] The AT showed a superior reinforcing effect when it was separated from nanocrystalline aggregates and dispersed at the nanoscale in the polymer matrix by *in situ* polymerization or direct blending after surface modification.[18–20]

In this chapter, we will introduce the composition, morphology, and physical and chemical characteristics of AT. Furthermore, we will summarize the preparation and interfacial design methods, dispersion and filler network of AT–rubber composites, and structure–properties relationship of AT–rubber composites, and the synergistic effect of AT and other microfibers on the mechanical properties of rubbers.

2.2 Fibrillar Attapulgite

2.2.1 Composition and Morphology of Fibrillar Attapulgite

As mentioned before, fibrillar AT is one type of hydrated magnesium aluminum silicate mineral composed of many nano-fibrillar single crystals. AT has the structural formula $Mg_5[A1]Si_8O_{20}(HO)_2(OH_2)_4 4H_2O$. Because of lattice replacement of the structure of AT, a crystal contains no quantitative Na^+, Ca^{2+}, Fe^{3+}.

The fibrillar single crystal is the smallest structural unit, with a length of 100–3000 nm and a diameter of 10–25 nm. Each single crystal consists of many unit layers. Each unit layer consists of two silicon–oxygen tetrahedrons.[21] In between adjacent unit layers are five aluminum–oxygen tetrahedrons.[22–24] There are three levels in the structure of AT: (1) the unit layers are connected by oxygen atoms to form tunnel-like structure single crystals, (2) the single crystals are compactly arranged in parallel and to form crystal bundles, and (3) the crystal bundles are then agglomerated into AT microparticles (Figure 2.1).[25] Figure 2.2 shows the three levels of structure. Because of the structure of the tunnel (Figure 2.3), a large void volume exists about AT, and it possesses a large specific surface area.

The contact area of AT is smaller than that of layered silicates because of the similar line–line contact between AT. Additionally, AT does not contain many cations like montmorillonite, so the interaction (van der Waals force)

Figure 2.1 Representative SEM images of as-received fibrillar silicate powder: (A) low magnification and (B) high magnification. (Reproduced from *Polymer*, **48**, M. Tian *et al.*, Bis-GMA/TEGDMA dental composites reinforced with electrospun nylon 6 nanocomposites nanofibers containing highly aligned fibrilar silicate single crystals, 2720–2728, Copyright 2007 with permission from Elsevier.[25])

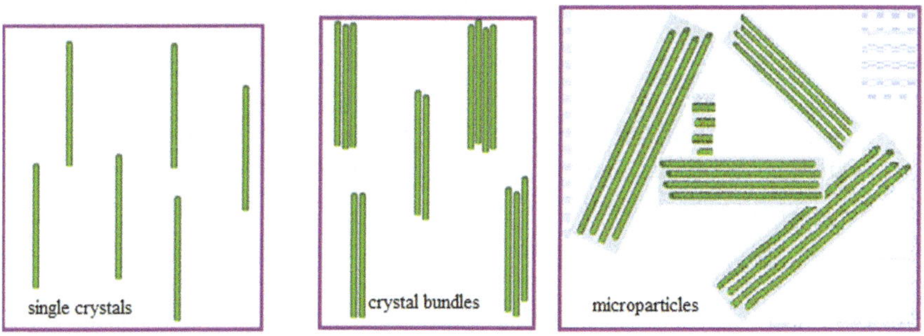

Figure 2.2 Schematic diagram of three levels of AT.

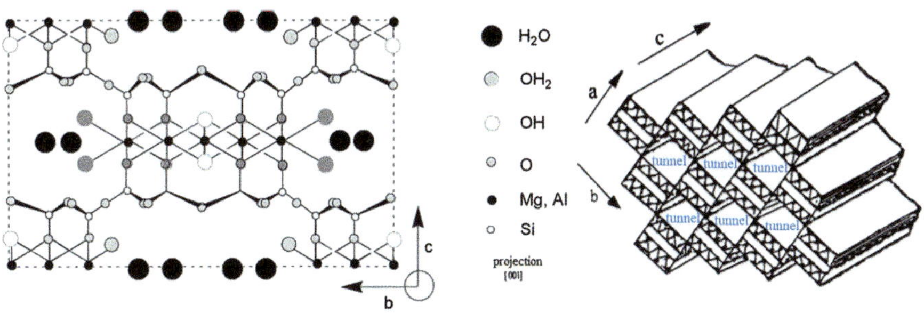

Figure 2.3 Projection of AT in the direction of [001] and a three-dimensional model of the crystal channels of AT. (With kind permission from Springer Science and Business Media: *Journal of Materials Science*, Structure and properties of fibrillar silicate/SBR composites by direct blend process, **38**, 2003, M. Tian.[26])

between the single crystals in AT is considerably weaker than that between the silicate layers in montmorillonite, resulting in easier separation between single nanocrystals of AT in polymer matrixes, thus showing an effective reinforcing effect on polymers.[26–28]

2.2.2 Physical Properties of Fibrillar Attapulgite

2.2.2.1 Surface Properties of Fibrillar Attapulgite

Fibrillar AT is a type of strongly polar inorganic filler with physical and chemical properties similar to those of silicates. Because of the large specific surface area and many active functional groups, many active sites exist on the surface of fibrillar AT. There are three types of hydroxyls in fibrillar AT, as demonstrated by the FTIR spectra (see Figure 2.4) of the AT. We can observe four absorption peaks at 3615 cm^{-1}, 3582 cm^{-1}, 3551 cm^{-1}, and 3398 cm^{-1}. The peaks at 3615 cm^{-1}, 3582 cm^{-1} are attributed to the hydroxyl stretching vibration connected with Mg, and the peaks at 3551 cm^{-1} and 3398 cm^{-1} are attributed to the stretching vibration of the water connected to Mg at the edge of the tunnel and physical water absorption, respectively.[29] In addition, the absorption peaks at 483 cm^{-1} and 446 cm^{-1} are the characteristic absorption peaks at the low wavenumbers of AT.[30,31] The abundant Si–OH groups on the surface of fibrillar AT single crystals can react with silane coupling agents such as (3-glycidyloxypropyl) trimethoxysilane.[32] The hydroxyl groups facilitate the surface modification of AT.

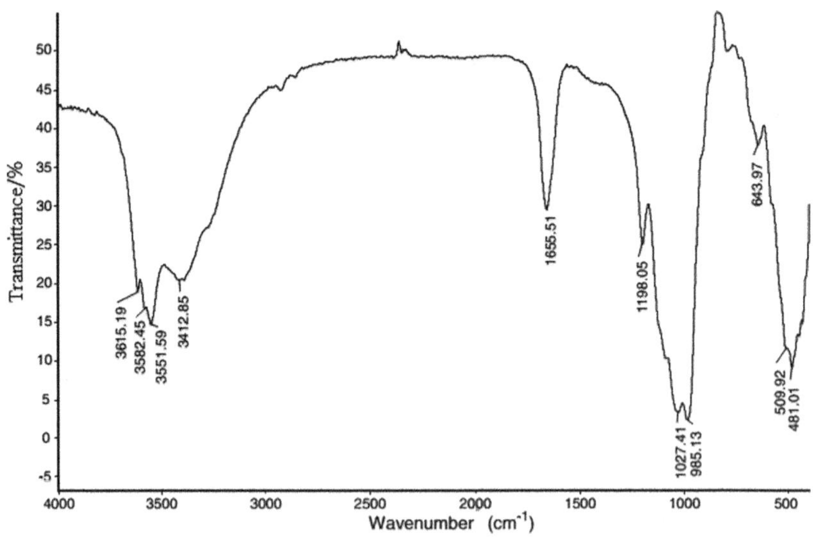

Figure 2.4 FTIR spectrum of AT at room temperature. (Reproduced from *Composites Science and Technology*, **65**, M. Tian *et al.*, Surface modification of fibrillar silicate and its reinforcing mechanism on FS/rubber composites, 1129–1138, Copyright 2005 with permission from Elsevier.[37])

2.2.2.2 Adsorption Characteristics

Because of the tunnel structure, AT has a large volume of void and a large specific surface area.[33] The adsorption of AT, depending on its surface area, surface void structure, and ionic state, can be physical or chemical. In physical adsorption, the adsorbate is adsorbed on the inside and outside surfaces of AT by van der Waals forces.[34] The size of the adsorbate particles is from the nanoscale to the microscale. The chemical adsorption of AT is caused by the structural hydroxyls of AT and the alkaline earth metal cations of AT. The structural hydroxyls of AT are active acidic hydroxyls and can interact with pyridine. Additionally, the alkaline earth metal cations can react with formic acid. Figure 2.5 shows the temperature-programmed desorption infrared (FTIR-TPD) spectra of AT after the adsorption of formic acid. Elemental analysis shows that AT contains alkaline earth metals. After the adsorption of formic acid for three hours, followed by blowing for one hour, some new absorption peaks appear in the spectra of AT. At room temperature, the absorption peak at 1700 cm^{-1} is for the carbonyl group on the formic acid adsorbed on AT by physical adsorption or by hydrogen bond. As the temperature increases to 150 °C, the formic acid adsorbed on AT by physical adsorption or by hydrogen bond is gradually disrupted, leaving only the strong

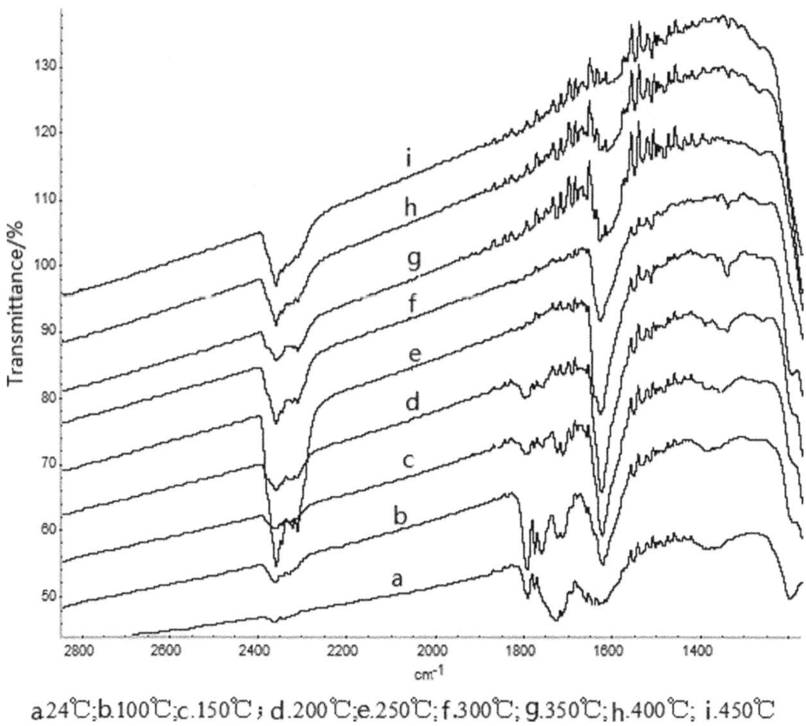

Figure 2.5 FTIR-TPD spectra of AT after adsorption of formic acid.

adsorption peak at 1628 cm^{-1}, which is attributed to the asymmetric stretching vibration of the $O{=}C{-}O^-$ of the formate that reacted with the alkaline earth metal cations in AT. Then, the adsorption peak at 1628 cm^{-1} indicates that the surface of AT is obviously alkaline. The intensity of the peak at 1628 cm^{-1} decreases with increasing temperature, but is still high at 400 °C, indicating that the strength of the basic center on AT is still high.

2.2.2.3 *Ion Exchange Capacity*

The cation exchange capacity (CEC) of most of the cations on AT is quite low, usually below 50 mmol 100 g^{-1} AT. Most of them were between 20 and 30 mmol 100 g^{-1} AT.[35] The CEC of Na^+ in AT is relatively high, up to 65 mmol 100 g^{-1}, higher than that of kaolin clay, but only one half to a third of those of montmorillonite and vermiculite. The value of CEC of AT increases slightly with the decrease of particle size.

2.2.3 Chemical Properties of Fibrillar Attapulgite

There are adsorbed water, zeolite water, structural water and crystalline water in AT. Heating can take off most of the water, except part of the structural water, causing internal lattice interruption and thus increasing lattice defects, which can be the active centers. Studies have shown that when AT is baked at above 350 °C, its crystal structure collapses. Thus, it is generally activated by heating at about 300 °C, which can not only maintain the crystal structure of AT, but also improve the activity. Additionally, acidification of AT can change the chemical properties and increase the activity of AT. From the edge of the tunnel to the center, the cations in AT, such as Mg^{2+}, Al^{3+}, and Fe^{3+}, are replaced in turn by H^+. After the replacement of these cations by H^+, the surface activity of AT can be increased. Importantly, the large number of hydroxyls on the surface of AT can react with silane coupling agents to ensure the good interfacial interaction of AT with the rubber matrix, thus making AT a good reinforcing filler for high-performance, low-cost rubber composites.

2.3 Fibrillar Attapulgite–Rubber Composites

The large surface area, high thermal resistance, and fibrous morphology of AT provide much potential for diverse applications such as absorbent, catalyst and carrier for catalysts, densifying agent, adhesive agent, nucleating agent in polymers, and cheap fillers to reduce cost. By conventional polymer melt blending, modified AT can be dissociated into single crystals or crystal bundles with diameters less than 100 nm, and these single crystals or crystal bundles can be uniformly dispersed in the rubber matrix to provide an excellent reinforcing effect. AT–rubber composites with low costs and excellent properties have been successfully prepared.

2.3.1 Preparation Methods for Fibrillar Attapulgite–Rubber Composites

Major preparation methods for polymer nanocomposites include *in situ* sol–gel polymerization and direct blending (of polymer melt, solution, and emulsion) methods. The polymer melt blending method and polymer emulsion blending method are commonly used to prepare elastomer composites because these methods are simple, readily find industrial applications, and require no organic solvent.

Silicate attapulgite–styrene butadiene rubber (SBR) composites with excellent properties and low cost were first prepared from a direct blending process (including polymer melt blending and emulsion co-coagulation) by Tian *et al.*[26] After unpurified AT and SBR were directly blended on a two-roll mill, most AT particles were exfoliated into fibrillar single crystals or crystal bundles with a diameter smaller than 100 nm under mechanical shear (the shallow dark domains in Figure 2.6a). This result further demonstrates that the interaction between AT single crystals is much weaker than that between the layers of layered silicates. It is well known that layered silicates cannot be directly separated into nano-units to disperse in rubber by direct blending. A disadvantage of these fibrillar single crystals or crystal bundles is that they are broken under shear and become shorter. According to TEM and SEM observations, some particles 0.2–0.5 μm long (the darker domains in Figure 2.6a and b) are shorter than raw AT. These smaller particles may be the impurities in AT and/or unseparated AT. Furthermore, the smooth and clear fracture surfaces of the composites indicate that the interfacial adhesion between raw AT and the rubber matrix is poor (Figure 2.6c).

After AT was purified by sodium polyacrylate (NaA) and dried with a spray drier, it was blended with SBR directly to prepare purified AT–SBR composites. Although some AT was exfoliated into AT single crystals or crystal bundles, most AT remained as large particles (Figure 2.7) in SBR because the purified AT was driven by thermodynamic force to re-agglomerate into a fibrillous

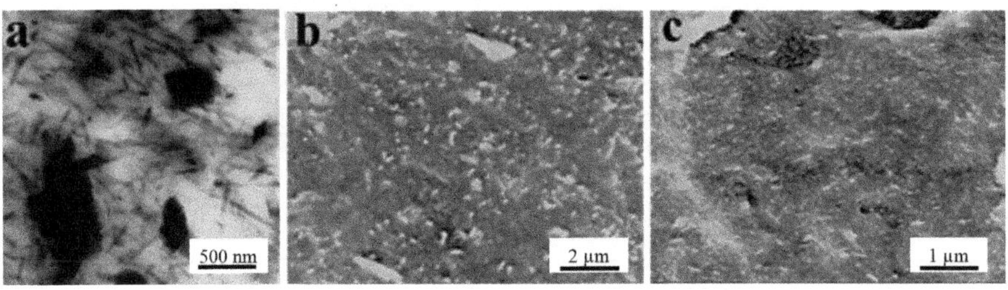

Figure 2.6 TEM image (a) and SEM images (b, c) of unpurified AT–SBR composites prepared by polymer melt blending. (With kind permission from Springer Science and Business Media: *Journal of Materials Science*, Structure and properties of fibrillar silicate/SBR composites by direct blend process, **38**, 2003, M. Tian.[26])

network during the drying process. This fibrillous network is a more compact structure than that of unpurified AT and is difficult to disperse in the SBR matrix directly.

The co-coagulation method can improve the dispersion of AT in SBR better than the direct blending method (Figure 2.8). AT fibers get shorter from NaA purification and the shorter fibers are dispersed uniformly in SBR (Figure 2.8a). AT fibers are relatively long after sodium salt of acrylic acid-*co*-maleic anhydride (PAA) purification, but the dispersion is less uniform (Figure 2.8b). The dispersion of unpurified AT fibers (Figure 2.8c) is worse than those of the two types of purified AT fibers just mentioned, but better than that of unpurified AT–SBR composites prepared by polymer melt blending.

It can be seen from Table 2.1 that the mechanical properties of purified AT–SBR composites are very close to those of the raw AT–SBR composite prepared by the melt blending method, indicating that purification does not achieve the desired effect. The key is that the dispersion of AT is not improved by purification.

Compared with the melt blending method (see Table 2.1), the emulsion co-precipitation method (see Table 2.2) can better improve the reinforcing effect of raw AT on rubber, as demonstrated by the increased tensile strength and tear strength of the composite. The reinforcing effect of the purification of AT by PAA is almost the same as that of raw AT, whereas the reinforcing effect of the purification of AT by NaA is slightly better than that of the raw AT.

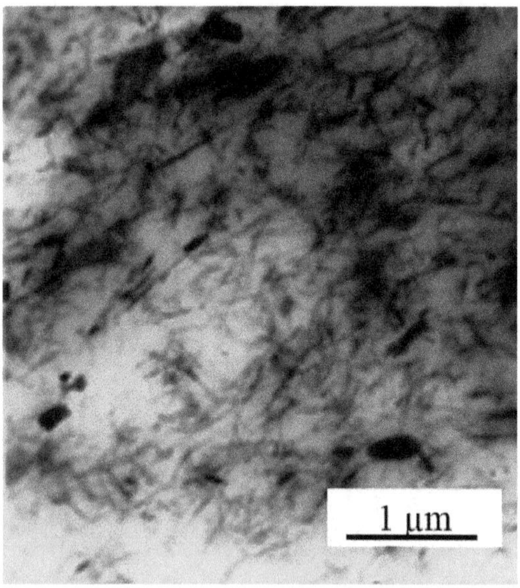

Figure 2.7 SEM image of purified AT–SBR composite prepared by polymer melt blending. (With kind permission from Springer Science and Business Media: *Journal of Materials Science*, Structure and properties of fibrillar silicate/SBR composites by direct blend process, **38**, 2003, M. Tian.[26])

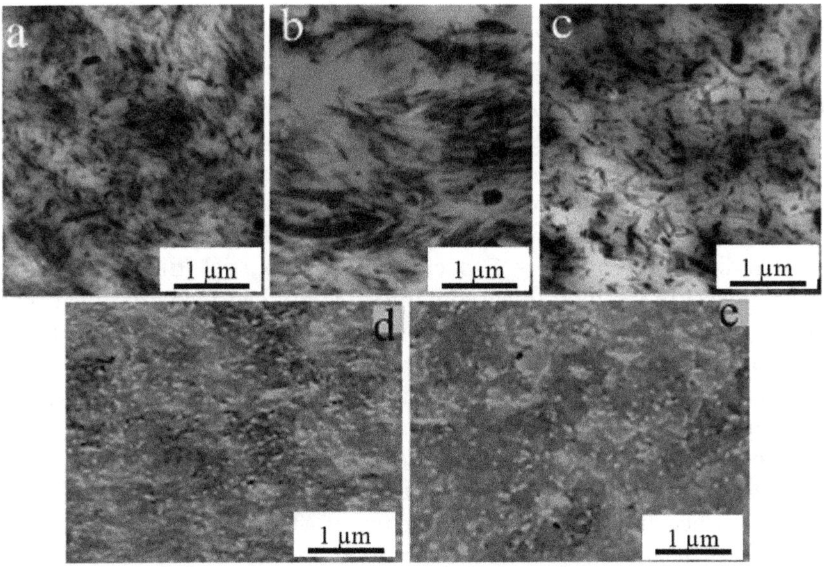

Figure 2.8 SEM images of AT–SBR composites prepared by the emulsion co-coagulation method. (a), (d): NaA-purified AT; (b), (e): PAA-purified AT; (c): unpurified AT. (With kind permission from Springer Science and Business Media: *Journal of Materials Science*, Structure and properties of fibrillar silicate/SBR composites by direct blend process, **38**, 2003, M. Tian.[26])

Table 2.1 Effect of purification of AT on properties of composite prepared by the melt blending method.

Properties	Raw AT	AT purified by NaA
Hardness/Shore A	56	56
Modulus at 300% elongation/MPa	2.1	1.9
Tensile strength/MPa	7.6	6.1
Elongation at break/%	900	972
Tear strength/kN m^{-1}	22.2	21.7

Table 2.2 Effect of purification of AT on properties of composite prepared by the emulsion co-precipitation method.

Properties	Raw AT	AT purified by PAA	AT purified by NaA
Hardness/Shore A	66	66	66
Modulus at 300% elongation/MPa	2.8	3.0	2.6
Tensile strength/MPa	9.2	9.6	11.8
Elongation at break/%	884	892	804
Tear strength/kN m^{-1}	33.0	30.4	30.8

2.3.2 Interface Design of Fibrillar Attapulgite–Rubber Composites

There are two key issues to be resolved in using inorganic fillers to reinforce a polymer: one is the dispersion of the fillers in the polymer, and the other is the filler–polymer interaction. Generally, the surface properties of inorganic fillers are very different from those of the polymer, which makes the inorganic fillers incompatible with the polymer and makes it difficult to form a uniform dispersion of fillers. Especially at small particle diameters (<1 μm), the particles are inclined to agglomerate. Organic surface modification can enhance the dispersion of the fillers in the polymer. The strong interfacial adhesion between the fillers and the polymer can be obtained by organic modification of the fillers. The abundant Si–OH groups on the surface of AT single crystals provide the possibility for surface modification.[37]

To determine the effect of surface treatment, direct methods and indirect methods are usually adopted. By measuring the hydrophobicity, coating ratio, specific surface area and other parameters of AT, direct methods can characterize the surface properties of the modified surface of AT. But there are few reports on the determination of surface properties by direct methods. The surface properties of the modified AT surface can also be characterized by indirect methods, depending on the effect of the application. The effect of modification can be characterized by means of various parameters of the composite, but this method is complicated. FTIR is an effective method to characterize the effect of modification.[37] The effect of modification can be analyzed by comparing the infrared spectra of the samples before and after modification.

2.3.2.1 Modification of Fibrillar Attapulgite by Coupling Agents

Because the surface of fibrillar AT is rich in polar Si–OH groups, a coupling agent can be used to treat the fibrillar AT. The coupling agents that can be used include the silane coupling agent, titanate coupling agent, and aluminate coupling agent.

2.3.2.1.1 Modification with Silane Coupling Agents. The silane coupling agents, including KH570, Si69, KH550, KH560, and A151 (see Table 2.3 for full chemical names and formulae), are widely used in the surface modification of inorganic fillers.

The dispersion of modified fibrillar AT depends on the type of silane coupling agent. Among the five coupling agents mentioned above, KH570, with a polar methacryloxy group, and KH560, with a polar epoxy group, have the highest surface tensions and the closest surface tensions to that of fibrillar AT (see Table 2.4).[38] As a result, fibrillar AT is well wetted by these coupling agents, which are then adsorbed onto the AT surface by Si–O–Si bonds.

Here, we only discuss the reaction between the silane coupling agent KH570 and the hydroxyls of AT. The same discussion applies to other silane coupling agents. The conventional modification method for AT is as follows:

Table 2.3 Various silane coupling agents.

Abbreviation	Chemical name	Chemical formula
KH550	γ-Aminopropyltriethoxy silane	$H_2NCH_2CH_2Si(OC_2H_5)_3$
KH560	γ-Glycidylpropyltrimethoxy silane	$CH_2(O)CHCH_2OCH_2CH_2CH_2Si(OCH_3)_3$
KH570	γ-(Methacryloxy)proxyltri-methoxy silane	$CH_2C(CH_3)COOCH_2CH_2CH_2Si(OCH_3)_3$
Si69	Bis(3-triethoxysilylpropyl) tetrasulfide	$(C_2H_5O)_3Si(CH_2)_3S4Si(CH_2)_3(OC_2H_5)_3$
A151	Vinyltriacetoxy silane	CH_2=CH–Si–$(OOCCH_3)_3$

Table 2.4 Surface tensions for various silane coupling agents.

Silane coupling agents	KH560	Si69	A151	KH550	KH570	Fibrillar attapulgite[a]
Surface tension/mN m^{-1}	49.4	24	28.6	35.7	44.8	58.2

[a]Fibrillar attapulgite modified by Si69.

first, AT is treated in a high-speed mixer at a relatively low speed for 3–5 min to make it fluffy. Then, a solution of the silane coupling agent KH570 in ethanol solution is sprayed onto the surface of AT. Meanwhile, the mixer is run at a relatively high speed for 10 min to form a uniform coating of KH570 on the AT surface. Finally, the as-obtained KH570-modified AT is dried in an oven at 80 °C for 4 h to remove the ethanol.

The ^{29}Si MAS NMR spectra of unmodified AT, heat-treated AT, and KH570-modified AT (see Figure 2.9) all show two sharp peaks at −97.6 and −91.7 ppm. The first peak is attributed to Q3 Si, which is bonded with one hydroxyl, and the second peak is attributed to Q2 Si, which is bonded with two hydroxyls.[39] The relative height of the two peaks is changed for the heat-treated AT and the KH570-modified AT compared to the unmodified AT. The relative height of peaks is changed for the heat-treated AT because some hydroxyls of absorbed water are lost after the heat treatment, while the change of the relative height of peaks for the KH570-modified AT was probably caused by chemical reaction between the methoxyls of KH570 and the hydroxyls of AT. More details can be found in the ^{29}Si CP-MAS NMR spectrum of KH570-modified AT (Figure 2.10). There are two weak and broad peaks at −58.8 ppm and −67.2 ppm, which are attributed to two different ways in which Si is bonded to KH570. KH570 possesses three methoxyls. The peak at −58 ppm indicates that only one or two methoxyls in KH570 have hydrolyzed and reacted with the hydroxyls of AT to form Si–O–Si bonds, while the peak at −67 ppm indicates that three methoxyls in KH570 have hydrolyzed and reacted with the hydroxyls of AT to form Si–O–Si bonds, as illustrated in Figure 2.10.[40] These results demonstrate that KH570 can be adsorbed on the surface of AT by chemical bonding.

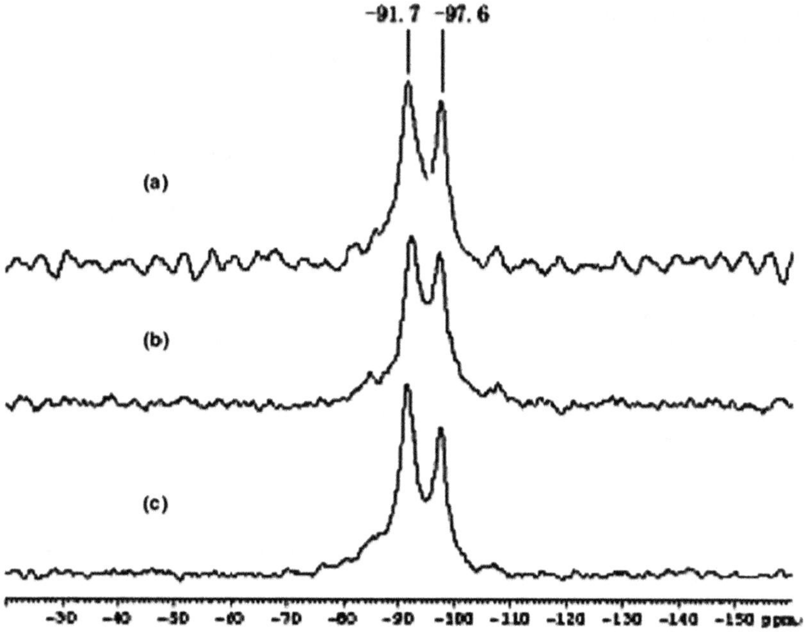

Figure 2.9 ^{29}Si MAS NMR spectra of AT: (a) unmodified AT; (b) AT heat-treated at 150 °C; (c) AT modified with KH570. (Reproduced from *Composites Science and Technology*, **65**, M. Tian *et al.*, Surface modification of fibrillar silicate and its reinforcing mechanism on FS/rubber composites, 1129–1138, Copyright 2005 with permission from Elsevier.[37])

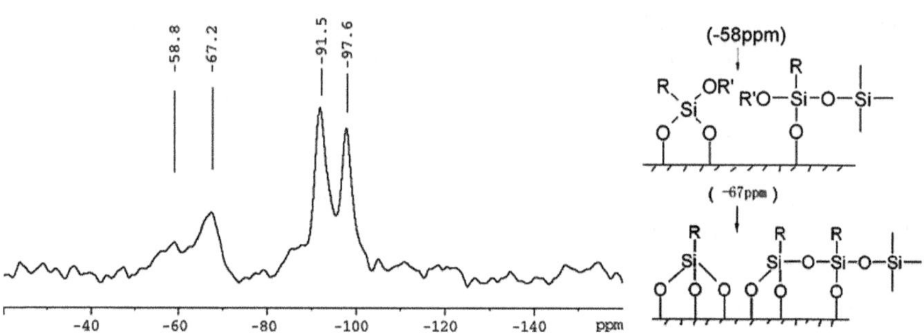

Figure 2.10 ^{29}Si MAS NMR spectra of AT modified with KH570. (Reproduced from *Composites Science and Technology*, **65**, M. Tian *et al.*, Surface modification of fibrillar silicate and its reinforcing mechanism on FS/rubber composites, 1129–1138, Copyright 2005 with permission from Elsevier.[37])

Figure 2.11 shows the FTIR spectra of AT and KH570-modified AT.[36] The spectrum of AT modified with KH570 (Figure 2.11b) has stronger absorption peaks at 2922 and 2851 cm^{-1} than the spectrum of unmodified AT (Figure 2.11a) because of the oscillation of the C–H bond. Meanwhile, the new absorption peak at 1721 cm^{-1} is ascribed to the carbonyl group

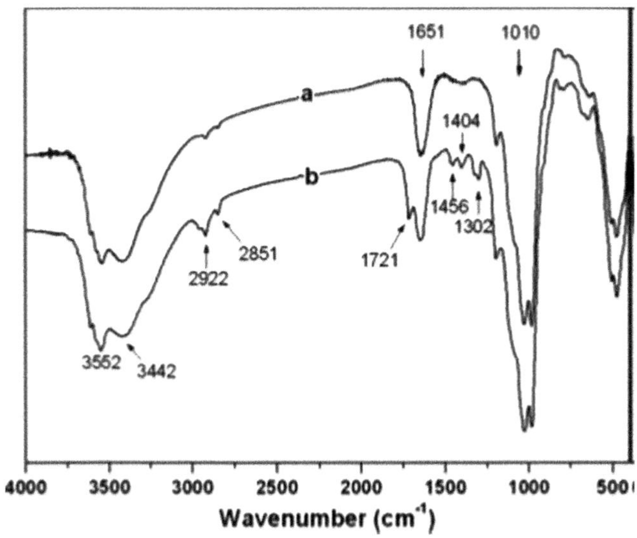

Figure 2.11 FTIR spectra of (a) AT and (b) AT modified with KH570. (Reproduced from ref. 36 with permission from John Wiley and Sons. Copyright © 2009 Society of Plastics Engineers.)

stretching vibration, and the one at 1456 cm^{-1} is due to $-CH_2-$ scissoring and $-CH_3$ asymmetric bending deformation, confirming the successful organification of AT. In addition, the absorption band at 3552 cm^{-1} is due to the surface hydroxy groups of AT.[41,42] The strong absorption band at 1010 cm^{-1} corresponds to the stretching vibration of the Si–O bond.[46]

The surface modification of AT by silane coupling agents may be carried out in supercritical CO_2 (sc-CO_2).[43] Figure 2.12 is the schematic representation of modification of AT in sc-CO_2. AT is first loaded loosely in a 300 mesh copper grid, dried in an oven at 80 °C, for 4 h and then moved onto the support of a reactor. The reactor is sealed after the addition of a solution of KH570 in ethanol at the bottom. The same mass ratio of AT to KH570 (AT/KH570 = 50/6) is used in both the conventional method and the sc-CO_2 method for the comparisons to be valid. Then, the pressure and temperature are elevated, and the contents of the vessel are continuously stirred. When the desired temperature and pressure (70 °C, 14 MPa) are attained, the reactor contents are stirred for 1 h to perform the modification in the sc-CO_2. After 1 h, the reactor is opened, and then the modified AT is dried in an oven at 80 °C for 4 h to remove any ethanol. The modified AT is dispersed in ethanol (the mass ratio of AT to ethanol is 1/10) in an ultrasonic bath for 15 min, and the dispersion is followed by centrifugation at 4000 rpm for 10 min and solvent separation to remove the modifier KH570 physically adsorbed on the surface of AT. After extraction, the modified AT is dried in the oven at 80 °C for 8 h to remove the ethanol.

The FTIR spectra of AT and the as-obtained AT modified by the conventional method and sc-CO_2 method are shown in Figure 2.13. It can be

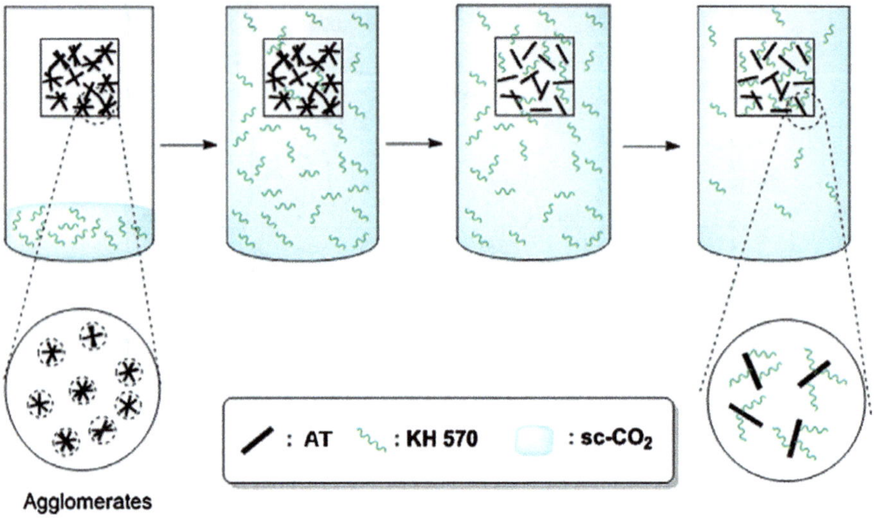

Figure 2.12 Schematic representation of modification of AT in sc-CO_2. (Reproduced from *Composites Science and Technology*, **79**, M. Tian *et al.*, Interfacial enhancement of fibrillar silicates (FSs)/EPDM nanocomposites by surface modification of FS in supercritical CO_2, 21–27, Copyright 2013 with permission from Elsevier.[43])

observed that the spectra of the AT modified by the two methods show new absorption bands at 1721 cm^{-1}, 1328 cm^{-1}, and 1304 cm^{-1}. The absorption band at 1721 cm^{-1} is ascribed to the stretching vibration of C=O, and the other two bands at 1328 cm^{-1} and 1304 cm^{-1} are ascribed to the stretching vibration of the C–O groups, indicating that the modifier KH570 is successfully adsorbed on the surface of AT using both modification methods. After extraction by ethanol, the absorption peaks at 1328 cm^{-1} and 1304 cm^{-1} of the C–O groups can still be seen, but they have become weaker than those of the AT modified using the conventional method because not only physical adsorption but also chemical adsorption of KH570 exists on the AT surface. The physically adsorbed KH570 is detached from the surface of AT after extraction, whereas the chemically adsorbed KH570 is still bonded to the surface of AT, as confirmed by the presence of the same characteristic absorption peaks after extraction. However, for the AT modified using the sc-CO_2 method, these characteristic absorption peaks show no obvious change after extraction because the chemical adsorption of KH570 on the AT surface predominates, suggesting a better modification effect using the sc-CO_2 method.

2.3.2.1.2 Modification with Organic Titanate Coupling Agents. The modification mechanism of titanate modification is similar to that of silane. The titanate coupling agent and the free protons on the AT surface form Ti–O bonds. After treatment with titanate, the surface of AT is covered with a layer of organic molecular film, which changes the surface properties of AT.

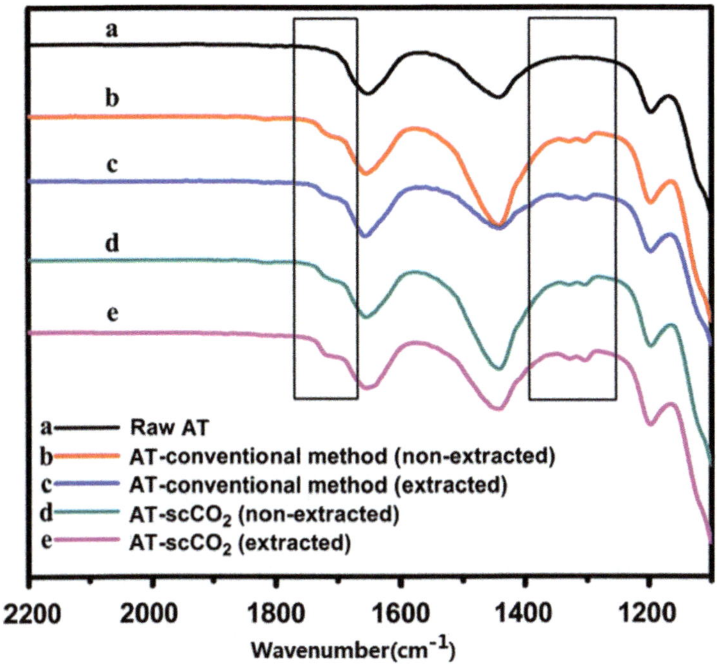

Figure 2.13 FTIR spectra of raw AT and various modified ATs. (Reproduced from *Composites Science and Technology*, **79**, M. Tian *et al.*, Interfacial enhancement of fibrillar silicates (FSs)/EPDM nanocomposites by surface modification of FS in supercritical CO_2, 21–27, Copyright 2013 with permission from Elsevier.[43])

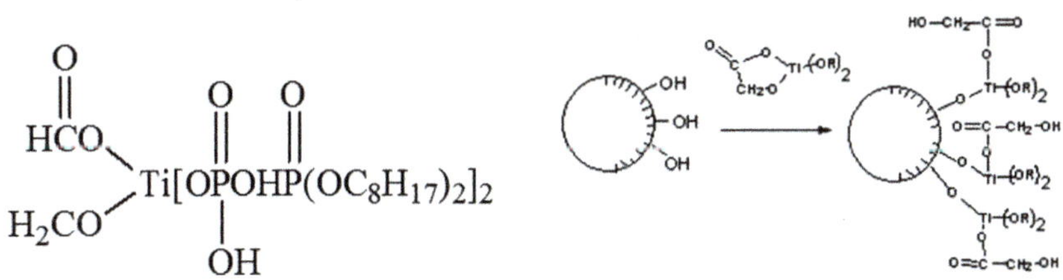

Figure 2.14 Structure of a chelated titanate (WDZ311) and schematic of the modification mechanism for AT.

WDZ311, a chelated titanate (Figure 2.14), is a commonly used titanate coupling agent. After modification with WDZ311, AT is vacuum treated at 150–170 °C to remove the titanate that is physically adsorbed on its surface. As can be seen from Figure 2.15, titanate modification of AT adds two new absorption peaks at 2960 cm^{-1} and 2932 cm^{-1} to the FTIR spectrum of AT. These two peaks are due to the stretching vibration of C–H belonging to $-CH_3$ and $-CH_2-$ in the titanate after AT and the titanate have reacted with each other. The intensity of the peak for the hydroxyl in titanate-modified

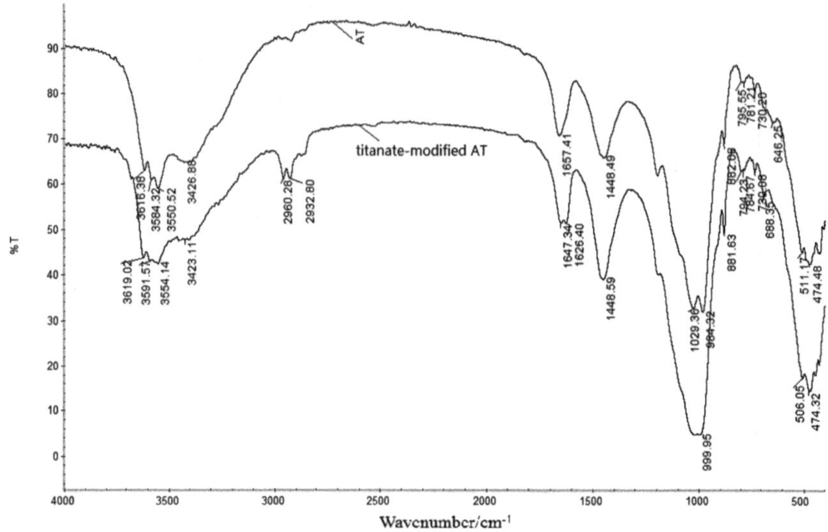

Figure 2.15 FTIR spectra of raw AT and titanate (WDZ311)-modified AT.

AT decreases greatly, confirming the reaction between the titanate and the hydroxyl in AT.

2.3.2.1.3 Modification with Organic Aluminate Coupling Agents. Organic aluminate coupling agents have unique advantages in the surface treatment of fibrillar AT. They can improve the processability and the physical and mechanical properties of the fibrillar AT, and are solid, color shallow, non-toxic at room temperature, easy to use, and highly thermally stable. The surface treatment mechanism for aluminate coupling agents is similar to that for titanate coupling agents. However, the use of aluminate coupling agents results in irreversible chemical bonds on the AT surface. Thus, the surface obtained by treatment with aluminate coupling agents is better than that obtained by treatment with titanate coupling agents.

2.3.2.2 Modification of Fibrillar Attapulgite by Surfactant

A surfactant molecule with both hydrophilic groups and hydrophobic groups can be used for the surface modification of AT. The inorganic AT surface changes from hydrophilic to hydrophobic due to the hydrophilic groups of surfactants attaching to the inorganic AT surface by physical and chemical interactions, endowing AT with a polymer affinity and improving the dispersion of AT in the polymer matrix.

AT has been modified by sodium stearate, which is a commonly used surfactant, and then vacuum treated at 150–170 °C. In Figure 2.16, there are peaks in the spectrum for the sodium stearate-modified AT that do not appear in the spectrum for unmodified AT. The peaks at 2919 cm^{-1} and 2849 cm^{-1} are attributed to the stretching vibration of the –CH_2– of the AT that

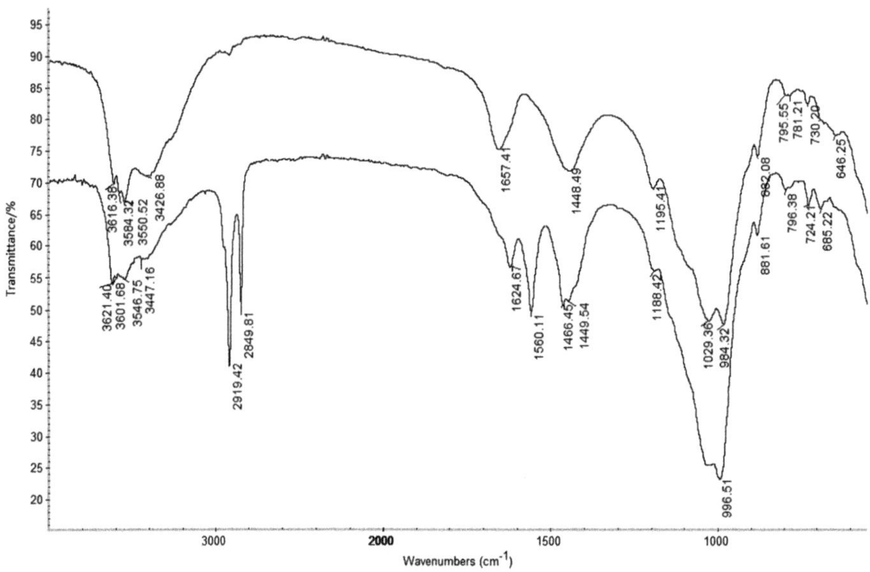

Figure 2.16 FTIR spectra of raw AT and sodium stearate-modified AT.

has reacted with the sodium stearate, and the peaks at 1560 cm^{-1}, 1466 cm^{-1}, and 724 cm^{-1} are due to the organic groups of sodium stearate. Meanwhile, the relative intensity of the hydroxyl groups in AT decreased significantly, indicating that the hydroxyl groups in AT have reacted with sodium stearate.

2.3.2.3 Acidification of Fibrillar Attapulgite

According to Gonzalez *et al.*, the acid activation of AT can adjust the unbroken silicate structure and the octahedron parts in AT are dissolved in the silicate minerals to produce colloidal SiO_2, thus improving the microporosity.[44] With the increase of the intensity of acid treatment, the AT structure changes significantly or is even destroyed. At the same time, along with the dissolution of the octahedral parts, the crystallinity of AT decreases, and the amount of SiO_2 phase and Si–OH groups increases. These changes greatly improve the adsorption performance of AT. After acid activation, both the micropore surface area and the specific area of AT significantly increase, but decrease when the acid concentration is too high. Analyses of adsorption–desorption isotherms show that the pore shape is approximately cylindrical and the pore volume increases after acid activation. In conclusion, the adsorption of AT is greatly increased after acid activation. AT was applied as a catalyst in the oxidation of alcohols in aqueous media by Wu *et al.*[45] Based on literature precedents, a plausible mechanism for the alcohol oxidation (Figure 2.17) was proposed. Naked $Si^{18}OH$ groups were formed by acid-catalyzed cleavage with $H_3{}^{18}O^+$. The $Si^{18}OH$ then prompted the oxidation of alcohols to the corresponding carbonyl compounds (Figure 2.18).

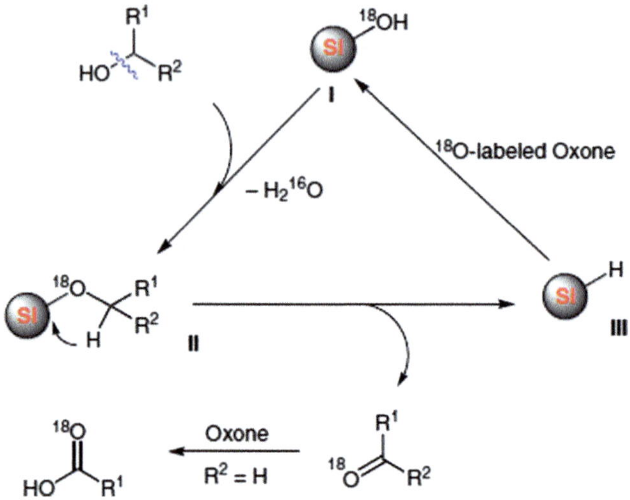

Figure 2.17 Proposed reaction mechanism for the alcohol oxidation. (Reproduced with permission from ref. 45. Copyright © Georg Thieme Verlag KG.)

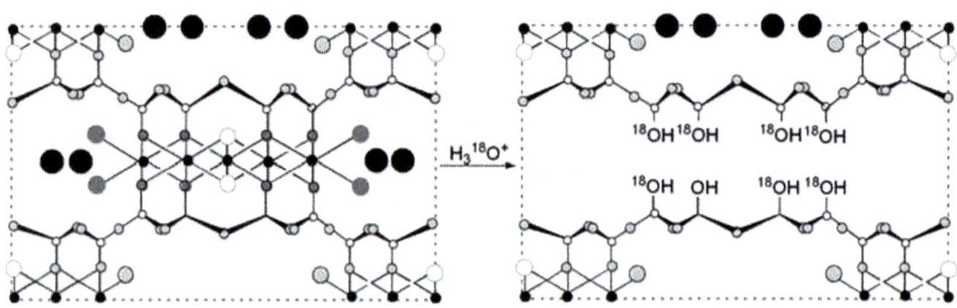

Figure 2.18 Proposed mechanism of acid leaching of AT. (Reproduced with permission from ref. 45. Copyright © Georg Thieme Verlag KG.)

2.3.2.4 *Silver Nanoparticle-Immobilized Fibrillar Attapulgite*

A chemical approach to construct one-dimensional inorganic nanomaterial-based hybrid nanostructures offers a promising method for functionalizing materials. Silver nanoparticles have superior optical, catalytic, and antibacterial properties; they are among the best substrates for surface-enhanced Raman spectroscopy and also exhibit antibacterial properties because of their large surface area and high reactivity.[46,47] The combination of antibacterial properties and low toxicity for silver nanoparticles has been attracting significant attention.[48–52] Because of the inactive surface of inorganic materials and the highly curved surface of this nanoscale substrate, with diameters of 10–20 nm, it is hard to modify and assemble silver nanoparticles on the surface of inorganic materials.

Poly(dopamine) (PDA) can be formed on a wide range of inorganic and organic materials by *in situ* spontaneous oxidative polymerization of

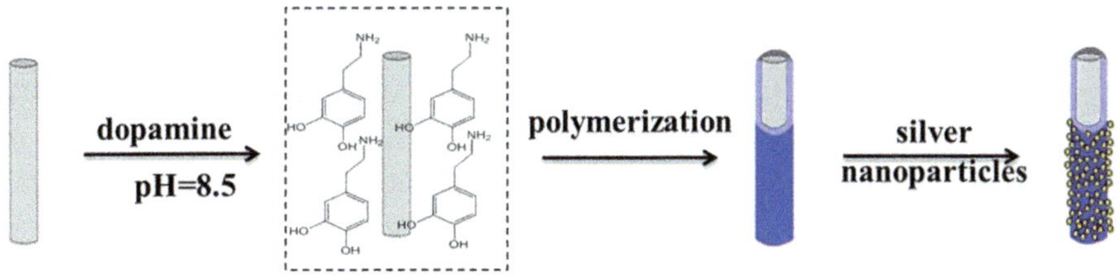

Figure 2.19 Schematic illustration of the procedure for preparation of AT–PDA–Ag hybrids. (Reproduced from ref. 55 with permission from John Wiley and Sons. Copyright © 2013 Wiley Periodicals, Inc.)

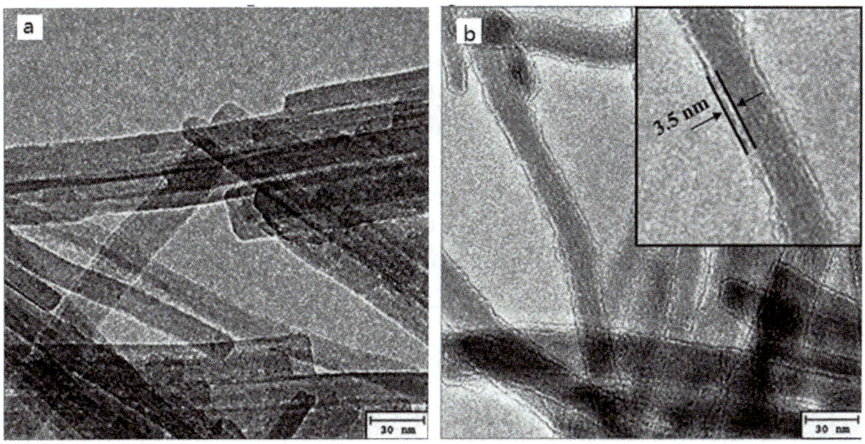

Figure 2.20 TEM images of (a) AT and (b) AT–poly(dopamine) (PDA). (Reproduced from ref. 55 with permission from John Wiley and Sons. Copyright © 2013 Wiley Periodicals, Inc.)

dopamine in alkaline solution at room temperature. The *in situ* spontaneous polymerization of dopamine provides the advantages of one-step surface functionalization with simple ingredients, mild reaction conditions, and applicability to various types of materials. More importantly, the PDA layer can be used as a versatile platform for secondary reactions.[53,54] Especially, the metal-binding ability of catechol and nitrogen-containing groups present in the PDA structure may be exploited to immobilize metal particles onto various substrates without conventional environmentally unfriendly oxidation and activation–sensitization processes.

The preparation of AT–PDA–Ag hybrids was carried out in a two-step procedure in which silver nanoparticle immobilization on the fibrillar AT surfaces is assisted by bound PDA (Figure 2.19).[55] The dopamine solution was colorless and transparent. During the *in situ* spontaneous oxidative polymerization of dopamine, the color of the solution turned to pink as soon as the catechol was oxidized to benzoquinone, and then the pink disappeared and the solution turned slowly into a deep dark colour, indicating that the polymerization may be followed by a reaction in the manner of melanin formation.

Figure 2.20 shows the TEM images of the pristine AT [Figure 2.20(a)] and AT–PDA with a dopamine concentration of 2 g L^{-1} [Figure 2.20(b)]. A fibrillar nanostructure of AT with diameters of 10–20 nm was observed. From Figure 2.20(b), we can see that AT is coated with a uniform PDA layer with a PDA thickness of about 3.5 nm at a concentration of dopamine of 2 g L^{-1}.

2.3.3 Dispersion and Filler Network of Fibrillar Attapulgite in Elastomers

2.3.3.1 Filler Network of Fibrillar Silicates in Elastomers

Generally, a network of filler is defined as an aggregate of fillers in a rubber matrix and is characterized by the Payne effect. The stronger the filler network is, the stronger the Payne effect is. In other words, a compound with a high Payne effect shows a high dynamic modulus (G') at very small strains, and its G' decreases remarkably fast along with the break-down of its filler network as the exerted strain increases. The dynamic modulus of the compound (unvulcanized) mainly depends on the filler network and the filler–polymer bonding, while that of the composite (vulcanized compound) depends also on the rubber network, as shown in eqn (2.1).

$$\text{Compound: } G'_{\text{compound}} = G'_{\text{filler–filler}} + G'_{\text{filler–polymer}}$$

$$\text{Composite: } G'_{\text{composite}} = G'_{\text{filler–filler}} + G'_{\text{filler–polymer}} + G'_{\text{rubber–rubber}} \tag{2.1}$$

The effect of the rubber's network on dynamic modulus is much smaller than that of the filler network and can be neglected. AT is a kind of inorganic filler with a strong polarity, and physical and chemical properties similar to those of silica. For unmodified AT, the AT–polymer bonding is so weak that its effect can also be neglected. Thus, the dynamic modulus of either the compound or the composite mainly depends on the filler network. Usually the modulus difference ($\Delta G'$) between the maximum and minimum of G' can be used to characterize the filler network. Herein, the relation between the $\Delta G'$ of the composite and that of the compound is used as an indication of change in the filler network during vulcanization.

2.3.3.1.1 Effect of Rubber Viscosity. Acrylonitrile–butadiene rubber (NBR) was used as the matrix, the Mooney viscosity of which is very high. Various amounts of the extended oil DOP were used to prepare NBR matrices with different Mooney viscosities (see Table 2.5).

As shown in Figure 2.21, for the compounds, the higher the Mooney viscosity of the NBR matrix the higher the Payne effect. However, for composites, the effect of the Mooney viscosity of the matrix on the Payne effect is much smaller, implying that the filler network formed by filler re-aggregation is weaker in an NBR matrix with a higher Mooney viscosity. As shown in Figure 2.21c, the change in filler network after vulcanization is significant for a composite with a low Mooney viscosity matrix. Based on the discussion

Table 2.5 Various polymer materials used as matrices.

	A1	A2	A3	B1	B2
Polymer matrices	NBR1051[a]	NBR1051[b]	NBR1051[c]	NBR3365	NBR1965
Molecular weight/g mol^{-1}	124674				
Mooney viscosity (ML_{1+4} 100 °C)	65	42	25	65	65
Acrylonitrile content/%	41	41	41	33	19

[a]Contain 0 phr, respectively, of dioctyl phthalate per 100 phr of NBR.
[b]Contain 20 phr, respectively, of dioctyl phthalate per 100 phr of NBR.
[c]Contain 40 phr, respectively, of dioctyl phthalate per 100 phr of NBR.

above, the dispersion and network formation of AT in a composite are determined by both the dispersion and the re-aggregating process. A high viscosity of the matrix is beneficial to the dispersion of AT, but prevents AT from re-aggregating. Thus, AT–NBR composites with different NBR Mooney viscosities show little difference in Payne effect. The results show that the Payne effect of an AT–NBR composite with a higher viscosity is more significant than an AT–NBR composite with a lower viscosity.

2.3.3.1.2 Effect of Rubber Polarity. As shown in Figure 2.22a, at the same NBR Mooney viscosity and NBR molecular weight, the higher the content of acrylonitrile in NBR is, *i.e.*, the higher the molecular polarity of NBR is, the stronger the Payne effect of the AT–NBR compound is. Among these three compounds, the Payne effect of AT–NBR1051 (acrylonitrile content of 41%) is the strongest, followed by that of AT–NBR3365 (acrylonitrile content of 33%), and that of AT–NBR1965 (acrylonitrile content of 19%) is the weakest. As Figure 2.22b shows, the Payne effects of these three AT–NBR composites follow a similar trend to those of the corresponding compounds. However, the largest change in filler network during vulcanization is found in the matrix NBR3365 (see Figure 2.22c), indicating that the filler network formed by the re-aggregation of filler is the strongest.

2.3.3.1.3 Effect of Rubber Matrix. As shown in Figure 2.23, the AT–SiR compound shows the lowest dynamic modulus due to the highly flexible molecules and weak interaction between the molecules, showing that no obvious Payne effect was observed. The Payne effect is the strongest for the AT–NBR compound because it has polar molecules, which lead to strong intermolecular forces. The Payne effects of the compounds with matrices of the nonpolar elastomers SBR and ethylene propylene diene monomer rubber (EPDM), both of which have a molecular structure with a number of side chains, are very similar. Figure 2.23b shows that the Payne effect of the AT–NBR composite is the strongest, followed by those of the AT–SBR composite and the AT–EPDM composite, and the AT–SiR composite shows no obvious Payne effect.

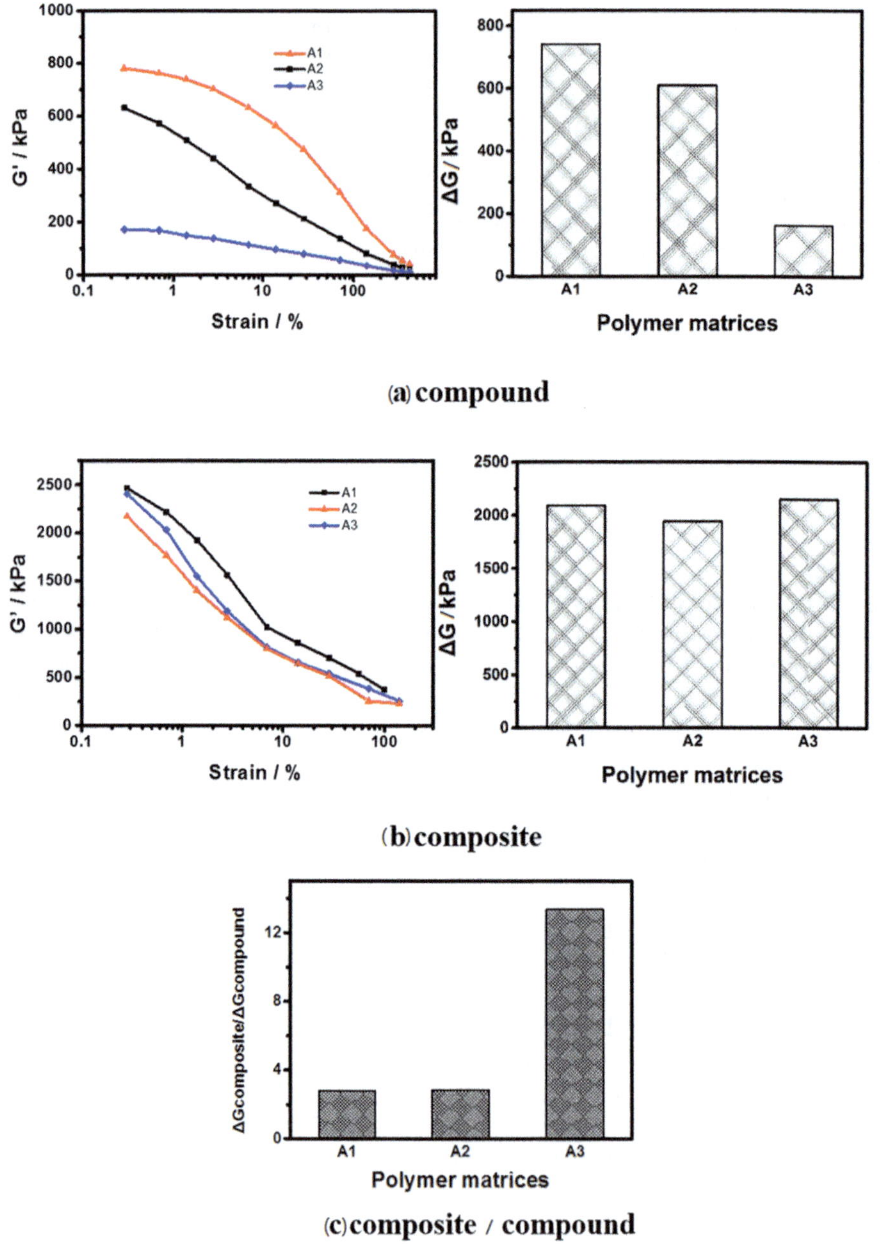

Figure 2.21 Payne effects of AT–NBR (a) compounds, (b) composites and (c) $\Delta G_{composite}/\Delta G_{compound}$ with different Mooney viscosities.

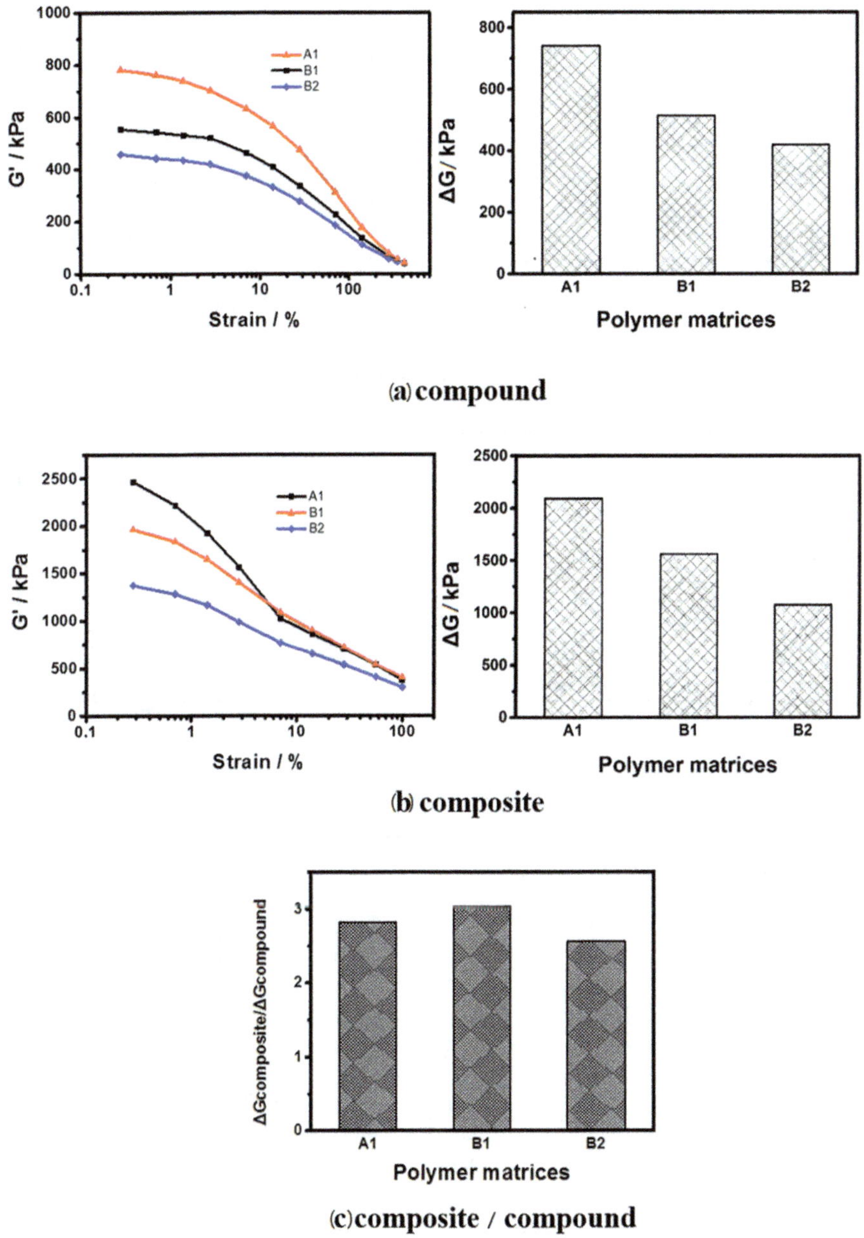

Figure 2.22 Payne effects of AT–NBR (a) compounds, (b) composites and (c) $\Delta G_{composite}/\Delta G_{compound}$ with different acrylonitrile contents: A1 (41%); B1 (33%); B2 (19%).

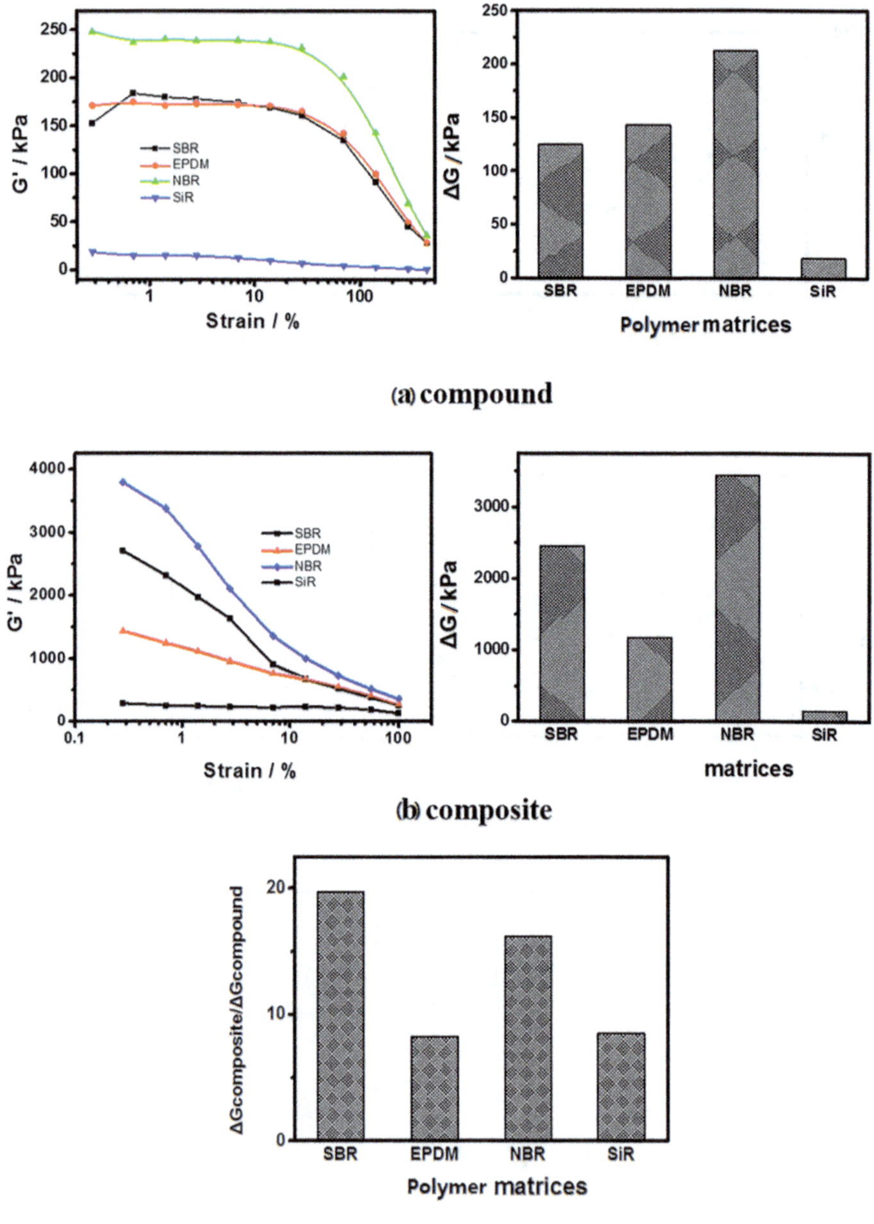

Figure 2.23 Payne effects of AT–NBR (a) compounds, (b) composites and (c) $\Delta G_{composite}/\Delta G_{compound}$ with different rubber matrices.

2.3.3.1.4 Effect of Silane Coupling Agent. As shown in Figure 2.24a, the Payne effects of modified AT–SBR compounds with different types of coupling agents are different. The Payne effects of AT–SBR compounds modified with KH550 or KH560 are the strongest, indicating that the filler networks in these two compounds are the strongest. The network of AT modified with Si69 is much weaker, and that of unmodified AT is the weakest (although close to that of AT modified with Si69). These results indicate that the dispersion of AT modified with KH550 or KH560 is better than those of AT modified with Si69 and unmodified AT in a SBR matrix. As shown in Figure 2.24b, the Payne effect of the unmodified AT–SBR composite is stronger than those of the modified AT–SBR composites. As shown in Figure 2.24c, the filler network of modified AT originating from the re-aggregation of AT is much weaker than that of unmodified AT. Moreover, the filler network of AT modified with KH560 is the strongest, followed by that of AT modified with Si69, and that of AT modified with KH550 is the weakest.

As shown in Figure 2.25, the Payne effects of the Si69-modified AT–SBR compounds are much stronger than that of the unmodified AT–SBR compound. As the loading of Si69 increases, the Payne effect of the compound becomes stronger and is the strongest at a loading of Si69 of 10% of the weight of AT. Further increases in Si69 loading will weaken the Payne effect of the compound because excessive Si69 absorbed onto the single crystals or crystal bundles of AT may lead to multimolecular layer absorption, which will weaken the filler network. On the other hand, the dispersion of AT is already very good, and excessive Si69 will not increase the amount of filler network constructing units. Thus, excessive Si69 will weaken the unit–unit bonding and hence the filler network.

2.3.3.2 Mechanism for Dispersion of Fibrillar Attapulgite in Elastomers

The dispersion of AT in elastomers is different from that of other particle fillers, which are dominated by erosion processes. A schematic of the dispersion and re-aggregation processes of AT in an elastomer matrix is shown in Figure 2.26. When added into the elastomer matrix, a shear force is exerted on AT, and due to the low interaction force between the single crystals, the large scale agglomerates erode into submicron particles (<1 μm). Then, through further rupture erosion these submicron particles exfoliate into single crystals and crystal bundles. The dispersion behavior of AT is quite different from that of silica, which exists in elastomers as spherical particles or particle agglomerates, and due to its poor compatibility with the elastomer matrix, modification is usually necessary. During the vulcanization process, these single crystals and crystal bundles absorb energy stemming from high temperature and tend to re-aggregate. However, the energy absorbed is not enough for single crystals and crystal bundles to re-crystalize back into large aggregates. Although the uniformity of the dispersion of AT is disrupted, the single crystals and crystal bundles can overlap and interact with one another to form a stronger filler network. Thus, the Payne effect is enhanced.

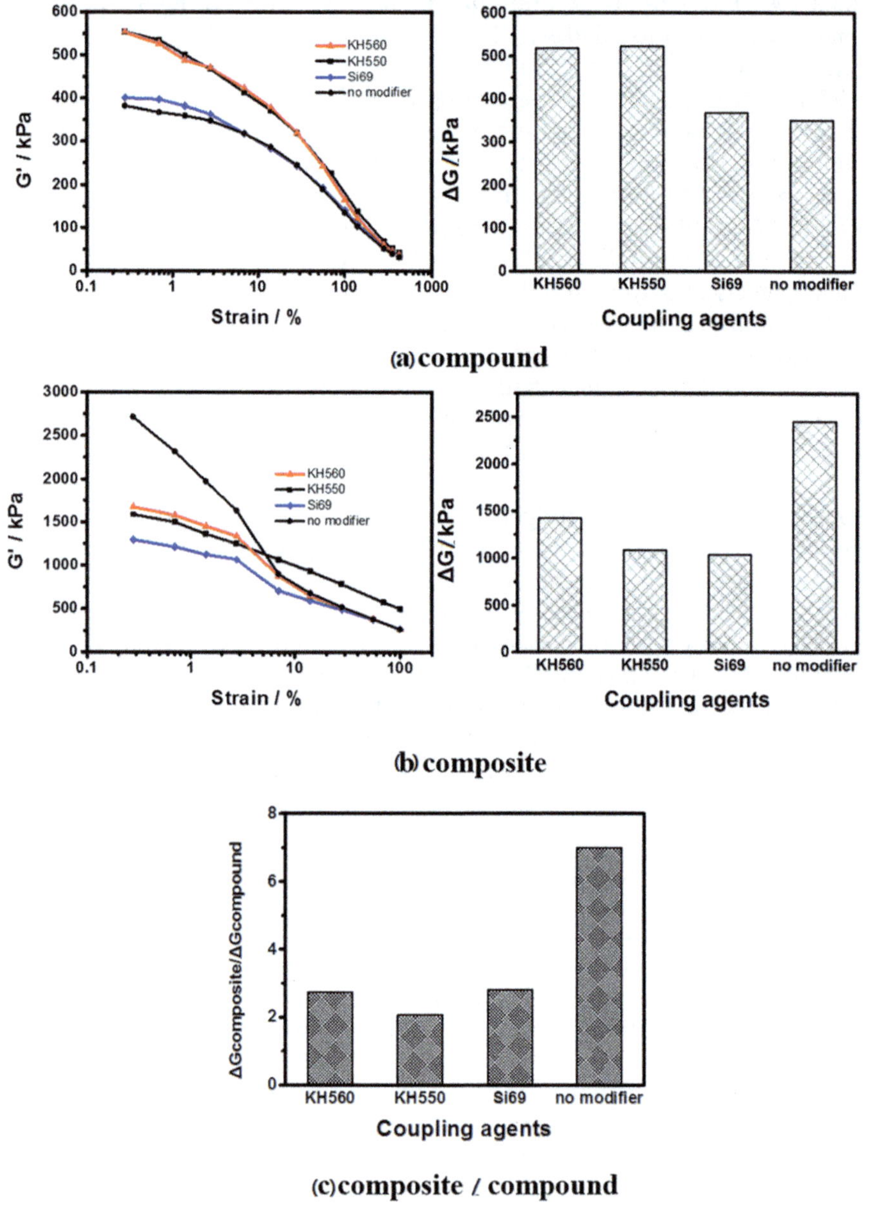

Figure 2.24 Payne effects of AT–NBR (a) compounds, (b) composites and (c) $\Delta G_{composite}/\Delta G_{compound}$ with different coupling agents.

2.3.4 Structure and Properties of Fibrillar Attapulgite–Rubber Composites

2.3.4.1 Anisotropy of Fibrillar Attapulgite–Rubber Composites

By conventional polymer melt blending, modified AT can be exfoliated into single crystals or crystal bundles with diameters smaller than 100 nm, and

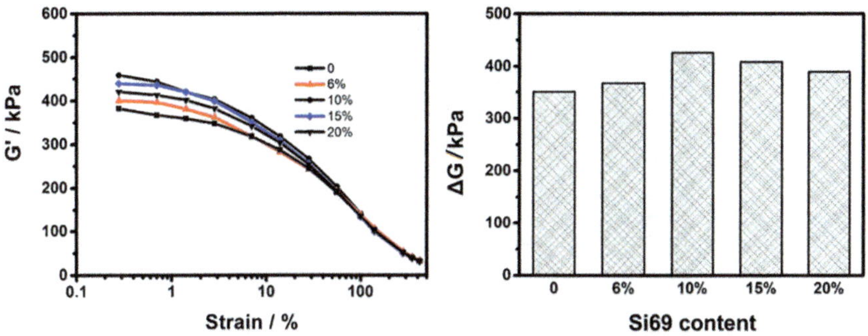

Figure 2.25 Payne effects of Si69-modified AT–SBR compounds with different Si69 contents.

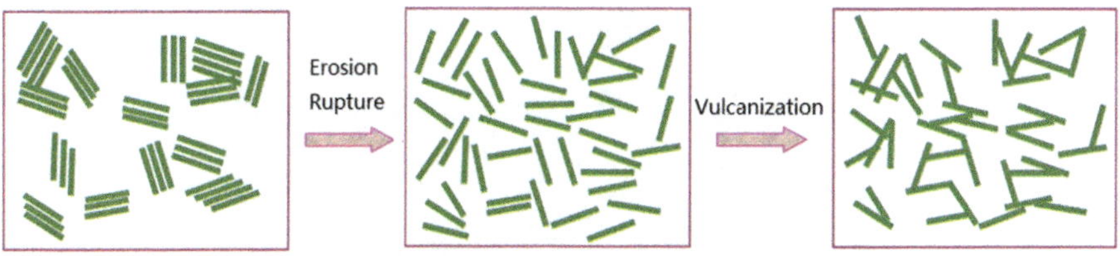

Figure 2.26 Dispersion and re-aggregation scheme of AT in elastomer matrix.

these single crystals or crystal bundles can be uniformly dispersed in the rubber matrix and provide an excellent reinforcing effect. In this way, AT–rubber composites with low cost and excellent properties can be prepared. The properties of the AT–rubber nanocomposites depend on many factors, such as the dispersion and dosage of AT, the type of modifier, and the type of matrix.

Tian *et al.* successfully prepared AT–rubber nanocomposites by directly mixing modified AT with the rubber matrix.[56] The differences in mechanical properties of the composites in two different directions showed that both AT–SBR and AT–NBR composites exhibit obvious anisotropy (see Figure 2.27), which strongly depends on the preparation process, AT concentration, and rubber matrix, and the authors concluded that the anisotropy of the composites is due to the orientation of AT.

2.3.4.1.1 Influence of Preparation Process. Three mixing processes (A, B, and C) were used to prepare AT–rubber composites.[56] In Process A, the two rolls of a two-roll mill were adjusted to the smallest gap, and when the rubber became fluidic, modified AT was added into the rubber step by step to ensure good dispersion. The vulcanizing agent and accelerator were added next. Finally, the compound sheet for vulcanization was obtained through a 1.5 mm gap. Process B is the same as Process A, except that the compound was mixed in the same direction all the time to align the nanofibrils. Process C is the same as Process A, except that the compound sheet prepared for

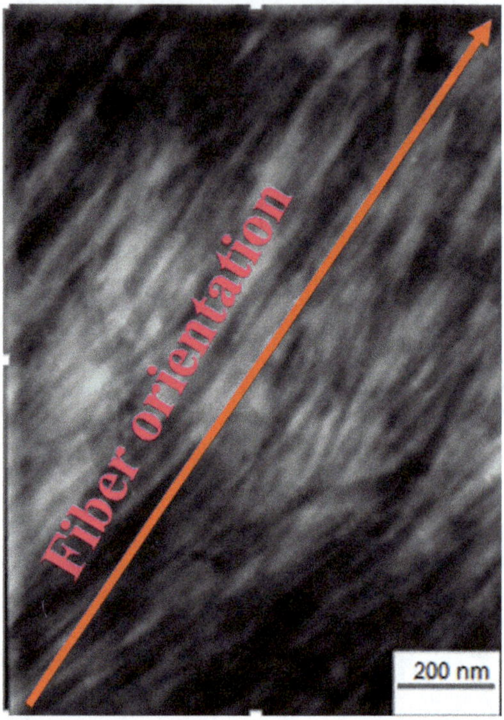

Figure 2.27 TEM image illustrating anisotropy of AT–rubber composite. (Reproduced from ref. 56 with permission from John Wiley and Sons. Copyright © 2005 WILEY-VCH Verlag GmbH & Co. KGaA, Weinheim.)

vulcanization was obtained through the smallest gap to make the nanofibrils oriented. As shown in Table 2.6, the AT–NBR composites prepared by Process B or C exhibit a higher stress at 100% strain and tensile strength, but lower elongation at break in the direction parallel to the fibers than in the direction perpendicular to the fibers, while the AT–NBR composite prepared by Process A does not show visible differences in these two directions. The authors concluded that the composites prepared by Processes B and C are anisotropic.

2.3.4.1.2 Influence of AT Concentration. Table 2.7 further demonstrates that the AT–NBR composites prepared by Process C exhibit anisotropy. On the whole, as the concentration of AT increases, the tensile stress and strength increase, but the elongation at break slightly decreases, especially in the direction parallel to the fibers. Moreover, the differences in the tensile stress at 100% strain or tensile strength between the directions parallel and perpendicular to the fibers increases with the increase in AT, indicating that the anisotropy of the composites becomes more evident. As seen from Table 2.8, the AT–NBR composites prepared by Process C have a significant increase in tensile strength as the AT concentration goes up,

Table 2.6 Effect of preparation process on anisotropy of AT (40 phr)–NBR composite.

Direction	L[a]	T[b]	L	T	L	T
Preparation process	A		B		C	
Tensile stress at 100% strain/MPa	9.7	8.9	15.3	8.3	12.9	7.8
Tensile strength/MPa	22.4	20.0	24.8	19.4	23.6	20.9
Elongation at break/%	288	288	216	236	246	277

[a]*L* stands for the direction parallel to the fibers.
[b]*T* stands for the direction perpendicular to the fibers.

Table 2.7 Effect of AT content on anisotropy of AT–NBR composites.

AT content/phr	40		50		70		80	
Direction	L[a]	T[b]	L	T	L	T	L	T
Tensile stress at 100% strain/MPa	12.9	7.8	13.8	7.4	20.5	9.9	32.5	15.0
Tensile strength/MPa	23.6	20.9	22.3	19.8	26.0	21.5	33.8	21.0
Elongation at break/%	246	277	221	345	149	257	101	192
Tear strength/kN m^{-1}	55.1	49.9	58.4	51.7	51.1	44.5	49.5	44.2

[a]*L* stands for the direction parallel to the fibers.
[b]*T* stands for the direction perpendicular to the fibers.

Table 2.8 Young's modulus of AT–NBR composites.

	Tensile Strength/MPa	
AT content/phr	L[a]	T[b]
30	13.76	13.11
40	15.24	15.14
50	34.49	22.81
70	51.61	23.20
80	61.08	29.00

[a]*L* stands for the direction parallel to the fibers.
[b]*T* stands for the direction perpendicular to the fibers.

indicating that the anisotropy of the nanocomposites is enhanced and providing further evidence that the concentration of AT affects the anisotropy of the composites.

2.3.4.1.3 Influence of Rubber Matrix.

As seen from Figure 2.28, the difference in tensile stress ($\sigma_L - \sigma_T$) at 100% strain between the parallel and perpendicular to the fibers directions for the AT–SBR and the AT–NBR composites increases as the concentration of AT increases. However, the difference for the AT–NBR composite is much larger than that for the AT–SBR composite at the same concentration of AT, demonstrating that AT–NBR composites are more anisotropic than AT–SBR composites. As shown in Figure 2.29, the Young's modulus difference ($E_L - E_T$) between the two different directions for the two composites is very small at low AT concentrations. The Young's modulus difference for the AT–NBR composites becomes significant at an AT concentration of 40 phr and increases as the concentration of AT increases, whereas the Young's modulus difference for the AT–SBR composites does not become significant until an AT concentration of 50 phr. As discussed above, the better dispersion of AT in NBR results in more nanofibrils and thus a higher rate of increase in the Young's modulus difference as the AT concentration increases. In conclusion, the type of rubber matrix plays an important role in the anisotropy of the composite because it greatly affects the dispersion of AT.

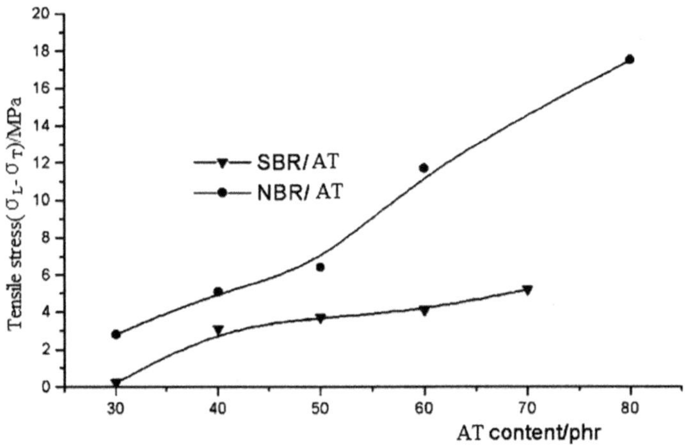

Figure 2.28 Effect of AT content on difference in tensile stress of composites. (Reproduced from ref. 56 with permission from John Wiley and Sons. Copyright © 2005 WILEY-VCH Verlag GmbH & Co. KGaA, Weinheim.)

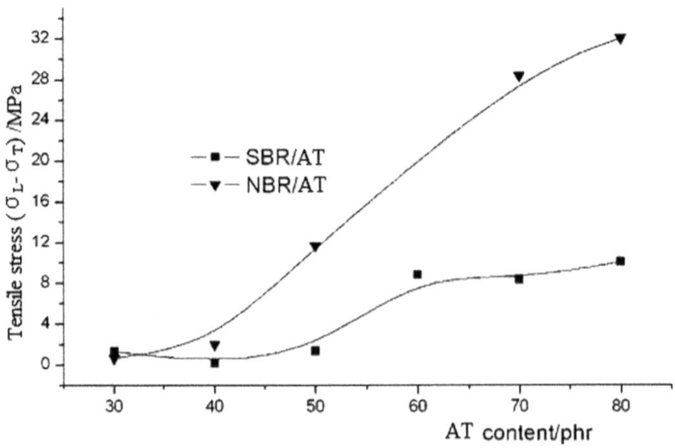

Figure 2.29 Effect of AT content on difference in Young's modulus of composites. (Reproduced from ref. 56 with permission from John Wiley and Sons. Copyright © 2005 WILEY-VCH Verlag GmbH & Co. KGaA, Weinheim.)

2.3.4.2 Mechanical Properties of Fibrillar Attapulgite–Rubber Composites with Different Modifiers

As seen from Table 2.9, the hardness, tensile stress at 300% elongation, tensile strength, and tear strength of Si69-modified AT–SBR composites are significantly higher, but the tensile set and elongation at break are significantly lower than those of the unmodified AT–SBR composite, further demonstrating that surface modification using Si69 can greatly improve the reinforcing ability of AT. The optimum loading level of Si69 was 2.4 g 40 g^{-1} AT, considering the cost.

The SEM micrographs of the composites in Figure 2.30 show some agglomerates in the size range 0.2–2 μm, much smaller than their original sizes. The finer the modified AT particles the better their dispersion,

Table 2.9 Effect of amount of Si69 on mechanical properties of AT–SBR composites.

Amount of Si69 (per 40 g of AT)	0	2.4	4	6	8
Hardness/Shore A	56 ± 1.0	71 ± 2.0	71 ± 1.0	73 ± 3.0	76 ± 2.0
Modulus at 300% elongation/MPa	2.1 ± 0.3	11.1 ± 0.4	11.2 ± 0.5	12.7 ± 0.6	14.8 ± 0.8
Tensile strength/MPa	7.6 ± 0.5	13.4 ± 0.6	12.9 ± 0.5	13.4 ± 0.4	14.8 ± 0.7
Elongation at break/%	900 ± 35	440 ± 23	360 ± 25	312 ± 15	300 ± 13
Tensile set/%	52 ± 2.0	16 ± 1.0	8 ± 1.0	8 ± 2.0	8 ± 0.5
Tear strength/kN m^{-1}	22.2 ± 1.5	52.4 ± 2.0	57.2 ± 2.5	56.9 ± 1.5	66.4 ± 1.3

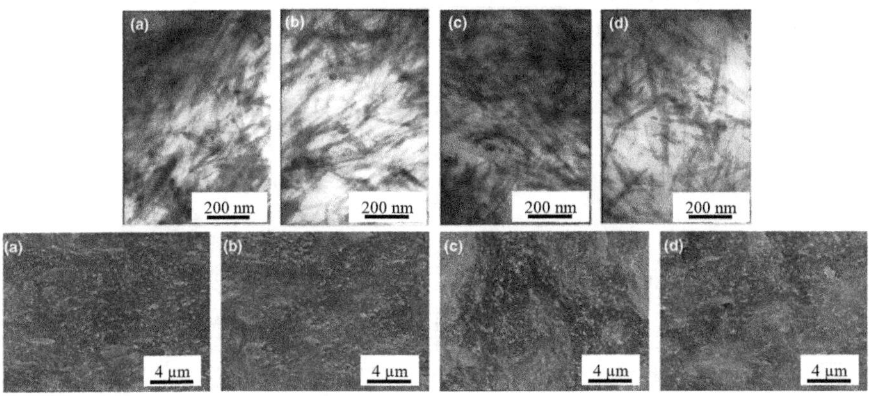

Figure 2.30 TEM and SEM images of AT–SBR composites: (a) unmodified AT; (b) KH570-modified AT; (c) Si69-modified AT; (d) KH560-modified AT. (Reproduced from *Composites Science and Technology*, **65**, M. Tian *et al.*, Surface modification of fibrillar silicate and its reinforcing mechanism on FS–rubber composites, 1129–1138, Copyright 2005 with permission from Elsevier.[37])

and the dispersion of AT modified with KH570 or KH560 is the best. The dispersion of unmodified AT is the worst because it forms many large agglomerates. The fracture surface of unmodified AT–SBR composite is smoother and has larger holes than modified AT–SBR composites due to the dislocation of agglomerates from the matrix, indicating that the interfacial adhesion between unmodified AT and SBR is the weakest. Similarly, the fracture surface of KH570-modified AT–SBR composite has fewer smaller holes than unmodified AT/SBR composite, indicating that the interfacial adhesion is still weak. The dispersion of Si69-modified AT is not as good as that of KH570-modified AT, but the less clear fracture of Si69-modified AT–SBR composite suggests that the interfacial adhesion is stronger. In addition, using KH560 as modifier, we can achieve a good dispersion of AT and improve the interfacial adhesion between AT and the matrix as well.

Dynamic mechanical thermal analysis (DMTA) is generally employed to evaluate the interfacial change of composites. Tian *et al.* purposely chose simpler composites vulcanized by dicumyl peroxide as samples for DMTA testing.[57] Figure 2.31 shows that the incorporation of AT into the rubber matrix greatly lowers the viscoelasticity of the rubber in the glassy transition region and hence the loss factor (tan σ) of the rubber in this region, but remarkably improves the storage modulus (E') of the rubber. Furthermore, with the help of a silane coupling agent, AT produces a stronger restricting effect to macromolecular motion than unmodified AT does due to the finer dispersion in the matrix and stronger interfacial interaction with the matrix. Also, KH570, with double bonds, shows a stronger effect on slowing down the thermal motion of rubber macromolecules than Si69, which contains sulfhydryl groups. In fact, this effect of the silane coupling agent is associated with what is used as vulcanizing agent. The silane coupling agent Si69 should be preferentially chosen, as sulfur is extensively used as the vulcanizing agent in the rubber industry.

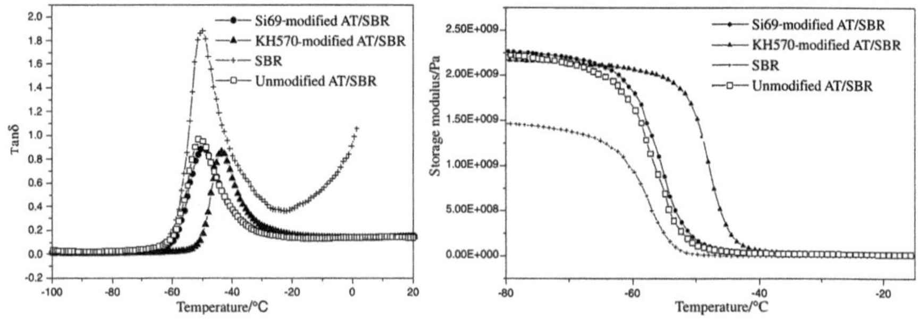

Figure 2.31 Dynamic properties of AT–SBR with different silane coupling agents. Generally, lower loss factor in the glassy transition region and higher storage modulus of rubber composite indicate a stronger interfacial interaction. (Reprinted by permission from Macmillan Publishers Ltd: *Polymer Journal*,[57] Copyright (2006).)

2.3.4.3 Mechanical Properties of Fibrillar Attapulgite–Rubber Composites with Different Curing Systems

Tian *et al.* explored the mechanical properties of AT–rubber composites under different curing systems.[37] From Tables 2.10 and 2.11, it can be seen that the vulcanizer has an important effect on the interfacial adhesion between AT and matrix. When the vulcanizer is sulfur, KH560 and Si69 are good choices to improve the interfacial adhesion because the active epoxy group in KH560 may take part in crosslinking reactions and the polysulfur bond in Si69 also takes part in crosslinking reactions. However, when the vulcanizer is peroxide, KH570 and A151 are good because chemical bonds may be formed between

Table 2.10 Effect of type of silane coupling agent on mechanical properties of AT–SBR composites cured by sulfur.[a]

Modified AT	AT0	AT1	AT2	AT3	AT4	AT5
Hardness/Shore A	56 ± 1.0	73 ± 2.0	71 ± 2.0	68 ± 1.0	71 ± 1.0	67 ± 1.0
Modulus at 300% elongation/MPa	2.1 ± 0.2	10.1 ± 0.5	11.1 ± 0.8	5.1 ± 0.3	8.4 ± 0.5	4.5 ± 0.3
Tensile strength/MPa	7.6 ± 0.5	12.9 ± 0.9	13.4 ± 0.8	11.7 ± 0.7	13.5 ± 0.8	14.4 ± 0.9
Elongation at break/%	900 ± 25	528 ± 20	440 ± 20	776 ± 30	492 ± 25	756 ± 20
Tensile set/%	52 ± 2.0	16 ± 3.0	16 ± 2.0	48 ± 3.0	36 ± 2.0	50 ± 3.0
Tear strength/kN m^{-1}	22.2 ± 1.5	34.6 ± 1.5	52.4 ± 2.0	28.0 ± 1.5	31.2 ± 1.8	26.2 ± 1.2

[a]AT0 stands for unmodified AT, and AT1, AT2, AT3, AT4, and AT5 stand for AT modified with KH560, Si69, A151, KH550, and KH570, respectively. Modifier amount is 6% of the weight of AT unless otherwise stated.

Table 2.11 Effect of type of silane coupling agent on mechanical properties of AT–SBR composites cured by DCP.

Modified AT	AT0	AT1	AT2	AT3	AT4	AT5
Hardness/Shore A	67 ± 1.0	70 ± 1.0	71 ± 1.0	72 ± 2.0	75 ± 1.0	75 ± 2.0
Modulus at 300% elongation/MPa	2.1 ± 0.2	4.6 ± 0.2	3.0 ± 0.2	5.8 ± 0.3	6.5 ± 0.3	10.3 ± 0.5
Tensile strength/MPa	3.9 ± 0.3	5.1 ± 0.4	9.5 ± 0.4	8.7 ± 0.5	6.9 ± 0.5	16.7 ± 0.8
Elongation at break/%	252 ± 15	128 ± 12	352 ± 14	188 ± 15	124 ± 10	136 ± 12
Tensile set/%	4 ± 0.5	0	4 ± 1.0	4 ± 0	4 ± 1.0	0
Tear strength/kN m^{-1}	16.0 ± 1.2	24.5 ± 1.5	40.9 ± 1.2	20.4 ± 1.4	17.4 ± 1.3	26.8 ± 1.0

the C=C of KH570 or A151 and the C=C of rubber. KH550 can improve the interfacial adhesion between the filler and the polymer in both curing systems.

2.3.4.4 *Effect of Fibrillar Attapulgite on the Mechanical Properties of Different Rubber Matrices*

Table 2.12 summarizes the mechanical properties of various rubber-based short-nanofibril silicate nanocomposites made by a traditional mechanical mixing technique in combination with an *in situ* modification method. Nanofibril silicates can endow various rubbers, such as SBR, NBR, NR, EPDM, chloride butadiene–acrylonitrile rubber (CNBR), hydrogenated butadiene–acrylonitrile rubber (HNBR), and chloroprene rubber (CR), with excellent properties. Compared with the neat rubber, the corresponding AT–rubber nanocomposites show a remarkable increase in hardness, stress at 100% strain, tensile strength, and tear strength, as well as a decrease in elongation at break. An exception is that the tensile strength of the AT–NR nanocomposite is slightly lower than that of neat NR, probably because the immobility of the rubber macromolecules as a result of the strong bonding between AT and the matrix hinders the strain-induced crystallization of NR at large tensile strains. Still, the tear strength and stress at 100% strain of the AT–NR composite are higher than those of neat NR, indicating the excellent reinforcement of AT. The excellent properties of AT–rubber composites are related to not only the dispersion of AT in the rubber matrix, but also the strong interfacial interactions between AT and the rubber. In summary, a series of novel AT–rubber composites were prepared by using a traditional mechanical mixing technique in combination with an *in situ* modification method. These composites have excellent mechanical properties similar to those of short-fiber-reinforced rubbers, excellent processing properties as expected, and low costs, and hence a very bright future in the rubber industry.

2.3.4.5 *Comparison of Fibrillar Attapulgite with Other Fillers to Enhance the Mechanical Properties of Rubber Composites*

The most important mechanical characteristics of short-fiber-reinforced composites (SFRC) are high modulus and apparent anisotropy. Figure 2.32 a shows the stress–strain behaviors of various kinds of filler-reinforced NBR composites.[57] Polyhexamethylene adipamide (PA66) short fiber is a kind of microfiber, while carbon black N330 is a kind of particulate filler that have been extensively used in the rubber industry for many years. At the same strain, the Young's modulus and stress of Si69 modified AT–NBR composites are higher than those of AT–NBR composites prepared by directly blending without *in situ* modification and traditional carbon black–NBR composites and almost as high as those of NBR composites reinforced by PA66 short fibers. Although the modulus of the AT–NBR composite is slightly lower than that of PA66 short fiber–NBR composites, the gentler linear stress–strain behavior displayed by the AT–NBR composite is actually more desirable for industrial applications. Nanofibers lead to lower stress concentrations in the composite and higher

Table 2.12 Mechanical properties of various rubbers and AT–rubber nanocomposites.[a]

Properties	SBR	AT/SBR	NR	AT/NR	NBR	AT/NBR	CNBR	AT/CNBR	EPDM	AT/EPDM	HNBR	AT/HNBR	CR	AT/CR
Hardness/Shore A	45	73	41	74	54	83	61	72	70	78	57	80	51	83
Modulus at 100% elongation/MPa	0.8	7.6	1.0	9.1	1.2	12.1	2.3	8.3	1.4	7.0	1.3	28.6	1.1	13.8
Tensile strength/MPa	2.3	14.7	29.3	24.5	3.2	22.7	9.7	22.7	4.7	10.7	6.1	30.4	6.6	14.6
Elongation at break/%	556	344	660	390	308	290	412	237	356	177	270	109	509	113
Tear strength/kN m^{-1}	11.6	60.8	30.3	65.4	14.3	58.8	22.9	48.5	18.1	36.1	13.9	26.4	—	—

[a]The AT loading is 40 phr. SBR, NR, NBR, CNBR, EPDM, HNBR, CR are the abbreviations of butadiene–styrene rubber, neutral rubber, butadiene–acrylonitrile rubber, chloride butadiene–acrylonitrile rubber, ethylene propylene diene monomer rubber, hydrogenated butadiene–acrylonitrile rubber, chloroprene rubber, respectively.

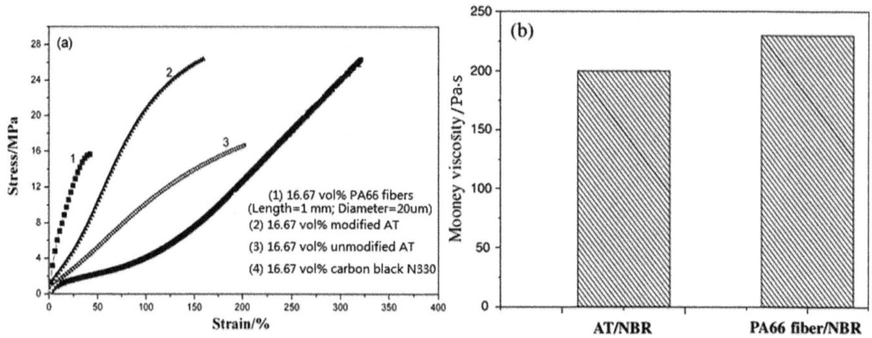

Figure 2.32 (a) Stress–strain characteristics and (b) Mooney viscosity of NBR composites reinforced with different fillers. (Reprinted by permission from Macmillan Publishers Ltd: *Polymer Journal*,[57] Copyright (2006).)

interfacial area but lower stiffness of the dispersion units than microfibers. Figure 2.32b compares the Mooney viscosity of nanofiber-filled NBR compound with that of microfiber-filled NBR compound. The temperature for the Mooney viscosity measurements was purposely set at 40 °C to prevent the PA66 fibers from softening. The AT–NBR compound, with a lower Mooney viscosity, presents better flow properties at the same fiber volume fraction, mainly because of the smaller absolute length of the nanofibrils. Furthermore, the extrudates of the AT–NBR compound from a capillary rheometer are more smooth than those of the short nylon fiber–NBR compound.

2.3.5 Synergistic Effect of Fibrillar Attapulgite and Other Microfibers on Mechanical Properties of Rubber

SFRCs have attracted growing interest because they have significantly mechanical properties and unique rubbery characteristics. When the short fibers in the composites are aligned, the SFRCs exhibit rigidity in the direction parallel to the fiber axes (the *L* direction) and viscoelasticity in the direction perpendicular to the fiber axes (the *T* direction); therefore, the mechanical properties of such SFRCs are anisotropic. In addition to microfibers (*e.g.*, conventional microfibers of polyester and/or nylon), numerous nanofibers, including carbon nanotubes,[58–60] carbon nanofibers,[61–63] halloysite nanotubes,[64] and electrospun nanofibers,[65] have also been studied for the development of SFRCs. In comparison with microfibers, nanofibers are more likely to impart high mechanical and/or functional properties, controlled degrees of anisotropy, and desirable appearances to SFRCs.

SFRCs with hydrogenated acrylonitrile butadiene rubber (HNBR) as the matrix and fibrillar silicate (AT) nanofibers and short aramid microfibers (DCAF) as the fillers were developed by Tian *et al.*[66] Table 2.13 shows the mechanical properties of the DCAF–AT–HNBR composites filled with various amounts of DCAF microfibers (the AT amount was kept constant at 40 phr). With increasing amount of DCAF microfibers, the tensile stress at low

Table 2.13 Mechanical properties of DCAF–AT–HNBR composites filled with various amounts of DCAF microfibers.[a]

	DCAF (phr)													
	0		3		6		9		12		15		18	
Property direction	L	T	L	T	L	T	L	T	L	T	L	T	L	T
Tensile stress at 5% strain/MPa	1.1	0.9	1.5	0.7	2.6	1.1	3.3	1.3	4.1	1.4	22.3	2.6	29.6	2.8
Tensile stress at 25% strain/MPa	3.6	2.8	9.1	2.1	13.1	3.1	17.5	3.2	21.4	3.4	—	5.6	—	6.5
Tensile strength at break/MPa	25.3	22.4	25.5	17.6	22.8	17.6	21.0	16.5	18.2	14.7	28.2	15.1	29.6	14.7
Elongation at break/%	157	213	197	279	220	230	206	241	137	254	12	197	5	176

[a]The AT loading is 40 phr.

strains and the anisotropy of the tensile stress increase, but the elongation at break decreases. Table 2.14 shows the mechanical properties of the DCAF–AT–HNBR composites filled with various amounts of KH570-modified AT (the amount of DCAF microfibers was kept constant at 3 phr). With increasing amount of KH570-modified AT, the tensile stress at low strains and the anisotropy of the tensile stress increase, but the effect is much smaller than that of DCAF microfibers on the tensile stress and the anisotropy at low strains. The increase in the AT amount also increases the tensile strength, but decreases the elongation at break of the composites. The results indicate that the properties of the DCAF–AT–HNBR composites are determined by the amount of AT nanofibers and DCAF microfibers. A small amount of microfibers combined with an appropriate amount of nanofibers results in synergistic reinforcement and imparts significantly improved mechanical properties to the SFRCs without substantially compromising the rubber characteristics.

Additionally, EPDM composites reinforced with micro- and nanofibers were made by mechanical blending of silanized fibrillar silicate (AT) nanofibers and PA66 microfibers by Tian *et al.*[67] A synergistic reinforcement of static and dynamic mechanical properties was observed. Figure 2.33 shows the tensile stress–strain curves of AT–PA66–EPDM composites at different volume ratios of AT nanofibers to PA66 microfibers. These stress–strain curves are all steep, with much higher tensile stress but lower elongation at break in the *L* direction than in the *T* direction, typical of SFRCs. Figure 2.34 displays the tensile stresses in the *L* direction and the ratios of the tensile stress in the *L* direction to that in the *T* direction, all at 25% elongation, of EPDM–AT–PA66 composites at different volume ratios of AT nanofibers to PA66 microfibers. According to Figures 2.33 and 2.34, the AT–EPDM composite reinforced with only AT nanofibers exhibits higher tensile strength, tensile stress at 25% elongation, and elongation at break in the *L* direction than the PA66–EPDM composite reinforced with only PA66 microfibers, indicating that AT nanofibers have a stronger reinforcement effect than PA66 microfibers on EPDM at the same fiber volume. AT–PA66–EPDM composites display steeper stress–strain curves and much higher tensile stresses at 25% elongation than EPDM composites reinforced with AT nanofibers or PA66 microfibers alone. At volume ratios of AT nanofibers to PA66 microfibers of 13/13 cm^3/cm^3 and 17/9 cm^3/cm^3, the composites exhibit the steepest stress–strain curves as well as the highest tensile stress at 25% elongation in the *L* direction, confirming the synergetic reinforcement effect of AT nanofibers and PA66 microfibers.

DMTA is used to study the viscoelasticity of polymers. The higher the storage modulus E', the greater the capacity of polymers to store elastic energy. A high storage modulus is an indication of high strength as a result of strong interactions among polymer chains. On the other hand, the energy loss increases with the increase of the loss factor ($\tan\delta$, the ratio of loss modulus to storage modulus). Tan δ is an indication of heat accumulation as a result of the friction among polymer chains (the viscosity of the polymer). As shown in Figure 2.35, the effect of the volume ratio of AT nanofibers to PA66 microfibers on dynamic properties is similar to that on tensile properties. EPDM reinforced by PA66 microfibers alone has the lowest storage modulus E', but

Table 2.14 Mechanical properties of DCAF–AT–HNBR composites filled with various amounts of KH570-modified AT and 3 phr of DCAF microfibers.

	AT (phr)													
	0		10		20		30		40		60		80	
Property direction	L	T	L	T	L	T	L	T	L	T	L	T	L	T
Tensile stress at 5% strain/MPa	0.5	0.4	0.7	0.6	1.2	0.7	2.8	1.2	1.8	1.1	2.8	1.3	4.9	1.6
Tensile stress at 25% strain/MPa	2.9	0.6	4.1	1.2	5.8	1.3	8.2	2	6.8	2.2	7.2	3	11.8	3.7
Tensile strength at break/MPa	4.1	5.7	11.7	11.6	16.5	13.8	21.5	18.3	23.3	18	26.2	20.2	30.0	21.4
Elongation at break/%	187	358	310	345	247	288	203	237	214	234	175	239	126	217

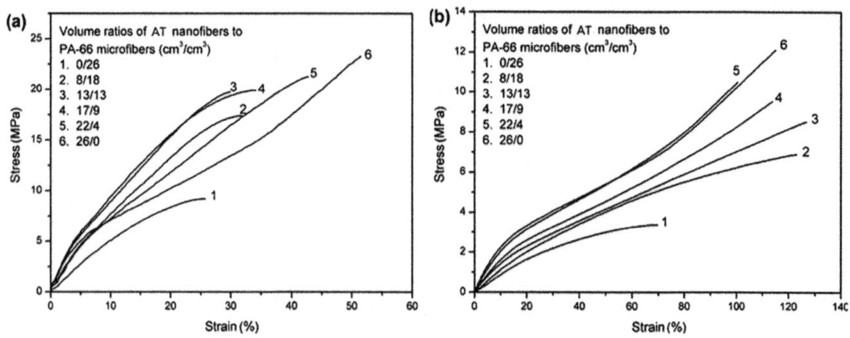

Figure 2.33 Tensile stress–strain curves of AT–PA66–EPDM composites at different volume ratios of AT nanofibers to PA66 microfibers: (a) in *L* direction and (b) in *T* direction. (Reproduced from *Composites Part B: Engineering*, **42**, M. Tian *et al.*, Static and dynamic mechanical properties and fracture morphology of EPDM composites containing silicate nanofibers and short PA-66 microfibers, 1937–1944, Copyright 2011 with permission from Elsevier.[67])

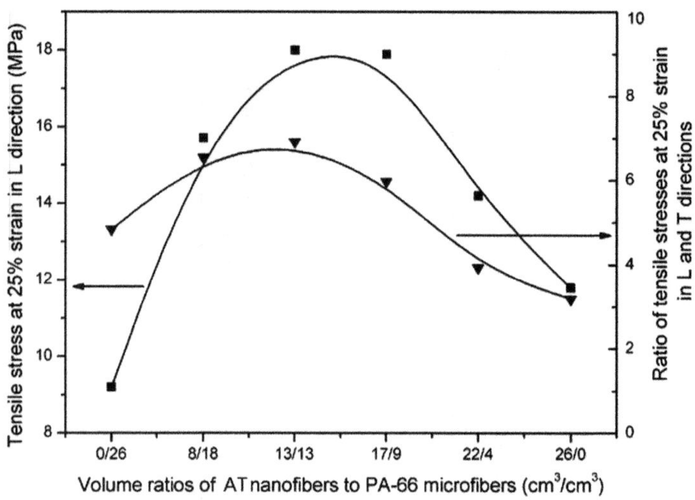

Figure 2.34 Tensile stress at 25% strain in *L* direction and ratio of tensile stresses in *L* and *T* directions at 25% strain of AT–PA66–EPDM–AT composites at different volume ratios of AT nanofibers to PA66 microfibers. (Reproduced from *Composites Part B: Engineering*, **42**, M. Tian *et al.*, Static and dynamic mechanical properties and fracture morphology of EPDM composites containing silicate nanofibers and short PA-66 microfibers, 1937–1944, Copyright 2011 with permission from Elsevier.[67])

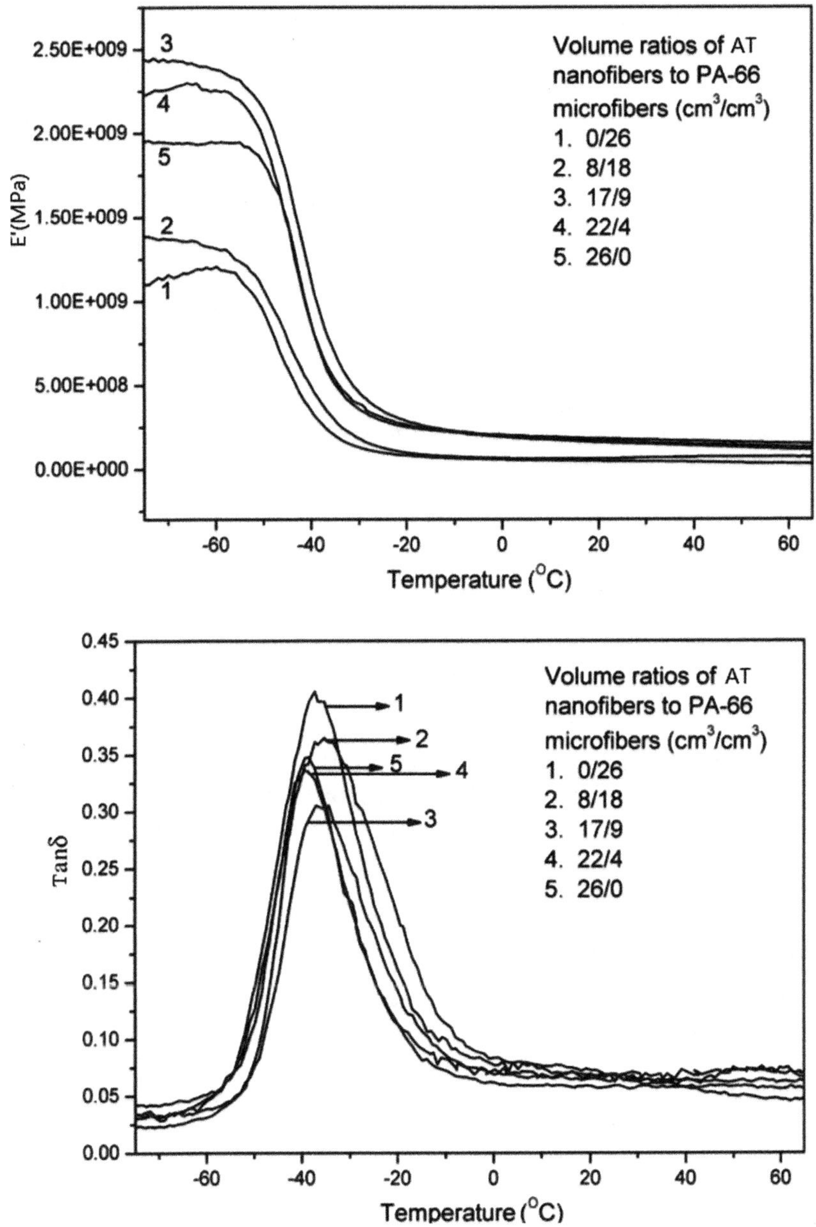

Figure 2.35 The results of DMTA temperature scanning of EPDM–FS–PA66 composites in *L* direction at different volume ratios of AT nanofibers to PA66 microfibers. (Reproduced from *Composites Part B: Engineering*, **42**, M. Tian *et al.*, Static and dynamic mechanical properties and fracture morphology of EPDM composites containing silicate nanofibers and short PA-66 microfibers, 1937–1944, Copyright 2011 with permission from Elsevier.[67])

the highest peak value of tan δ because the interfacial adhesion between the PA66 microfibers and the rubber is very weak. At a volume ratio of AT nanofibers to PA66 microfibers of 17/9 cm^3/cm^3, the storage modulus E' is the highest, but the peak value of tan δ is the lowest, reconfirming the synergetic reinforcement effect of AT nanofibers and PA66 microfibers. However, the volume ratio of AT nanofibers to PA66 microfibers hardly affects the glass transition temperature T_g (the temperature corresponding to the tan δ peak).

In summary, when AT nanofibers and PA66 microfibers are used together in an appropriate volume ratio, both the static and the dynamic mechanical properties (such as tensile stress and dynamic storage modulus) of the resulting composites are substantially improved. The synergetic reinforcement effect on the dynamic fatigue properties is attributed to the result of competitive factors including stress transfer, crack initiation and propagation, and matrix reinforcement, whereas the synergetic reinforcement effect on the static tensile properties is due to the strong interfacial adhesion between the AT nanofibers and rubber.

2.4 Conclusion

A series of novel nanofibril silicate–rubber composites can be prepared by using a traditional mechanical mixing technique. The enhancement effect of AT on AT–rubber composites may be affected by many factors, such as the preparation method, the amount of AT and modifiers, the type of modifiers, and the type of vulcanizers. Based on the special characteristics of AT, many high performance AT–rubber composites have been prepared, and this may be an important way to obtain low-cost, high-performance rubber composites.

Acknowledgements

We would like to express our sincere thanks to the National Science Fund for Distinguished Young Scholars of China (Grant No. 51525301), the National Natural Science Foundation of China (Grant No. 51373012), the National Basic Research Program of China (2015CB654704 (2015CB654704)) for financial support.

References

1. L. Q. Zhang and Y. P. Wu, *China Synth. Rubber Ind.*, 2000, **2**, 71–77.
2. L. Q. Zhang, *China Synth. Rubber Ind.*, 1996, **1**, 5.
3. E. P. Giannelis, *Adv Mater.*, 1996, **8**, 29–35.
4. M. Alexandre and P. Dubois, *Mater. Sci. Eng.*, 2000, **28**, 1–63.
5. Y. C. Ke, C. F Long and Z. N. Qi, *Appl. Polym. Sci.*, 1999, **71**, 1139–1146.
6. J. Heinemann, H. Reichert, R. Thomann and R. Muelhaupt, *Macromol. Rapid Commun.*, 1999, **20**, 423–430.

7. L. M. Liu, Z. N. Qi and X. G. Zhu, *J. Appl. Polym. Sci.*, 1999, **71**, 1133–1138.
8. G. Carotenuto, *Polym. News*, 2000, **25**, 191–193.
9. G. Carotenuto, *Polym. News*, 2000, **25**, 265–267.
10. C. D. Qu, M. Tian and L. Q. Zhang, *China Synth. Rubber Ind.*, 2003, **1**, 49.
11. H. H. Murray, *Appl. Clay Sci.*, 2000, **17**, 207–221.
12. Q. Tang, G. Qing, F. Wang, H. Guo, Y. Yang, Y. L. Du, J. S. Liang and F. Q. Zhang, *Powder Technol.*, 2015, **270**, 92–97.
13. M. Tian, W. L. Liang, G. Y. Rao, L. Q. Zhang and C. X. Guo, *Compos. Sci. Technol.*, 2005, **65**, 1129–1138.
14. H. H. Murray, *Appl. Clay Sci.*, 2000, **17**, 207–221.
15. P. Cerezoa, C. Viseras Iborraa, A. Lopez-Galindob, F. Ferraric and C. Carmellac, *Appl. Clay Sci.*, 2001, **20**, 81–86.
16. D. M. Araújo Melo, J. A. C. Ruiz, M. A. F. Melo, E. V. Sobrinho and A. E. Martinelli, *J. Alloys Compd.*, 2002, **344**, 352–355.
17. J. Herrero, J. A. Pajares and C. Blanco, *Clays Clay Miner.*, 1991, **39**, 651–657.
18. Q. P. Guo, L. Xiang, J. Y. Huang, T. L. Chen and K. R. Wang, *Eur. Polym. J.*, 1990, **26**, 355–358.
19. K. Song and G. Sandí, *Clays Clay Miner.*, 2001, **49**, 119–125.
20. Y. J. Deng, B. Joe Dixon and G. Norman White, *Clays Clay Miner.*, 2003, **51**, 150–161.
21. M. S. Augsburger, E. Starssera, E. Perinoa, R. C. Mercaderb and J. C. Pedregosa, *J. Phys. Chem. Solids*, 1998, **59**, 175–180.
22. J. Gu, N. Liu and Y. Li, *Silic. Bull.*, 1999, **6**, 50–54.
23. Y. S. Fang and J. Q. Xu, *et al.*, *J. Nanjing Univ. Earth Sci.*, 1990, **1**, 15–23.
24. S. H. Yan, *et al.*, *Translation. Clay mineralogy*, Beijing, China, Beijing press, 1981, p. 9.
25. M. Tian, Y. Gao, Y. Liu, Y. L. Liao, R. W. Xu, N. E. Hedin and H. Fong, *Polymer*, 2007, **48**, 2720–2728.
26. M. Tian, C. D. Qu, L. Q. Zhang and Y. Feng, *J. Mater. Sci.*, 2003, **38**, 4917–4924.
27. Z. J. Du, J. F. Rong, H. Q. Li, W. Zhang and Z. Jing, *J. Mater. Sci.*, 2003, **38**, 4863–4868.
28. E. N. Brown, A. K. Davis, K. D. Jonnalagadda and N. R. Sottos, *Compos. Sci. Technol.*, 2005, **65**, 2466–2473.
29. R. L. Frost, O. B. Locos, H. d. Ruan and J. Theo Kloprogge, *Vib. Spectrosc.*, 2001, **27**, 1–13.
30. T. H. Chen, *Anhui Geol.*, 1999, **3**, 199–203.
31. M. S. Wang and Y. M. He, *Sepi. Infrastruct. Spectrosc. Res. North Jiangsu*, 1999, **6**, 19–24.
32. M. Tian, L. L. Su, W. T. Cai, S. Yin, H. Fong and L. Q. Zhang, *J. Appl. Polym. Sci.*, 2011, **120**, 1439–1447.
33. M. S. Augsburger, E. Strasser, E. Perino, R. C. Mercader and J. C. Pedregosa, *J. Phys. Chem. Solids*, 1998, **2**, 175–180.
34. H. C. Gu and H. R. Xu, *Chem. Eng. Process.*, 1999, **4**, 5–7.
35. L. Y. Yang and J. Sheng, *Jiangsu Chem.*, 2001, **6**, 33–37.

36. L. Song, S. L. Fang, J. Zhan, Y. Hu and J. Wu, *Polym. Compos.*, 2010, **31**, 405–410.
37. M. Tian, W. L. Liang, G. Y. Rao, L. Q. Zhang and C. X. Guo, *Compos. Sci. Technol.*, 2005, **65**, 1129–1138.
38. J. Brandrup, H. Immergut, E. A. Grulke and D. Bloch, *Polymer handbook*, 4th edn, 1999, pp. 521–541.
39. G. E. Maciel and D. W. Sindorf, *J. Am. Chem. Soc.*, 1980, **102**, 7606–7607.
40. G. S. Caravajal, D. E. Leyden, G. R. Quinting and G. E. Maciel, *Anal. Chem.*, 1988, **60**, 1776–1786.
41. R. L. Frost, O. B. Locos, H. Ruan and J. T. Kloprogge, *Vib. Spectrosc.*, 2001, **27**, 1–13.
42. D. A. McKeown, J. E. Post and E. S. Etz, *Clay Mineral.*, 2002, **50**, 667–680.
43. M. Tian, X. L. Zhang, L. Q. Zhang, S. Yin, T. Nishi and N. Y. Ning, *Compos. Sci. Technol.*, 2013, **79**, 21–27.
44. E. Gonzalez, C. Pesquear and I. Benite, *Clay and Clay Miner.*, 1989, **3**, 258–262.
45. S. Wu, H. C. Ma and Z. Q. Lei, *Synlett*, 2010, **18**, 2818–2822.
46. P. A. Schueler, J. T. Ives, F. DeLaCroix, W. B. Lacy, P. A. Becker, J. Li, K. D. Caldwell, B. Drake and J. M. Harris, *Anal. Chem.*, 1993, **65**, 3177–3186.
47. L. H. Lu, H. S. Wang, Y. H. Zhou, S. Q. Xi, H. J. Zhang, J. W. Hu and B. Zhao, *Chem. Commun.*, 2002, **2**, 144–145.
48. N. Nino-Martınez, G. A. Martınez-Castanon, A. Arag on-Pina, F. Martınez-Gutierrez, J. R. Martınez-Mendoza and R. Facundo, *Nanotechnology*, 2008, **19**, 065711/1–065711/8.
49. V. Alt, T. Bechert, P. Steinrucke, M. Wagenerc, P. Seideld, E. Dingeldeind, E. Domanne and R. Schnettlera, *Biomaterials*, 2004, **25**, 4383–4391.
50. P. Sukdeb, K. T. Yu and J. M. Song, *Appl. Environ. Microbiol.*, 2007, **73**, 1712–1716.
51. P. V. AshaRani, L. K. M. Grace, P. H. Manoor and V. Suresh, *ACS Nano*, 2009, **3**, 279–290.
52. S. Varun, M. M. Megan, R. P. Blake and S. J. Ayusman, *Am. Chem. Soc.*, 2006, **128**, 9798–9808.
53. K. Ni, H. Lu, C. Wang, K. L. C. Black, D. Wei, Y. Ren and P. B. Messersmith, *Biotechnol. Bioeng.*, 2012, **109**, 2970–2977.
54. I. C. Sun, D. K. Eun, J. K. Na, S. Lee, I. J. Kim, I. C. Youn, C. Y. Ko, H. S. Kim, D. Lim, K. Choi, P. B. Messersmith, T. G. Park, C. Y. Kim, I. C. Kwon, K. Kim and C. H. Ahn, *Chem.–Eur. J.*, 2009, **15**, 13341–13347.
55. Y. Fu, G. f. Li, M. Tian, X. Wang, L. Q. Zhang and W. C. Wang, *J. Appl. Polym. Sci.*, 2014, **131**, 39859(1–7).
56. M. Tian, L. J. Cheng, W. L. Liang and L. Q. Zhang, *Macromol. Mater. Eng.*, 2005, **290**, 681–687.
57. M. Tian, Y. L. Lu, W. L. Liang, L. J. Cheng and L. Q. Zhang, *Polym. J.*, 2006, **38**, 1105–1113.
58. L. Bokobza and M. Kolodziej, *Polym. Int.*, 2006, **55**, 1090–1098.
59. L. Bokobza and C. Belin, *J. Appl. Polym. Sci.*, 2007, **105**, 2054–2061.
60. A. De-Falco, S. Goyanes, G. H. Rubiolo, I. Mondragon and A. Marzocca, *Appl. Surf. Sci.*, 2007, **254**, 262–265.

61. A. Kelarakis, K. Yoon, I. Sics, R. H. Somani, B. S. Hsiao and B. Chu, *Polymer*, 2005, **46**, 5103–5117.
62. C. Gauthier, L. Chazeau, T. Prasse and J. Y. Cavaille, *Compos. Sci. Technol.*, 2005, **65**, 335–343.
63. J. Karger-Kocsis, D. Felhoes and R. Thomann, *J. Appl. Polym. Sci.*, 2008, **108**, 724–730.
64. M. L. Du, B. C. Guo, Y. D. Lei, M. Liu and D. M. Jia, *Polymer*, 2008, **49**, 4871–4876.
65. J. S. Kim and D. H. Reneker, *Polym. Compos.*, 1999, **20**, 124–131.
66. M. Tian, L. L. Su, W. T. Cai, S. Yin, Q. Chen, H. Fong and L. Q. Zhang, *J. Appl. Polym. Sci.*, 2011, **120**, 1439–1447.
67. M. Tian, S. Yin, H. Zou, L. L. Su and L. Q. Zhang, *Composites, Part B*, 2011, **42**, 1937–1944.

CHAPTER 3

Rubber–Rectorite Composites with High Gas Barrier Properties

SHAOJIAN HE[a] AND LIQUN ZHANG*[b]

[a]School of Renewable Energy, North China Electric Power University, Beijing, 102206, China; [b]Key Laboratory for Nanomaterials, Ministry of Education, Beijing University of Chemical Technology, Beijing, 100029, China
*E-mail: zhanglq@mail.buct.edu.cn

3.1 Introduction

Traditionally, polymeric materials are filled with artificial or natural inorganic compounds so as to improve their properties, or simply to reduce cost. For rubber, carbon black continues to be the most important reinforcing agent in the rubber industry, but the polluting nature and dependence on petroleum feedstock of carbon black cause researchers to look out for other 'white' reinforcing agents. Since silica is not quite as reactive to rubber as carbon black, silane coupling agents are used for surface modification of silica, but the curing time of rubber with silica is longer than that with carbon black, and silane coupling agent-modified silica is expensive. In contrast, clay is an abundant resource and the cost of pristine clay is very low; besides, clays do not present an environmental pollution threat or health risks to humans. Rubber molecules can enter the galleries in clay layers, pushing

RSC Smart Materials No. 22
Functional Polymer Composites with Nanoclays
Edited by Yuri Lvov, Baochun Guo and Rawil F. Fakhrullin

Published by the Royal Society of Chemistry, www.rsc.org

the layers apart and creating nanocomposites. Such nanocomposites exhibit high performance,[1–4] such as excellent gas barrier properties, improved thermal stability, enhanced flame retardance and mechanical properties, *etc.* that are dramatically different from and often superior to those of their bulk counterparts. Therefore, rubber–clay nanocomposites have attracted considerable attention in the field of materials science and engineering.

The term nanocomposite is used because the clay platelets dispersed in the rubber matrix for reinforcement have nanometer dimensions (dozens to hundreds of nanometers in length and several to dozens of nanometers in thickness).[5] The significance of nanocomposites is that considerable property enhancements are achievable at very low filler loadings, whereas much higher filler loadings of conventional fillers in a composite can be required to obtain the same result. Importantly, the low content of filler makes nanocomposites accessible to the same processing techniques as the rubber matrix. Therefore, the rubber–clay nanocomposite with high performance is a novel and promising material for application over a wide field.

3.1.1 Structure of Clays

Clay is a natural, fine grained ceramic material, composed mainly of silica, alumina and water. There are many kinds of clay minerals,[5] such as montmorillonite, rectorite, kaolinite, halloysite, illite, vermiculite, attapulgite, sepiolite and mixed layered minerals. Two particular characteristics of clay are generally considered for rubber–clay nanocomposites. The first is the ability of the clay particles to disperse into individual layers. The second characteristic is the ability to fine-tune their surface chemistry through ion exchange reactions with organic and inorganic cations. These two characteristics are, of course, interrelated since the degree of dispersion of clay layers in a particular rubber matrix depends on the interlayer cation. Herein, two different kinds of clay, montmorillonite and rectorite, are introduced because they possess very good swelling properties, high cation exchange capacities, high aspect ratio and large surface area.

The most studied clay for the preparation of rubber–clay nanocomposites is montmorillonite, which belongs to the general family of 2 : 1 layered silicates.[6,7] Montmorillonite consists of layers made up of two tetrahedrally coordinated silicon atoms fused to an edge-shared octahedral sheet of either aluminum or magnesium hydroxide, as shown in Figure 3.1. The clay layers are coupled through relatively weak dipolar and van der Waals forces, and the interlayer distance is about 1 nm. The exchangeable cations residing between the clay layers can be replaced by either inorganic or organic cations,[8] which makes it possible for the clay layers to be separated to prepare polymeric nanocomposites.

Rectorite, discovered early in 1891 by E. W. Rector, is another kind of clay mineral, with a structure and characteristics much like those of montmorillonite. It is a sort of regularly interstratified clay mineral with alternate pairs of dioctahedral mica-like layers and dioctahedral smectite-like layers in a 1 : 1

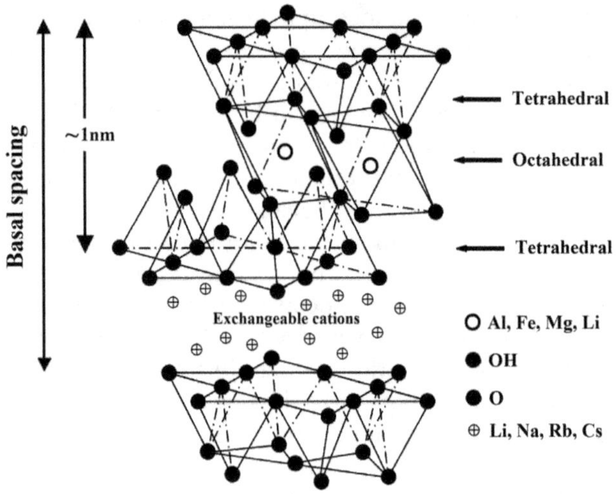

Figure 3.1 Structure of montorillonite (d_{001} refers to basal plane spacing). (Reproduced from *Progress in Polymer Science*, **28**, S. S. Ray and M. Okamoto, Polymer/layered silicate noncomposites: a review from preparation to processing, 1539–1641, Copyright 2003 with permission from Elsevier.[7])

ratio.[9,10] The structure comparison between montmorillonite and rectorite is shown in Figure 3.2. The general chemical formulae of the mica-like layer and the smectite-like layer are $(Ca_{0.55}Na_{0.02}K_{0.01}Mg_{0.03})_{0.61}(Al_{4.1}Fe^{2+}{}_{0.09}Mg_{0.07})_{4.26}$-$(Si_{6.46}Al_{1.54})_8O_{22}$ and $(Na_{0.79}K_{0.39}Ca_{0.26})_{1.44}Al_4[Si_6Al_2]_8O_{22}$, respectively.[11] The non-exchangeable cations lie in the interlayer region of 2:1 mica-like layers, whereas the exchangeable hydrated cations lie in the interlayer region of 2:1 smectite-like layers. Like montmorillonite, the interlayer cations in a smectite-like layer of rectorite can also be replaced easily by either organic or inorganic cations. The thickness of a single rectorite layer is equal to the sum of the thickness of the mica-like layer and the smectite-like layer (~2 nm); thicker than that of the montmorillonite layer (~1 nm). A single sheet of pristine rectorite is about 40 μm in length and 5 μm in width, whereas that of pristine montmorillonite is about 1 μm in both length and width.[11,12] Thus, compared with montmorillonite, rectorite has a higher aspect ratio of width to thickness and is expected to endow the rubber–clay nanocomposites with a larger enhancement in composite performance.

3.1.2 Types of Rubber–Clay Nanocomposites

Any physical mixture of rubber and inorganic clay (or inorganic material in general) does not necessarily form a nanocomposite. Specifically, when the rubber is unable to intercalate between the clay layers, a phase-separated composite is obtained, whose properties are in the same range as the traditional micro-composites. Beyond these conventional rubber–clay composites (micro-composites), depending on the dispersion status of the clay layers, the rubber–clay

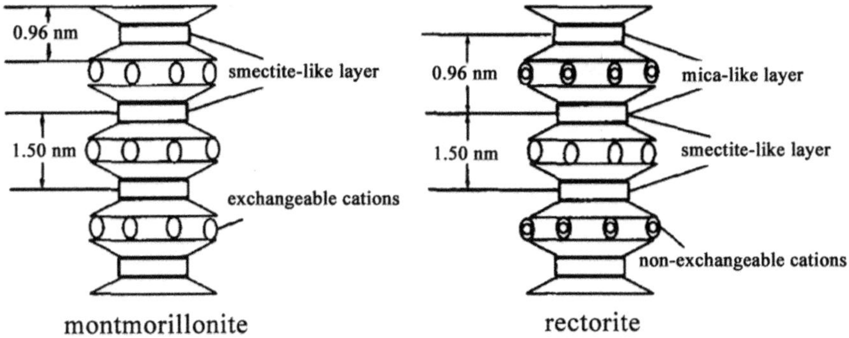

Figure 3.2 Structure comparison between montorillonite and rectorite. (Reproduced from *Microporous and Mesoporous Materials*, **57**, J. P. Olivier and M. L. Occelli, Surface area and microporisity of pollared rectorite catalysts from a hybrid density functional theory method, 291–296, Copyright 2003 with permission from Elsevier.[9])

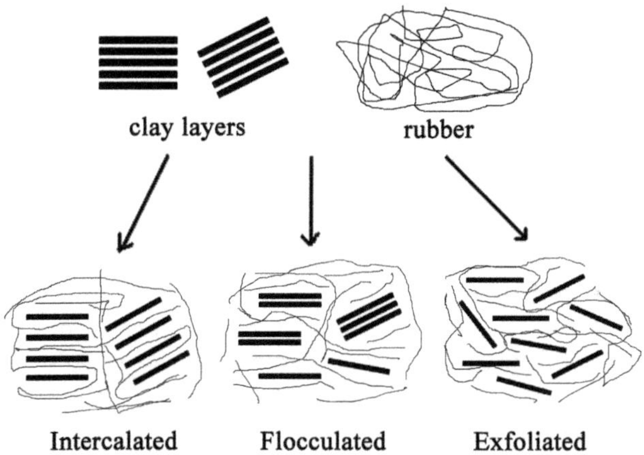

Figure 3.3 Schematic representation of intercalated, flocculated and exfoliated rubber–clay nanocomposites. (Reproduced from ref. 3 with permission from John Wiley and Sons. Copyright © 2007 Society of Plastic Engineers.)

nanocomposites can be classified into three types:[13–16] intercalated, flocculated (or separated) and exfoliated nanocomposites. The schematic diagram of these three types of rubber–clay nanocomposite is shown in Figure 3.3.

XRD is usually used to identify the structure of rubber–clay composites,[17] and the typical XRD curves of conventional, intercalated, flocculated and exfoliated rubber–clay nanocomposites are shown in Figure 3.4. The basal plane spacing d_{001} related to the interlayer spacing of clay layers (the distance between a plane in the unit layer and the corresponding plane in the next unit layer)[5] can be characterized by XRD. If the rubber molecules do not enter the clay galleries, d_{001} remains the same as the pristine clay, and the composite

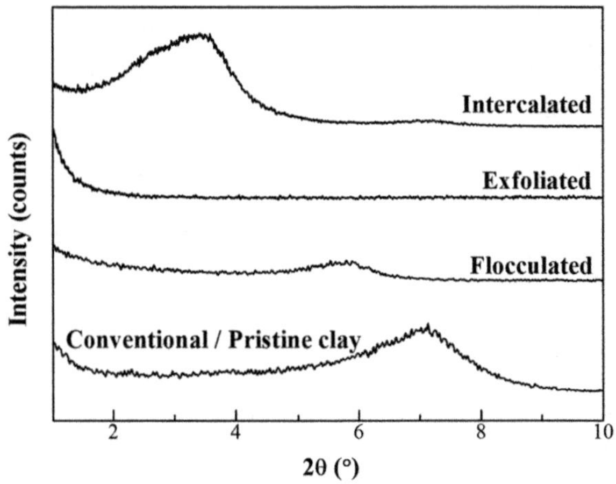

Figure 3.4 XRD curves of intercalated, flocculated and exfoliated rubber–clay nanocomposites. (Reproduced from ref. 48 with permission from John Wiley and Sons. Copyright © 2006 WILEY-VCH Verlag GmbH & Co. KGaA, Weinheim.)

is 'conventional' (or micro-composite). The properties of these conventional composites are very poor due to the obvious phase separation and the weak filler–rubber interaction.

In the intercalated nanocomposite, the repetitive multilayer structure is well preserved, allowing the interlayer spacing to be determined. The intercalation of the rubber chains usually increases the interlayer spacing, in comparison with the interlayer spacing of the organically modified clay used, leading to a shift of the diffraction peak towards lower angle values.

In comparison, the exfoliated nanocomposite is formed with the individual clay layers that are separated in a continuous rubber matrix, and no more diffraction peaks are visible in the XRD curve, either because of a much too large interlayer spacing between the layers (*i.e.* exceeding 8 nm in the case of ordered exfoliated structure) or because the nanocomposite does not present ordering anymore. Usually, the clay content of an exfoliated nanocomposite is much lower than that of an intercalated nanocomposite.

In the flocculated nanocomposites, there is no intercalation of rubber molecules into the clay galleries, and the clay layers exhibit a partly exfoliated structure with the forms of either individual layers or nanometer-thick layer aggregates.[18] Therefore, a small diffraction peak indicating the existence of nanometer-thick layer aggregates can be observed. However, the interlayer spacing is a little larger than the initial value of pristine clay resulting from the cations of flocculating agent entering the clay intergallery,[19] because the flocculated nanocomposites are always prepared by the latex compounding method that will be mentioned in the following section.

Transmission electron spectroscopy (TEM) is used to characterize the nanocomposite morphology. In some cases, both intercalated and exfoliated structures can co-exist in the nanocomposites, so one must rely on TEM observation combined with XRD to define the overall structure. Additionally, atomic force microscopy (AFM), small-angle X-ray scattering (SAXS), nuclear magnetic resonance (NMR), dielectric relaxation spectroscopy and inelastic neutron scattering have also been employed to characterize the structure of the nanocomposites.

3.1.3 Preparation of Rubber–Clay Nanocomposites

The main goal for the successful development of clay-based nanocomposites is to achieve complete exfoliation of the clay layers in the rubber matrix. To date, there are mainly three strategies reported to prepare rubber–clay nanocomposites: melt processing,[20] solution blending[21] and latex compounding.[22] Although the route of *in situ* polymerization is viable for the preparation of plastics-based nanocomposites, it has not been used for rubbers so far.

From an industrial point of view, melt processing is a direct, time-saving, and environmentally friendly method to prepare rubber–clay nanocomposites, and the existing conventional rubber compounding lines of high output can still be used. Melt intercalation of rubbers in organically modified clay is primarily driven by enthalpic interactions between rubber macromolecules and clay layers.[23] The role of the alkyl ammonium cations in organically modified clay is to lower the surface energy of the clay so as to improve the wettability by rubber. By appropriate choice of the alkyl ammonium cation, the interactions between the clay layers and the intercalated rubber chains can be tailored to prepare rubber–clay nanocomposites with high performance.

In the solution blending method, dry rubber is dissolved in a suitable solvent along with the organically modified clays or mixed together after dissolution in a respective suitable solvent. The curatives are usually mixed with the rubber–clay compound after partial or complete solvent removal, or may be added before solvent removal. The driving force for rubber intercalation from the solution is the entropy gained by desorption of small molecules from the clay galleries.[23]

Before the process of either melt processing or solution blending, the clays usually need to be modified by organic components, such as ion exchange reaction with quaternary ammonium salts, to ensure that the clay can dissolve in the solvent for solution blending, or to increase the interlayer spacing of the clay so that the rubber macromolecules can easily intercalate into the clay galleries during melt processing.

Unfortunately, the extensive use of solvent in solution blending may also pose serious risks to human health and the environment. In addition, the commercial organically modified clay is usually quite expensive compared with widely used carbon black or silica. In comparison, the pristine clay, without organic modification, is usually used in the latex compounding method. Generally, most of the commercial rubbers are available in the form of latex, which is an aqueous dispersion of rubber particles with a size from hundreds of nanometers to several micrometers. The clay layers can be easily dispersed in water because water acts as a swelling agent owing to hydration of the interlayer cations, usually Na^+ or K The water swelling capability of natural clays is not the same, depending on the type of clay and its cation exchange capacity. Therefore, rubber–clay nanocomposite can be prepared by co-coagulating the mixture of rubber latex and aqueous suspension of clay using an appropriate flocculating agent. Compared with the melt processing or solution blending methods, the latex compounding method is promising for industrialization because of its environmental friendliness, simplicity of the preparation process and superior performance/cost ratio.[18]

3.2 Gas Transport Mechanism

The exploitation of rubber–clay nanocomposites with high performance has become an area of intensive research activity, and application fields of these nanocomposites have been increasingly developed. The gas barrier properties of rubbers are of particular importance for airtight applications, and the addition of clay layers in rubbers would be of great help. When clay with a high aspect ratio (width/thickness) is nanodispersed in the rubber matrix in a planar oriented fashion, there is a great increase of the diffusion distance because a much more tortuous path is created for the diffusing gas,[14] and the schematic diagram of the gas diffusion paths in different composites is shown in Figure 3.5. Therefore, the gas barrier properties of rubber–clay nanocomposites are highly improved when compared with gum rubber.

Generally, the thickness of a single rectorite layer is larger than that of a montmorillonite layer, and the width and length vary from a micron to several microns, much larger than that of a montmorillonite layer. As a result, the separable layer thickness and layer aspect ratio (width/thickness) of rectorite are larger than those of regular montmorillonite, which make it more effective in the enhancement of the gas barrier properties of rubber–clay nanocomposites. Therefore, in this chapter, the dominant mechanisms for the transport of gas molecules in rubber and rubber–clay nanocomposites

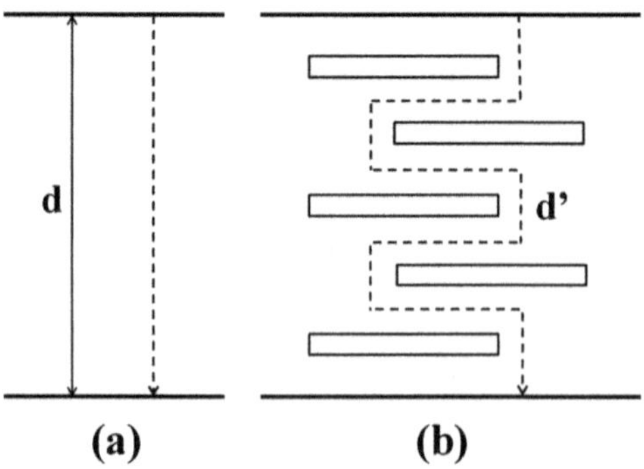

Figure 3.5 Gas diffusion path in (a) gum rubber and (b) rubber–clay nanocomposite. (Reproduced from *Applied Clay Science*, **15**, P. C. LeBaron, Z. Wang and T. J. Pinnavaia, Polymer-layered silicate nanocomposites: an overview, 11–29, Copyright 1999 with permission from Elsevier.[14])

are presented, along with the experimental measurement of gas permeability, and the influence of filler characteristics on the gas barrier properties of rubber–clay nanocomposites is discussed; then, the emphasis is given to the preparation and characterization of rubber–rectorite nanocomposites with high gas barrier properties; finally, recent progress in potential approaches to further improve the gas barrier properties is introduced.

3.2.1 Gas Transport in Rubber

The permeation of gases through a rubber film is a complex process that contains the following four processes:[1] initially, the gas molecules are absorbed on the surface of the rubber film; next, the gas molecules are dissolved inside the rubber; then the gas molecules diffuse through the rubber; finally, the gas molecules are desorbed from the other surface of the rubber film. Among these four processes, the gas sorption and diffusion are the two main stages that determine the gas permeability of the rubber.

During the sorption process, the gas molecules are assumed to be located in the free volume holes of the rubber, which are derived either from thermal perturbations or from micro-Brownian motions of the rubber molecules, such as the motion of chain segments, side groups and chain units.

The diffusion process occurs by gas molecules jumping through neighboring free volume holes, and depends on the number and the size of these holes and on the frequency of the jumps. The number and the size of free volume holes usually correspond to the static free volume that is related to the gas solubility, while the frequency of the jumps corresponds to the dynamic free volume that is related to the gas diffusivity.

The solubility, S, is a thermodynamic factor related to the interactions between the rubber and the gas molecules, and is associated with the specific free volume, ν_f:[24]

$$\nu_f \equiv \nu - \nu_0 = \frac{S}{\rho_g} \tag{3.1}$$

where ν is the specific volume, ν_0 is the occupied specific volume and ρ_g is the density of the gas. The fractional free volume (ν_f/ν) can be determined by positron annihilation lifetime spectroscopy (PALS).[25]

The diffusion coefficient (or diffusivity), D, is a kinetic factor that reflects the mobility of the gas molecules in the rubber, correlating with the free volume in the matrix, the molecular diameter of the gas, Φ, and the velocity of the gas molecules, u:[26]

$$D = g\Phi u \exp\left(-\frac{\gamma\nu_0}{\nu_f}\right) \tag{3.2}$$

where g is a geometric factor and γ is a factor related to the overlap of free volume, *i.e.* the degree to which more than one molecule may access the same free volume.

Permeability, P, is the product of the diffusivity and the solubility:

$$P = DS \tag{3.3}$$

3.2.2 Gas Transport in Rubber–Clay Nanocomposites

In a filler filled rubber composite, the gas permeation process is correlated with the size, shape, dispersion status and loading of the fillers.[1] The rubber–clay nanocomposite could be considered to consist of a permeable phase (rubber matrix) in which impermeable clay layers are dispersed. Therefore, the permeability of a rubber–clay nanocomposite is mainly influenced by the following four factors: (1) the volume fraction of the clay layers; (2) the orientation of the clay layers relative to the diffusion direction; (3) the aspect ratio of the clay layers; (4) the free volume in the rubber matrix.

Nielsen[27] developed a simple model to describe the permeability in filler filled polymers. The model strives to predict the observed permeability based strictly on tortuosity arguments. The presence of a filler (spherical, plate, cylindrical, *etc.*) introduces a tortuous path for a diffusing penetrant, as shown in Figure 3.5. The matrix in a rubber–clay nanocomposite maintains the same characteristics as the neat rubber, so a decrease of the solubility resulting from the reduced volume fraction of the rubber matrix is expected:

$$S = S_0(1 - \phi) \tag{3.4}$$

where S_0 is the solubility coefficient of the neat rubber, and ϕ is the volume fraction of the clay in the rubber matrix. The increased clay loading results in the reduction of the matrix volume fraction. As a result, according to eqn (3.3), the permeability of rubber–clay nanocomposites usually decreases with the increase of clay loading.

Besides, the clay layers act as impermeable barriers to the diffusing gas molecules, forcing them to travel longer paths, so a decrease in diffusivity is also expected in the nanocomposite:

$$D = \frac{D_0}{\tau} \tag{3.5}$$

where D_0 is the diffusion coefficient of the matrix, τ is the tortuosity factor, which is directly related to the aspect ratio, the shape and the orientation of the clay layers in the rubber matrix.[27] Generally, the reduction of the diffusion coefficient is higher than that of the solubility coefficient when the nanodispersed clay layers are introduced in the rubber matrix.

Therefore, the effect of tortuosity on the permeability is expressed as:

$$P = P_0\frac{1 - \phi}{\tau} \tag{3.6}$$

where P_0 is the permeability of the matrix.

The model described by Nielson makes a key assumption that the sheets are placed such that the normal of the sheet is coincident with the direction of diffusion. Clearly, this arrangement of the sheets results in the highest tortuosity, and any deviation from the arrangement where the sheets lie perpendicular to the film plane would in fact lead to deterioration in the barrier properties. Therefore, Bharadwaj[28] discussed the influence of relative orientation of the clay layers in the polymer matrix on the gas permeability by modeling. A range of relative orientations of the sheets with respect to each other and the plane of the film is envisaged as illustrated in Figure 3.6(a), and the modeling results of the effect of sheet orientation on the relative permeability are shown in Figure 3.6(b). As is expected, the barrier properties are enhanced when $F = 1$ (F is the order parameter) and decay to nearly that of the pure polymer when $F = -1/2$, where the sheets are arranged such that there is negligible increase in the tortuosity. Besides, shorter sheets, by virtue of the small aspect ratio, are very sensitive to orientational order, and a small random rotation of the sheet normals from the direction of preferred orientation ($F = 1$) is sufficient to degrade the barrier properties significantly.

Furthermore, the degree of dispersion of the nanoplatelets is determined by the degree of delamination of the clay. Hence, the author also modeled the influence of the clay delamination status in the polymer matrix on the permeability.[28] Figure 3.7(a) shows the schematics of nanocomposites having one, two and four sheet aggregates dispersed throughout the matrix, and the curves in Figure 3.7(b) show the relative permeability as a function

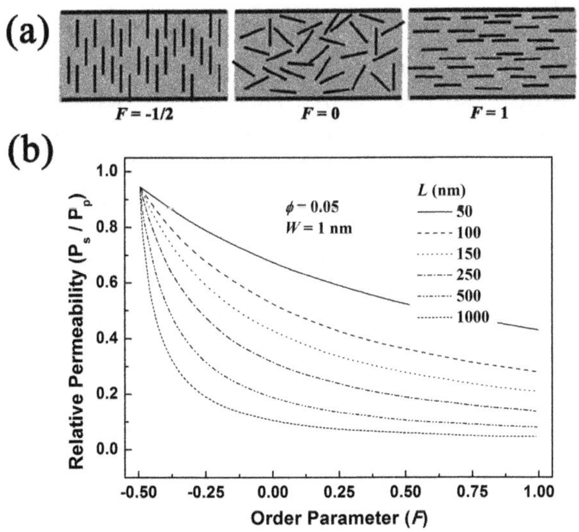

Figure 3.6 (a) Illustrations for three order parameter (F) values of −1/2, 0 and 1; (b) effect of sheet orientation on the relative permeability in exfoliated nanocomposites at a filler volume fraction of 0.05 and layer thickness of 1 nm. (Reproduced with permission from R. K. Bharadwaj, *Macromolecules*, 2001, **34**, 9189. Copyright (2001) American Chemical Society.[28])

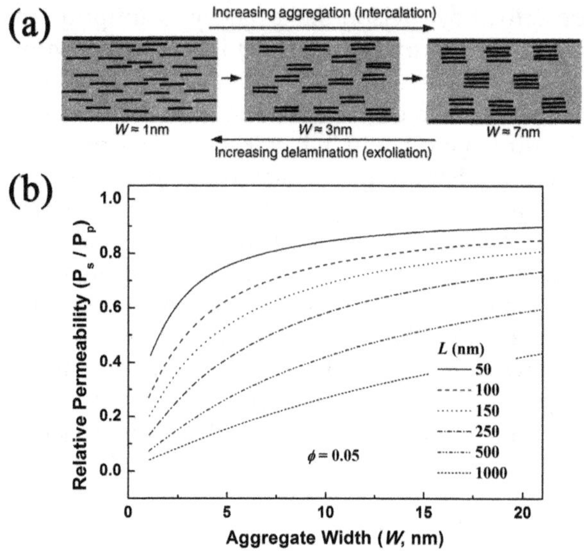

Figure 3.7 (a) Illustrations for one, two and four sheet aggregates dispersed throughout the matrix; (b) effect of aggregate thickness on the relative permeability with several different sheet lengths at a filler volume fraction of 0.05. (Reproduced with permission from R. K. Bharadwaj, *Macromolecules*, 2001, **34**, 9189. Copyright (2001) American Chemical Society.[28])

of the aggregate thickness (or width, *W*) at several different lengths of the sheets with a filler volume fraction of 0.05. In the first case, single sheets are dispersed throughout the polymer showing an ideal structure, and the thickness of the single sheet is ~1 nm, which serves to maximize the aspect ratio. In contrast, consider a case where aggregates comprised of two silicate sheets with a layer of polymer in between are dispersed throughout the matrix. Assuming that the polymer layer is 1 nm thick, aggregates of two and four silicate sheets with one and three layers of polymer chains in between lead to an effective thickness of ~3 nm and ~7 nm, respectively. As the modeling results show in Figure 3.7(b), dispersing the same total number of sheets as two-sheet aggregates results in a dramatic decrease in the tortuosity, and aggregates of four sheets begin to show a percolating path for the diffusing penetrant. The above argument can be extended to include the lateral displacement of the sheets with respect to each other while maintaining a regular stacking arrangement, leading to a variability in the length of the aggregate in addition to the thickness. Clearly, exfoliation is the critical factor in determining the maximum performance of polymer nanocomposites for barrier applications. Moreover, at a length of 1000 nm, the expected relative permeability is ~0.05 for a thickness of 1 nm and degrades to ~0.2 for a thickness of 7 nm, but for a length of 500 nm in the same thickness, the relative permeability degrades from ~0.1 to ~0.4. This indicates that the relative permeability is extremely sensitive to incomplete delamination, especially for the short sheet lengths.

The modeling results show that the fully delaminated (exfoliated) nanocomposite presents much higher values for the tortuosity factor and the aspect ratio in comparison with the partially delaminated (intercalated) nanocomposite and it is much more effective at decreasing permeability. It was found experimentally that[17] the nitrogen permeability of a rubber–montmorillonite nanocomposite with an exfoliated structure decreased by 33% with 10 phr montmorillonite, as compared to only 25% decrease in permeability for the nanocomposite with a flocculated structure at the same filler loading. However, at high clay loadings, complete exfoliation is very difficult to achieve, because the formation of aggregated clay layers becomes easier when the filler loading increases. And the increased aggregation of clay layers causes a decrease of the aspect ratio of fillers, leading to insufficient enhancement of the gas barrier properties.

Moreover, as mentioned in eqn (3.2), the increased free volume in a rubber should increase the diffusion coefficient, resulting in an increase in permeability. In many cases, the presence of the inorganic phase may induce an increase in the size and the number of the free volume holes in a rubber–clay nanocomposite, especially at the interfaces, due to the incompatibility between the rubber matrix and the inorganic fillers. Besides, a higher number of clay agglomerates may induce more possibilities to form large holes (pores) in the matrix, which can facilitate gas transport in the nanocomposite.[29] Therefore, it is desirable to prepare rubber–clay nanocomposites with as little clay aggregation as possible in order to achieve much improved gas barrier properties.

3.2.3 Characterization of Gas Barrier Properties

The gas barrier properties of a rubber are characterized by the permeability coefficient. The experimental methods for permeability measurement have been developed for many years, with the variable-pressure (manometric) method and the variable-volume (volumetric) method according to ASTM D 1434 (standard test method for determining gas permeability characteristics of plastic film and sheeting) as the two main methods.

In a typical experimental setup,[30] a rubber film is positioned between two compartments in a permeation cell. The gas is introduced in the one compartment (compartment A) under constant pressure p_0. At constant temperature, the concentration of the gas in the other compartment (compartment B) is monitored by a certain instrument, such as a thermal conductivity detector, coulometric detector, infrared spectrometer, gas chromatograph or mass spectrometer. The flux density of the gas, $J(t)$, is defined as follows:

$$J(t) = A\frac{\mathrm{d}c}{\mathrm{d}t} \tag{3.7}$$

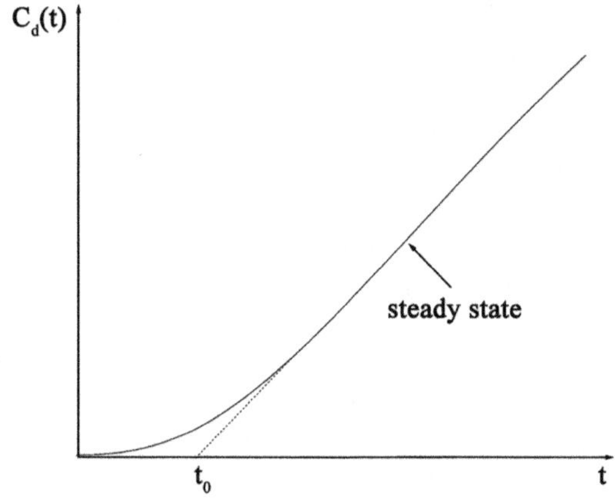

Figure 3.8 Typical diagram of experimental permeability measurements. (Reproduced from *European Polymer Journal*, **45**, G. Choudalakis and A. D. Gotsis, Permeability of polymer/clay nanocomposites: a review, 967–984, Copyright 2009 with permission from Elsevier.[1])

where A is the test area of the rubber film, c is the concentration of the gas in compartment B, and t is the time.

If c is plotted *vs.* t for a sufficiently long time, a typical diagram of experimental results could be obtained, as shown in Figure 3.8 When the concentration of the gas in the rubber film reaches a steady distribution, the plot of c *vs.* t becomes a linear line and $J(t)$ reaches the steady state flux, J_∞:

$$J_\infty = P\frac{p_0}{d} \tag{3.8}$$

where d is the thickness of the rubber film.

From the slope of the linear part of the c *vs.* t curve, the permeability coefficient can be calculated.

3.3 Rubber–Rectorite Composites with High Gas Barrier Properties

In recent years, rectorite has been used in different rubber or elastomer matrices to form rubber–rectorite nanocomposites with excellent properties and a unique phase structure, such as thermoplastic polyurethane rubber (TPUR),[10,31,32] natural rubber (NR),[12,33] styrene–butadiene rubber (SBR),[11,34–36] acrylonitrile-butadiene rubber (NBR),[37] hydrogenated acrylonitrile–butadiene rubber (HNBR)[38] and poly(di-isoamyl itaconateco-isoprene) (PDII) bio-based elastomer.[39] Compared with montmorillonite, rectorite has a higher aspect ratio of the width to the thickness, and is expected to endow the rubber–clay nanocomposites with larger enhancement in composite performance, especially for the gas barrier properties. In addition, since rectorite layers are thick, they may be able to endure the strong shear force during processing and maintain the high aspect ratio after processing,[11] which makes them more promising and valuable for application in rubber–clay nanocomposites compared with montmorillonite. Therefore, in this section, the emphasis is placed on the gas permeability of the various reported rubber–rectorite nanocomposites.

3.3.1 Rectorite Filled Non-Polar Elastomer Composites

Butyl rubber and halogenated butyl rubber are the commonly used rubbers for inner tubes and liners of tires. However, they are difficult to process and hard to co-crosslink with body rubbers, which are generally NR, SBR and butadiene rubber (BR), due to the significant difference in the saturation degree. In addition, they are more expensive than the commonly used body rubbers. Thus, SBR–rectorite or NR–rectorite nanocomposites are expected to be good candidates for tire tube or inner liner materials due to their comparable gas barrier properties, improved co-crosslinking properties and low cost.

In addition to application as the candidate materials for tire tubes and inner liners, the NR–rectorite nanocomposites may also be used for the rubber balloon. Wang *et al.*[12] obtained NR–rectorite nanocomposites *via* latex compounding. Figure 3.9 shows a schematic of the latex compounding method. Firstly, rectorite was dispersed in water with vigorous stirring and an aqueous suspension of clay layers was achieved. In the presence of water, the cations existing in the space of smectite-like layers have a strong tendency to hydrate, thereby forcing the layers apart. The change in the interlayer spacing

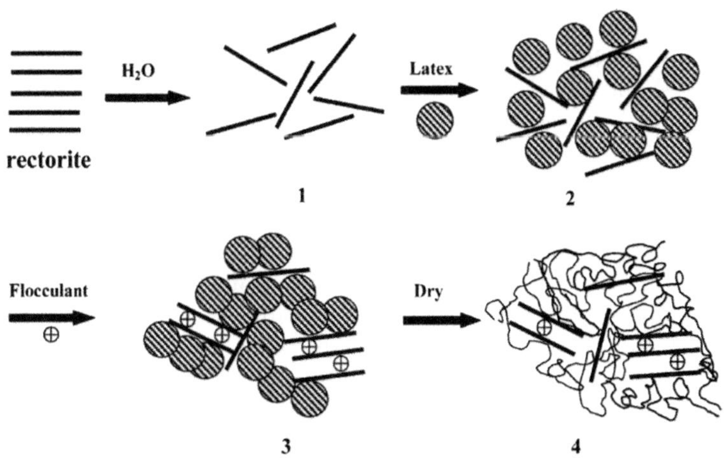

Figure 3.9 Schematic of the latex compounding method for the preparation of rubber–rectorite nanocomposite. (Reproduced from *European Polymer Journal*, **41**, Y. Wang *et al.*, Preparation and properties of natural rubber/rectorite nanocomposites, 2776–2783, Copyright 2005 with permission from Elsevier.[12])

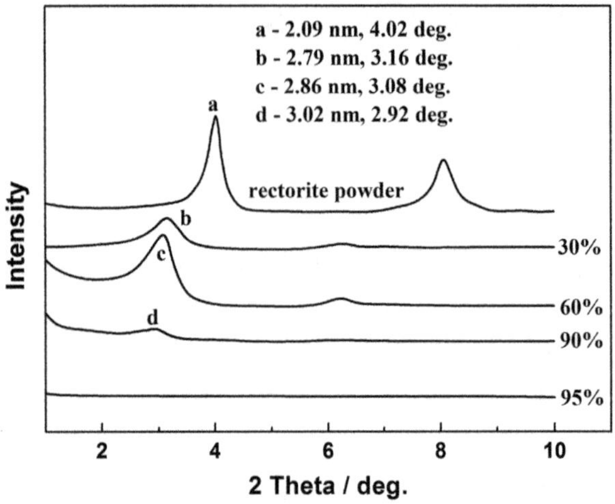

Figure 3.10 XRD patterns of rectorite and water mixture with different water contents. (Reproduced from *European Polymer Journal*, **41**, Y. Wang *et al.*, Preparation and properties of natural rubber/rectorite nanocomposites, 2776–2783, Copyright 2005 with permission from Elsevier.[12])

of rectorite layers in rectorite–water mixtures with various water contents can be characterized by XRD, as shown in Figure 3.10. With increasing water content, the d_{001} diffraction peaks of rectorite gradually shift to a smaller angle, which means the interlayer spacing is gradually enlarged by the introduction of water. When the water content is more than 90%, no diffraction peak can be observed, indicating that rectorite can be separated into individual layers in the dilute aqueous solution. Secondly, rubber latex was mixed with the aqueous suspension of rectorite. Since both the surfaces of rubber latex particles and rectorite layers are hydrophilic, the introduction of rubber latex does not cause the aggregation of rectorite layers. A uniform and stable mixture is obtained. Thirdly, the mixture was co-coagulated by the addition of a flocculating agent. The structure of the nanocomposite is formed in this process. On the one hand, the existence of latex particles between the rectorite layers in the water medium should result in a complete separation (exfoliation) of rectorite layers; on the other hand, cations in the flocculating agent probably cause some of the separated rectorite layers to reaggregate so that the rubber latex particles between the rectorite layers may be expelled. As a result, there are some non-exfoliated (stacking) layers (still on the nanometer scale) in the nanocomposite, and the rubber–rectorite nanocomposite prepared by this method is usually flocculated (separated) nanocomposite. Finally, the mixture was dried to remove the residual water.

The nitrogen permeability of the NR–rectorite nanocomposites as a function of rectorite volume fraction was measured, followed by analysis of the permeability data with the Nielsen model, and the results are shown in Figure 3.11. Nielsen model allows the gas permeability to be predicted as a function

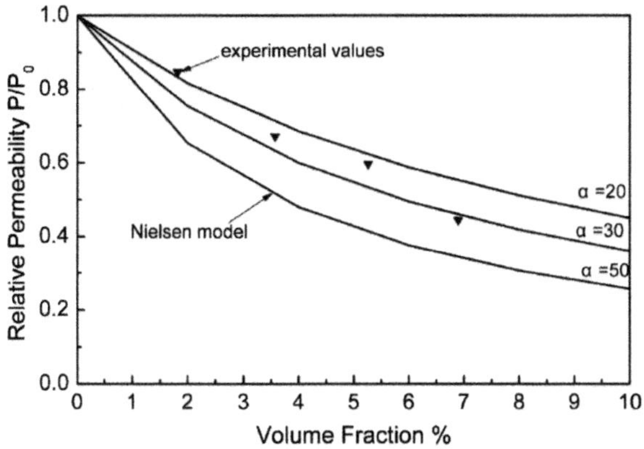

Figure 3.11 Correlation of experimental values of NR–rectorite nanocomposites and the Nielsen model (solid lines are the Nielsen model). (Reproduced from *European Polymer Journal*, **41**, Y. Wang *et al.*, Preparation and properties of natural rubber/rectorite nanocomposites, 2776–2783, Copyright 2005 with permission from Elsevier.[12])

of aspect ratio of the filler as well as its volume fraction in the matrix, and the model is described by the following formula:[27]

$$P_c = P_0\left(\frac{1-\phi}{1+\alpha\phi/2}\right) \tag{3.9}$$

where P_c and P_0 are the permeability of the composite and the pure polymer, respectively, ϕ is the volume fraction of the filler in the matrix, and α is the aspect ratio of filler particles.

As can be seen from Figure 3.11, the nitrogen permeability of the nanocomposites rapidly decreased with the increase of rectorite volume fraction. The permeability decreased by ~50% when the rectorite loading reached 7 vol%. The remarkable improvement in barrier properties is attributed to the reduction of permeable amorphous rubber content in the nanocomposites and the increase of tortuosity of the diffusion path for penetrant molecules, resulting in the reduction of the solubility coefficient and diffusion coefficient, as described in eqn (3.4) and (3.5), respectively. The Nielsen model was found to well describe the effect of rectorite volume fraction on the permeability of the nanocomposites when an aspect ratio of 20–30 for the dispersed rectorite was assumed.

Moreover, the authors[11] prepared the SBR–rectorite nanocomposites *via* the latex compounding method as well. Rectorite layers dispersed in SBR with a thickness of 30–50 nm and a length of ~1 μm were observed, as shown in Figure 3.12(a), while those of montmorillonite in SBR–montmorillonite nanocomposite obtained by the same method were 10–20 nm and 200–300 nm, respectively, as shown in Figure 3.12(b). The results indicated that the SBR–rectorite nanocomposite exhibited a higher aspect ratio of clay layers

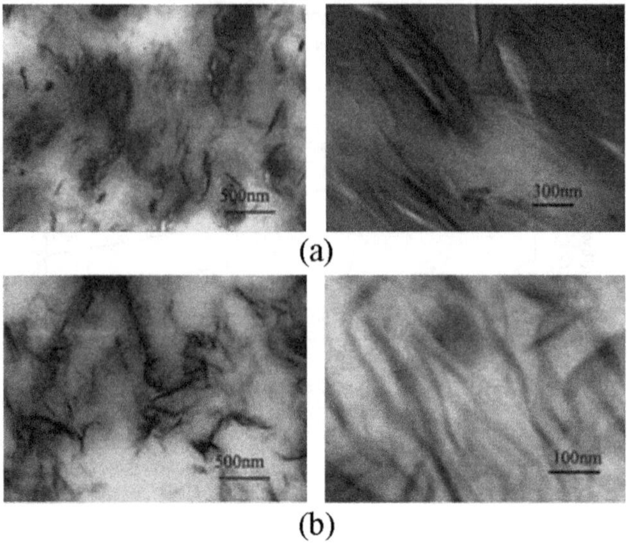

Figure 3.12 TEM micrographs of (a) SBR–rectorite and (b) SBR–montorillonite nanocomposites. (Reproduced from ref. 11 with permission from John Wiley and Sons. Copyright © 2005 Wiley Periodicals, Inc.)

compared with SBR–montmorillonite nanocomposite. It is believed that serious breakage of nanodispersed clay layers may occur during rubber processing due to the high viscosity of rubber, which could result in the reduction of the aspect ratios of clay layers in rubber. Therefore, it is important to keep in mind that the high aspect ratio of rectorite needs to be maintained so that high gas barrier properties can be achieved.

PALS was used to investigate the free volume hole properties of SBR–rectorite nanocomposites and conventional SBR–carbon black (N326) composites.[35] Combined with differential scanning calorimeter (DSC) results, it was found that rectorite had a stronger effect on restraining rubber chain mobility than carbon black, which resulted in the decrease of fractional free volume and gas permeability. Figure 3.13 shows that the nitrogen permeability of both SBR–rectorite nanocomposites and SBR–carbon black composites are gradually reduced with the increase of filler volume fraction, and the reduction in the permeability of SBR–rectorite nanocomposites is much faster than that of carbon black filled SBR composites. Incorporation of either rectorite or carbon black can improve gas barrier properties, due to the increase of tortuous diffusion path and the decrease of fractional free volume. The big difference in permeability reduction between rectorite and carbon black filled SBR is mainly attributed to the different filler effects on the tortuousity of the diffusion path.

After fitting the experimental results to eqn (3.9), aspect ratios of ~30 for the rectorite and ~1 for the carbon black in SBR matrix were obtained. Both experimental and fitting results suggest that the addition of fillers with high

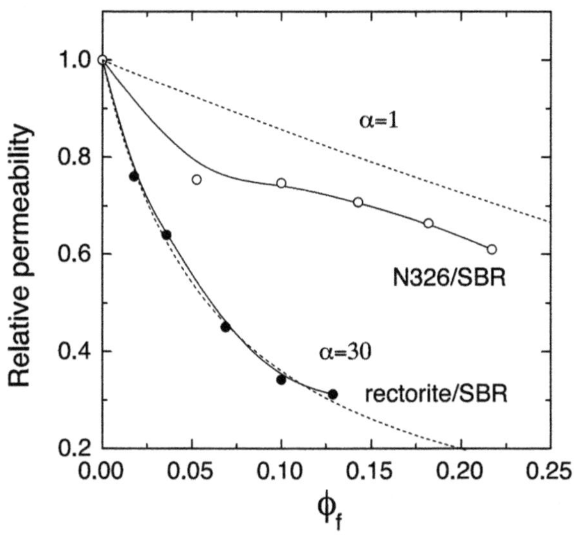

Figure 3.13 Effects of the filler loading on relative nitrogen permeability of SBR–rectorite nanocomposites and SBR–carbon black composites. The dash lines represent predictions from eqn (3.9) for aspect ratios of 1 and 30, respectively. (Reproduced from *Polymer*, **46**, Z. F. Wang *et al.*, Influence of fillers on free volume and gas barrier properties in styrene-butadienne rubber studied by positrons, 719–724, Copyright 2005 with permission from Elsevier.[35])

aspect ratio into the polymer matrix is a promising and practical way to prepare rubber materials with high gas barrier properties.

Since the addition of nanodispersed clay layers into rubber could well decrease the gas permeation, one would expect that a clay loading as high as possible may be added into the rubber matrix to maximize the improvement of gas barrier properties. However, unfortunately, the processability of the filler filled rubber composite would usually become worse with the increase of filler loading because of the strengthening of filler network interaction. Therefore, He *et al.*[34] introduced glycerin into SBR–rectorite nanocomposite to reduce the filler network interaction. With the addition of glycerin, the mechanical properties of the nanocomposites were improved, along with the disappearance of delayed vulcanization, and, most important of all, the high gas barrier properties were maintained. In comparison, with the use of aromatic oil, the widely used processing oil to improve processability of rubber composites, a reduction of the gas barrier properties occurred in the SBR–rectorite nanocomposites. The addition of aromatic oil can improve the mobility of the rubber chains and increase the free volume, causing an increase of solubility and diffusivity, as described in eqn (3.1) and (3.2), respectively, so as to increase the permeability. In contrast, the formation of hydrogen bond interactions between rectorite and glycerin, as shown in Figure 3.14, can reduce the possibilities to form large holes (pores) in the

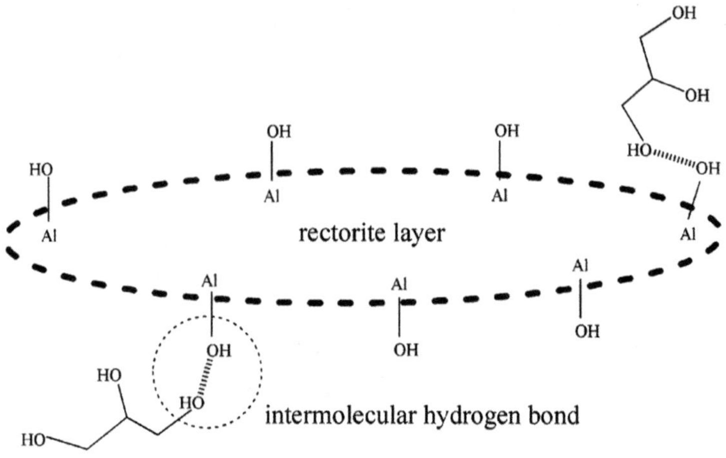

Figure 3.14 Sketch of the interaction between glycerin and rectorite. (Reproduced from ref. 34 with permission from John Wiley and Sons. Copyright © 2012 Society of Plastic Engineers.)

rubber matrix, and thus prevent the gas barrier properties of SBR–rectorite nanocomposites from worsening.

3.3.2 Rectorite Filled Polar Elastomer Composites

The gas barrier properties of the NBR–rectorite nanocomposites obtained from the latex compounding method have been reported.[37] Figure 3.15 presents the relative permeability of rectorite, carbon black and silica filled NBR composites as a function of filler volume fraction. Similar to the SBR–rectorite and NR–rectorite nanocomposites, the permeability of the NBR–rectorite nanocomposites gradually decreased with increasing filler volume fraction. Compared with carbon black or silica filled NBR, NBR–rectorite nanocomposites exhibited much better gas barrier properties. The authors also found that the nitrogen permeability of NBR–rectorite nanocomposites decreased by 75% with 10 vol% rectorite, as compared to only 48% decrease of nitrogen permeability for NBR–montmorillonite nanocomposites with the same filler volume fraction in the previous report.[40] An aspect ratio of 25–50 for the rectorite layers in NBR–rectorite nanocomposites was obtained by fitting the data using the Nielsen model, which was almost the same as that estimated by TEM observation. In comparison, a smaller aspect ratio of ~15 was obtained for NBR–montmorillonite nanocomposites. Hence, it is the higher aspect ratio of rectorite than that of montmorillonite in NBR matrix that leads to the better gas barrier properties of NBR–rectorite nanocomposites.

PDII is a type of newly invented and promising bio-based elastomer.[41] However, the barrier properties of PDII and its composites with carbon black and silica are not satisfying at all. By applying latex compounding, Yao *et al.*[39] prepared PDII–rectorite and PDII–montmorillonite nanocomposites

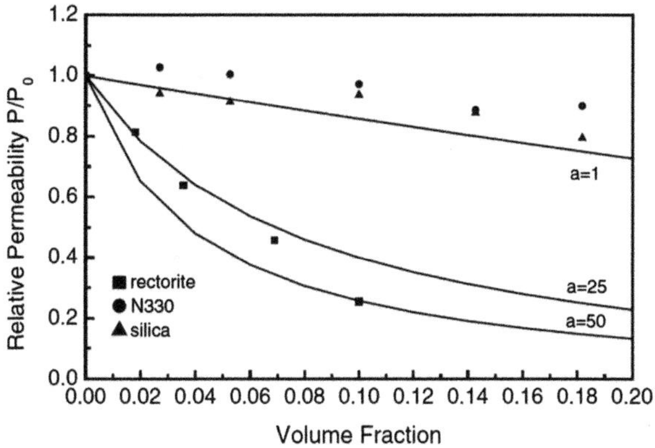

Figure 3.15 The relative permeability of rectorite, carbon black and silica filled NBR composites as a function of filler volume fraction. The solid lines represent predictions from eqn (3.9) for aspect ratios of 1, 25 and 50, respectively. (Reprinted by permission from Macmillan Publishers Ltd: *Polymer Journal* (ref. 37), Copyright (2005)).

with much improved gas barrier properties. The nitrogen permeabilities of the PDII–rectorite and PDII–montmorillonite nanocomposites are shown in Figure 3.16. The permeability of the pure PDII elastomer was 11.7×10^{-17} $m^2\ Pa^{-1}\ s^{-1}$, while the PDII filled with 20 phr montmorillonite had a permeability of 5.2×10^{-17} $m^2\ Pa^{-1}\ s^{-1}$, less than half of that of unfilled PDII. The permeability decreased only slightly with further increase in the loading of montmorillonite, which showed that the effect of the filler content on the gas impermeability decreased. The authors attributed such behavior to the poor distribution of the montmorillonite resulting from the formation of a mass of clay layer aggregates at high loading, which was confirmed by XRD and TEM characterization. In contrast to that of PDII–montorillonite, the permeability of PDII–rectorite decreased continuously with the increasing filler loading, and the lowest air permeability reached 1.7×10^{-17} $m^2\ Pa^{-1}\ s^{-1}$ at a filler loading of 80 phr, a reduction of 85.5% compared with that of the unfilled PDII. At a given content of clay, such as 10 phr, PDII–montmorillonite showed better air barrier properties than PDII–rectorite. In contrast, as mentioned above in the NBR matrix,[37,40] compared with montmorillonite, rectorite endowed the NBR–clay nanocomposites with a larger enhancement of the nitrogen barrier properties with the same filler loading. The effect of montmorillonite and rectorite on the gas barrier properties of PDII elastomer is quite different from that in NBR matrix. This may be attributed to the good compatibility of polar ester groups of PDII with clay layers through relatively strong boundary interactions, as shown in Figure 3.16(a). In addition, the authors also stated that the thicker layers of rectorite were less likely to be crushed during the mill process.

In the above examples, the latex compounding method was utilized for the preparation of rubber–rectorite nanocomposites, and pristine rectorite

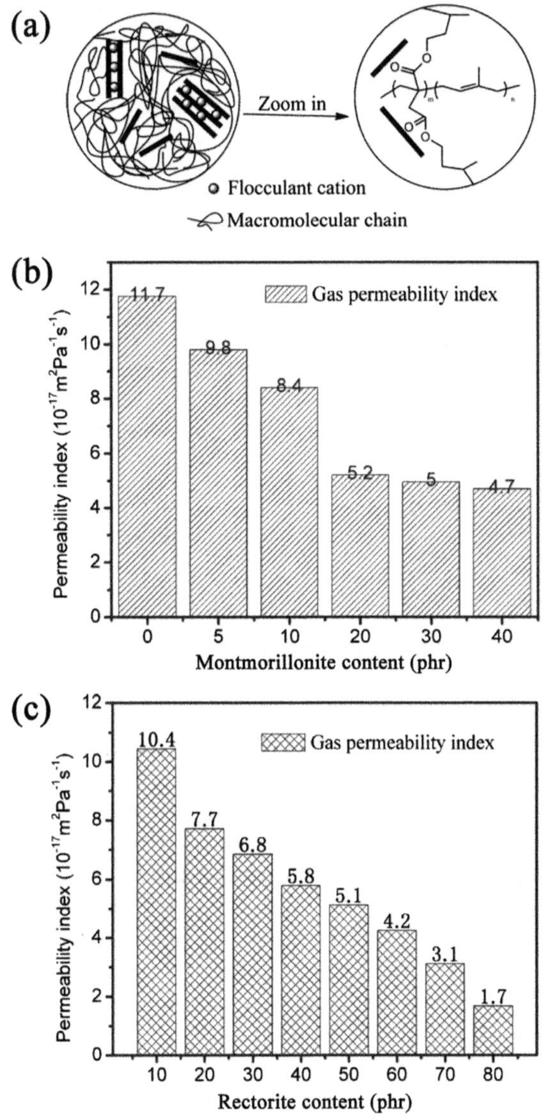

Figure 3.16 (a) Sketch of the designed PDII–clay nanocomposite; gas permeability of (b) PDII–montorillonite and (c) PDII–rectorite nanocomposites. (Reproduced from ref. 39 with permission from John Wiley and Sons. Copyright © 2004 Wiley Periodicals, Inc.)

instead of organically modified rectorite was used in the nanocomposites. In another work, Zhang *et al.*[38] used two different types of commercial organically modified clay to prepare HNBR–clay nanocomposites by the melt processing method. The clays used in their work are montmorillonite and rectorite, modified by dimethyl di(hydrogenated tallow) ammonium salt and octadecyl ammonium salt, respectively. The mechanism for forming the nanocomposite structure by melt processing is totally different from that by latex compounding. During melt processing, rubber macromolecules first intercalate into the intergalleries of organically modified clay, leading to decreased interaction between layers, and then the intercalated organically modified clay is exfoliated under shear force or during the vulcanization process. Consequently, the nanocomposites prepared by the melt processing method bear intercalated clay layers, and sometimes both intercalated and exfoliated clay layers simultaneously. Remarkable improvement was achieved in tensile strength, tear strength, aging resistance and oxygen permeation properties of HNBR–clay nanocomposites compared with that of unfilled HNBR. It is noteworthy that the oxygen permeability of HNBR–rectorite nanocomposites decreased by ~25% with 10 phr rectorite, as compared to ~30% decrease for HNBR–montmorillonite nanocomposites with the same filler volume fraction.

Generally, it is expected that rectorite can endow the rubber–clay nanocomposites with a larger enhancement in the gas barrier properties due to the higher aspect ratio compared with montmorillonite. However, sometimes the results are unexpected, and rubber–montmorillonite nanocomposite may exhibit better gas barrier properties than rubber–rectorite nanocomposite, as shown in the above mentioned PDII and HNBR matrix. Besides, compared with the SBR–montmorillonite nanocomposite prepared by the same method,[19] SBR–rectorite nanocomposite exhibits just a little higher nitrogen permeability. Therefore, much research still remains in order to understand the complex structure–property relationships in various rubber–clay nanocomposites.

3.4 Potential Approaches to Further Improve Gas Barrier Properties

There have been reports on preparation of rubber–rectorite nanocomposites with relatively high gas barrier properties, as described in the last section. However, the rubber–rectorite nanocomposites with extremely high gas barrier properties would need to possess the fully exfoliated clay structure, with the planes of all clay layers oriented parallel to that of the nanocomposite films. Therefore, much work still remains to further improve the gas barrier properties of rubber–rectorite nanocomposites. In this section, the potential approaches to further improve the gas barrier properties are presented.

3.4.1 Improvement in Interfacial Interaction and Filler Dispersion

Clearly, exfoliation is the critical factor in determining the maximum performance of rubber–clay nanocomposites for gas barrier applications. Theoretically, based on the view of thermodynamics, when the interaction between the clay layers and the rubber matrix during processing is strong enough to slow down the tendency of the clay self-aggregation, it is possible to improve the dispersion of clay layers in the nanocomposites. In fact, the rectorite layers are hydrophilic whereas the rubber macromolecules are hydrophobic, so rectorite needs to be modified by organic molecules to improve the compatibility with rubber.

The hydrated cations between the rectorite layers can be exchanged with cationic surfactants such as organic quaternary ammonium salts. Cationic surfactants have been reported to modify rectorite, including dodecyl benzyl dimethyl ammonium salt,[10,31,32,42] dimethyl dioctadecyl ammonium salt,[43] hexadecyl trimethyl ammonium salt,[10,32,44] *N*,*N*-dimethyl-*N*-dodecyl benzyl ammonium salt,[45] benzidine,[10] cetyltrimethyl ammonium salt[46] and octadecyl dimethyl benzyl ammonium salt.[46,47] Although the modification of rectorite by organic alkyl ammonium salts can greatly improve the compatibility with rubber and the dispersion of rectorite layers in the rubber matrix, the interfacial interaction between the two phases is still weak due to the nature of van der Waals forces. The solution for this is to build a strong interfacial interaction or chemical bonding between the two phases.

One important way to build a strong interfacial interaction is to apply the reactive modifier in place of traditional organic alkylammonium salts. Jia *et al.*[48] introduced an unsaturated organic ammonium salt (*N*-allyl-*N*,*N*-dimethyl-octadecyl ammonium chloride) into SBR–montmorillonite nanocompound, and the chemical bonding between nanodispersed montmorillonite layers and SBR macromolecules was formed after vulcanization. In another approach,[49] rubber–montmorillonite nanocomposite was prepared by a novel two-step method. The proposed schematic of nanocomposite preparation by means of the latex compounding method is shown in Figure 3.17. A silane coupling agent with amine groups, 3-aminopropyl triethoxy silane (KH550), was firstly intercalated into the nanodispersed montmorillonite to form reactive clay; then they were co-coagulated with rubber latex; finally, another silane coupling agent with sulfide groups, bis(triethoxy silyl propyl)tetra sulfide (Si69), was incorporated into the nanocompound to induce a strong interaction between KH550 modified montmorillonite and the rubber matrix after vulcanization.

The other way to build a strong interfacial interaction is to introduce an ion exchangeable group directly on the rubber macromolecules. By applying latex compounding, He *et al.*[17] successfully prepared a rubber–montmorillonite nanocompound with an exfoliated structure by co-coagulating butadiene-styrene-vinyl pyridine rubber (VPR) latex and an aqueous suspension of montmorillonite with dilute sulfuric acid solution as flocculating agent. During co-coagulation, the released acidified VPR macromolecules could be strongly absorbed onto the surfaces of the clay layers and the expected strong interfacial interaction between rubber and montmorillonite layers could be obtained. Obviously, in comparison with a partially delaminated (intercalated or flocculated) nanocomposite, a fully delaminated (exfoliated) nanocomposite is much more effective in decreasing permeability.

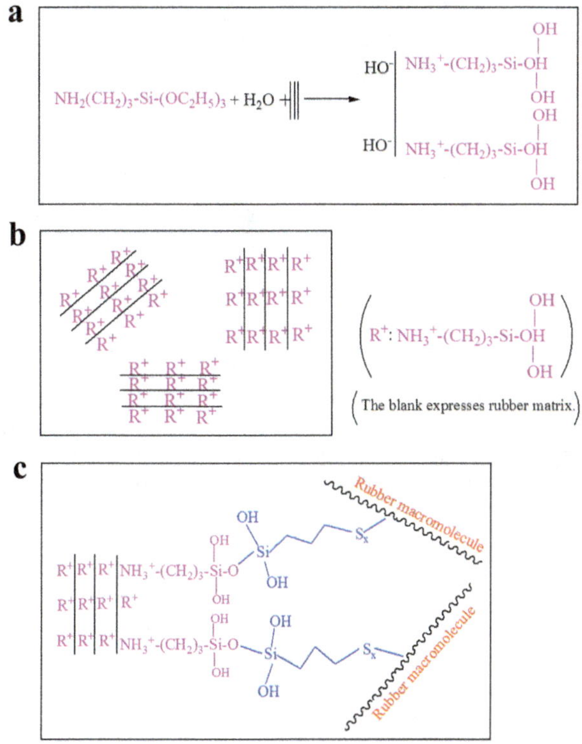

Figure 3.17 Proposed schematic of the modification mechanism: (a) cation exchange in clay suspension with KH550 added; (b) the structure of the nanocompound after co-coagulation; (c) the structure of the nanocomposite after vulcanization. (Reproduced from *Composites Science and Technology*, **68**, Q-X. Jia *et al.*, Enhanced interfacial interaction of rubber/clay nanocomposites by a novel two-step method, 1050–1056, Copyright 2008 with permission from Elsevier.[49])

3.4.2 Optimization of Filler Arrangement

The formation of the aggregation of clay layers during processing is hard to avoid; nevertheless, it is encouraging that the formation of edge to edge aggregates rather than face to face aggregates is beneficial for the gas barrier properties due to the increased aspect ratio of the clay layers. Nazarenko and coworkers[50] employed layer multiplying coextrusion to create films consisting of alternating layers of LDPE and LLDPE-*g*-MA–organically modified montmorillonite nanocomposite, and found an additional gas barrier enhancement in the nanocomposite by the formation of longer skewed aggregates (as shown in Figure 3.18) upon annealing and the contraction of the filled layers.

In addition, well dispersed platelet particles with a high aspect ratio, especially when they are aligned in the direction parallel to the film surface, can result in a significant decrease of the gas permeability. Attention has been

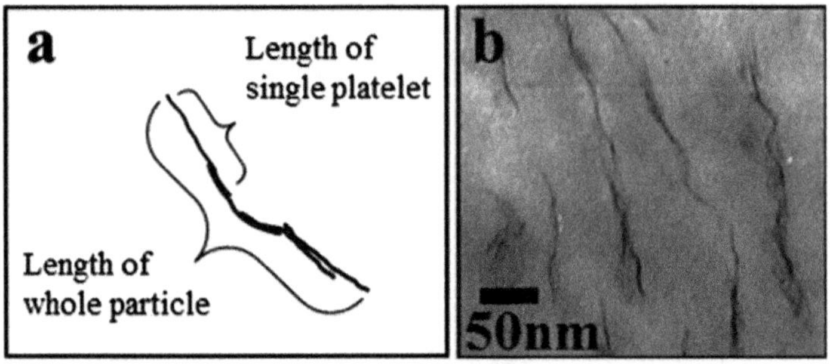

Figure 3.18 Skewed agglomerate consisting of several single clay platelets (a) schematic; (b) as revealed by TEM at higher magnification in a nanocomposite with 1.0 vol% montmorillonite. (Reproduced from *Polymer*, **61**, J. D. Decker *et al.*, Polyethylene-based nanocomposites containing organoclay: A new approach to enhance gas barrier *via* multilayer coextrusion and interdiffusion, 42–54, Copyright 2015 with permission from Elsevier.[50])

paid to the preparation of nanocomposites with highly oriented nanosheets parallel to that of the nanocomposite membranes so as to improve the properties in a certain direction. It has been reported that highly oriented boron nitride (BN) nanosheets in a polymer matrix can be achieved by oscillatory shearing on a two-roll mill or by stretching orientation,[51] direct current electric field orientation[52–54] and magnetic field orientation.[55] Nanocomposites with a significant boost of performance in a certain direction were obtained. It is believed that these techniques can be applied to prepare highly oriented rectorite layers in a rubber matrix, so as to further improve the gas barrier properties of rubber–rectorite nanocomposites.

3.5 Other Properties of Rubber–Rectorite Nanocomposites

The rubber–rectorite nanocomposites exhibit remarkable gas barrier properties due to the high aspect ratio of rectorite layers. In addition, attention could be paid to some other properties of rubber–rectorite nanocomposites, such as mechanical properties, thermal stability, aging resistance and oil resistance.

3.5.1 Mechanical Properties

Nano-strengthening by employing nanoparticles is necessary for high-efficiency strengthening of rubbers, which has already been validated by numerous research and industrial applications. The main reinforcement mechanism can generally be attributed to rubber–filler and filler–filler network interactions as well as strain amplification effects.[56] The rubber–filler interactions directly depend on the detailed structure and surface activity of the filler in the rubber matrix. Rubber–rectorite nanocomposites prepared *via* the latex compounding method could result in the formation of the flocculated nanocomposite. There is no intercalation of rubber molecules into the clay galleries, and the clay layers exhibit a partly exfoliated structure, where rectorite layers are totally separated into very fine units in the rubber matrix, *i.e.* individual sheets and some nanoscale stacking sheets with a thickness of 30–50 nm and a length less than 1 μm.[11,12,37] When the nanocomposite is stretched, a strong restricting effect is applied on the rubber chains by the fillers due to the high aspect ratio of rectorite layers and huge contact areas between rubber macromolecules and rectorite layers. Therefore, the introduction of pristine rectorite into a rubber matrix could lead to an increase in mechanical properties, including stress at 300% strain, Shore A hardness, tear strength, elongation at break and tensile strength.

For example,[11] SBR–rectorite nanocomposite with 20 phr rectorite using $C_4H_9N^+(CH_2CH_2OH)_3Br^-$ as interfacial agent exhibits a tensile strength of 12.7 MPa, which is 6.3 times that of pure SBR. Compared with SBR–carbon black (N330) and SBR–calcium carbonate composites with the same filler loading, SBR–rectorite nanocomposite achieved higher values on the stress at 300% strain, tensile strength, elongation at break and tear strength. This substantial enhancement in mechanical properties is ascribed to the nanodispersion of rectorite in the SBR matrix and the high aspect ratio of the rectorite layers. The improvement in mechanical properties was also found in NBR–rectorite nanocomposites.[37] Compared with NBR–carbon black (N330) and NBR–silica nanocomposites at a filler loading of 5, 10 and 20 phr, NBR–rectorite nanocomposites exhibit the highest hardness, stress at 300% strain and tear strength. But the tensile strength of the NBR–rectorite nanocomposite at 20 phr is slightly lower than that of carbon black filled NBR. This behavior may be attributed to the existence of a filler network that is actually an inhomogeneous microstructure, and a higher stress concentration exists in the nanocomposite caused by rectorite layers with a high aspect ratio. It is noticed that the elongation at break of NBR–rectorite nanocomposites is lower than that of carbon black or silica filled NBR.

For pure NR, the stress–strain curve demonstrates a typical stress-induced crystallization behavior, as shown in Figure 3.19(a). In the lower strain region, the modulus of rubber is low and slowly increases with the increase of strain, and after the strain approaches a certain value, the stress will sharply increase within a small strain range due to the occurrence and subsequent rapid development of tensile crystallization. After the introduction of nanodispersed rectorite layers, the NR–rectorite nanocomposites exhibit a unique stress–strain behavior due to the reinforcement and the hindrance of rectorite layers to the stress-induced crystallization of NR.[12] As shown in Figure 3.19, the modulus of NR is significantly improved with the increase of filler loading when the strain is lower than 400% because of reinforcement of nanodispersed rectorite layers. Correspondingly, the stress transition caused by tensile crystallization gradually weakens, and even disappears, at the filler loading of 20 phr, as shown in Figure 3.19(d), but the tensile strength of NR–rectorite nanocomposites is still higher than that of pure NR. When the filler

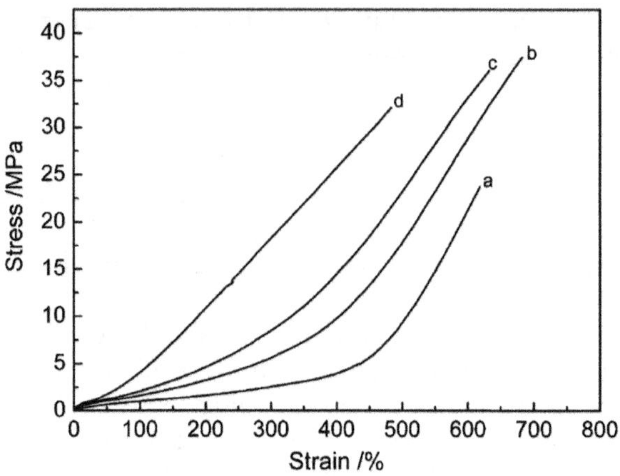

Figure 3.19 Stress–strain curves of NR–rectorite nanocomposites with different filler loadings: (a) 0 phr, (b) 5 phr, (c) 10 phr and (d) 20 phr. (Reproduced from *European Polymer Journal*, **41**, Y. Wang *et al.*, Preparation and properties of natural rubber/rectorite nanocomposites, 2776–2783, Copyright 2005 with permission from Elsevier.[12])

loading is more than 10 phr, the tensile strength and elongation at break of NR decrease with the increase of filler loading, which is attributed to the poor dispersion of rectorite layers when introducing more rectorite layers and the impedance of the stress-induced crystallization of NR.

Moreover, the modification of rectorite by organic alkyl ammonium salts can improve the compatibility with rubber and the dispersion of rectorite layers in the rubber matrix; thus, the rubber–rectorite nanocomposites can be obtained by the introduction of organo-modified rectorite into the rubber matrix *via* melt processing. The addition of a small amount of organo-modified rectorite to TPUR matrix can significantly improve the mechanical properties of the composites.[10] The highest tensile strength of the TPUR–rectorite nanocomposite is 57.9 MPa with only 2 wt% filler loading, about 52.0% higher than that of pure TPUR, and the biggest enhancement in tear strength of TPUR is 40.0% by adding only 2 wt% rectorite.

3.5.2 Thermal Stability

Thermogravimetric analysis (TGA) is commonly used to evaluate the thermal stability of a polymer. The thermal stability of polymer composites is affected by the structure of polymer, the type of additive, the dispersion of filler, and the interaction between polymer and filler. For polymer–clay nanocomposite, a nanocomposite with an intercalated structure is usually more stable than that with an exfoliated structure due to restriction of the polymer thermal motion in the clay gallery,[57] and a nanocomposite with stronger interfacial interaction would result in better thermal stability. It has been reported that the thermal stability of TPUR improves by adding only 2 wt% organo-modified rectorite.[31] The decomposition temperature at 10% weight loss increases from 315 °C to 330 °C, and the weight loss at 500 °C decreases from ~83% to ~72%. This result may be attributed to the nanodispersion of organo-modified rectorite in TPUR matrix and the stable structure of high temperature-resistant mica layers in rectorite. Besides, the presence of well-dispersed rectorite layers in TPUR–rectorite nanocomposite will reduce the permeability of volatile decomposed products out of the composites in comparison with the unfilled TPUR. In this case, the byproducts and heat of decomposition will be difficult to penetrate into the composites by adding nanodispersed rectorite layers so as to reduce the weigh loss of the nanocomposite.

3.5.3 Aging Resistance and Oil Resistance

Oxidative aging of rubber is one of the most serious problems in the rubber industry because the absorption of a small amount of oxygen causes considerable deterioration in the physicomechanical properties of rubber. Depending on the structure of rubber materials, the deterioration leads to chain scission, crosslinking, or formation of oxygen containing functional groups in the polymer chain or as degradation products. The thermal aging process involves the oxidation of rubber chains by molecular oxygen in a free radical process which can be catalyzed by oxygen, ozone, heat or mechanical stress, *etc.* The clay layers in the rubber matrix could slow down the rate of oxygen diffusing into the rubber, or the active clay may play a role as a free radical catching agent; thus, the incorporation of clay can improve the thermal aging resistance of rubber. Besides, the clay layers could act as impermeable barriers to not only the oxygen but also oil molecules diffusing in the rubber matrix. Therefore, the rubber–rectorite nanocomposites exhibit an improvement in thermal aging resistance and oil aging resistance. For instance, the addition of a small amount of organo-modified rectorite (less than 10 wt%) into TPUR matrix can improve the thermal aging resistance and oil resistant properties,[31] and the oil aging resistance of HNBR–rectorite nanocomposite with 10 phr rectorite is better than that of unfilled HNBR.[38] The presence of well-dispersed rectorite layers can make the penetrant travel a long tortuous path in the composites so as to improve oil resistance, and the improvement in thermal aging resistance may result from the slowing down of heat transmission by the rectorite layers. However, much work still remains to better explain the results.

3.6 Conclusions and Prospects

Rectorite, with a structure much like montmorillonite, is a promising filler for the preparation of nanocomposites. It is expected that rectorite can endow rubber–clay nanocomposites with a more effective enhancement in the performance, especially the gas barrier properties, due to the higher aspect ratio compared with montmorillonite. The high gas barrier properties of rubber–rectorite nanocomposites are attributed to a reduction of the volume fraction of the permeable rubber phase and an increase of the tortuous diffusion path, resulting in a decrease of solubility and diffusivity, respectively. The increase of the tortuous path can retard the diffusion of the gas molecules through rubber, which is the main reason for the reduced permeability.

To date, latex compounding is one of the most cost effective methods for controlling the dispersion of the rectorite layers in a rubber matrix. However, the completely exfoliated rubber–rectorite nanocomposite is still very difficult to achieve at high clay loadings. Although there has already been research work in the preparation of rubber–rectorite nanocomposites with relatively high gas barrier properties, challenges still remain in order to fully understand the complex structure–property relationships.

Moreover, the rubber–rectorite nanocomposites with extremely high gas barrier properties would need to possess the fully exfoliated clay structure, with the planes of most clay layers oriented parallel to that of the nanocomposite films. Much work remains to further improve the gas barrier properties of the rubber–rectorite nanocomposite. The increase of the free volume in the rubber matrix can facilitate gas transport in the nanocomposite, so the reasonable design and control of the interfacial interactions between rubber and rectorite is very important for a composite achieving high gas barrier properties. The formation of aggregation of clay layers leads to a decrease of the aspect ratio of fillers, and thus the actual aspect ratios are much lower than that needed for efficient enhancement of the barrier properties. Optimization of the arrangement of rectorite layers in the rubber matrix is the key to achieving a high aspect ratio.

Rubber–rectorite nanocomposites are promising for application to airtight materials due to their high gas barrier properties. However, further development is limited by reasons like a high cost/performance ratio and deficiency in some properties. To break through the bottleneck, we should develop preparation techniques with low cost, and exploit more excellent properties and find out more optimum application fields of the rubber–rectorite nanocomposites.

Acknowledgements

The authors would like to thank the National Basic Research Program of China (grant number: 2015CB654700), the National Natural Science Foundation of China (grant numbers: 51303051, 51320105012 and 51221002), and the Fundamental Research Funds for the Central Universities (grant numbers: 2014MS34) for financial support.

References

1. G. Choudalakis and A. D. Gotsis, *Eur. Polym. J.*, 2009, **45**, 967.
2. A. B. Morgan, *Polym. Adv. Technol.*, 2006, **17**, 206.
3. R. Sengupta, S. Chakraborty, S. Bandyopadhyay, S. Dasgupta, R. Mukhopadhyay, K. Auddy and A. S. Deuri, *Polym. Eng. Sci.*, 2007, **47**, 1956.

4. S. Pavlidou and C. D. Papaspyrides, *Prog. Polym. Sci.*, 2008, **33**, 1119.
5. B. Chen, *Br. Ceram. Trans.*, 2004, **103**, 241.
6. M. Alexandre and P. Dubois, *Mater. Sci. Eng., R*, 2000, **28**, 1.
7. S. S. Ray and M. Okamoto, *Prog. Polym. Sci.*, 2003, **28**, 1539.
8. P. Liu, *Appl. Clay Sci.*, 2007, **38**, 64.
9. J. P. Olivier and M. L. Occelli, *Microporous Mesoporous Mater.*, 2003, **57**, 291.
10. X. Y. Ma, H. J. Lu, G. Z. Liang and H. X. Yan, *J. Appl. Polym. Sci.*, 2004, **93**, 608.
11. Y. Q. Wang, H. F. Zhang, Y. P. Wu, J. Yang and L. Q. Zhang, *J. Appl. Polym. Sci.*, 2005, **96**, 324.
12. Y. Q. Wang, H. F. Zhang, Y. P. Wu, J. Yang and L. Q. Zhang, *Eur. Polym. J.*, 2005, **41**, 2776.
13. M. Okamoto, *J. Ind. Eng. Chem.*, 2004, **10**, 1156.
14. P. C. Lebaron, Z. Wang and T. J. Pinnavaia, *Appl. Clay Sci.*, 1999, **15**, 11.
15. R. Krishnamoorti and K. Yurekli, *Curr. Opin. Colloid Interface Sci.*, 2001, **6**, 464.
16. V. Mittal, *Materials*, 2009, **2**, 992.
17. S. J. He, Y. Q. Wang, Y. P. Feng, Q. S. Liu and L. Q. Zhang, *Nanotechnology*, 2010, **21**, 115601.
18. S. J. He, Y. Q. Wang, Y. P. Wu, X. H. Wu, Y. L. Lu and L. Q. Zhang, *Plast., Rubber Compos.*, 2010, **39**, 33.
19. Y. P. Wu, Y. Q. Wang, H. F. Zhang, Y. Z. Wang, D. S. Yu, L. Q. Zhang and J. Yang, *Compos. Sci. Technol.*, 2005, **65**, 1195.
20. A. Usuki, A. Tukigase and M. Kato, *Polymer*, 2002, **43**, 2185.
21. M. Ganter, W. Gronski, P. Reichert and R. Mülhaupt, *Rubber Chem. Technol.*, 2001, **74**, 221.
22. L. Q. Zhang, Y. Z. Wang, Y. Q. Wang, Y. Sui and D. S. Yu, *J. Appl. Polym. Sci.*, 2000, **78**, 1873.
23. R. A. Vaia, H. Ishii and E. P. Giannelis, *Chem. Mater.*, 1993, **5**, 1694.
24. A. Hiltner, R. Liu, Y. Hu and E. Baer, *J. Polym. Sci., Part B: Polym. Phys.*, 2005, **43**, 1047.
25. R. A. Pethrick, *Prog. Polym. Sci.*, 1997, **22**, 1.
26. M. H. Cohen and D. Turnbull, *J. Chem. Phys.*, 1959, **31**, 1164.
27. L. E. Nielsen, *J. Macromol. Sci., Part A*, 1967, **1**, 929.
28. R. K. Bharadwaj, *Macromolecules*, 2001, **34**, 9189.
29. S. Nazarenko, P. Meneghetti, P. Julmon, B. Olson and S. Qutubuddin, *J. Polym. Sci., Part B: Polym. Phys.*, 2007, **45**, 1733.
30. G. J. V. Amerongen, *Rubber Chem. Technol.*, 1947, **20**, 494.
31. X. Y. Ma, H. J. Lu, G. Z. Lian, J. C. Zhao and T. L. Lu, *J. Appl. Polym. Sci.*, 2005, **96**, 1165.
32. X. Y. Ma, X. H. Qu, Q. Zhang and F. Chen, *Polymer*, 2008, **49**, 3590.
33. L. Wang, P. Y. Xiang, L. Q. Zhang and Y. P. Wu, *J. Appl. Polym. Sci.*, 2013, **128**, 2578.
34. S. J. He, L. Q. Zhang, X. H. Wu, J. Lin, M. M. Xi and Y. Q. Wang, *Polym. Compos.*, 2012, **33**, 336.

35. Z. F. Wang, B. Wang, N. Qi, H. F. Zhang and L. Q. Zhang, *Polymer*, 2005, **46**, 719.
36. L. Wang, Y. Q. Wang, L. Q. Zhang and Y. P. Wu, *J. Appl. Polym. Sci.*, 2013, **127**, 765.
37. Y. Q. Wang, Y. P. Wu, H. F. Zhang, W. Zhao, C. X. Wang and L. Q. Zhang, *Polym. J.*, 2005, **37**, 154.
38. R. L. Zhang, L. F. Zhao, L. Liu and Y. D. Huang, *Plast., Rubber Compos.*, 2014, **43**, 235.
39. H. Yao, H. Qiao, Q. Y. Xing, N. Gao, J. F. Chu, L. Q. Zhang, K. C. Hua, J. Kulig and R. G. Wang, *J. Appl. Polym. Sci.*, 2014, **131**, 40682.
40. Y. P. Wu, Q. X. Jia, D. S. Yu and L. Q. Zhang, *J. Appl. Polym. Sci.*, 2003, **89**, 3855.
41. R. G. Wang, H. Yao, W. W. Lei, X. X. Zhou, L. Q. Zhang, K. C. Hua and J. Kulig, *J. Appl. Polym. Sci.*, 2013, **129**, 1546.
42. Y. Huang, X. Y. Ma, G. Z. Liang, H. X. Yan, X. Qu and F. Chen, *Clay Miner.*, 2007, **42**, 59.
43. L. Yuan, X. Y. Ma, G. Z. Liang and Y. Huang, *J. Compos. Mater.*, 2007, **41**, 1051.
44. J. H. Liu and A. Q. Wang, *J. Appl. Polym. Sci.*, 2008, **110**, 678.
45. L. Yuan, X. Y. Ma, A. J. Gu, H. X. Yan, G. Z. Liang, W. Wang and J. Y. Wu, *Polym. Adv. Technol.*, 2009, **20**, 826.
46. H. L. Zhang, C. Z. Zhu, K. Z. Yan and J. Y. Yu, *J. Mater. Civ. Eng.*, 2014, **27**, C4014002.
47. B. Li, F. X. Dong, X. L. Wang, J. Yang, D. Y. Wang and Y. Z. Wang, *Eur. Polym. J.*, 2009, **45**, 2996.
48. Q. X. Jia, Y. P. Wu, Y. L. Xu, H. H. Mao and L. Q. Zhang, *Macromol. Mater. Eng.*, 2006, **291**, 218.
49. Q. X. Jia, Y. P. Wu, Y. Q. Wang, M. Lu and L. Q. Zhang, *Compos. Sci. Technol.*, 2008, **68**, 1050.
50. J. J. Decker, K. P. Meyers, D. R. Paul, D. A. Schiraldi, A. Hiltner and S. Nazarenko, *Polymer*, 2015, **61**, 42.
51. Z. Q. Kuang, Y. L. Chen, Y. L. Lu, L. Liu, S. Hu, S. P. Wen, Y. Y. Mao and L. Q. Zhang, *Small*, 2015, **11**, 1655.
52. H. B. Cho, T. Nakayama, T. Suzuki, S. Tanaka, W. Jiang, H. Suematsu and K. Niihara, *Jpn. J. Appl. Phys.*, 2011, **50**, 01BJ05.
53. H. B. Cho, N. C. Tu, T. Fujihara, S. Endo, T. Suzuki, S. Tanaka, W. Jiang, H. Suematsu, K. Niihara and T. Nakayama, *Mater. Lett.*, 2011, **65**, 2426.
54. H. B. Cho, T. Nakayama, Y. Tokoi, S. Endo, S. Tanaka, T. Suzuki, W. Jiang, H. Suematsu and K. Niihara, *Compos. Sci. Technol.*, 2010, **70**, 1681.
55. H. B. Cho, Y. Tokoi, S. Tanaka, H. Suematsu, T. Suzuki, W. Jiang, K. Niihara and T. Nakayama, *Compos. Sci. Technol.*, 2011, **71**, 1046.
56. Z. H. Wang, J. Liu, S. Z. Wu, W. C. Wang and L. Q. Zhang, *Phys. Chem. Chem. Phys.*, 2010, **12**, 3014.
57. J. W. Gilman, *Appl. Clay Sci.*, 1999, **15**, 31.

CHAPTER 4

Design and Physicochemical Characterization of Novel Organic–Inorganic Hybrids from Natural Aluminosilicate Nanotubes

A. TAKAHARA*[a,b] AND Y. HIGAKI[a,b]

[a]Institute for Materials Chemistry and Engineering, Kyushu University, 744 Motooka, Nishi-ku, Fukuoka 819-0395, Japan; [b]International Institute for Carbon-Neutral Energy Research (WPI-I2CNER), Kyushu University, 744 Motooka, Nishi-ku, Fukuoka 819-0395, Japan
*E-mail: takahara@cstf.kyushu-u.ac.jp

4.1 Imogolite and Halloysite

Nanotubular materials are fundamental building blocks of future nanodevices for novel applications.[1] One of the well studied nanotubular materials from the fullerene family is the carbon nanotube, which was discovered by Iijima *et al.* in 1991.[2] The carbon nanotube is regarded as an ideal reinforcement nanofiller for polymers, and has been actively investigated for potential nanotechnological applications. However, the conventional techniques to produce a sizeable amount of carbon nanotubes require energy

RSC Smart Materials No. 22
Functional Polymer Composites with Nanoclays
Edited by Yuri Lvov, Baochun Guo and Rawil F. Fakhrullin

Published by the Royal Society of Chemistry, www.rsc.org

intensive processes, and the extreme synthetic conditions are not feasible for widespread application from the industrial point of view. Therefore, researchers are eager to find a competent substitution for carbon nanotubes to fully exploit the advantages of nanotubular materials. Naturally occurring single wall nanotube materials consisting of hydrous aluminosilicate were found in 1962 from the weathered pumice bed of volcanic ash soil in Kyushu, Japan.[3] This hydrous aluminosilicate material with general formula [Al_2O_3·SiO_2·$2H_2O$] is known as imogolite, named after the imogo-soil of the Kumamoto prefecture. Imogolite has a high aspect ratio and surface area with a length from several hundred nanometers to micrometers, like carbon nanotubes.[4] The unique physicochemical properties of the imogolite nanotube are expected to have potential nanotechnological applications in catalysts, membranes, and adsorbents.[5–7] However, there are some problems with imogolite nanotubes that remain to be overcome: for instance, the uniform dispersion in the polymer matrix, tunability of the imogolite dimensions (diameter and length), and interfacial adhesion between the nanotube and polymer matrix.

The dispersion of imogolite is probably the most pressing issue to be resolved for composite application.[8] The uniform dispersion of imogolite in the polymer matrix is very important from the perspective of nanomaterial engineering since it governs the macroscopic performance of polymeric nanocomposites. Imogolite nanotubes should be individually distributed in the polymer matrix without aggregation to ensure efficient load transfer to the imogolite network. Well dispersed imogolite imparts transparency to nanocomposites since an evenly dispersed imogolite nanotube will offer better transmission of light, while the aggregates induce light scattering.[9] There are a number of studies which report that surface treatment or chemical functionalization of inorganic nanotubes is an indispensable step toward better interfacial adhesion between the hydrophilic inorganic nanofiller and hydrophobic polymer matrix.[10–13] The surface of the inorganic nanofiller usually consists of hydrophilic groups which should be modified with organic moieties to increase the miscibility with the polymer matrix. For a better entanglement effect, inorganic nanofillers are *in situ* synthesized in the polymer matrix.

Apart from imogolite, halloysite is also a naturally occurring clay mineral with submicron sized hollow cylindrical morphology (Figure 4.1). Halloysite was discovered in the 18th century and was named in honor of O. d'Halloy who first studied the mineral.[14] Since then, halloysite has attracted much attention due to its tubular structure, and has been suggested for use in plastic composites, controlled release of chemicals,[15,16] and as a conductive filler by coating with metal using electroless plating.[17] Halloysite exhibits different surface chemical properties at the inner and outer sides of the tubes.[18] In halloysite, the silica layer is located on the outer surface of the tube, while the alumina layer is located in the inner lumen surface. Aluminum and silicon oxides have different dielectric and ionization properties, which are evident from observations of the electrical zeta-potentials of these oxide colloids in

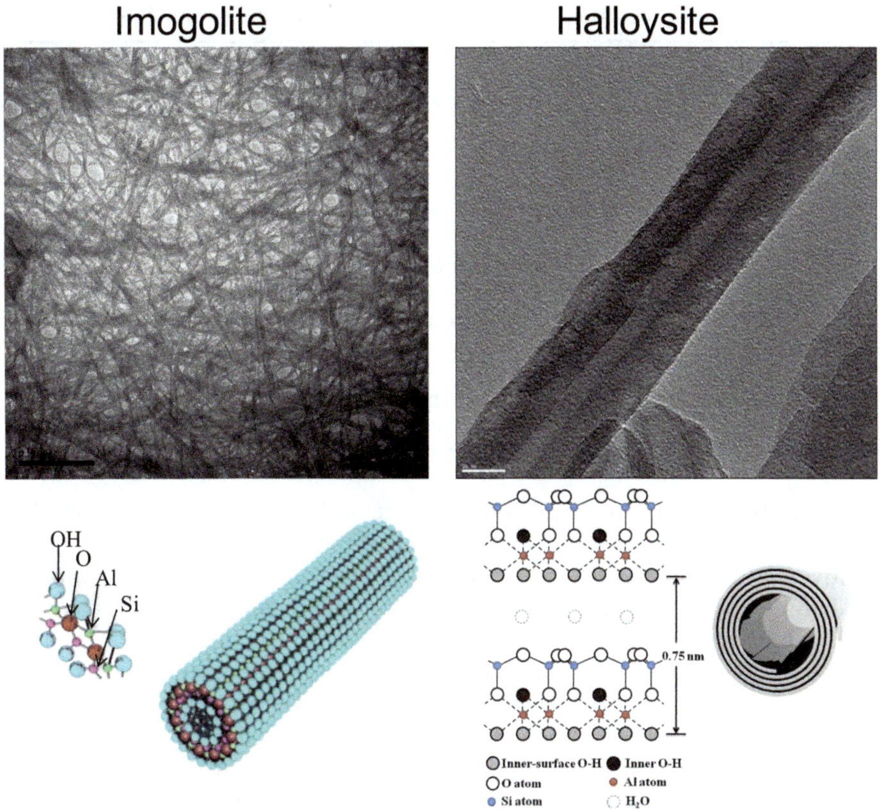

Figure 4.1 TEM images and chemical structures of (left) imogolite and (right) halloysite nanotubes.

water. The ability to have different charges at the inner and outer parts of the halloysite nanotubes allows for the selective loading of negatively charged molecules into their lumen.[19] The different chemistry of the inner and outer surfaces of the halloysite nanotubes can also be exploited for selective modification.

Up to now, nanotube reinforced nanocomposites have mostly been based on carbon nanotubes. However, carbon nanotubes and some inorganic nanotubes reported are all artificial materials involving accuracy-required and time-consuming preparation processes. Generally, these materials are opaque, which might not be suitable for certain optical applications. On top of that, issues concerning their potential toxicity to humans and the environment have become a stumbling block to the development and application of these nanotubes. In this respect, imogolite and halloysite, low cost and naturally available green nanotubes with a high aspect ratio, surface area and inherited transparent properties, are worth exploring as perspective nanomaterials.[20]

Table 4.1 Summary of structures and properties of imogolite and halloysite.

	Imogolite	Halloysite
Morphology	Tubular thread-like units	Mostly tubular, sometimes spherical or platy
Ideal unit formula	$Al_4Si_2O_6(OH)_8$	$Al_2Si_2O_5(OH)_4$
Dimensions	1 nm inner diameter 2 nm external diameter 1–5 micron length	10–50 nm inner diameter 50–200 nm external diameter 0.5–2 micron length
Tube exterior chemistry	Alumina octahedrons	Silica tetrahedrons
Tube interior chemistry	Silica tetrahedrons	Alumina octahedrons
Surface charge	Positive	Negative
Density	2.65 g cm^{-1}	2.55 g cm^{-1}
Specific surface area	Theoretical: 530 $m^2\ g^{-1}$ Observed: 200–400 $m^2\ g^{-1}$[37]	Theoretical: 40 $m^2\ g^{-1}$ Observed: 80–150 $m^2\ g^{-1}$

Comparison of the structures of imogolite and halloysite minerals might be interesting and reveal some facts about their structure (Table 4.1).[21,22] The positive surface charge of the imogolite suggests that the internal alumina octahedrons of halloysite are also positively charged, since both of these alumina octahedral sheets are chemically identical. Halloysite tubes are rolled like spirals, while imogolite is a single walled concentric tube.[23] The external and internal diameters of imogolite are much smaller than the ones of halloysite. The single walled structure of the imogolite provides a larger specific surface area compared to halloysite. It was expected that imogolite would have a larger cation exchange capacity (CEC) than halloysite due to its larger specific surface area, but this is not the case.[24]

4.2 Surface Modification of Imogolite and Application to Nanohybrids

A method for modifying imogolite at the nanotube level was developed. Imogolite nanotubes can be dispersed only in weakly acidic water by electrostatic repulsion.[25] In order to achieve modification at an individual tube level rather than at a bundle level, a process conducted in an aqueous solution is favorable. Unfortunately, most of the coupling agents that can provide a sufficiently hydrophobic surface are insoluble in water or sensitive to water. Ammonium salts of alkyl phosphate, however, were proven to be an exception. Various alkyl phosphates can be dissolved in water by converting into the corresponding ammonium salts.[26] Phosphate groups show excellent affinity with metal oxide surfaces,[27–30] and are stable in water over a wide pH range, and are rather insensitive to nucleophilic substitution and prone to heterocondensation (M–O–P bond formation) as compared to homocondensation (P–O–P). Thus, surface modification with organophosphorus compounds has the advantage of being operable in a wide range of solvents from aprotic to protic, even in water. According to the above consideration, artificial synthesis of imogolite and modification with alkyl phosphate were examined. The application of the modified imogolite nanotubes to polymer based nanohybrids was also described.

4.2.1 Modification of Imogolite Nanotubes with Alkyl Phosphate

Figure 4.2 shows the surface modification procedure of imogolite by alkyl phosphate compounds.[31] The ammonium salt of dodecyl phosphate ($DDPO_4(NH_4)_2$) was used for surface modification of imogolite. The interaction between imogolite and $DDPO_4$ was confirmed by IR measurements and X-ray photoelectron spectroscopy (XPS). Figure 4.3 shows FT-IR spectra of the pristine imogolite, $DDPO_4H_2$, $DDPO_4(NH_4)_2$, and $DDPO_4$-imogolite. For $DDPO_4H_2$, the absorption of the PO–H vibration around 3420 cm^{-1}, the absorption of the P–O stretching vibration around 1030 cm^{-1}, and the absorption of the P=O vibration at 1239 cm^{-1} were observed. The absorption spectrum of $DDPO_4(NH_4)_2$ shows some differences from those of $DDPO_4H_2$, as the absorption band around 1030 cm^{-1} splits into several sharp absorptions and the P=O vibration shifts to 1201 cm^{-1} as well. The absorption around 3130 cm^{-1} is attributed to N–H vibration of the ammonium groups. For

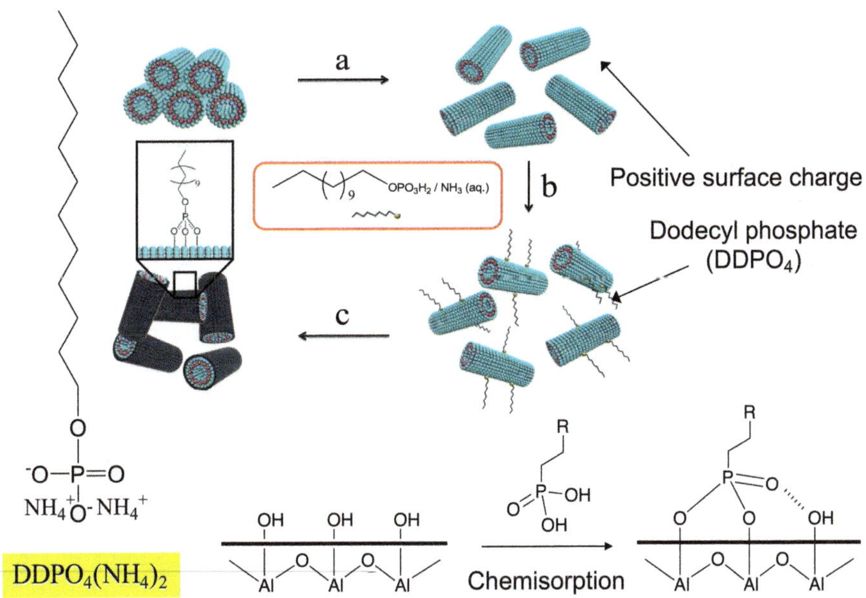

Figure 4.2 Surface modification procedure of imogolite with the ammonium salt of dodecylphosphate ($DDPO_4(NH_4)_2$). (a) Dispersion of imogolite in water, (b) $DDPO_4$ adsorption onto imogolite nanotubes, (c) purification. Adapted with permission from Ma *et al.*, *Chem. Lett.* 2011, **40**, 159–161. Copyright 2011 The Chemical Society of Japan.[31]

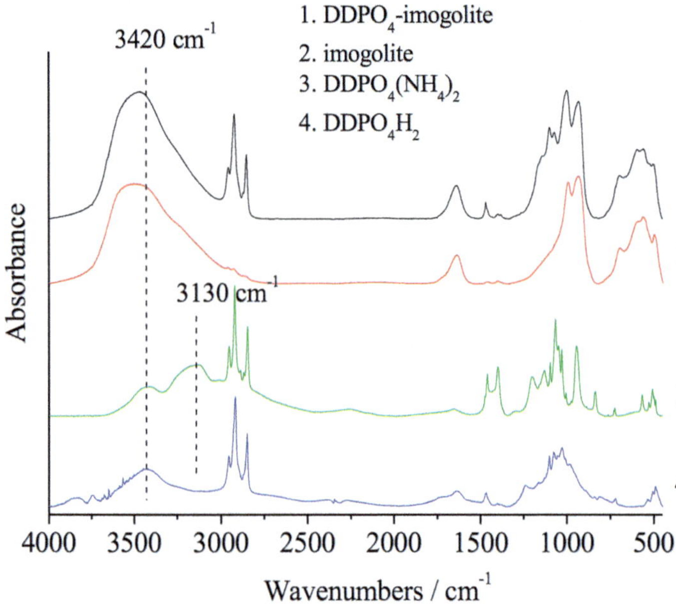

Figure 4.3 FT-IR spectra of the (1) $DDPO_4$-imogolite, (2) pristine imogolite, (3) $DDPO_4(NH_4)_2$, and (4) $DDPO_4H_2$. Adapted with permission from Ma *et al.*, *Chem. Lett.* 2011, **40**, 159–161. Copyright 2011 The Chemical Society of Japan.

$DDPO_4$-imogolite, the absence of the N–H absorption at 3130 cm^{-1} indicates that ammonium cations do not adsorb on the surface of imogolite. The typical absorption bands of imogolite at 995, 935, and 560 cm^{-1} still exist, suggesting the retention of the Si–O–Al skeleton in imogolite nanotubes, while the absorption at 995 cm^{-1} is strengthened by the coexistent absorption of the phosphate groups. The absorption of P=O at 1239 cm^{-1} for $DDPO_4H_2$ and 1201 cm^{-1} for $DDPO_4(NH_4)_2$ disappears from the spectrum of $DDPO_4$-imogolite, presumably due to the condensation between the phosphate groups and the aluminol groups. The PO–H vibration around 3420 cm^{-1} was still found, suggesting there are some POH groups in $DDPO_4$-imogolite. These POH groups might be due to either the unbound $DDPO_4H_2$ or the bonded ones that still have unreacted POH groups. In the thermogravimetric analysis (TGA), no weight loss due to the unbound $DDPO_4H_2$ was observed, thus it is reasonable that those POH groups in $DDPO_4$-imogolite are due to the bonded $DDPO_4H_2$. The adsorption of the $DDPO_4$ onto imogolite was further confirmed by XPS spectra of the Al_{2p}. For the pristine imogolite, the Al_{2p} signal is found around 74.3 eV with a symmetric peak; while for $DDPO_4$-imogolite, the Al_{2p} peak becomes wide and asymmetric, which can be fit with two Gaussian curves that consist of a signal at 74.3 eV and one at 76.3 eV. The second component is ascribed to an increase in the positive charge on Al atoms because of the formation of Al–O–P bonds at the surface of imogolite, while the first indicates the unreacted Al–OH. Thus, it can be concluded that dodecyl phosphate attaches to the surface of imogolite through covalent interaction.

The wettability of the dodecyl phosphate modified imogolite nanotube was confirmed by the water contact angle (CA) of the $DDPO_4$-imogolite film. $DDPO_4$-imogolite was dispersed in ethanol with a concentration of 2 mg mL^{-1} and cast onto a silicon wafer by spin coating. For contrast, an imogolite aqueous solution was also cast onto a silicon wafer by the same procedure. The static contact angle for the pristine imogolite cast surface was 22°. In contrast, for the $DDPO_4$-imogolite cast surface, the contact angle was 93°. This result indicates that the hydrophilicity of the external surface of imogolite changed by the adsorption of $DDPO_4$, which converts the hydrophilic surface of imogolite nanotubes to hydrophobic.

4.2.2 Polystyrene–Imogolite Nanohybrids

Surface modified $DDPO_4$-imogolite nanotubes were dispersed in a polymer matrix to prepare polymer–imogolite hybrids. Polystyrene (PS)–imogolite nanohybrids were prepared by solution blending followed by compression molding. The temperature dependence of the dynamic viscoelasticity of the PS–imogolite nanohybrids and the pristine PS was evaluated by dynamic mechanical analysis. As shown in Figure 4.4(a), the inclusion of imogolite led to a slight increase in the storage modulus below 100 °C, while the storage modulus showed a significant increase above 100 °C, compared with the pristine PS. PS is in a glassy state below 100 °C, and in a rubbery state above 100 °C. The reinforcement effects depend on the thermal dynamic state of the polymer matrix. On the other hand, an increase in the α_a-absorption temperature was observed with the inclusion of imogolite, as shown in Figure 4.4(b). Moreover, the peak intensity of α_a-absorption gradually decreased with the increase of imogolite loading amounts, suggesting that the mobility of PS

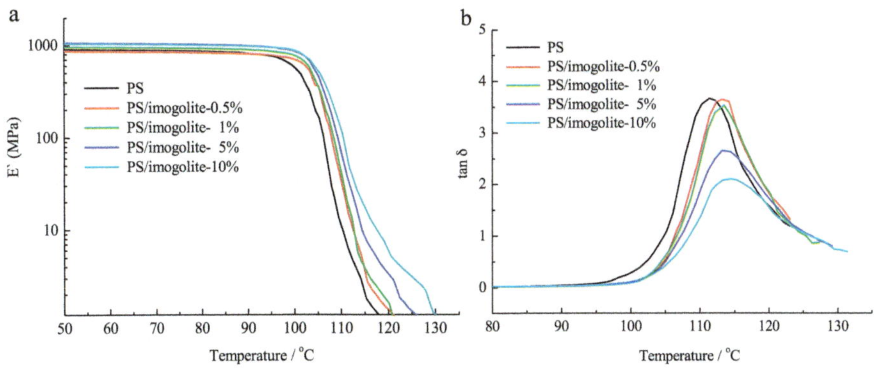

Figure 4.4 Dynamic storage modulus E' (a) and $\tan\delta$ peaks (b) of PS and PS–imogolite nanohybrid films at 1 Hz.

chain segments was restricted by the presence of rigid imogolite. Enhancement in the thermal stabilities was observed in PS–imogolite nanohybrids. Obvious increases in the thermal decomposition temperature of PS were observed with the inclusion of the modified imogolite. The thermal stability would be attributed to the restriction in the micro-Brownian motion of PS chain segments by imogolite. Polymer chains in a restricted environment need a more thermal energy to obtain the same degree of thermal motion for thermal degradation than non-restricted chains. The mechanical properties of the PS–modified imogolite hybrid were measured by tensile test. The elastic modulus of PS–imogolite nanohybrids increased with imogolite content, whereas the strength and elongation at break decreased, particularly when the loading amount of modified imogolite exceeded 1 wt%.

4.2.3 Surface-Initiated Atom Transfer Radical Polymerization (SI-ATRP) from Imogolite Surface

Imogolite has been successfully modified with dodecyl phosphate at the nanotube level by chemisorption. However, the dodecyl phosphate modified imogolite cannot homogeneously disperse in organic solvents, as confirmed by the appearance of a white dispersion. As a result, PS–imogolite nanohybrids based on dodecyl phosphate modified imogolite become brittle, because of the aggregates with weak filler–filler interaction. In order to obtain a better imogolite nanotube dispersion state, surface modification with a long organic chain is preferred, because this is favorable for the effective steric repulsion between nanofillers, as well as an affinity between nanofillers and the circumference media, *i.e.* solvents or polymer matrices. A "grafting from" method by surface-initiated activator regenerated by electron transfer (ARGET) atom transfer radical polymerization (ATRP) was applied for grafting poly(methyl methacrylate) (PMMA) from the imogolite surface.[32,33] Figure 4.5 shows the preparation procedure for a PMMA grafted imogolite nanotube (PMMA-*g*-imogolite).

Various ATRP initiators that can adsorb onto inorganic surfaces have been synthesized, however, most of the surface-attachable groups have almost exclusively been alkoxy- or chlorosilanes. Unfortunately, organosilanes are not suitable for the modification of imogolite, because surface modification with organosilanes usually needs a dried condition to prevent unfavorable side reactions. In addition, the surface bonds between organosilane and the external surface of imogolite were reported to be labile in humid atmosphere. On the other hand, organophosphorus compounds appear complementary of organosilanes, as they show an excellent affinity toward metal oxides. We synthesized an initiator carrying a phosphoric acid group, 8-(2-bromo-2-methylpropanoyloxy) octyl phosphoric acid ($BMPOPO_4H_2$), which was further converted to a water soluble ammonium salt [$BMPOPO_4(NH_4)_2$]. Homogeneous modification of the imogolite surface can be achieved by using a water soluble initiator carrying a surface-attachable phosphoric acid group.

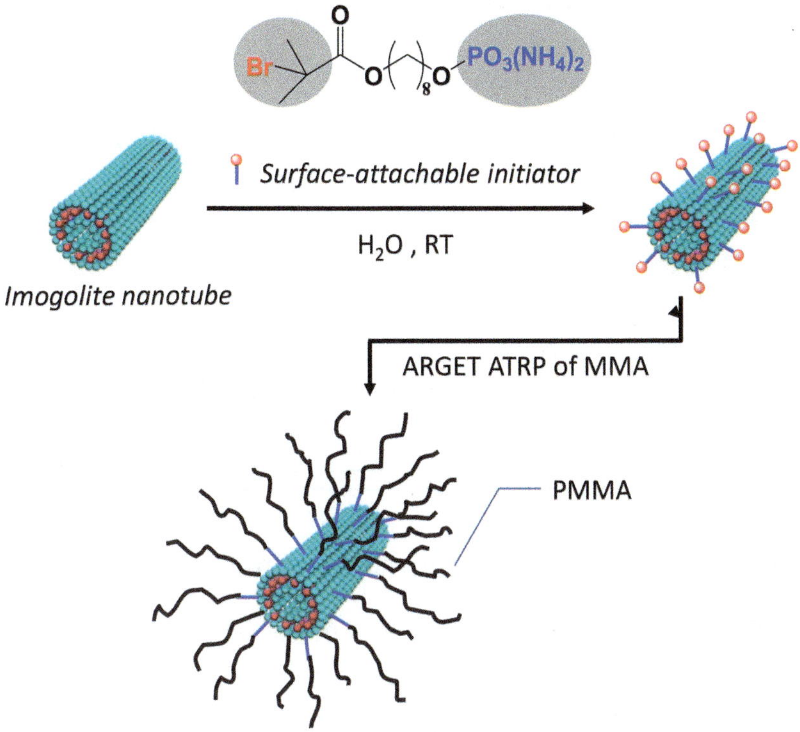

Figure 4.5 Schematic representation of the preparation of a PMMA grafted imogolite nanotube. Adapted from ref. 32 with permission from The Royal Society of Chemistry.

The initiator we designed seems capable of providing the modified imogolite with adequate hydrophobicity, because it contains a relatively long hydrophobic chain. The hydrophobicity provides affinity and dispersibility in common organic solvent and monomers for the modified imogolites.

The ATRP initiator $BMPOPO_4$ was fixed onto the imogolite surface in aqueous solution at room temperature. The acidity of the initial reaction mixture was controlled to be pH 5 with acetate buffer in order to disperse the imogolite fibers. The adsorption of $BMPOPO_4$ onto the imogolite surface was confirmed by FT-IR measurement and XPS analysis. FT-IR spectra in Figure 4.6 revealed that the Al–O–Si vibration of imogolite at 995, 935, and 560 cm^{-1} still existed after modification, indicating that this reaction was a safe process for the imogolite structure. The absorbance bands at 2932 cm^{-1} (C–H) and 1735 cm^{-1} (C=O) confirmed the adsorption of the ATRP initiator onto the imogolite surface, while the absence of an N–H vibration band at 3130 cm^{-1} indicated that ammonium cations did not adsorb onto imogolite. XPS spectra provided more information on the interaction between the imogolite surface and the ATRP initiator. In the wide scan XPS spectra of $BMPOPO_4$-imogolite, the characteristic peaks of P_{2p}, P_{2s}, and Br_{3d} were found at around 134.5, 191.8,

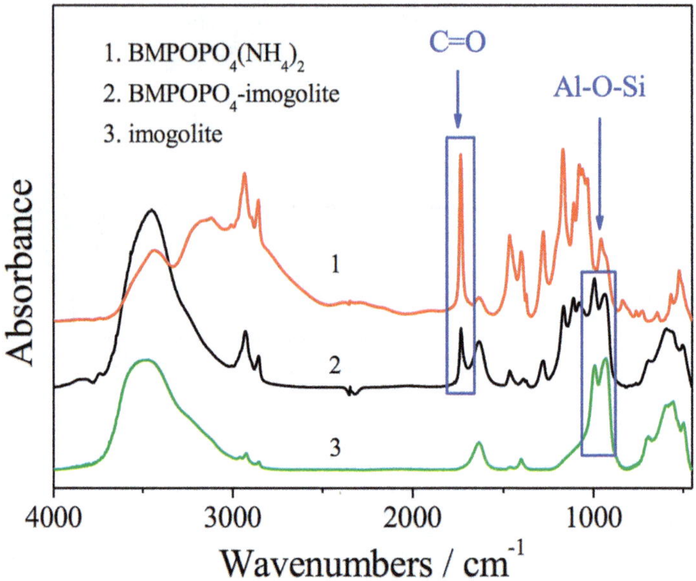

Figure 4.6 FT-IR spectra of the (1) $BMPOPO_4(NH_4)_2$, (2) $BMPOPO_4$-imogolite, and (3) pristine imogolite.

and 70.0 eV, respectively. In addition, no signal of nitrogen was found, further confirming that the ammonium cations did not adsorb onto imogolite. The high resolution XPS spectra of Al_{2p} showed that the position of Al_{2p} was unchanged after the immobilization of the ATRP initiator. Thus $BMPOPO_4$ attached onto the imogolite surface, possibly through electrostatic adsorption rather than covalent bonding. The contrast of the interaction type in between the $BMPOPO_4$ and dodecyl phosphate was likely due to the difference in the hydrophilicity. The amount of adsorbed $BMPOPO_4$ was estimated to be 49 wt% by TGA.

The surface-initiated ARGET ATRP of MMA in the presence of initiator immobilized imogolite was conducted. The GPC profile showed that grafted PMMA with a molecular weight of M_n = 26 600 and 32 700, and a molecular weight distribution of M_w/M_n = 1.22 and 1.33 were obtained after a polymerization time of 50 and 90 min, respectively. After $BMPOPO_4$ modification, the modified imogolite can disperse in various solvents, while the pristine imogolite cannot be dispersed in any organic solvent. The dispersion was neither homogeneous nor stable. However, when PMMA was grafted to the surface of imogolite nanotubes, the PMMA-*g*-imogolite showed good dispersibility in organic solvents including THF, chloroform, and toluene. As shown in Figure 4.7(a), the homogeneous dispersion of PMMA-*g*-imogolite in THF with a concentration of 10 mg mL^{-1} was stable for more than two months.

The morphology of the PMMA-*g*-imogolite nanotubes was observed with atomic force microscopy (AFM) (Figure 4.7(b)). Discrete nanostructures were randomly distributed on the mica surface and no aggregation was observed,

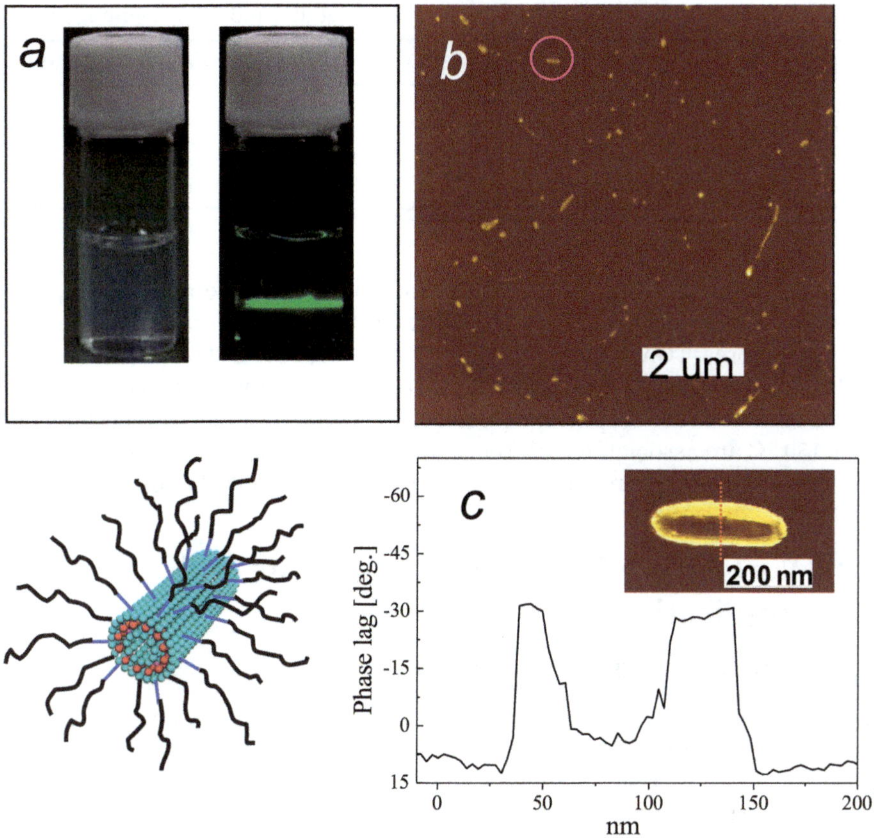

Figure 4.7 (a) Photographs of THF dispersion of PMMA-*g*-imogolite (PMMA M_n: 32 700, M_w/M_n: 1.33) with a concentration of 10 mg mL^{-1} (the one on the right shows scattering of a green light beam by dispersion due to the Tyndall effect). (b) AFM phase image (wide scan) and (c) cross-sectional analysis of a PMMA-*g*-imogolite (AFM high resolution phase image (insert)). Adapted from ref. 32 with permission from The Royal Society of Chemistry.

indicating excellent dispersibility of the PMMA-*g*-imogolite. The high resolution phase image and the corresponding cross-section analysis in Figure 4.7(c) indicate that the PMMA-*g*-imogolite render a hard middle part and soft edge. This further confirms the core–shell structure of the PMMA-*g*-imogolite. The hardness distribution along the objects corresponds to the different thermal dynamic states of PMMA, and suggests that the mobility of the PMMA chain segments near imogolite is lower compared with those far from imogolite. Individual bare imogolite has a height of *ca.* 3.0 nm. For the PMMA-*g*-imogolite, the height is not as uniform as that of bare imogolite. The average height value of PMMA-*g*-imogolite analyzed from 60 AFM images was estimated to be 10.6 ± 2.5 nm. If we consider the object with the smallest

Table 4.2 TGA results of imogolite, $BMPOPO_4$-imogolite, and PMMA-*g*-imogolite.

Sample	Weight loss at 150 °C (%)	Weight loss at 600 °C (%)
Imogolite	12.6	31.3
$BMPOPO_4$-imogolite	6.7	50.9
PMMA-*g*-imogolite-50 min	2.1	85.6
PMMA-*g*-imogolite-90 min	1.1	93.7

height value (6 nm) contains a single imogolite nanotube as the core, the ones with larger height values are expected to have nanotube bundles as their rigid cores. TGA results of pristine imogolite, $BMPOPO_4$-imogolite, and PMMA-*g*-imogolite are summarized in Table 4.2. The weight losses before 150 °C are due to the evaporation of adsorbed solvents, and the weight losses after 150 °C are associated with the sample decomposition. The weight loss profiles in TGA clearly indicate the successful grafting of PMMA on imogolite.

4.3 Surface Modification of Halloysite and Application to Nanohybrids

4.3.1 Surface Modification of Halloysite

Halloysites have rapidly garnered significant interest in the fields of material science due to their versatile features of high surface area, large pore size, and tunable surface chemistry, which have enabled this nanomaterial to be utilized as catalysts, in electronic devices, for entrapment of hydrophilic and lipophilic active agents, and as a nanofiller for polymers. Halloysite is a naturally occurring aluminosilicate nanotube with a stoichiometry of $Al_2Si_2O_5(OH)_4{\cdot}nH_2O$. Layered halloysite occurs mainly in two polymorphs; the anhydrous form with interlayer of 7 Å and the hydrated form with expanded interlayer spacing of 10 Å, as a result of the incorporation of water in the interlamellar space. Attempting to use organosilanes for selective surface modification on halloysite appears futile as silanes bind to both surfaces. The use of organophosphonic acid in surface modification is very attractive for preparation of highly hydrolytic stable hybrid material, and surface modification can be performed in water at ambient conditions in contrary to organosilane coupling agents. The preparation of inorganic micelles with a hydrophobic lumen and hydrophilic shell is achieved by selective modification on the Al_2O_3 surface of a halloysite lumen using alkylphosphonic acid [octadecyl phosphonic acid (ODP)] (Figure 4.8).[34] The inorganic micelles of modified halloysite are associated with their unique architecture, in which the hydrophobic lumen creates a space for encapsulation of neutral and hydrophobic guest molecules by partitioning from a polar solvent, while the solid shell provides stability to the micellar structure to stay intact and retain the guest molecules. The stability of conventional organic micelles, which are characterized as a hydrophobic core and hydrophilic palisade

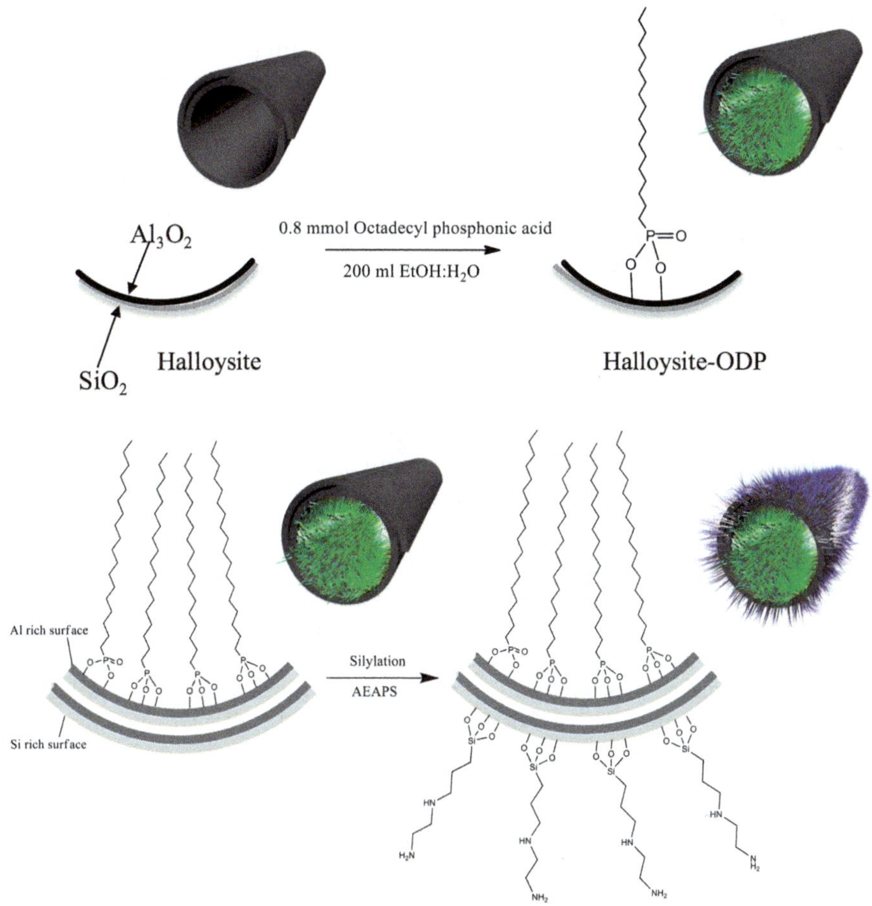

Figure 4.8 Lumen modification of halloysite and preparation of inorganic micelles through subsequent exterior surface modification of halloysite. Adapted with permission from Yah *et al.*, *J. Am. Chem. Soc.* 2012, **134**, 1853–1859. Copyright 2012 American Chemical Society.[34]

architecture, is largely dependent on their critical micelle concentration and temperature. Introduction of cross linkers may increase micelle stability by forming covalent bonds in the core; however, they may cause safety concerns as most of the cross linkers and synthetic surfactants are toxic. The construction of inorganic micelles as appropriate absorbents and carriers for delivery offers some advantages over the organic micelles in this regard. Effort has been directed toward engineering of the micellar core to increase the loading capacity of hydrophobic guest molecules and achieve controlled release.

The TEM images revealed that halloysite possesses thick walls with a cylindrical shape and contains open ended channels that run longitudinally along the nanotube (Figure 4.1). The samples contain agglomerates of short and broken nanotubes with irregularity in diameter, wall thickness, and morphology. As evaluated from TEM images, the halloysites have an external diameter in the range of 30–70 nm with an inner diameter of less than 20 nm, while the thickness of the wall is about 25 nm. After modification of the halloysite, the transparent central channel that runs longitudinally along the nanotube became less resolved, indicating that the lumen has been covered with a thin layer of grafted ODP. Nevertheless, judging from the TEM images, the pore is not completely filled or blocked by ODP, which may be benefit for inclusion and controlled release applications.

The porosity of halloysite before and after ODP modification was measured by nitrogen adsorption–desorption at 77 K (Figure 4.9). There is a noticeable increase in nitrogen uptake that corresponds to monolayer–multilayer adsorption of pore walls. The BET specific surface area (S_{BET}) and pore volume (V_{pore}) of pristine halloysite are 43 cm^2 g^{-1} and 0.53 cm^3 g^{-1}, respectively. The isotherm of halloysite-ODP is similar to that of pristine halloysite with relatively lower S_{BET} and V_{pore} of 27.6 cm^2 g^{-1} and 0.38 cm^3 g^{-1}, respectively (Table 4.3). The pore distribution curve shown in Figure 4.9 displays one primary population of pores at 31.7 and 27.7 nm, for halloysite and halloysite-ODP, respectively. A slight reduction in pore diameter of halloysite after modification strongly indicates the formation of a single grafted ODP layer. Although the values calculated by nitrogen adsorption are biased toward the maximum diameter compared to values observed by TEM images, it is clear that these values are of more statistical significance for larger size samples and should be recommended for evaluation of natural halloysite pore systems whose pore size and shape are generally not uniform.

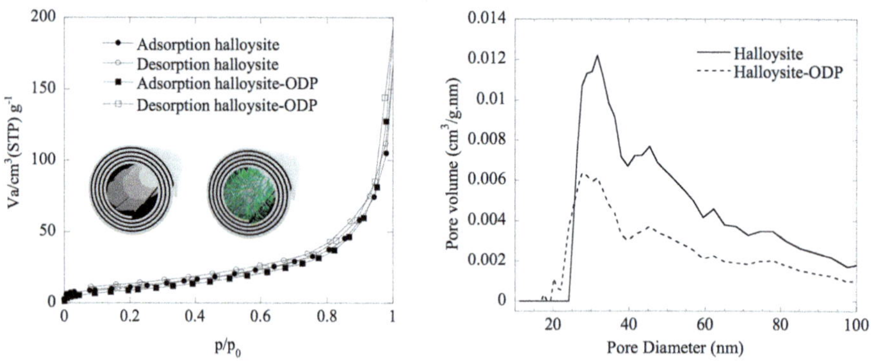

Figure 4.9 Nitrogen adsorption isotherms (left side) and pore size distribution curves (right side) of halloysite and halloysite-ODP.

Table 4.3 Structural data for pristine halloysite and halloysite-ODP obtained from the BET nitrogen adsorption test.

Sample	S_{BET} (m^2 g^{-1})	V_{pore} (cm^3 g^{-1})	d_{peak} (nm)
Halloysite	43.0	0.53	32 ± 1
Halloysite-ODP	27.6	0.38	28 ± 1

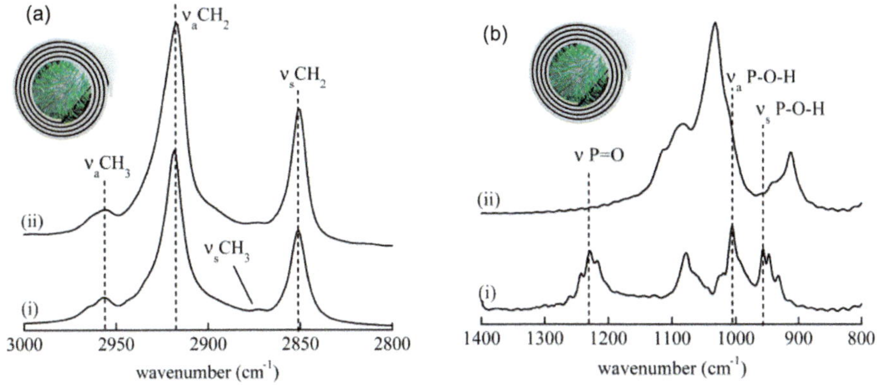

Figure 4.10 FT-IR spectra of (a) methylene stretching region (3000 to 2800 cm^{-1}) and (b) P–O stretching region (1400 to 800 cm^{-1}) of (i) ODP and (ii) halloysite-ODP. Adapted with permission from Yah *et al.*, *J. Am. Chem. Soc.* 2012, **134**, 1853–1859. Copyright 2012 American Chemical Society.[34]

Figure 4.10 shows the methylene stretching region of ODP and halloysite-ODP. The methylene groups are known to give two bands corresponding to symmetric and asymmetric C–H stretching at 2850–2852 and 2916–2924 cm^{-1}, respectively. In the same region, terminal methyl groups of symmetric and asymmetric stretching modes are detected at 2872 and 2958 cm^{-1}, respectively. Owing to the fact that there are fewer CH_3 terminal groups than CH_2 groups in their molecular structure, the terminal methyl bands are weaker than those of methylene groups. Comparing the frequency value, it can be noted that all three frequencies (symmetry CH_3 stretching at 2873 cm^{-1} is too weak to look for shifting) of ODP are shifted toward lower values, which usually indicates a less free vibration of alkyl groups inside the confined spaces of the halloysite lumen. Also, the location of the CH_2 stretching band in halloysite-ODP at a lower wavenumber compared with ODP indicated that ODP in the halloysite lumen is relatively ordered.

In Figure 4.10(b), the P–O stretching in the region between 1300 and 800 cm^{-1} is broad and complex making it difficult to interpret, but the changes observed show that a strong interaction of the phosphonate headgroup with the alumina surface is present. The P–O stretches of ODP at 1228 and 956 cm^{-1}, which are assigned to P=O and P–O–H groups in ODP, are absent in halloysite-ODP. Also, the PO–H band at 2500–2000 cm^{-1} for free ODP is not detected in the halloysite-ODP indicating that the ODP binds to the aluminol site in the deprotonated form. However, P–O–H and P–O–Al vibrate at similar frequencies of 950 cm^{-1}, whereas the intense Si–O stretching band at 1100–1000 cm^{-1} overlaps with the R-PO_3^{2-} symmetric and asymmetric stretches. Hence, it is hard to conclude from these data alone whether the binding mode is mono-, bi- or tridentate. Nevertheless, the drastic decrease of P–O–H at 2375–2200 cm^{-1} and P=O vibration at 1228 cm^{-1} could possibly indicate that the binding mode is mostly tridentate. The definite assignment of these bands is difficult since the

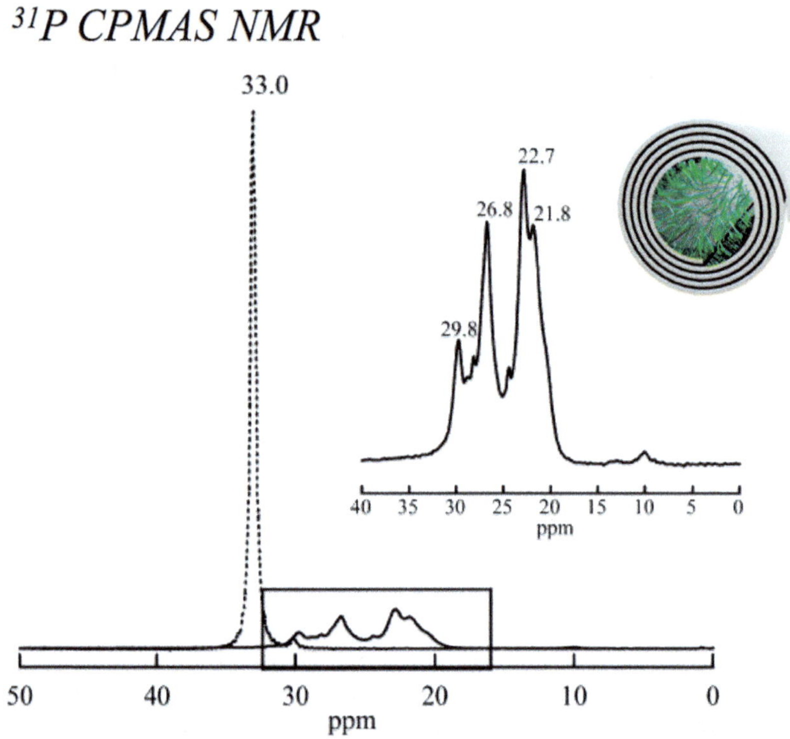

Figure 4.11 ^{31}P CPMAS NMR spectra of ODP (dotted line) and halloysite-ODP (solid line). Adapted with permission from Yah *et al.*, *J. Am. Chem. Soc.* 2012, **134**, 1853–1859. Copyright 2012 American Chemical Society.[34]

ranges for the different P–O stretching peaks greatly overlap and depend on the degree of hydrogen bonding or metal binding.

The ^{31}P solid-state NMR spectra, shown in Figure 4.11, are more useful for revealing the nature of the interaction of the phosphonic acid headgroup with the metal oxide surface. ^{31}P chemical shifts are sensitive to variations in the O–P–O bond angle as well as the degree of electronegativity of the nearest neighbor atoms. The surface of the alumina site is complex, as the bonding involves both the coordination of the phosphoryl oxygen to the Lewis acid sites, and the condensation of the P–OH functions with surface Al–OH groups. Accordingly, the ODP headgroup may act as a mono-, bi-, or tridentate ligand on the surface. Significant change is observed for halloysite-ODP where the phosphorus resonance is shifted further upfield and is inhomogeneously broadened. The peaks were greatly shifted from the pure ODP indicating a strong interaction with the alumina surface of halloysite lumen. The broad distribution of chemical shifts is due to different types of surface bonds and bonding sites on the aluminol surface. The spectrum displayed multiple peaks between 30 and 20 ppm, with distinguishable peaks centered on 29.8, 26.8, and 22.7 ppm. These peaks are, respectively, attributed to phosphates bound to the internal surface through one, two, and three Al–O–P bonds. From the average upfield shift of 6.2 and 10.3 ppm from free acid, the primary mode of attachment to the aluminol surface is probably bidentate and tridentate with a fraction of less associated monodentate configuration as well ($\Delta\delta = 3.3$ ppm). The FT-IR results support the ^{31}P solid-state NMR results, which show a significant decrease in the intensity of the P=O stretching band in halloysite-ODP. Similar ^{31}P resonance shifts have been observed in numerous studies where phosphonic acids were reacted with silica, titania, alumina, SnO_2, and zirconia surfaces. This indicates that it is not differences in the surface that cause the shift but more likely the binding modes of the phosphonic acid.

4.3.2 Adsorption of Ferrocene into the Modified Halloysite and the Controlled Release

The controlled loading and releasing properties of the modified halloysite nanotubes as an inorganic nano-container are demonstrated by using ferrocene as a representative hydrophobic molecule.[34] In an equilibrium adsorption study, 5 mg of halloysite sample and 1 mL of ferrocene solution were placed in a tube and stirred for two days in a bioshaker at 25 °C. After centrifugation at 7000 rpm for 10 min, the liquid phase concentrations of ferrocene were analyzed by UV-vis absorption spectroscopy. The adsorption amount at the equilibrium state Q_e (mg g^{-1}) was obtained using the following equation:

$$Q_e = (C_0 - C_e)V/w$$

where C_0 and C_e are the initial and equilibrium concentrations of ferrocene (μM), V is volume of the liquid phase (mL), and w is weight of adsorbent (pristine or modified halloysite) (mg). The amount of ferrocene included in the halloysite is plotted as a function of the molar concentration of ferrocene. The equilibrium adsorption of ferrocene by halloysite and halloysite-ODP are shown in Figure 4.12. Both adsorption isotherms are linear, with correlation coefficients of 0.88 and 0.95 for halloysite and halloysite-ODP, respectively. The linearity of the adsorption isotherms is consistent with the idea that inclusion of ferrocene by the halloysite is a partitioning process. In the partitioning process, the hydrophobic ferrocene distributes itself between the organic phase and aqueous phase. The dominant driving forces for the adsorption of ferrocene are hydrophobic interaction (a kind of van der Waals attraction) and capillary forces, where hydrophobic guests, such as ferrocene, are driven into the less polar lumen of halloysite-ODP. The slope of the isotherms, the partition coefficient, clearly show that the partitioning equilibria favor the less polar environment of halloysite-ODP for hydrophobic ferrocene.

Figure 4.13 shows the weight (mg) of ferrocene released *versus* time from samples that were saturated with 20 mM ferrocene solution at room temperature. The common trend showed a faster release rate of ferrocene initially, followed by slower release. For an unmodified halloysite system, a small initial burst release up to one hour of assay is observed, which is attributed to immediate dissolution of the portion of ferrocene on the surface of halloysite, and the release has halted completely in the following hours. Overall,

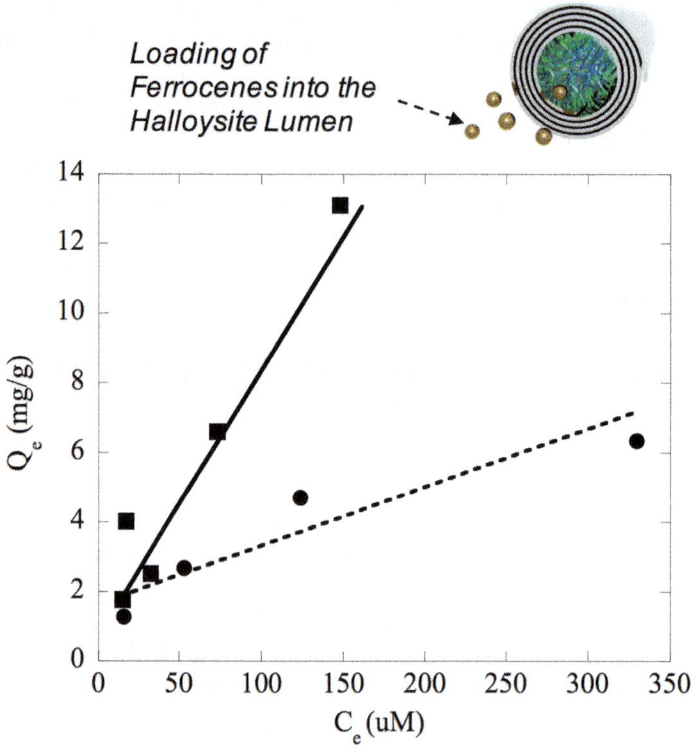

Figure 4.12 Equilibrium adsorption of ferrocene by halloysite (●) and halloysite-ODP (■) at room temperature from 1:1 THF:water solution. Adapted with permission from Yah *et al.*, *J. Am. Chem. Soc.* 2012, **134**, 1853–1859. Copyright 2012 American Chemical Society.[34]

the ODP modified halloysite sample was found to release more ferrocene in the same period of time in comparison to unmodified halloysite. The main factor may be that there was a very little amount of ferrocene adsorbed in the unmodified halloysite because of its lower adsorption capacity. The hydrophobic modification of the inner lumen by ODP caused not only the adsorption capacity increment but long-term slow release. The hydrophobic interaction between the ferrocene and alkyl chain in ODP make the ferrocene molecules stay inside the halloysite-ODP lumen, and the ferrocenes are released slowly after the initial rapid release.

4.3.3 Preparation of Inorganic Micelles by Bifunctionalization of Halloysite Nanotubes

Further reaction of halloysite-ODP with *N*-(2-aminoethyl)-3-aminopropyltrimethoxysilane (AEAPS) resulted in the grafting of silyl groups, as evidenced by the signal at −64.6 ppm, characteristic of hydrolyzed AEAPS, shown in the ^{29}Si CPMAS NMR spectrum (Figure 4.14). Note that the ^{31}P CPMAS NMR

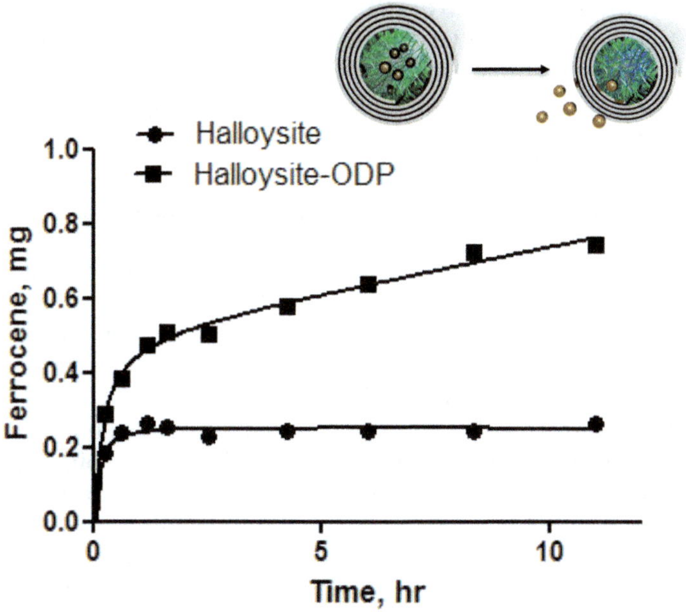

Figure 4.13 Release profiles of ferrocene from halloysite (•) and halloysite-ODP (■). Adapted with permission from Yah *et al.*, *J. Am. Chem. Soc.* 2012, **134**, 1853–1859. Copyright 2012 American Chemical Society.[34]

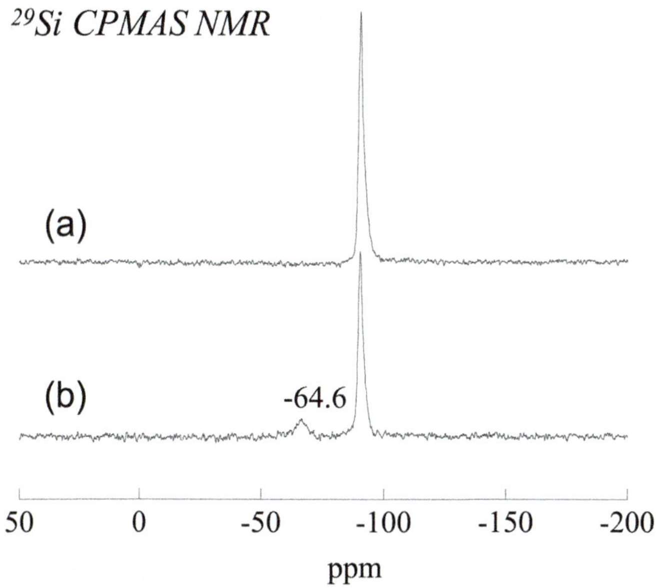

Figure 4.14 ^{29}Si CPMAS NMR spectra of the halloysite-ODP (a), and halloysite-ODP reacted successively with AEAPS (b).

spectrum shown is identical to the ^{31}P CPMAS NMR of halloysite-ODP shown in Figure 4.11, implying that ODP grafted in the lumen of halloysite remains intact after AEAPS treatment. Modification of the silica rich external surface of halloysite by silyl groups demonstrates the possibility of bifunctionalization of halloysite, as illustrated in Figure 4.8.

4.3.4 Polymer Growth from the Lumen of Halloysite

A radical polymerization initiator with a catecholic anchor group that can modify the lumen of halloysite was designed.[35] The catechol group is a new addition to the toolbox of selective modification schemes for nano-clays with silica–metal oxide surfaces, inspired by strong, wet adhesion systems developed by nature. 3,4-Dihydroxyphenylalanine (dopamine, DOPA) is found in mussel adhesive proteins that contribute to the high adhesion ability of the byssus to a wide variety of marine substrates. Though the binding mechanism of catechol is still under extensive debate, the chemisorption to metal oxide surfaces was proposed to comprise a monodentate–bidentate bond. The binding of catechol to silica is noncovalent; its binding energy is estimated to be 14 kcal mol^{-1}, which is smaller than that for covalent binding with a metal oxide.

The modified halloysite with a dopamine bearing molecule with an ATRP-initiating bromo group was further used as an initiator for surface-initiated ATRP (SI-ATRP) for selective grafting of a PMMA brush layer at the lumen (Figure 4.15).[35] The lumen of the halloysite was modified with 2-bromo-*N*-[2-(3,4-dihydroxyphenyl)ethyl]-isobutyryl amide (DOPA-initiator), which serves as an ATRP initiating site as well. The attachment of the DOPA-initiator to the halloysite surfaces was confirmed from the characteristic C–H and carbonyl stretching vibrations at 2930 cm^{-1} and 1722 cm^{-1} in

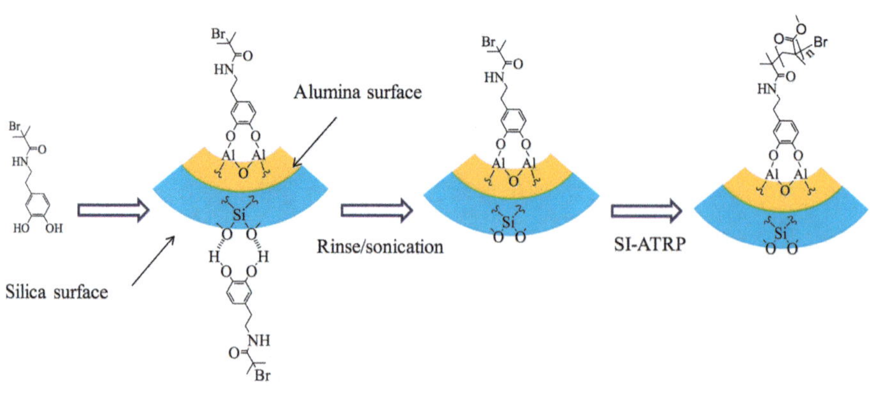

Figure 4.15 Selective modification of halloysite by a biomimetic DOPA bearing initiator and subsequent SI-ATRP. Adapted with permission from Yah *et al.*, *J. Am. Chem. Soc.* 2012, **134**, 12134–12137. Copyright 2012 American Chemical Society.[35]

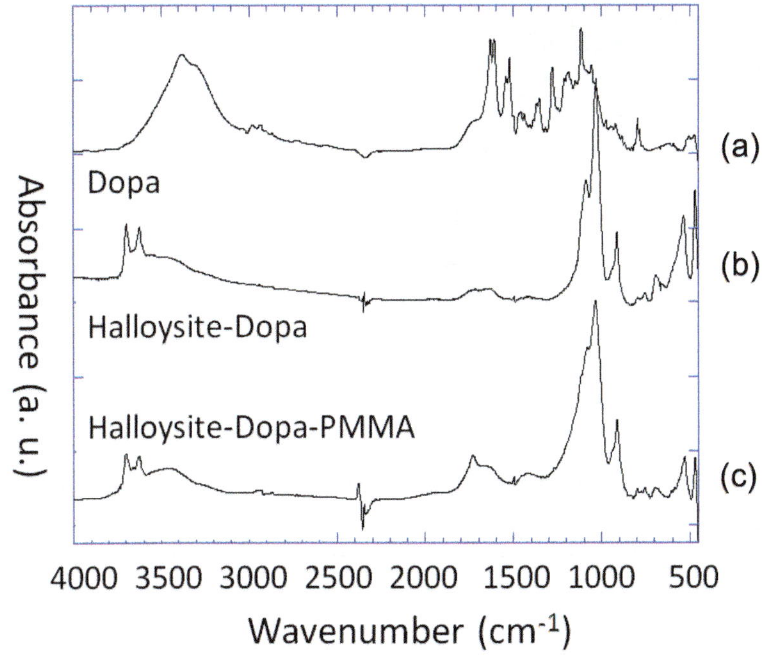

Figure 4.16 IR spectra of (a) DOPA-initiator, (b) DOPA-initiator modified halloysite, and (c) PMMA grafted halloysite.

the FT-IR spectrum, respectively (Figure 4.16). The amount of the DOPA-initiator attached to the halloysite surfaces was determined by TGA analysis to be 1.1% relative weight loss. The selective adsorption nature of the DOPA-initiator on metal oxide over silica was evidenced from solid-state ^{13}C NMR of silica and alumina nanoparticles before and after treatment. For alumina particles modified with DOPA-initiator, the ^{13}C NMR spectra displayed two characteristic resonances. The broad resonances at 20–50 ppm (aliphatic) and 181.1 ppm were attributed to the grafted DOPA species through formation of alumina–catechol covalent bonds. However, when silica particles were treated with DOPA-initiator, silica–catechol species were hardly detected. This result may likely be ascribed to weak a silica–catechol bond, which can be readily removed by a sonication process. Alumina–catechol bonds are much more stable than silica–catechol bonds, allowing the selective surface modification of aluminol groups of the halloysite lumen by catechol groups under aqueous conditions.

The growth of a PMMA brush from the tube lumen can be ascertained from the TEM images. Figure 4.17 shows the TEM images of halloysite (a) and PMMA grafted halloysite (b and c). For the purpose of easy visualization of the polymer portion of the sample, halloysite underwent a prolonged polymerization time. After being polymerized for 24 h, the interior cavity was completely filled with a PMMA brush. The globular structures, which

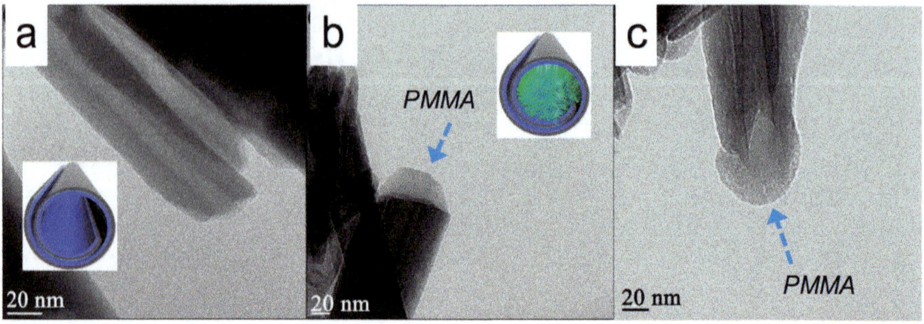

Figure 4.17 TEM images of halloysite (a) and after SI-ATRP (b and c). Adapted with permission from Yah *et al.*, *J. Am. Chem. Soc.* 2012, **134**, 12134–12137. Copyright 2012 American Chemical Society.[35]

presumably correspond to the polymer portion and protruded from the opening of the tube, were clearly observed. TEM images provide further evidence that PMMA brushes grow through selectively from grafted DOPA in the halloysite lumen.

4.3.5 Intelligent Polymer Nanocomposites with Flame Retardant Loaded Halloysite

Halloysite is expected to be used as a nano-container for molecular adsorption and molecular storage applications, in addition to serving as a reinforcing filler for functional nanocomposite applications. Because the halloysite surface is covered with hydrophilic silanol groups, surface modification is necessary to incorporate it into hydrophobic polymer matrices. Surface modification of halloysite with organosilane compounds improves the dispersion state and interfacial interaction between the halloysite and polymer resin.

Segmented polyurethanes (PU) are multi-block polymers that have been widely used due to their excellent biocompatibility, abrasion resistance, and elastic mechanical properties. Various nanofillers, such as minerals, sheets, or fibers, have been incorporated into PU to enhance the mechanical performance and thermal stability. Organic phosphorus based flame retardants are known as effective and environmentally benign flame retardants for polymers, including flexible PU foams. The char forming activity is supported by foaming or intumescence, which forms a protective top layer on the polymer surface. However, corrosion induced by phosphorus flame retardants remains the challenge for applications.

PU nanocomposites with surface modified halloysite were investigated.[36] The bifunctionally modified halloysites, which consist of a hydrophobized exterior and lumen loaded with a hydrophobic flame retardant, bisphenol-A bis(diphenyl phosphate) (BDP), (A-HNTs-ODP-BDP) were prepared as shown in Figure 4.18.[37] Pellethane 2363-80AE, which consists of poly(tetramethylene glycol) (PTMG), 4,4′-methylenebis(phenylisocyanate) (MDI), and

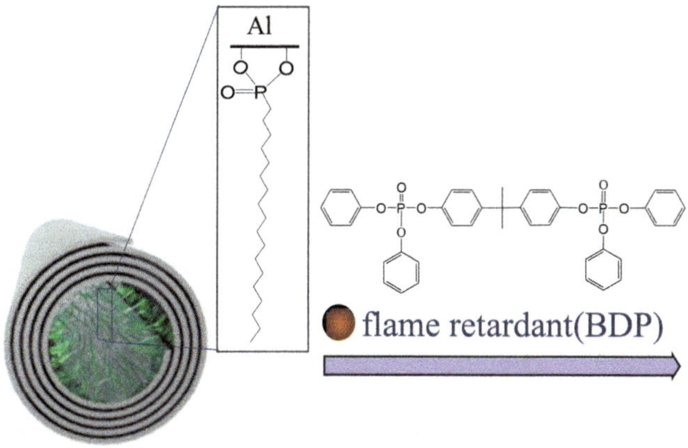

Figure 4.18 Preparation of bifunctionally modified halloysite that consists of a hydrophobized surface and lumen loaded with a hydrophobic flame retardant, bisphenol-A bis(diphenyl phosphate) (BDP), (A-HNTs-ODP-BDP). Adapted with permission from Jing *et al.*, *Chem. Lett.* 2013, **42**, 121–123. Copyright 2013 The Chemical Society of Japan.[37]

1,4-butanediol (BDO), was used as a matrix PU. A-HNTs-ODP-BDP and PU were blended by solution blending, and the PU–A-HNTs-ODP-BDP nanocomposite film was prepared by a bar coating process.

The strength of PU depends on the phase morphology of the soft and hard segments. For thermo-plastic segmented PU, the strength mainly depends on the hard segment structure and content. The introduction of halloysite as the reinforcing filler in PU led to a remarkable improvement of tensile strength, elongation at break, and Young's modulus of the nanocomposites, as listed in Table 4.4. The tensile strength of the PU nanocomposites was increased from 6.7 MPa for neat PU to 9.7 MPa for PU–neat halloysite (HNTs)_0.5 wt%, to 25.5 MPa for PU–AEAPTMS modified halloysite (A-HNTs)_0.5 wt% and to 28.4 MPa for PU–*n*-octadecyltrimethoxysilane modified halloysite (O-HNTs)_0.5 wt%. The Young's modulus increased with halloysite loading, reaching 39.5 MPa for PU–O-HNTs_0.5 wt%, four times higher than that of neat PU. The tensile strength in the 5 wt% halloysite content nanocomposites was lower than that in the 0.5 wt% halloysite content nanocomposites, although the Young's modulus in the 5 wt% halloysite content nanocomposites was larger than that in the 0.5 wt% halloysite content nanocomposites. These results show that the introduction of HNTs has remarkably enhanced the tensile modulus and toughness of the PU nanocomposites; especially for PU–O-HNTs_0.5 wt% nanocomposites to exhibit a higher elongation at break combined with much higher tensile strength and modulus than other nanocomposites. However, excess halloysite loading results in a reduction mechanical properties. The halloysite loading increases the pseudo cross-linking density, leading to an effective continuum network. As a result, the modulus and strength of the nanocomposites are increased.

Table 4.4 Mechanical and thermal properties of PU–A-HNTs-ODP-BDP hybrid.

Sample type	Young's modulus/MPa	Elongation at break/%	Tensile strength/MPa	$T_{-40\,\mathrm{wt\%}}$ [a] K^{-1}
Neat PU	10.0 ± 0.1	720 ± 10	6.7 ± 0.1	631
PU–A-HNTs_5 wt%	35.5 ± 0.9	590 ± 40	13.7 ± 0.2	640
PU–A-HNTs-ODP-BDP_5 wt%	56.3 ± 0.2	702 ± 20	14.2 ± 0.3	681

[a]Temperature at 40% weight loss in TGA measurement.

Strong interaction of modified halloysite with hard domains and increased surface area allow the ductility to be improved. At higher halloysite contents, the halloysites are supposed to aggregate, leading to the formation of large aggregates. The coarse aggregates cause the stiffness distribution to induce a local stress concentration in the PU–halloysite hybrid films. The tensile mechanical characteristics of the high halloysite content composite films, less fracture stress and elongation at break, would be attributed to defect generation along with the failure progress based on the stress concentration.

Thermal decomposition behavior of PU, PU–A-HNTs_5 wt%, and PU–A-HNTs-ODP-BDP_5 wt% nanocomposites under nitrogen atmosphere was analyzed by TGA. The temperature at 40% weight loss ($T_{-40\,\mathrm{wt\%}}$) for neat PU is 631 K, whereas $T_{-40\,\mathrm{wt\%}}$ for the nanocomposite filled with 5 wt% A-HNTs is 640 K. The $T_{-40\,\mathrm{wt\%}}$ further increased to 681 K for the PU–A-HNTs-ODP-BDP_5 wt% nanocomposite, which is 41 K higher than that of PU–A-HNTs_5 wt%, indicating an effective thermal stability enhancement and degradation inhibition by BDP loading. Neat BDP starts to degrade from 510 K; the main weight loss of BDP reflects the generation of metaphosphoric acid and dehydration of BDP. In the PU–A-HNTs-ODP-BDP_5 wt% nanocomposite system, BDP would be released from the nanotube lumen during the heating process, and effectively work after 600 K, leading to a generation of metaphosphoric acid and dehydration by following a type of solid-phase mechanism to provide the cross-linked intumescence layer supporting char formation and less release of flammable gases. Furthermore, a kind of synergy effect by the combination of BDP and halloysite would exist to suppress thermal degradation. The mechanical properties could also be improved without sacrificing thermal stability (Table 4.4). Therefore, the A-HNTs-ODP-BDP complex remains the reinforcing effect, while it is effective for thermal stability enhancement due to the release of BDP on heating.

4.4 Conclusion

Surface functionalization of tubular nano-clays of imogolite and halloysite using the selective binding of organophosphonic acids and organosilane compounds, and the use of the surface modified nanotubes in polymer hybrids were summarized. Surface modification of imogolite with alkyl phosphonic acid salt through the specific interaction of phosphonic acid and the exterior alumina sites of imogolite was presented. SI-ATRP was performed with the selectively adsorbed phosphonic acid functionalized ATRP-initiator to prepare polymer brushes on the imogolite surface. Selective modification of halloysite nanotube exterior and inner surfaces was demonstrated. Aqueous phosphonic acid binds to alumina sites at the tube lumen to make the lumen hydrophobic. Subsequent modification with organosilane affords bifunctionalized halloysites with both the lumen and exterior surface modified. Loading of hydrophobic organic compound to the modified lumen was confirmed. SI-ATRP was performed through the selectively adsorbed DOPA-functionalized ATRP-initiator to prepare polymer brushes on the nanotube lumen. Also, surface modified halloysite was applied for preparation of novel intelligent polyurethane nanocomposites with improved thermal stability and mechanical properties.

Acknowledgements

We gratefully acknowledge the financial support of a Grant-in-Aid for Scientific Research (A) (No. 19205031) and (No: 26248053) from Japan Society for the Promotion of Science.

References

1. C. M. Lieber, *Solid State Commun.*, 1998, **107**, 607.
2. S. Iijima, *Nature*, 1991, **354**, 56.
3. N. Yoshinaga and S. Aomine, *Soil Sci. Plant Nutr.*, 1962, **8**, 22.
4. P. D. G. Cradwick, V. C. Farmer, J. D. Russell, C. R. Masson, K. Wada and N. Yoshinaga, *Nature (London), Phys. Sci.*, 1972, **240**, 187.
5. C. Su, J. B. Harsh and P. M. Bertsch, *Clays Clay Miner.*, 1992, **40**, 280.
6. S. Imamura, T. Kokubu, T. Yamashita, Y. Okamoto, K. Kajiwara and H. Kanai, *Catalyst*, 1996, **160**, 137.
7. F. Ohashi, S. Tomura, K. Akaku, S. Hayashi and S.-I. Wada, *J. Mater. Sci.*, 2004, **39**, 1799.
8. W. Ma, W.-O. Yah, H. Otsuka and A. Takahara, *J. Mater. Chem.*, 2012, **22**, 11887–11892.
9. K. Yamamoto, H. Otsuka, S.-I. Wada, D.-W. Sohn and A. Takahara, *Polymer*, 2005, **46**, 12386–12392.
10. R. H. Baughman, A. A. Zakhidov and W. A. D. Heer, *Science*, 2002, **297**, 787.
11. D. Sukanta, P. E. Lyons, S. Sorel, E. M. Doherty, P. J. King, W. J. Blau, P. N. Nirmalraj, J. J. Boland, V. Scardaci, J. Joimel and J. N. Coleman, *ACS Nano*, 2009, **3**, 714.
12. S. M. Khaled, R. Sui, P. A. Charpentier and A. S. Rizkall, *Langmuir*, 2007, **23**, 3988.
13. A. A. Koval'chuk, A. N. Shchegolikhin, V. G. Shevchenko, P. M. Nedorezova, A. N. Klyamkina and A. M. Aladyshev, *Macromolecules*, 2008, **41**, 3149.

14. M. K. Singh, T. Shokuhfar, J. J. A. Gracio, A. C. M. Sousa, J. M. D. F. Fereira, H. Garmestani and S. Ahzi, *Adv. Funct. Mater.*, 2008, **18**, 694.
15. P. Berthier, *Ann. Chim. Phys.*, 1826, **32**, 332.
16. R. Price, B. Gaber and Y. Lvov, *J. Microencapsulation*, 2001, **18**, 713.
17. Y. M. Lvov, D. G. Shchukin, H. Mohwald and R. R. Price, *ACS Nano*, 2008, **2**, 814.
18. S. Baral, S. Brandow and B. P. Gaber, *Chem. Mater.*, 1993, **5**, 1227.
19. V. Vergaro, E. Abdullayev, Y. M. Lvov, R. Cingolani, R. Rinaldi and S. Leporatti, *Biomacromolecules*, 2010, **11**, 820.
20. L. Guimarães, A. N. Enyashin, J. Frenzel, T. Heine, H. A. Duarte and G. Seifert, *ACS Nano*, 2007, **1**, 362.
21. T. Bates, F. Hildebrand and A. Swineford, *Am. Mineral.*, 1950, **35**, 463.
22. G. Tari, I. Bobos, C. S. F. Gomes and J. M. F. Ferreira, *J. Colloid Interface Sci.*, 1999, **210**, 360.
23. S. Krivovichev, *Minerals and Advanced Materials I*, Springer-Verlag, Berlin, Heidelberg, 2008.
24. S. Aomine and N. Yoshinaga, *Soil Sci.*, 1955, **79**, 349.
25. K. Wada, *Am. Mineral.*, 1967, **52**, 690.
26. R. Hofer, M. Textor and N. D. Spencer, *Langmuir*, 2001, **17**, 4014–4020.
27. P. H. Mutin, G. Guerrero and A. Vioux, *J. Mater. Chem.*, 2005, **15**, 3761–3768.
28. R. Matsuno, K. Yamamoto, H. Otsuka and A. Takahara, *Chem. Mater.*, 2003, **15**, 3–5.
29. R. Matsuno, K. Yamamoto, H. Otsuka and A. Takahara, *Macromolecules*, 2004, **37**, 2203–2209.
30. K. Yamamoto, H. Otsuka, S.-I. Wada and A. Takahara, *Chem. Lett.*, 2001, **30**(11), 1162–1163.
31. W. Ma, J. Kim, H. Otsuka and A. Takahara, *Chem. Lett.*, 2011, **40**, 159–161.
32. W. Ma, H. Otsuka and A. Takahara, *Chem. Commun.*, 2011, **47**, 5813–5815.
33. W. Ma, H. Otsuka and A. Takahara, *Polymer*, 2011, **52**, 5543–5550.
34. W.-O. Yah, A. Takahara and Y. M. Lvov, *J. Am. Chem. Soc.*, 2012, **134**, 1853–1859.
35. W.-O. Yah, H. Xu, H. Soejima, W. Ma, Y. Lvov and A. Takahara, *J. Am. Chem. Soc.*, 2012, **134**, 12134–12137.
36. H. Jing, Y. Higaki, T. Ishikawa, K. L. White, H. Otsuka and A. Takahara, *Sci. Adv. Mater.*, 2015, **7**, 974–980.
37. H. Jing, Y. Higaki, W. Ma, H. Hui, W.-O. Yah, H. Otsuka, Y. Lvov and A. Takahara, *Chem. Lett.*, 2013, **42**, 121–123.

CHAPTER 5

Surface Modification of Halloysite

JING HUANG[a], ZHENGHAI TANG[a] AND BAOCHUN GUO*[a]

[a]Department of Polymer Materials and Engineering, South China University of Technology, Guangzhou 510540, China
*E-mail: psbcguo@scut.edu.cn

5.1 Introduction

Surface modification of halloysite nanotubes (HNTs) is the essential consideration when fabricating HNT-based composites. For example, the mechanical properties of polymer–HNTs composites are determined by the dispersion and interfacial interaction of the composites, which are governed by the surface properties of the modified HNTs. Additionally, surface modification alters the hydrophilicity of HNTs and permits the loading of hydrophobic substances in the lumen and the subsequent sustainable release. Although there arc many choices for surface modification of HNTs, selecting the modification method is dependent on the target application and the nature of the matrix. Generally, there are three expected targets for the surface modification of HNTs.

(i) To improve the interfacial compatibility between HNTs and polymer matrices or the dispersion of HNTs in the organic solvents.
(ii) To immobilize functional particles or molecules for different applications, such as catalysis or adsorption (removal) of pollutants such as metal ions and dyes.

RSC Smart Materials No. 22
Functional Polymer Composites with Nanoclays
Edited by Yuri Lvov, Baochun Guo and Rawil F. Fakhrullin

Published by the Royal Society of Chemistry, www.rsc.org

(iii) To load hydrophobic molecules or enlarge the lumen volume/surface area for better loading efficiency.

In this chapter, the authors review the recent progress in modification methods for HNTs. Regarding modification mechanisms, silanization, grafting, addition reactions, non-covalent interactions, and selective modification of the inner/outer side of the lumen are summarized and discussed. Finally, concluding remarks are made.

5.2 Structural Characteristics and Surface Chemistry of HNTs

5.2.1 Composition and Structure of HNTs

Halloysite was first reported by Berthier in 1826 as a dioctahedral 1:1 clay mineral of the kaolin group.[1] In many places in China, such as Hubei, Hunan, Henan, and Sichuan, there exist large-scale deposits of high-purity halloysite, which are also found abundantly in many other countries, for example France and New Zealand.[2] Halloysite collected from different sites can have widely varying morphologies, such as spheroidals, short tubes, and long tubes.[3] Pure halloysite is a white mineral; however, certain deposits are slightly red as a result of impurities from ferric (Fe^{3+}) ions. Generally, the stone-like raw halloysite is easily ground into powder.[3,4]

As a type of naturally occurring clay with nanotubular structures, HNTs have the idealized molecular formula of $Al_2Si_2O_5(OH)_4{\cdot}nH_2O$, which is similar to kaolinite except there is an additional water monolayer between the adjacent clay layers. When $n = 2$, its layer periodicity is 10 Å, and thus the mineral is called halloysite-10 Å. When heating halloysite-10 Å to 120 °C, the water between the layers will be lost, which irreversibly changes it into halloysite-7 Å with $n = 0$. Ideally, an HNT crystal consists of a layered structure that contains octahedrally coordinated aluminium (Al^{3+}) and tetrahedrally coordinated silicon (Si^{4+}) in a 1:1 stoichiometric ratio. In reality, chemical analyses of many HNTs reveal that trace metal ions such as ferric (Fe^{3+}), chromium (Cr^{3+}), and titanium (Ti^{4+}), may be substituted for Al^{3+}, and Al^{3+} may be substituted for Si^{4+}. Morphologically, HNTs are multi-walled inorganic nanotubes. The inner and outer diameters of the tubes range from 10 to 40 nm and 40 to 70 nm, respectively.[3–8] The structure of HNTs is schematically illustrated in Figure 5.1.

Different from kaolinite and montmorillonite, most of the aluminols are located in the interior of the HNTs, whereas the exterior of HNTs consists of primary siloxanes and a few silanols/aluminols that are exposed at the edges and defects of the sheet. Consequently, the density of the surface hydroxyl groups of HNTs is relatively low. HNTs exhibit many interesting characteristics, such as a tubular nanostructure with a large aspect ratio, natural

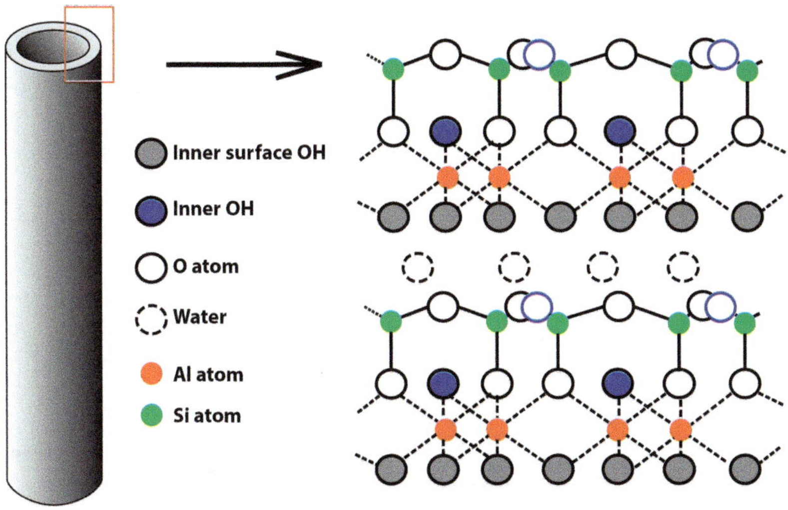

Figure 5.1 Schematic illustration of the crystalline structure of HNTs.

availability, rich functionality, good biocompatibility, and high mechanical strength; therefore, HNTs are found in various applications, such as ceramic raw materials, nanocontainers for active compounds, catalysts, adsorbents, and polymer fillers.[2,4,8] However, the direct application of pure HNTs in the aforementioned areas is hindered due to their highly hydrophilic surfaces and limited number of active sites, which make their surface modification essential.

5.2.2 General Considerations on the Modification Chemistry of HNTs

The hydroxyl groups located on the inner surfaces and at the edges or defects of the sheets provide reactive sites to covalently attach chemical species. Therefore, a considerable amount of research has focused on the covalent functionalization of HNTs, such as silanization and addition reactions with (iso)cyanates. However, covalent modification is essentially hindered by the limited number of hydroxyl groups on the surface of HNTs. Therefore, more and more attention has been given to non-covalent modification. Moreover, the polarity of hydroxyl groups, the existence of few transition metal elements, and the charges on the surface offer favourable conditions for non-covalent functionalization. In addition, etching the lumen of HNTs with an inorganic acid or an alkali is a simple but extremely effective method to enlarge the volume and surface area of the lumen. Specially, selective modification of the inner lumen can also be obtained according to the different surface properties of the HNTs. The general chemistry modification of HNTs is categorized in Table 5.1.

Table 5.1 Categorized surface modification methods for HNTs.

Mechanism	Modifiers	Typical interaction scheme	Ref.
Silanization	Silanes	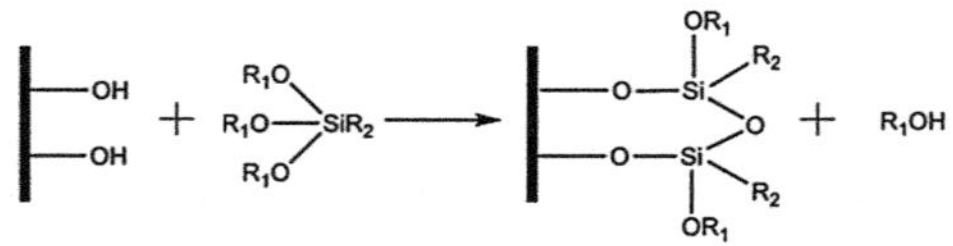	10–47
Addition	(Iso)Cyanate esters or isocyanates	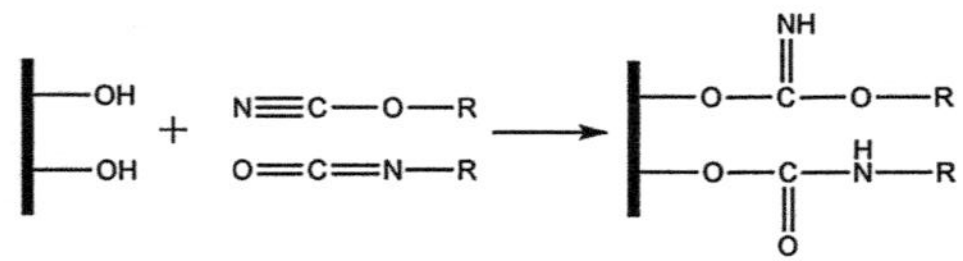	59–61
Esterification	Phosphoric acid or lactic acid	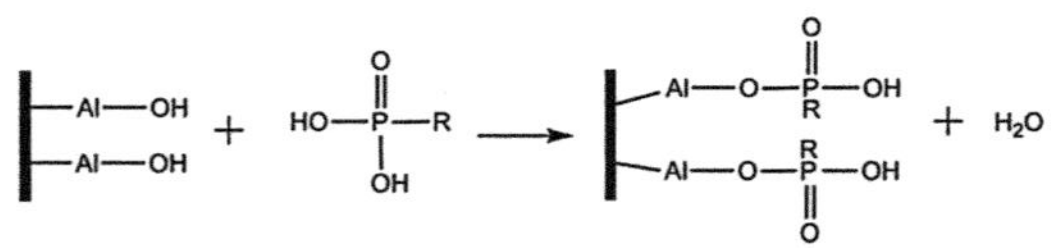	66 and 105
Hydrogen bonding	Hydrogen bonding donors or acceptors	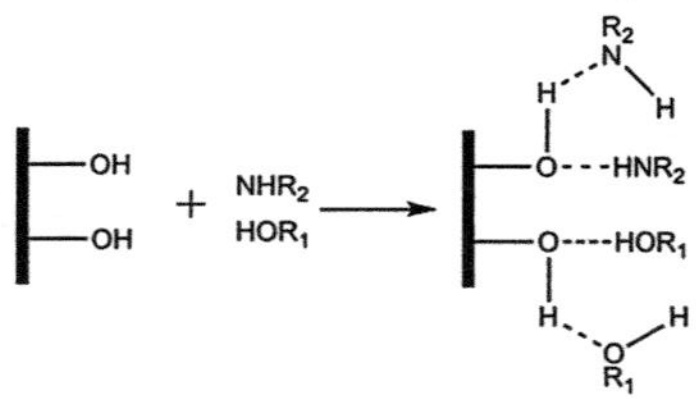	63–68,76–84

Charge transferring	Electron donors	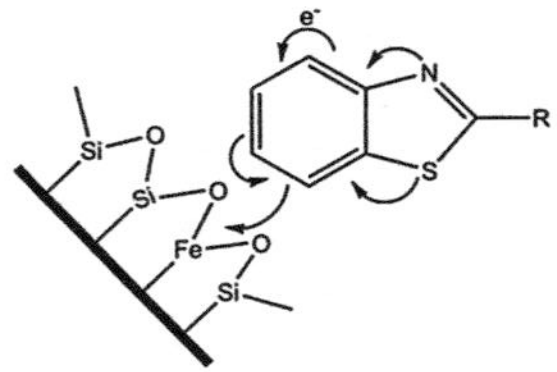	73–75
Electrostatic	Surfactants or polyelectrolyte	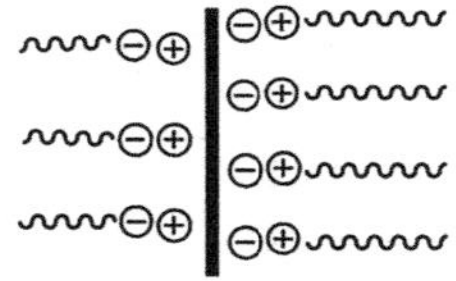	70–72,85–94
Bidentate coordination	Catechol	inner/outer 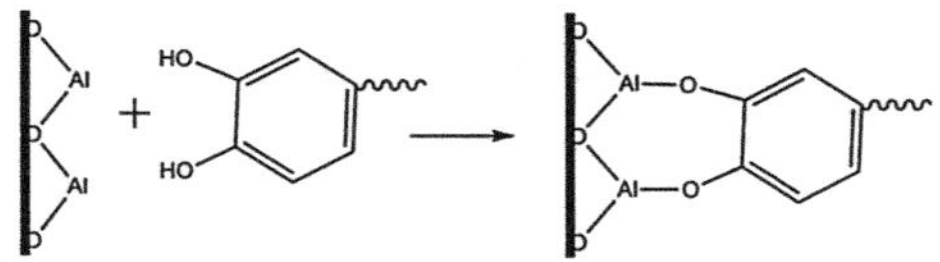	107
Etching	Alkali or acids	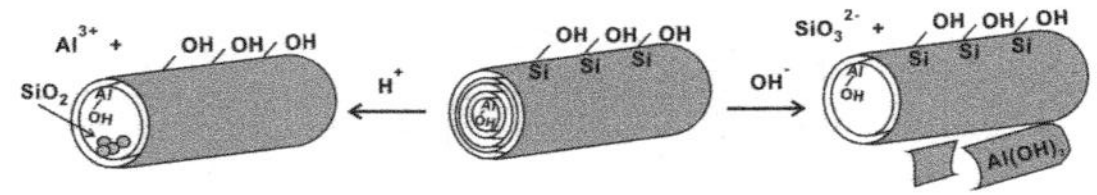	101–104

5.3 Modification Chemistry and Techniques of HNTs

5.3.1 Surface Silanization

Grafting silanes *via* condensation between the hydrolysed silanes and the surface hydroxyl groups of HNTs, including the aluminol groups and silanol groups, is the most common covalent modification for HNTs. The modification can be performed both *via* an *in situ* process and pretreatment in solution mediums such as toluene and water–alcohol mixtures.

5.3.1.1 Reactivity of Silanes Towards HNTs

Most silicates, including HNTs, are chemically reactive towards various silanes. Although the covalent linkage between surface silanols on silicate and those hydrolysed silanes have been confirmed, the actual mechanisms and processes are extremely complicated and influenced by a number of factors, such as the reactant (silicate and silane) structure and reaction conditions, *i.e.*, temperature, reaction medium, and catalysis. Before a reaction begins, HNTs sometimes require treatment by aqueous hydrogen peroxide to remove inorganic salt impurities on the surface.[9]

A reaction mechanism was suggested by Yuan *et al.* The authors modified HNTs by grafting γ-aminopropyltriethoxysilane (APTES) and found that the modification mechanism included not only the direct grafting of APTES onto the hydroxyl groups of the internal walls, edges, and external surfaces of the nanotubes but also oligomerized APTES *via* condensing with the already grafted APTES to form a cross-linked structure. An analysis of the APTES-treated HNTs (APTES-HNTs) showed that the HNTs preserve their characteristic interlayer spacing of 7 Å, which indicates that silane grafting did not occur at the interlayer spacing. Moreover, the modification mechanism was strongly affected by thermal and evacuation pretreatments. Thermal pretreatment affected the modification by controlling the water content, which further determined the amount of oligomerization, where as evacuation pretreatment could dramatically enhance the loading of hydrolysed APTES into the lumen of the HNTs and increase the grafting ratio.[10]

While drying the HNTs can reduce the amount of grafted APTES, Vahedi and Pasbakhsh showed that a low concentration of sodium hydroxide (NaOH) treatment of HNTs can considerably increase the amount of grafted APTES on the surface of HNTs due to the increase in the amount of available –OH groups on the surface of the HNTs.[11] The higher density of external silanol groups of hydroxylated HNTs (h-HNTs) by NaOH treatment compared with that of pure HNTs (p-HNTs) was explicitly confirmed by Zeng *et al.* (Figure 5.2).[12] Actually, it has been reported that acid or alkali pretreatments are useful for introducing more active sites and improving the reactivity between the nanolayers and coupling agents.[13–15]

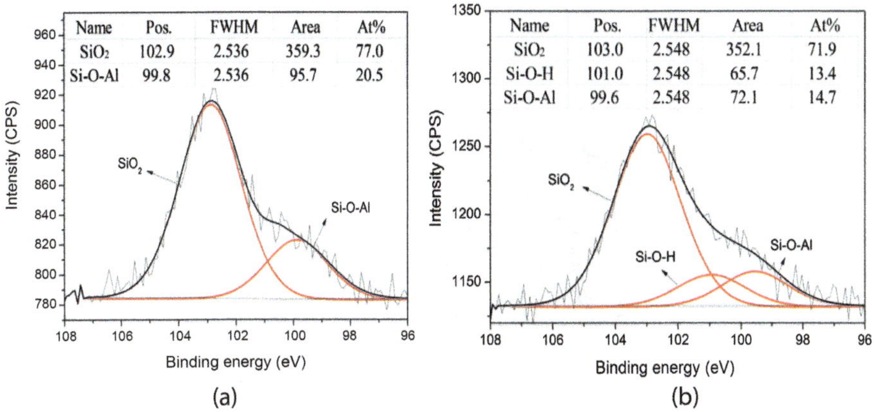

Figure 5.2 High-resolution XPS spectra of Si 2p of (a) p-HNTs and (b) h-HNTs. (Reproduced from *Polymer*, **55**, S. Zeng *et al.*, Facile hydroxylation of halloysite nanotubes for epoxy nanocomposite applications, 6519–6528, Copyright (2014) with permission from Elsevier.[12])

5.3.1.2 Silanization of HNTs in Polymer-Based Composites

Modification by silanes can efficiently improve HNT dispersion and interfacial interaction between HNTs and the polymer matrix, and therefore enhance the performance of the composites. Various silanes have been grafted onto HNTs when preparing polymer composites because the structures of silanes can be designed according to the nature and reactivity of the polymer matrix to optimize target effects. In general, the other side end group of silanes can be divided into different categories, such as along-chain alkyl[16,17] to improve the hydrophobicity of HNTs; an unsaturated double bond[18–20] to bond with the polymer covalently; an amino group[16,17,21,22] to construct a hydrogen bond or covalent bond with the polymer; an epoxy group[23] to react with epoxy resin; and a polysulfide bond ($C–S_x–C$)[24,25] to act as a sulfur-donor and bond with rubber. The following examples give more detailed information regarding the selection of silanes to fabricate polymer–HNTs composites.

Guo *et al.* treated HNTs with the silane 3-(trimethoxysilyl)propyl methacrylate (MAPTS) to improve both the dispersion of HNTs and the interaction between HNTs and polyamide 6 (PA6) chains. Both excess silanes and pre-hydrolysing in an ethanol (95%) solution with a pH of 4 were used to ensure the grafting ratio of silanes. The graft weight was calculated as 4.9 wt% relative to the weight of HNTs (Figure 5.3). The hydrophobicity of modified HNTs (m-HNTs) was dramatically improved according to the extraction experiment, which can be distinctly observed in the inset picture. Unsurprisingly, the nanocomposites consisting of PA6 and m-HNTs exhibited significantly improved mechanical properties and a heat distortion temperature, which were attributed to the covalent interfacial bonding and the excellent dispersion state of the m-HNTs.[26]

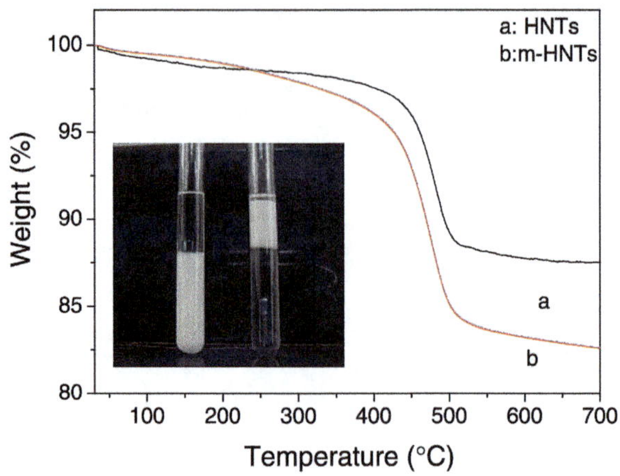

Figure 5.3 Thermal gravimetric analyses on (a) HNTs and (b) m-HNTs; the inset is the extraction of HNTs and m-HNTs in toluene–water. (Reproduced from *Polym. J.*, **41**, Guo *et al.*, Structure and Performance of Polyamide 6/Halloysite Nanotubes Nanocomposites, 835–842, Copyright (2009) with permission from Macmillan Publishing Ltd: *Nature*.[26])

The effects of silanes on the properties of poly(hydroxybutyrate-*co*-hydroxyvalerate) (PHBV)–HNTs composites were revealed by Carli *et al.* Four silanes were used: APTES and 3-methylaminopropyltrimethoxysilane (APTMS) that contained amino groups, which were able to form hydrogen bonds with the carbonyl group on PHBV chains; 3-glycidyloxypropyltrimethoxysilane (GOPTMS) with an ether functional group, which subsequently gave a polar character that could interact with the polymer matrix through secondary interactions; and octyltriethoxysilane (OTES) with long alkyl chains that could interact with PHBV through van der Waals forces. The characterizations confirmed the grafting of the silanes on the HNTs. However, the authors found that the HNTs modified with aminosilanes did not exhibit the expected reinforcing effect, which was attributed to the catalysed degradation of PHBV in the presence of aminosilanes. For the PHBV composites with m-HNTs using GOPTMS and OTES, the secondary interactions prevented such an adverse effect. Despite the poor dispersion of the m-HNTs, the m-HNTs improved the mechanical performance of the PHBV. Consequently, in summary, the mechanical properties of the nanocomposite may be controlled by the interaction between the functional group of the silane and the polymer.[16]

5.3.1.3 Silanization of HNTs in HNT-Based Functional Composites

When treated with silanes containing special functional groups, HNTs can be better applied in functional areas. For example, aminosilane and thiolsilane functionalized HNTs have a strong coordination interaction with metal ions such as copper (Cu^{2+}), mercury (Hg^{2+}), silver (Ag^{+}), gold (Au^{2+}),

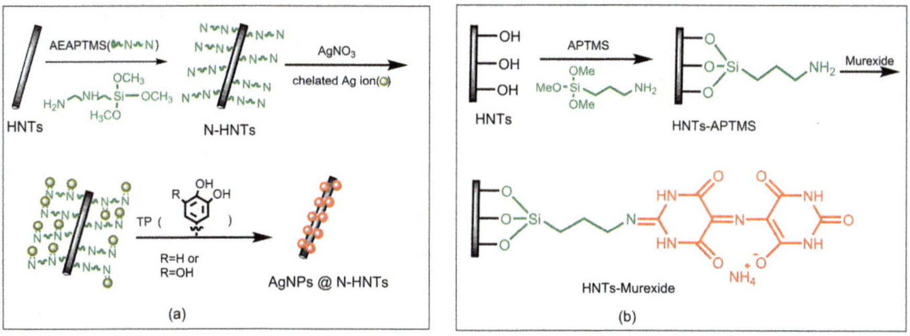

Figure 5.4 Schematic illustration of the synthesis of HNT supported AgNPs (a) and HNTs-murexide (b).

platinum (Pt^{2+}), *etc.* On the one hand, the functionalized HNTs adsorb heavy metal ions directly and act as heavy metal sorbents in an aqueous solution;[27,28] on the other hand, the functionalized HNTs can immobilize precious metal nanoparticles by an additional *in situ* reduction of these ions (Figure 5.4a) applied as highly effective catalysts,[29–34] antibacterial materials,[35] plasmonic materials,[36] and surface-enhanced Raman scattering substrates.[37] Moreover, APTES-HNTs have been found to substantially affect the loading and subsequent release of a model dye molecule. The functional aminopropyl groups make the pH value act as an external trigger to control the loading of the guest on HNTs and the subsequent controlled release.[38] Long-chain alkyl-modified HNTs formed pincushion agglomerates on the surface of the liquid droplets, which create a superhydrophobic surface similar to that of the plant gall surface prepared by aphids. Moreover, the liquid marbles exhibited high mechanical strength upon impact without the use of low surface energy fluoroalkyl or fluorine-modified materials.[39]

However, a few functional applications have to rely on silanes as reaction intermediates. Massaro *et al.* first modified HNTs with 3-mercaptopropyl trimethoxysilane (MPTMS) and then anchored vinylimidazolium ionic liquids (ILs) by a thiol–ene reaction under microwave irradiation. The final functionalized HNTs were used as support for palladium (Pd) particles as a catalyst for the Suzuki reaction with the catalytic efficiency higher than conventionally used supports.[40,41] Moreover, HNTs that attached with different hydrophobic ILs were used as a drug delivery system for cardanol.[42] Another example reveals that murexide-modified HNTs were obtained by APTMS as a reaction intermediate (Figure 5.4b). Then, the functionalized HNTs could be applied as a selective sorbent in solid phase extraction of trace Pd in solution samples.[43] Similarly, 2-hydroxybenzoic acid[44] and *N*-2-pyridylsuccinamic acid[45] were developed to modify HNTs for solid phase extraction of trace Fe and lead (Pb), respectively. If APTES-HNTs react with succinic anhydride in dimethyl formamide, modified HNTs with carboxylic acid groups (HNTs-COOH) will be generated.[46] By adjusting the pH value of the solution, the aggregation–dispersion properties of HNTs-COOH can be controlled, which are extremely important for applications that use HNTs, such as nanocomposite materials, catalysts, capillaries, drug delivery systems, and cosmetics. In addition, APTES-HNTs would be converted into a positively charged surface under acidic conditions due to the existence of amine-tailed short organic chains. A combination then occurs that is driven by the electrostatic force between the negative GO sheets and positive APTES-HNTs. The HNTs@rGO composite was fabricated after the reduction of GO. Such composites were investigated as an adsorbent and electrode material, which highlights their potential applications in waste water treatment and energy storage issues.[47]

5.3.2 Grafting Polymers onto HNTs

HNTs grafting silanes have wide applications, as mentioned above. Nevertheless, surface grafting polymers are occasionally better choices to apply HNTs in polymer-based composites. Surface grafting polymers enable a covalent linkage between HNTs and polymer chains, which improves the compatibility between HNTs and the polymer matrix. Furthermore, it is a convenient yet practical way to prepare polymer–HNTs composites that have strong interfaces. Polymers can be grafted onto HNTs in two ways. Directly grafting the polymers onto the surface of HNTs is typically referred to as the "grafting to" method, whereas pre-anchoring the initiators or polymerizable monomers (vinyl or condensation types) onto HNTs and then polymerizing *in situ* is referred to as the "grafting from" method. These processes are generally performed in solutions. However, the reactions between polymers or monomers and HNTs are hindered by the limited number of reactive groups on HNTs; as a result, silanes are often used to introduce various reactive groups or initiators.

5.3.2.1 Grafting To

Many types of polymers can be grafted onto HNTs *via* the grafting methods to optimize their application targets. To increase the compatibility between polypropylene (PP) and HNTs, PP chains were first grafted onto the surface of HNTs *via* a two-step method. First, APTES-HNTs were prepared. Second, APTES-HNTs reacted with maleic anhydride grafted PP (PP-*g*-MAH) by amidation, and a PP-*g*-HNTs composite was successfully obtained (Figure 5.5). Characterization of the model compound showed the weight loss of the grafted PP was 2.8 wt%. The PP-*g*-HNTs composite performed much better due to the better interfacial bonding of the composite. Thus, this work provides an effective method to modify inorganics used in polyolefin composites.[48]

If APTES-HNTs were chemically linked with polyethyleneimine (PEI) using glutaraldehyde as a coupling agent, PEI-*g*-HNTs composites would be obtained. It was demonstrated that PEI-*g*-HNTs composites were a low cost, high adsorption capacity, and environmentally friendly Cr(VI) adsorbent. The Cr(VI) uptake capacity of the composite was approximately 64 times and

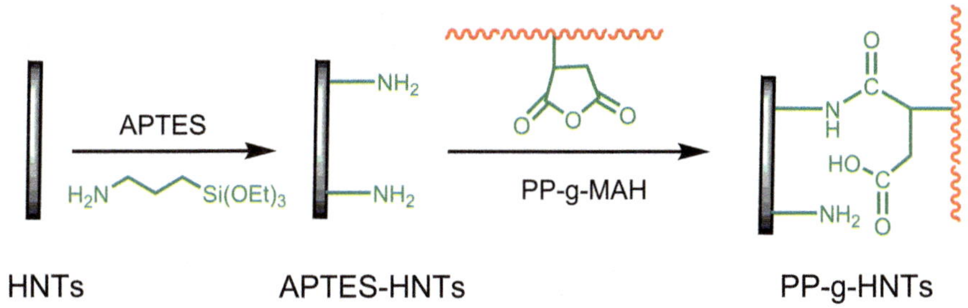

Figure 5.5 Schematic diagram of the preparation process and structure of PP-*g*-HNTs.[49]

two times greater than that of the original HNTs and APTES-HNTs,[27] respectively. The excellent performances of the PEI-*g*-HNTs were attributed to the presence of a large number of primary and secondary amine groups per molecule of PEI.[49] High-efficiency grafting of HNTs using *p*-conjugated polyfluorenes was conducted successfully by Zhang *et al.* Conjugated polyfluorenes with –C≡C–H end groups could be covalently grafted onto the azide-modified HNTs *via* "click" chemistry. The results showed the grafting degree of the HNTs reached up to 150%, which suggests a high level in the field of organic–inorganic nanocomposite materials. These composites will lead to a new generation of donor–acceptor nanohybrid materials that have promising applications in photoelectric fields.[50]

5.3.2.2 Grafting From

Polymers that grow from the surface of HNTs would follow different mechanisms of polymerization. The most common mechanism is free radical polymerization. For example, polymethylmethacrylate (PMMA) was grafted on HNTs through MMA free radical polymerization. The surface of the HNTs was first functionalized with MAPTS. Afterwards, the polymerization of MMA was conducted in the presence of MAPTS-HNTs. The final modified HNTs were used as a reinforcing agent in polar polymers, such as PVC[51] and epoxy acrylate resin.[52] In another example, poly(4-vinylpyridine) (P4VP) was first grafted onto the surface of HNTs *via in situ* free radical polymerization, and then, silver nanoparticles were immobilized on P4VP *via* a complex reaction and reduction. The final product blended with polyethersulfone (PES) formed antibacterial ultrafiltration membranes.[53] However, the aforementioned free radical polymerization was performed in an uncontrolled fashion. Therefore, polymers grafted onto HNTs are random and not well-defined.

Polycondensation between polyfunctional monomers is another common polymerization mechanism. For example, polyhydroxyurethane–HNTs nanocomposites were prepared by *in situ* surface-initiated polymerization of a five-membered cyclic carbonate 2,2-bis[*p*-(1,3-dioxolan-2-one-4-yl-methoxy)phenyl] propane and hexamethylenediamine from the surfaces of aminosilane-modified

HNTs. The grafting percentage and grafting efficiency were determined to be 41% and 23%, respectively. The grafted HNTs had better dispersibility than those of pristine HNTs in polymer or organic solvents.[54]

Controlled/"living" polymerizations from the HNT surface that allow better control over the target molecular weight, and the molecular weight distribution of the target polymer have attracted more and more attention. The two common types are atom transfer radical polymerization (ATRP) and reversible addition–fragmentation chain transfer polymerization (RAFT).

Different types of polymer brushes can be grown onto the surface of HNTs *via* ATRP by immobilizing different monomers. Polystyrene (PS) and polyacrylonitrile (PAN) were grown from both the interior and exterior surfaces of the HNTs bearing an ATRP initiator (2-bromoisobutyrylbromide), which was introduced by silanization. The shell thickness and interior cavity size could be continuously tuned by polymerization time and the monomer feeding amount (Figure 5.6). Polymeric nanotubes or nanowires were further derived after dissolution of the HNTs template.[55] Furthermore, the authors prepared a sulfonated PS layer. Thus, by specific interactions of desired materials with the sulfonated PS gel, favourable growth of other materials was allowed, and the composition of the composite layers could be broadly controlled, ranging from metal or metallic compounds, polymer, inorganic, and carbon. Consequently, a general synthesis approach for HNT-based composites has been proposed.[56] Specifically, Wang *et al.* realized controlled polymerization of 2-methacryloyloxyethylphosphorylcholine (MPC) on the HNTs surface *via* reverse atom transfer radical polymerization (RATRP). Similarly, HNTs were first modified by 3-chloropropyltriethoxy silane before polymerization to introduce the active sites. The final modified HNTs were used to improve the antifouling performance of polyethersulfone ultrafiltration.[57]

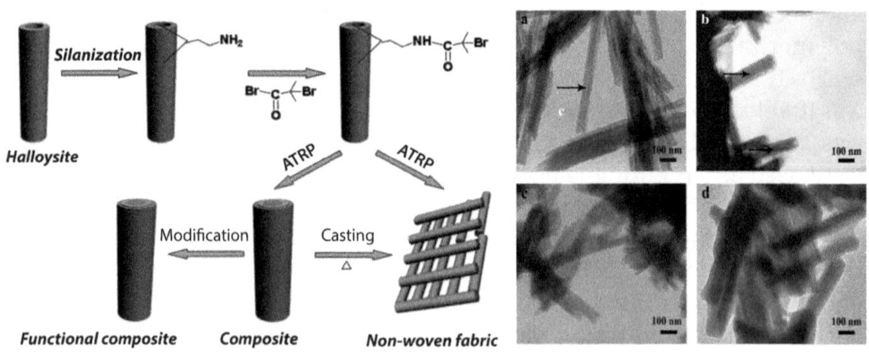

Figure 5.6 Illustrative synthesis of polymer–HNTs composite nanotubes and nonwoven fabric and TEM images of PS–HNTs composite nanotubes obtained at different polymerization times *via* the ATRP process. The polymerization times were 1.5 h, 3 h, 4.5 h, and 6 h for a, b, c, and d, respectively. (Reproduced from *J. Appl. Polym. Sci.* **110**, Li *et al.*, Polymer-Modified Halloysite Composite Nanotubes, 3638–3646, Copyright (2008) with permission from Wiley Periodicals, Inc.[55])

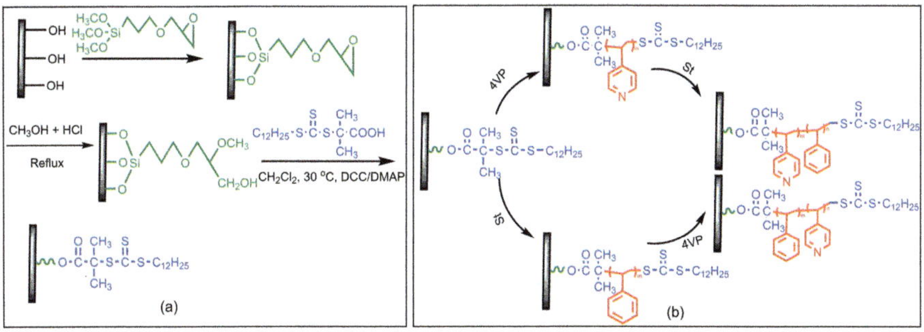

Figure 5.7 Schematic illustration of (a) anchoring of RAFT agent onto HNTs (b) grafting of amphiphilic brushes onto HNTs *via* a surface living RAFT polymerization.[58]

Amphiphilic brushes of poly(4-vinylpyridine)-*block*-polystyrene (P4VP-*b*-PS) and polystyrene-*block*-poly(4-vinylpyridine) (PS-*b*-P4VP) were grafted onto HNTs *via* RAFT (Figure 5.7). In the subsequent block copolymerization, the homopolymer brushes grafted onto HNTs as a chain transfer agent could effectively control polymerization, and the characterization showed that amphiphilic brushes were successfully grafted onto HNTs in a living manner. Therefore, the segment size in block copolymer brushes could be adjusted by the polymerization time. The Pickering emulsification behaviour of amphiphilic HNTs in a water–soybean oil diphase mixture demonstrated that modified HNTs could emulsify the water–soybean oil diphase mixture, and the emulsification performance was dependent on the microstructure of the amphiphilic brushes such as the hydrophilic–hydrophobic segment size and sequence.[58] This provides a new point for the functional application of HNTs.

5.3.3 Addition Reactions Between HNTs and (Iso)Cyanates

In addition to silanes, (iso)cyanates can react with hydroxyls on the surface of HNTs directly *via* the addition mechanism. Taking advantage of this, covalently linked organic–inorganic hybrids were formed by HNTs co-cured with epoxy/cyanate resins (Figure 5.8). The characterizations confirmed that silanols and aluminols were reactive towards the cyanate ester, where the principal products were iminocarbonates. However, the reactions do not deplete all of the hydroxyl groups of the HNTs, and the HNTs are partially wrapped after thermal treatment. Clearly, these hybrids exhibited improved performance and morphological characteristics due to the covalently linked interface.[59] Unlike that mentioned above, Zhang *et al.* first modified HNTs with isophoronediisocyanate (IPDI), and then the rest of the isocyanato group reacted with diethylenetriamine (DETA), which can act as a curing agent and facilitate the primary amine–epoxide reaction. A covalently linked interface and better performance of HNT-filled

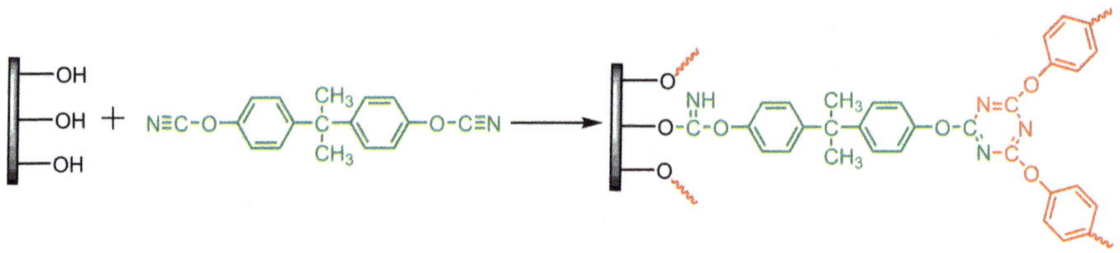

Figure 5.8 Schematic illustration of the reaction mechanism between silanols/aluminols and cyanate ester.

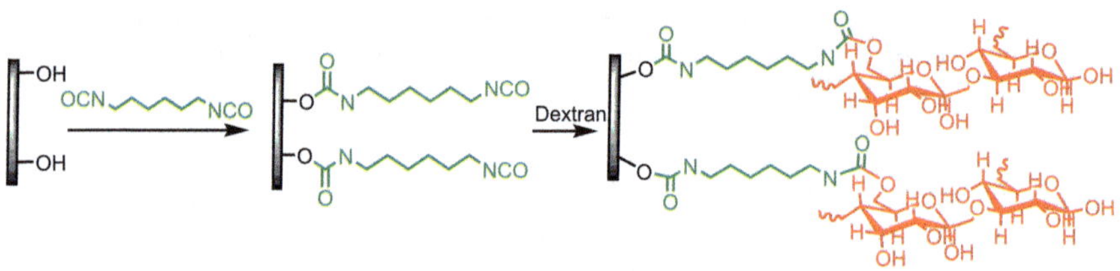

Figure 5.9 Schematic illustration of the synthesis route of dextran-grafted HNTs.[61]

epoxy nanocomposites were also obtained.[60] Similarly, Yu *et al.* used an isocyanato group on HNTs *via* 1,6-hexamethylene diisocyanate (HDI) and then further grafted dextran (Figure 5.9). Finally, the modified HNTs were incorporated into a PES ultrafiltration membrane, which greatly improved the antifouling properties.[61] In addition, it should be mentioned that the surface aluminols were claimed to be reactive to the carboxyl of L-lactic acid and formed aluminium lactate by a condensation reaction. The characterizations suggested that the grafting reaction may have only occurred in the Al–OH groups, and essentially no Si–O–Si bonds were formed during the reaction. Therefore, high-performance polylactic acid (PLA–HNTs) composites were prepared by incorporating L-lactic acid-grafted HNTs and PLA-grafted HNTs into the PLA matrix.[62]

5.3.4 Non-Covalent Interactions Between HNTs and Organic Molecules

Due to the relatively lower density of hydroxyl groups on the HNT surface, covalent modification for HNTs is somewhat limited. Fortunately, several special properties of HNTs, including the polarity of hydroxyls and the existence of few transition metal elements and surface charges, provide favourable conditions for non-covalent functionalization of HNTs. Moreover, much attention has been given to non-covalent modification of nanoparticles. In this chapter, we divided the non-covalent modification of HNTs into two types, *i.e.*, modification with small organic molecules and modification with polymers.

5.3.4.1 Hydrogen Bonding

The invariable presence of hydroxyl groups on the HNT surface provides the possibility of forming direct hydrogen bonding with many polar groups, such as carboxylates, hydroxyls, aminos, *etc.* Therefore, molecules containing these polar groups can be used to modify the surface of HNTs. Low-molecular weight organics, such as melamine (MEL), melamine cyanurate (MCA), and diphenyl guanidine (DPG), which are summarized in Table 5.2,[63] were chosen to interact with HNTs and subtly construct a

Table 5.2 Organics used for the formation of hydrogen bonds.

Organics	Molecular structures
Melamine (MEL)	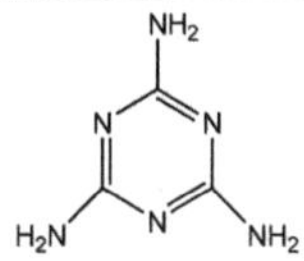
Diphenyl guanine (DPG)	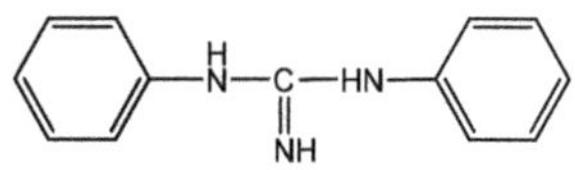
2,4,6-Trimercapto-*s*-triazine (TCY)	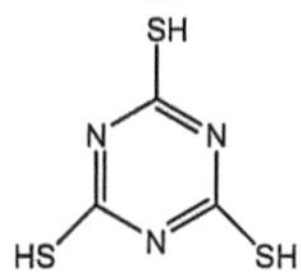
Melamine cyanuric acid (MCA)	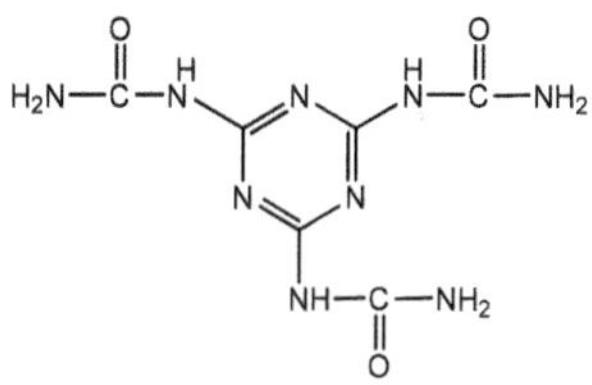
Tri-(2-hydroxyethyl) isocyanurete (THEIC)	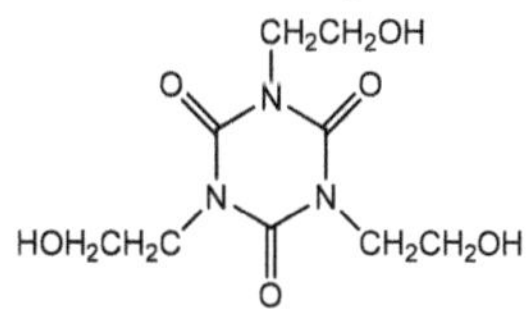
Beta cyclodextrin (Beta-CD)	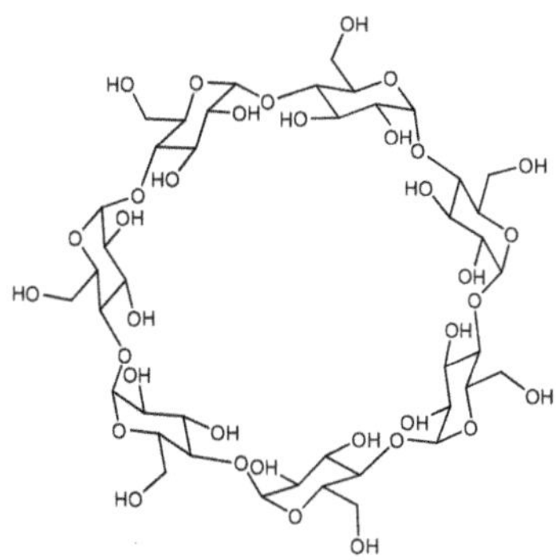

filler network in crystalline plastics *via* a hydrogen bonding self-assembly process; this directly facilitated the improved flexural modulus of the PP nanocomposites by up to 80%. This hydrogen bonding-induced assembled filler network also reinforced polar PA6; this process could be used with other inorganic materials, such as nanosilica. Functionalizing HNTs in this manner provides a general method to reinforce polymers that has many advantages, such as "green processing", unsophistication, facile scale-up, cost effectiveness, *etc.*[63,64]

In rubber–HNTs composites, an unsaturated carboxylic acid, such as methacrylic acid[65] or sorbic acid,[66] is a good candidate as a modifier for HNTs because it can cross-link with rubber chains simultaneously. Moreover, the *in situ* formed zinc salts, which were produced by the reaction between zinc oxide and carboxylic acid, also showed unusual reinforcing effects for the rubber compounds due to intermediate linkages connecting SBR and HNTs *via* a grafting/complexation mechanism. Moreover, certain functional ILs, such as 1-methylimidazolium mercaptopropionate (MimMP) and bis(1-methylimidazolium) mercaptosuccinate (BMimMS), could also form intermediate linkages that connect SBR and HNTs *via* grafting/hydrogen bonding (Figure 5.10).[67,68] It has been found that the addition of ILs can not only strengthen the interfacial interaction but also promote vulcanization and effectively facilitate HNT dispersion in a rubber matrix, which significantly improves the mechanical properties and lowers the mechanical loss.

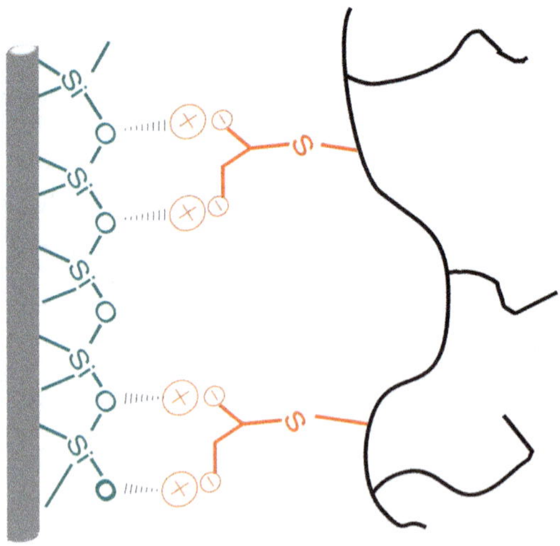

Figure 5.10 Schematic illustration of the interfacial interactions *via* grafting/hydrogen bonding between HNTs and SBR. (Reproduced from *Polymer*, **52**, Y. Lei *et al.*, Functional thiol ionic liquids as novel interfacial modifiers in SBR/HNTs composites, Copyright (2011) with permission from Elsevier.[68])

5.3.4.2 Electrostatic Interactions Between HNTs and Surfactants

Different compositions of the external surface and internal surface of HNTs confer a negative and positive charge at the outer and inner surfaces in a wide pH range in water.[69] Consequently, ionic surfactants and polyelectrolytes are often used to selectively modify HNTs *via* electrostatic interaction to improve HNT dispersal and stability in an aqueous medium, which are crucial for certain functional applications.

Two oppositely charged surfactants, sodium dodecanoate (NaL) and decyltrimethylammonium bromide (DeTAB), were exploited by Cavallaro *et al.* to functionalize HNTs, where the colloidal stability and solubilization ability of functionalized HNTs in the aqueous phase were studied. It was demonstrated that the adsorption of anionic surfactant into the lumen of the HNTs increased the net negative charge of the nanotubes, which enhanced the electrostatic repulsions and, consequently, the dispersion stability. On the contrary, the adsorption of the cationic surfactant on the external surface of HNTs decreased the net negative charge of the nanotubes, which resulted in aggregation (Figure 5.11). In addition, the solubilization capability of these functionalized nanotubes towards hydrophobic compounds was improved. This paper proposed a simple strategy to prepare hybrid materials, such as inorganic micelles, that could be used in water for solubilization and delivery of a hydrophobic compound by taking advantage of the sustainable and biocompatible properties.[70] Furthermore, the degree of hydrophobicity of the nanotube core could be tuned and manipulated by changing the length of the alkyl on the surfactants. Additionally, these dispersions were found to be non-foaming but extremely efficient in encapsulating oils that could be exploited for industrial applications on a large scale.[71] Moreover, the common surfactant, sodium dodecyl sulfate (SDS), has also been reported to enhance the dispersion of HNTs.[72]

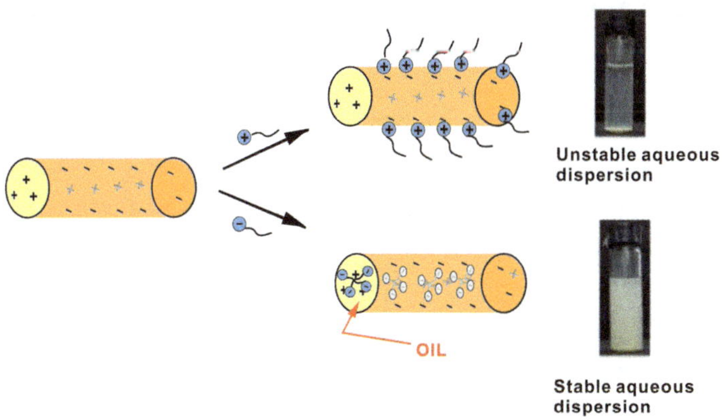

Figure 5.11 Illustration of the hybrid surfactant–HNTs materials and their colloidal stability. (Reproduced with permission from G. Cavallaro, G. Lazzara and S. Milioto, *J. Phys. Chem. C*, 2012, **116**, 21932.[70] Copyright (2012) American Chemical Society.)

5.3.4.3 Charge Transferring

HNTs are electrophilic due to the presence of several metal atoms on HNTs, such as aluminium, iron, and transition metals with unoccupied orbitals, which offers great opportunities to modify HNTs *via* electron transfer interactions. For example, 2,5-bis(2-benzoxazolyl) thiophene (BBT),[73] 2,2,0-(1,2-ethenediyldi-4,1-phenylene) bisbenzoxazole (EPB),[74] and benzothiazolesulfide (CBS),[75] which are capable of donating electrons, were selected as the interfacial modifier for PP–HNT composites. Their structures are shown in Figure 5.12. Hybrid fibrils composed of largely BBT and a small amount of HNTs were formed due to electron transfer interaction, which led to a much higher crystallinity compared with previously reported PP nanocomposites. As a result, the nanocomposites with BBT exhibited substantially better tensile and flexural properties.[73] EPB and HNTs formed aggregates, the cracking of which was responsible for the improved mechanical properties of the composites during fracture of the PP–HNTs composites.[74] Unlike the aforementioned, CBS, which is a commonly used accelerator in the tire industry, was used as a compatibilizer between the HNTs and PP *via* interfacial electron transferring and the grafting mechanism, which provides a new point for interfacial design.[75]

5.3.5 Non-Covalent Interactions Between HNTs and Polymers

Compared with small organic molecules, coatings are generally formed when polymers non-covalently modify HNTs. Furthermore, it is also a simple way to fabricate polymer–HNTs composites directly.

5.3.5.1 Hydrogen Bonding with Polymers

There are many environmentally friendly and commercially available polymers, such as polyethylene glycol (PEG) and polyvinyl alcohol (PVA), which can interact with HNTs *via* hydrogen bonding. For example, PEG

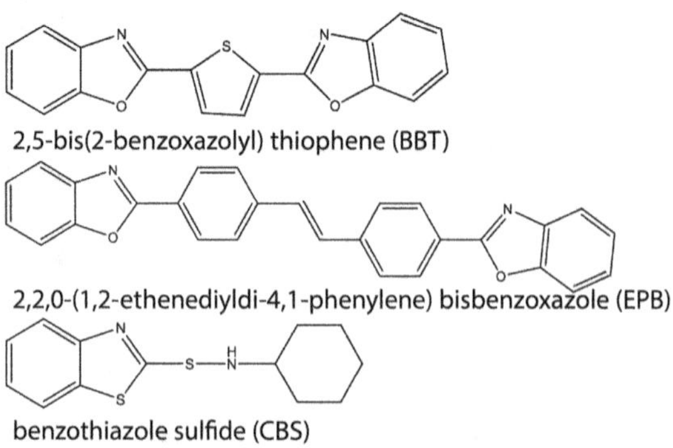

Figure 5.12 Several conjugated substances used to form charge transfer pairs with HNTs.

was used as a dispersing agent for HNTs–potato starch film.[76] Due to the good distribution of the modified HNTs, the tensile strength of the films was clearly enhanced. In polymer-based composites, hydrogen bonding between HNTs and polymers can be used to construct a relatively stronger interface and improve the mechanical properties of the composites. The reported matrices with hydrogen bonding capability include carboxylated styrene–butadiene rubber (xSBR),[77] polyamide (PA),[26] polyvinyl alcohol (PVA),[78] *etc.* For example, the xSBR–HNTs composite obtained by co-coagulation exhibited an extremely high reinforcing efficiency. With the inclusion of only 5 phr of HNTs, the modulus (at 100% elongation) was almost doubled compared with that of neat xSBR vulcanizate. The striking reinforcing efficiency was attributed to the strong interfacial hydrogen bonding between the pendent carboxylic groups on the rubber chains and the hydroxyls on the HNTs.[77]

Many natural macromolecules contain diverse oxygenic and nitrogenous groups that can also form hydrogen bonds with HNTs. Consequently, there are many reports on HNT nanocomposites based on natural polymers.[79–84] These nanocomposites typically exhibit high aqueous dispersity and long-term stability. Various biomedical applications of these systems can be found, for example, human select in protein–HNTs composites to capture flowing cancer cells,[80] chitosan–HNTs composites,[81] alginate–HNTs composites[84] for scaffolds in tissue engineering, deoxyribonucleic acid (DNA)–HNTs composites for controlled release of drugs,[82] *etc.* The good dispersibility of HNTs in these matrices and the hydrogen bonding interface contribute to the large reinforcement in these systems, which is important in scaffold application. It should be noted that chitosan is also a type of natural polycation electrolyte, and thus, the primary interaction mechanism between chitosan and HNTs is an electrostatic interaction. More details will be found in the following section.

5.3.5.2 *Electrostatic Interaction with Polymeric Electrolytes*

In contrast with negatively charged surfactants that increase the net negative charge of HNTs and hence improve the colloidal stability of HNTs, the polymeric electrolytes chosen to modify HNTs are typically polycation electrolytes. Thus, the electrolytes primarily coat the negatively charged HNT surface *via* electrostatic interaction, generally along with hydrogen bond interaction.[81] Additionally, the coated HNTs are often used in the biological medicine field. For example, Levis and Deasy used chitosan and PEI to coat HNTs and achieved a significant delayed drug release.[85] Zhai *et al.* synthesized chitosan–HNTs hybrid nanotubes through the electrostatic assembly of chitosan onto HNTs. The results indicated that the hybrid nanotubes could form three-dimensional (3D) nanocomposites with a hierarchically porous structure. After being cross-linked by glutaraldehyde, the hybrids exhibited an excellent capacity for horseradish peroxidase (HRP) immobilization (Figure 5.13).[86]

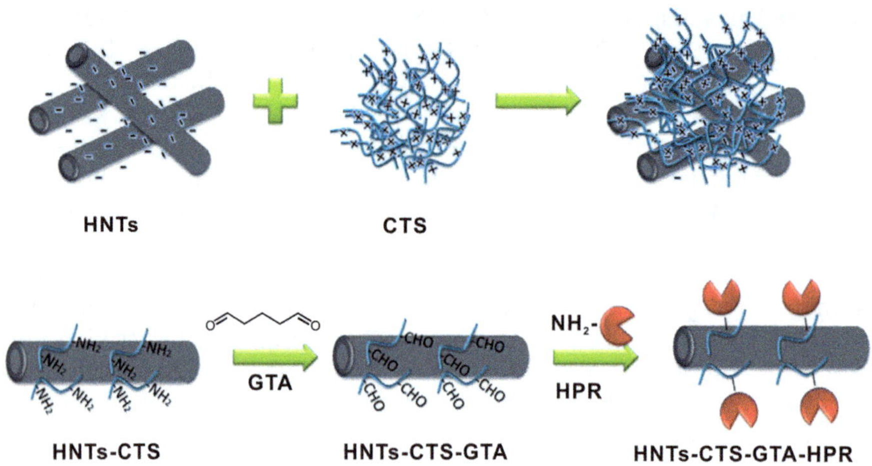

Figure 5.13 Schematic illustration of the preparation of chitosan–HNTs hybrid nanotubes and their immobilization mechanism. (Reproduced from *Chemical Engineering Journal*, **214**, R. Zhai *et al.*, Chitosan-halloysite hybrid-nanotubes: Horseradish peroxidase immobilisation and applications in phenol removal, 304–309, Copyright (2013) with permission from Elsevier.[86])

Moreover, the layer-by-layer self-assembly technique (LbL) based on electrostatic attraction can fabricate organized polyelectrolyte nanoshells on the HNT surface. During this process, oppositely charged polyelectrolytes are deposited on the nanoparticles alternately through electrostatic attraction. Much work has been summarized by Lvov *et al.*[7,69] These LbL nanoshells allow additional control for the sustained release of drug-loaded HNTs. The shell of polycation–polyanion multilayers on the HNTs provides a diffusion barrier, and therefore slows or even prevents release. Additionally, the release can be controlled by the type of polyelectrolyte, their molecular weight, number of polymeric layers, degree of ionization, pH of the solution, *etc.*[6,87–90] For example, a large molecular weight polyallylamine hydrochloride (PAH) (M_w 70 000)–polystyrene sulfonate (PSS) (M_w 70 000) polyelectrolyte shell can effectively decrease the release rate of the dexamethasone drug from 6 to 15 h.[6] Multilayers of PAH–PSS prevented undesirable leakage of the loaded inhibitor from the HNTs' interior. The schematic illustration of the fabrication of 2-mercaptobenzothiazole-loaded HNTs–polyelectrolyte nanocontainers and the zeta potential changes for sequential deposition of PAH and PSS polyelectrolytes on HNTs are shown in Figure 5.14. Furthermore, the polyelectrolyte shell provides the release of the encapsulated inhibitor controlled by pH changes in the environment surrounding the HNTs, which will ensure inhibitor release triggered by pH changes directly in the corrosion pit.[88]

In addition, monomers or initiators may be anchored on the HNT surface by electrostatic interaction and then polymerized to form coatings. For example, the *in situ* oxidative polymerization of anilinium[91] and pyrrole[92,93]

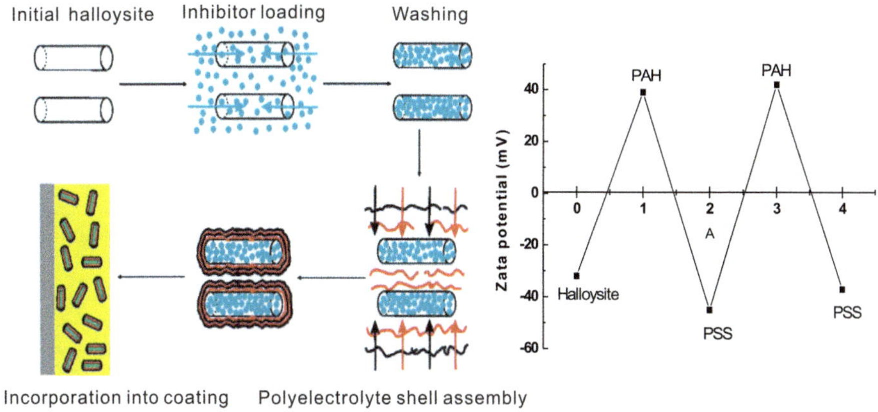

Figure 5.14 (Left) Schematic illustration of the fabrication of 2-mercaptobenzothiazole-loaded HNT–polyelectrolyte nanocontainers. (Right) Zeta potential data for sequential deposition of PAH and PSS polyelectrolytes on halloysite nanotubes, pH 7.5. (Reproduced with permission from D. G. Shchukin, S. V. Lamaka, K. A. Yasakau, M. L. Zheludkevich, M. G. S. Ferreira and H. Mohwald, *J. Phys. Chem., C*, 2008, **112**, 958.[88] Copyright (2008) American Chemical Society.)

on the HNTs surface is based on the electrostatic interaction between their cations and HNTs in an HCl aqueous solution. Thus, anilinium or pyrrole chloride salts act as both the monomer and the surface modifier in the oxidative polymerization. When the oxidant was added, polymerization occurred and conductive composites based on HNTs were obtained. Moreover, both the polypyrrole (PPy) nanotubes and carbon–HNTs composites could be prepared by etching HNTs[92] or pyrolysis of PPy.[93] Furthermore, like silica and carbon nanofibres, HNTs were also successfully modified with a hyperbranched polymer or copolymer shell by the self-condensing vinyl polymerization (SCVP) of 2-((bromoacetyl)oxy)ethyl acrylate (BAEA) and the self-condensing vinyl copolymerization of *n*-butyl acrylate (BA) and BAEA with BAEA as the inimer (AB*), respectively, from the surfaces of the 2-bromoisobutyric acid-modified HNTs *via* the ATRP technique.[94] Additionally, active sites were introduced by electrostatic adsorption of negatively charged 2-bromoisobutyric acid (Figure 5.15).

5.3.5.3 Plasma Polymerization

Plasma polymerization has attracted considerable attention because it is an environmentally friendly surface modification technology, which can easily alter the hydrophilic surfaces of layered silicate clays.[95–98] In the plasma polymerization process, plasma is generated by electrical discharge. The monomer gases are then activated and form electrons, ions, and radicals with the help of plasma energy, which are able to polymerize and form a deposit

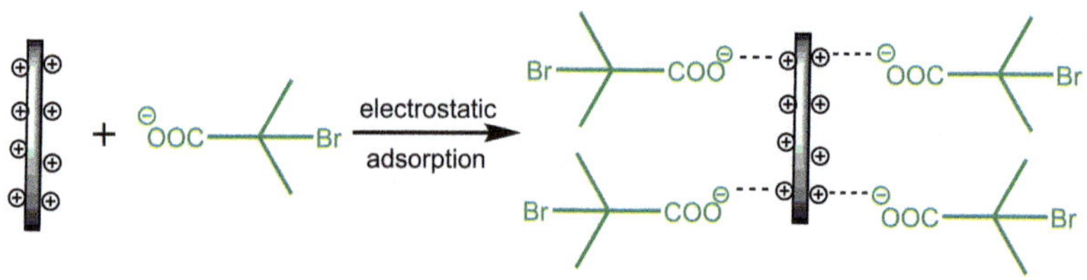

Figure 5.15 Schematic illustration of absorption of 2-bromoisobutyric acid on the HNTs.

on the surface of the filler; this changes their surface chemistry. However, the final modification effect largely depends on the plasma-treating parameters, such as radio frequency power, pressure, and time.[95,96,99]

Plasma polymerization of styrene was successfully used to modify the hydrophilic surfaces of HNTs and the modified HNTs were further used to reinforce PS. The characterizations confirmed that plasma polymerization did not form covalent linkages between the polymer coating and HNTs in this case because no C–O and Si–C stretching was recorded. However, compared to pristine HNTs, modified HNTs exhibited a substantially greater reinforcement effect due to the compatibility of the new hydrophobic HNT surfaces with the PS polymer matrix.[100] Poikelispää *et al.* used pyrrole and thiophene as monomers for plasma polymerization on HNTs to change their surface polarity and chemistry, and the modified HNTs were used in a natural rubber–butadiene rubber blend in the presence of carbon black. The surface modification resulted in a higher bound rubber content and lower Payne effect, which indicates a better filler–polymer interaction and good dispersion.[99]

5.3.6 Enlargement of the Lumen by Controlled Etching

Many applications of HNTs are conditioned by their lumen size. Therefore, enlarging the lumen using a chemical treatment is extremely important. Both sulfuric acid (H_2SO_4) and NaOH treatments are common means. Zhang *et al.* first systematically investigated the effects of H_2SO_4 treatment on the physico-chemical and pore characteristics of HNTs. The results indicated that H_2SO_4 would destroy the crystal structure of HNTs. The [AlO_6] octahedral layers would be dissolved, which induced rupture and collapse of [SiO_4] tetrahedral layers, resulting in porous nanorods. The BET surface area and the total pore volumes of the HNTs increased to 267.1 $m^2\ g^{-1}$ and 0.975 $cm^3\ g^{-1}$, respectively, after acid treatment for 13 h.[101]

Later, White *et al.* further revealed the mechanism of nanotube transformation in both alkaline and acid environments by studying the long-term stability of HNTs in acidic and basic aqueous suspensions at room temperature. It was found that extensive dissolution was initiated by dissolution of the inner Al–OH surface catalysed by acidic or basic species, which led to progressive thinning of the tubes from the inside without significant alteration of their shape. In a strong acid solution, the greater solubility of Al(III) *versus* Si(IV) leads to the formation of small SiO_2 nanoparticles inside the tubes, whereas in a NaOH solution, the greater solubility of Si(IV) *versus* Al(III) leads to the formation of fragmented $Al(OH)_3$ nanosheets.[102] It should be noted that the degree of the process can be controlled by acid or alkali concentration, time, and temperature of the treatment.

HNT applications, such as adsorbents, drug carriers, and fillers, greatly benefit from acid or alkali treatment. For example, acid-treated HNTs with fixed micropore and controllable mesopore size distributions were confirmed to be low cost yet effective adsorbents for methylene blue. The maximum adsorption capacity of acid-treated HNTs for methylene blue is greater than 60 mg g^{-1} and can be controlled by pH values.[101] With an acid etching technique under proper conditions, the specific surface area of the tubes increased by more than six-fold (40–250 m^2 g^{-1}), and the HNTs loading efficiency for benzotriazole increased four-fold, which is comparable with that using the conventional polymeric microencapsulation technique.[103] PVA–HNTs films with improved tensile strength, Young's modulus, and optical transmittance were prepared with sulfuric acid-treated HNTs within a short time (less than 2 h) because the increased surface areas significantly enhanced the interfacial adhesion.[104]

5.3.7 Selective Modification of the Lumen

As mentioned in Section 5.3.4.2, the oppositely charged innerside and outside of HNTs make it possible for selective modification on the HNT surface. Differently, a few types of modifiers, such as phosphonic acid and dopamine or its derivatives, which will coordinate with aluminium oxide but not silicon oxide and selectively modify HNTs, are summarized in particular in this section. Interestingly and importantly, bifunctionalization of the silica–alumina oxide surface of HNTs can be achieved in this way. Much work has been done by Yah *et al.* First, octadecylphosphonic acid (ODP) was chosen to selectively modify the lumen of HNTs. The characterizations of selectively modified tubes showed that ODP bounded within the HNT lumen through the bidentate and tridentate P–O–Al linkage. The ODP-modified HNTs were modified further by coupling silane *N*-(2-aminoethyl)-3-aminopropyltrimethoxysilane (AEAPS) on their outer surfaces, which created an inorganic micelle-like architecture with a hydrophobic aliphatic chain core and a hydrophilic silicate shell (Figure 5.16). These modified HNTs could be used for water purification as well as for drug immobilization and controlled release.[105] For example, ODP-modified HNTs were loaded with hydrophobic bisphenol-A bis(diphenylphosphate) (BDP) (a flame retardant for polymers). The modified HNTs had a high BDP loading ability. Additionally, this flame retardant was linearly released from the nanotubes over 60 h, which demonstrates their potential utility as effective nanocontainers for smart composite systems.[106]

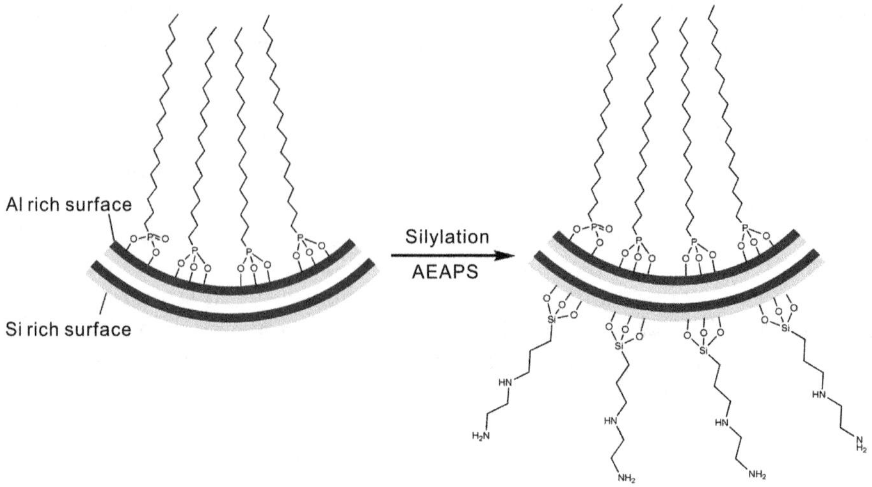

Figure 5.16 Schematic illustration of bifunctionalization of the silica–alumina oxide surface of halloysite by ODP and subsequent AEAPS silylation. (Reproduced with permission from W. O. Yah, A. Takahara and Y. M. Lvov, *J. Am. Chem. Soc.*, 2012, **134**, 1853.[105] Copyright (2011) American Chemical Society.)

Following this, a biomimetic dopamine derivative was chosen as the selective modifier. However, an additional step of sonication and tetrahydrofuran–water rinsing was required to remove the dopamine derivative on the external surface linked *via* weak silica–catechol bonds. However, alumina–catechol bonds were much more stable than silica–catechol bonds, which allowed selective modification of the aluminol group of the HNT lumen by the catechol group under aqueous conditions. Then, bromo groups in the dopamine derivative allowed the surface initiated-ATRP polymerization to selectively graft a layer of PMMA brush in the lumen (Figure 5.17). Additionally, the outermost surface of the silica of the HNTs–dopa composite can be covalently bonded with a second class of organic groups *via* silylation reaction.[107] However, if dopamine-modified HNTs are used for biomacromolecule immobilization, there is no need to remove the dopamine adhered on the outer surface, though it is necessary to provide proper conditions for dopamine self-polymerization to improve the number of immobilization sites.[108] All of this depends on the application of the modified HNTs.

5.4 Concluding Remarks

The tubular morphology of HNTs provides extraordinary opportunities to design various functional systems, including sustained drug release, immobilization of metal nanoparticles, polymer-based composites, *etc.* However, these applications are largely determined by the surface modification of HNTs, which is one of the key factors that governs the composite structure and functions. To

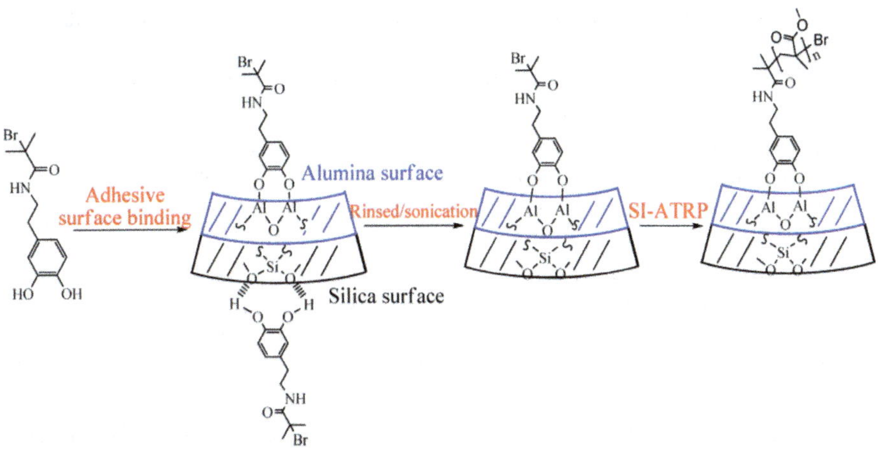

Figure 5.17 Selective modification of halloysite lumen with dopabiomimetic initiator and subsequent polymerization. (Reproduced with permission from W. O. Yah, H. Xu, H. Soejima, W. Ma, Y. Lvov and A. Takahara, *J. Am. Chem. Soc.*, 2012, **134**, 12134.[107] Copyright (2012) American Chemical Society.)

realize the functions or optimize the performance of HNT composites, the modification conditions must be judiciously chosen. For example, the silanization of HNTs is strongly dependent on the reaction conditions and the activation of HNTs. Also the modification *via* electrostatic interaction is greatly affected by the pH value because it determines the zeta potential. Furthermore, except for certain commonly known methods, such as silanization and wrapping *via* hydrogen bonding/electrostatic interaction, several interesting modifications, including controlled enlargement and selective inside/outside modification of the lumen, are worthy of attention because they provide new methods towards functional materials based on the specific lumen structure.

Acknowledgements

The authors gratefully acknowledge the National Basic Research Program of China (Grant No. 2015CB654703), and National Natural Science Foundation (Grant No. 51222301, 51333003, 51473050 and U1462116) for financial supports.

References

1. P. Berthier, *Ann. Chim. Phys.*, 1826, **32**, 332.
2. M. L. Du, B. C. Guo and D. M. Jia, *Polym. Int.*, 2010, **59**, 574.
3. E. Joussein, S. Petit, J. Churchman, B. Theng, D. Righi and B. Delvaux, *Clay Miner.*, 2005, **40**, 383.
4. M. X. Liu, Z. X. Jia, D. M. Jia and C. R. Zhou, *Prog. Polym. Sci.*, 2014, **39**, 1498.

5. T. F. Bates, F. A. Hildebrand and A. Swineford, *Am. Mineral.*, 1950, **35**, 463.
6. N. G. Veerabadran, D. Mongayt, V. Torchilin, R. R. Price and Y. M. Lvov, *Macromol. Rapid Commun.*, 2009, **30**, 99.
7. Y. Lvov and E. Abdullayev, *Prog. Polym. Sci.*, 2013, **38**, 1690.
8. E. Abdullayev and Y. Lvov, *J. Mater. Chem. B*, 2013, **1**, 2894.
9. K. M. Ng, Y. T. R. Lau, C. M. Chan, L. T. Weng and J. S. Wu, *Surf. Interface Anal.*, 2011, **43**, 795.
10. P. Yuan, P. D. Southon, Z. W. Liu, M. E. R. Green, J. M. Hook, S. J. Antill and C. J. Kepert, *J. Phys. Chem. C*, 2008, **112**, 15742.
11. V. Vahedi and P. Pasbakhsh, *Polym. Test.*, 2014, **39**, 101.
12. S. S. Zeng, C. Reyes, J. J. Liu, P. A. Rodgers, S. H. Wentworth and L. Y. Sun, *Polymer*, 2014, **55**, 6519.
13. J. Madejová, H. Pálková, M. Pentrák and P. Komadel, *Clays Clay Miner.*, 2009, **57**, 392.
14. B. Paul, W. N. Martens and R. L. Frost, *Appl. Surf. Sci.*, 2011, **257**, 5552.
15. C. Zha, W. C. Wang, Y. L. Lu and L. Q. Zhang, *ACS Appl. Mater. Interfaces*, 2014, **6**, 18769.
16. L. N. Carli, T. S. Daitx, G. V. Soares, J. S. Crespo and R. S. Mauler, *Appl. Clay Sci.*, 2014, **87**, 311.
17. H. Jing, Y. Higaki, T. Ishikawa, K. L. White, H. Otsuka and A. Takahara, *Sci. Adv. Mater.*, 2015, **7**, 974.
18. M. L. Du, B. C. Guo and D. M. Jia, *Eur. Polym. J.*, 2006, **42**, 1362.
19. P. Pasbakhsh, H. Ismail, M. N. A. Fauzi and A. A. Bakar, *Appl. Clay Sci.*, 2010, **48**, 405.
20. M. T. Albdiry and B. F. Yousif, *Mater. Des.*, 2013, **48**, 68.
21. H. J. Haroosh, Y. Dong, D. S. Chaudhary, G. D. Ingram and S.-i. Yusa, *Appl. Phys. A*, 2012, **110**, 433.
22. P. Pal, M. K. Kundu, A. Malas and C. K. Das, *J. Appl. Polym. Sci.*, 2014, **131**, 1.
23. M. X. Liu, B. C. Guo, M. L. Du, Y. D. Lei and D. Jia, *J. Polym. Res.*, 2007, **15**, 205.
24. S. Rooj, A. Das, V. Thakur, R. N. Mahaling, A. K. Bhowmick and G. Heinrich, *Mater. Des.*, 2010, **31**, 2151.
25. V. S. Raman, S. Rooj, A. Das, K. W. Stöckelhuber, F. Simon, G. B. Nando and G. Heinrich, *J. Macromol. Sci., Part A*, 2013, **50**, 1091.
26. B. C. Guo, Q. L. Zou, Y. D. Lei and D. Jia, *Polym. J. (Tokyo, Jpn.)*, 2009, **41**, 835.
27. P. Luo, J. S. Zhang, B. Zhang, J. H. Wang, Y. F. Zhao and J. D. Liu, *Ind. Eng. Chem. Res.*, 2011, **50**, 10246.
28. S. Jana and S. Das, *RSC Adv.*, 2014, **4**, 34435.
29. S. Barrientos-Ramirez, G. M. de Oca-Ramirez, E. V. Ramos-Fernandez, A. Sepulveda-Escribano, M. M. Pastor-Blas and A. Gonzalez-Montiel, *Appl. Catal., A*, 2011, **406**, 22.
30. M. L. Zou, M. L. Du, H. Zhu, C. S. Xu and Y. Q. Fu, *J. Phys. D: Appl. Phys.*, 2012, **45**, 325302.

31. Y. Zhang, X. He, J. Ouyang and H. Yang, *Sci. Rep.*, 2013, **3**, 2948.
32. Y. Zhang, Y. L. Xie, A. D. Tang, Y. H. Zhou, J. Ouyang and H. M. Yang, *Ind. Eng. Chem. Res.*, 2014, **53**, 5507.
33. T. T. Yang, M. L. Du, M. Zhang, H. Zhu, P. Wang and M. L. Zou, *Nanomater. Nanotechnol.*, 2015, **61**, 357.
34. M. L. Zou, M. L. Du, M. Zhang, T. T. Yang, H. Zhu, P. Wang and S. Y. Bao, *Mater. Res. Bull.*, 2015, **61**, 375.
35. Y. T. Zhang, Y. F. Chen, H. Q. Zhang, B. Zhang and J. D. Liu, *J. Inorg. Biochem.*, 2013, **118**, 59.
36. M. Zieba, J. L. Hueso, M. Arruebo, G. Martínez and J. Santamaría, *New J. Chem.*, 2014, **38**, 2037.
37. H. Zhu, M. L. Du, M. L. Zou, C. S. Xu and Y. Q. Fu, *Dalton Trans.*, 2012, **41**, 10465.
38. P. Yuan, P. D. Southon, Z. Liu and C. J. Kepert, *Nanotechnology*, 2012, **23**, 375705.
39. H. Wu, H. Watanabe, W. Ma, A. Fujimoto, T. Higuchi, K. Uesugi, A. Takeuchi, Y. Suzuki, H. Jinnai and A. Takahara, *Langmuir*, 2013, **29**, 14971.
40. M. Massaro, S. Riela, G. Cavallaro, M. Gruttadauria, S. Milioto, R. Noto and G. Lazzara, *J. Organomet. Chem.*, 2014, **749**, 410.
41. M. Massaro, S. Riela, G. Lazzara, M. Gruttadauria, S. Milioto and R. Noto, *Appl. Organomet. Chem.*, 2014, **28**, 234.
42. M. Massaro, C. G. Colletti, R. Noto, S. Riela, P. Poma, S. Guernelli, F. Parisi, S. Milioto and G. Lazzara, *Int. J. Pharm.*, 2015, **478**, 476.
43. R. J. Li, Q. He, Z. Hu, S. R. Zhang, L. J. Zhang and X. J. Chang, *Anal. Chim. Acta*, 2012, **713**, 136.
44. R. J. Li, Z. Hu, S. R. Zhang, Z. H. Li and X. J. Chang, *Int. J. Environ. Anal. Chem.*, 2013, **93**, 767.
45. Q. He, D. Yang, X. L. Deng, Q. Wu, R. J. Li, Y. H. Zhai and L. J. Zhang, *Water Res.*, 2013, **47**, 3976.
46. Y. Joo, Y. Jeon, S. U. Lee, J. H. Sim, J. Ryu, S. Lee, H. Lee and D. Sohn, *J. Phys. Chem. C*, 2012, **116**, 18230.
47. Y. S. Liu, X. Q. Jiang, B. J. Li, X. D. Zhang, T. Z. Liu, X. S. Yan, J. Ding, Q. Cai and J. M. Zhang, *J. Mater. Chem. A*, 2014, **2**, 4264.
48. M. L. Du, B. C. Guo, M. X. Liu and D. M. Jia, *Polym. J. (Tokyo, Jpn.)*, 2006, **38**, 1198.
49. X. K. Tian, W. W. Wang, Y. X. Wang, S. Komarneni and C. Yang, *Microporous Mesoporous Mater.*, 2015, **207**, 46.
50. H. L. Zhang, X. Y. Zhu, Y. G. Wu, H. Z. Song and X. W. Ba, *J. Mater. Sci.*, 2015, **50**, 4387.
51. C. Liu, Y. F. Luo, Z. X. Jia, B. C. Zhong, S. Q. Li, B. C. Guo and D. M. Jia, *eXPRESS Polym. Lett.*, 2011, **5**, 591.
52. J. H. Zhang, D. H. Zhang, A. Q. Zhang, Z. X. Jia and D. M. Jia, *J. Reinf. Plast. Compos.*, 2013, **32**, 713.
53. J. Y. Zhang, Y. T. Zhang, Y. F. Chen, L. Du, B. Zhang, H. Q. Zhang, J. D. Liu and K. J. Wang, *Ind. Eng. Chem. Res.*, 2012, **51**, 3081.

54. Z. B. Tang, P. Liu, J. S. Guo and Z. X. Su, *e-Polym.*, 2008, **8**, 1919.
55. C. P. Li, J. G. Liu, X. Z. Qu, B. C. Guo and Z. Z. Yang, *J. Appl. Polym. Sci.*, 2008, **110**, 3638.
56. C. P. Li, J. G. Liu, X. Z. Qu and Z. Z. Yang, *J. Appl. Polym. Sci.*, 2009, **112**, 2647.
57. Z. T. Wang, H. X. Wang, J. D. Liu and Y. T. Zhang, *Desalination*, 2014, **344**, 313.
58. Y. F. Hou, J. Q. Jiang, K. Li, Y. W. Zhang and J. D. Liu, *J. Phys. Chem. B*, 2014, **118**, 1962.
59. M. X. Liu, B. C. Guo, M. L. Du, X. J. Cai and D. M. Jia, *Nanotechnology*, 2007, **18**, 455703.
60. J. H. Zhang, Z. X. Jia, D. M. Jia, D. H. Zhang and A. Q. Zhang, *High Perform. Polym.*, 2014, **26**, 734.
61. H. X. Yu, Y. T. Zhang, X. B. Sun, J. D. Liu and H. Q. Zhang, *Chem. Eng. J.*, 2014, **237**, 322.
62. W. Xu, B. H. Luo, W. Wen, W. J. Xie, X. Y. Wang, M. X. Liu and C. R. Zhou, *J. Appl. Polym. Sci.*, 2015, **132**, 41451.
63. M. L. Du, B. C. Guo, M. X. Liu, X. J. Cai and D. M. Jia, *Phys. B*, 2010, **405**, 655.
64. M. L. Du, B. C. Guo, M. X. Liu and D. M. Jia, *Polym. J. (Tokyo, Jpn.)*, 2007, **39**, 208.
65. B. C. Guo, Y. D. Lei, F. Chen, X. L. Liu, M. L. Du and D. Jia, *Appl. Surf. Sci.*, 2008, **255**, 2715.
66. B. C. Guo, F. Chen, Y. D. Lei, X. L. Liu, J. J. Wan and D. M. Jia, *Appl. Surf. Sci.*, 2009, **255**, 7329.
67. B. C. Guo, X. L. Liu, W. Y. Zhou, Y. D. Lei and D. M. Jia, *J. Macromol. Sci., Phys.*, 2010, **49**, 1029.
68. Y. D. Lei, Z. H. Tang, L. X. Zhu, B. C. Guo and D. M. Jia, *Polymer*, 2011, **52**, 1337.
69. Y. M. Lvov, D. G. Shchukin, H. Mohwald and R. R. Price, *ACS Nano*, 2008, **2**, 814.
70. G. Cavallaro, G. Lazzara and S. Milioto, *J. Phys. Chem. C*, 2012, **116**, 21932.
71. G. Cavallaro, G. Lazzara, S. Milioto, F. Parisi and V. Sanzillo, *ACS Appl. Mater. Interfaces*, 2014, **6**, 606.
72. H. L. Lun, J. Ouyang and H. M. Yang, *Phys. Chem. Miner.*, 2013, **41**, 281.
73. M. X. Liu, B. C. Guo, Q. L. Zou, M. L. Du and D. M. Jia, *Nanotechnology*, 2008, **19**, 205709.
74. M. X. Liu, B. C. Guo, M. L. Du and D. M. Jia, *Polym. J. (Tokyo, Jpn.)*, 2008, **40**, 1087.
75. M. X. Liu, B. C. Guo, Y. D. Lei, M. L. Du and D. M. Jia, *Appl. Surf. Sci.*, 2009, **255**, 4961.
76. Y. Q. He, W. N. Kong, W. C. Wang, T. L. Liu, Y. Liu, Q. J. Gong and J. P. Gao, *Carbohydr. Polym.*, 2012, **87**, 2706.
77. M. L. Du, B. C. Guo, Y. D. Lei, M. X. Liu and D. M. Jia, *Polymer*, 2008, **49**, 4871.

78. W. Y. Zhou, B. C. Guo, M. X. Liu, R. J. Liao, A. B. M. Rabie and D. M. Jia, *J. Biomed. Mater. Res., Part A*, 2010, **93A**, 1574.
79. M. H. Shamsi and K. E. Geckeler, *Nanotechnology*, 2008, **19**, 075604.
80. A. D. Hughes and M. R. King, *Langmuir*, 2010, **26**, 12155.
81. M. X. Liu, Y. Zhang, C. C. Wu, S. Xiong and C. R. Zhou, *Int. J. Biol. Macromol.*, 2012, **51**, 566.
82. Y. Lee, G. E. Jung, S. J. Cho, K. E. Geckeler and H. Fuchs, *Nanoscale*, 2013, **5**, 8577.
83. R. De Silva, P. Pasbakhsh, K. Goh, S. P. Chai and J. Chen, *J. Compos. Mater.*, 2013, **48**, 3705.
84. M. X. Liu, L. B. Dai, H. Z. Shi, S. Xiong and C. R. Zhou, *Mater. Sci. Eng., C*, 2015, **49**, 700.
85. S. R. Levis and P. B. Deasy, *Int. J. Pharm.*, 2003, **253**, 145.
86. R. Zhai, B. Zhang, Y. Z. Wan, C. C. Li, J. T. Wang and J. D. Liu, *Chem. Eng. J.*, 2013, **214**, 304.
87. D. G. Shchukin and H. Möhwald, *Adv. Funct. Mater.*, 2007, **17**, 1451.
88. D. G. Shchukin, S. V. Lamaka, K. A. Yasakau, M. L. Zheludkevich, M. G. S. Ferreira and H. Mohwald, *J. Phys. Chem. C*, 2008, **112**, 958.
89. V. Vergaro, Y. M. Lvov and S. Leporatti, *Macromol. Biosci.*, 2012, **12**, 1265.
90. T. G. Shutava, R. F. Fakhrullin and Y. M. Lvov, *Curr. Opin. Pharmacol.*, 2014, **18**, 141.
91. L. Zhang, T. M. Wang and P. Liu, *Appl. Surf. Sci.*, 2008, **255**, 2091.
92. Y. S. Liu, H. M. Nan, Q. Cai and H. D. Li, *J. Appl. Polym. Sci.*, 2012, **125**, 638.
93. Y. S. Liu, Q. Cai, H. D. Li and J. M. Zhang, *J. Appl. Polym. Sci.*, 2013, **128**, 517.
94. B. Mu, M. F. Zhao and P. Liu, *J. Nanopart. Res.*, 2007, **10**, 831.
95. C. Nah, M. Y. Huh, J. M. Rhee and T. H. Yoon, *Polym. Int.*, 2002, **51**, 510.
96. N. Célini, F. Bergaya and F. Poncin-Epaillard, *Polymer*, 2007, **48**, 58.
97. G. Le Dû, N. Celini, F. Bergaya and F. Poncin-Epaillard, *Surf. Coat. Technol.*, 2007, **201**, 5815.
98. M. Tiwari, J. W. M. Noordermeer, W. J. van Ooij and W. K. Dierkes, *Polym. Adv. Technol.*, 2008, **19**, 1672.
99. M. Poikelispää, A. Das, W. Dierkes and J. Vuorinen, *J. Appl. Polym. Sci.*, 2013, **127**, 4688.
100. M. H. Shamsi, M. Luqman, F. Basarir, J.-S. Kim, T.-H. Yoon and K. E. Geckeler, *Polym. Int.*, 2010, **59**, 1492.
101. A. B. Zhang, L. Pan, H. Y. Zhang, S. T. Liu, Y. Ye, M. S. Xia and X. G. Chen, *Colloids Surf., A*, 2012, **396**, 182.
102. R. D. White, D. V. Bavykin and F. C. Walsh, *Nanotechnology*, 2012, **23**, 065705.
103. E. Abdullayev, A. Joshi, W. B. Wei, Y. F. Zhao and Y. Lvov, *ACS Nano*, 2012, **6**, 7216.
104. K. Fujii, A. N. Nakagaito, H. Takagi and D. Yonekura, *Compos. Interfaces*, 2014, **21**, 319.

105. W. O. Yah, A. Takahara and Y. M. Lvov, *J. Am. Chem. Soc.*, 2012, **134**, 1853.
106. H. Jing, Y. Higaki, W. Ma, H. Wu, W. O. Yah, H. Otsuka, Y. M. Lvov and A. Takahara, *Chem. Lett.*, 2013, **42**, 121.
107. W. O. Yah, H. Xu, H. Soejima, W. Ma, Y. Lvov and A. Takahara, *J. Am. Chem. Soc.*, 2012, **134**, 12134.
108. C. Chao, J. D. Liu, J. T. Wang, Y. W. Zhang, B. Zhang, Y. T. Zhang, X. Xiang and R. F. Chen, *ACS Appl. Mater. Interfaces*, 2013, **5**, 10559.

CHAPTER 6

Halloysite Based Smart Hybrid Nanomaterials for the Solubilization of Hydrophobic Compounds in Aqueous Media

G. CAVALLARO[a], G. LAZZARA*[a] AND S. MILOTO[a]

[a]University of Palermo, Department of Physics and Chemistry, Viale delle Scienze pad. 17, Palermo, 90128, Italy
*E-mail: giuseppe.lazzara@unipa.it

6.1 Introduction

Chemical modification of clay nanoparticles is a well-known strategy for designing sustainable nanomaterials with target properties suitable for several technological purposes.[1] Within this issue, promising nanoclays are halloysite nanotubes (HNTs) because of their versatile morphological and chemical characteristics, such as large specific area, hollow tubular shape and tunable surface chemistry.[2,3] Different deposits worldwide offered HNTs with peculiar morphology, purity and size distribution.[4] Moreover, HNTs are non-toxic, cheap and abundantly available over the world.[2] Due to these features, a growing scientific and industrial interest has been addressed to HNT based nanomaterials in the last decade.[1] Particularly, both covalent and non-covalent modification of HNT surfaces have been newly proposed in order

RSC Smart Materials No. 22
Functional Polymer Composites with Nanoclays
Edited by Yuri Lvov, Baochun Guo and Rawil F. Fakhrullin

Published by the Royal Society of Chemistry, www.rsc.org

to obtain ecocompatible nanostructures that can be used in life and materials science. Recent studies proved that chemically functionalized HNTs are prospective efficient catalytic supports[5–7] as well as adsorbent nanomaterials for water decontamination.[8–10] Filling polymers with HNTs represents an alternative route to generate nanocomposite films with desirable mechanical performance and humidity control ability.[11–13] In addition, both pristine and modified HNTs are proper nanocarriers for the encapsulation and controlled release of antibacterial,[14–16] anticorrosive[17,18] and self-healing compounds.[19]

HNTs are naturally occurring rolled and hydrated kaolinite multilayers. Their basic structure and properties can be found in Chapter 5. HNTs present two different interaction sites (inner side and outer side) that can be selectively involved in specific chemical reactions with the aim to obtain hybrid nanomaterials with target properties. The following reports are recommended for further reading.[2,3] The cycloaddition of azides and alkynes onto the lumen of HNTs produced microfibers that can be used as fillers to reinforce biopolymeric materials.[20] On the other hand, efficient nanocontainers for enzyme immobilization were obtained by coating the outer surface of HNTs with polydopamine.[21] The grafting of poly(*N*-isopropylacrylamide) onto the external surface of HNTs enabled the synthesis of thermoresponsive drug carriers for curcumin delivery.[22] Supports with high photocatalytic activity for methanol degradation were successfully prepared by the deposition of TiO_2 onto the shell of HNTs.[6] Furthermore, selective modification of HNT surfaces can be easily achieved through electrostatic interactions with ionic molecules, such as surfactants and polymers. The selectivity of the functionalization is based on the surface charge of the HNTs, which possess a positively charged cavity and a negatively charged outer surface in the pH range between 2 and 8 as a consequence of different ionization properties of aluminum and silicon oxides.[23] This peculiarity affects the liquid crystalline phase behavior of HNT aqueous dispersions[24] and, consequently, the self-assembly of the nanotubes in the meniscus of evaporating droplets.[25] Accordingly, it was demonstrated[25] that the modification of the lumen of HNTs with anionic poly(styrene) sulfonate (PSS) generates a better ordering. Consequently, PSS–HNTs hybrids can be very useful for applications in capillary orientation.

The adsorption of ionic molecules onto HNT surfaces represents a tremendous tool to control the hydrophobic–hydrophilic interface of the nanotubes as well as their colloidal stability in solvents with different polarities.

Here, we describe several strategies for designing HNT based hybrid nanomaterials suitable for the adsorption of hydrophobic compounds. The attained knowledge represents a starting point for the development of smart nanoparticles that can be used for water decontamination as well as pharmaceutical purposes on a large scale.

6.2 Experimental

In this paragraph we review the experimental procedures for surfactant–HNTs hybrid material preparations and the methods which are generally reported for the characterization of the prepared materials.

6.2.1 HNT Functionalization with Surfactants

Hybrid nanomaterials based on HNTs and surfactants were prepared using several approaches, which can be classified on the basis of the interactions between the two components. Particularly, HNT functionalization was obtained by:

- electrostatic interactions between ionic surfactants and HNT surfaces;
- covalent HNT surface modification.

HNT functionalization can be easily achieved by mixing ionic surfactants and HNTs in aqueous media at a pH ranging between 2 and 8. The formation of the hybrid material is due to the ionic exchange involving the head polar group of the surfactant and the HNT surfaces. Anionic surfactants selectively interact with the positively charged inner surface, driving hydrophobization of the HNT cavity. Oppositely, cationic surfactants are selectively adsorbed onto the HNT shell, which possesses a negative surface charge (Figure 6.1).

As Table 6.1 illustrates, several ionic surfactants were used for the functionalization of HNTs by means of electrostatic interactions.

In detail, the hybrid materials were prepared as follows. A certain amount of HNTs were added to saturated aqueous surfactant solutions in order to obtain dispersions with a surfactant/HNT weight ratio ≥ 1, which ensures the HNT surface modification by surfactant. The obtained dispersions were

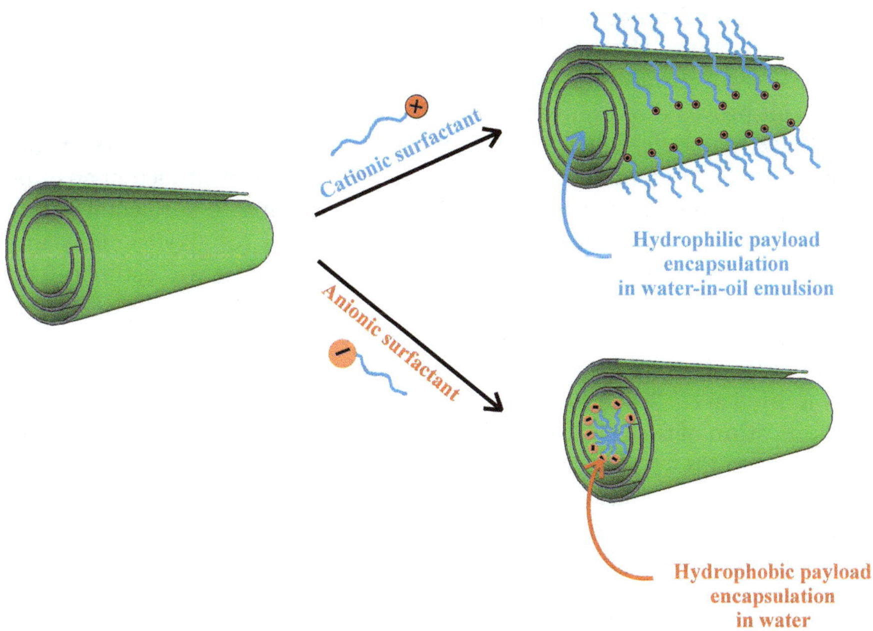

Figure 6.1 A schematic illustration of HNT selective surface modification to enhance the loading of the active cargo onto the lumen or tune the hydrophobicity of the HNT external surface.

Table 6.1 Ionic surfactants for the preparation of hybrid materials based on HNTs.

Surfactant	Symbol	Polar head group	References
Sodium undecanoate	NaC11	Anionic	8
Sodium dodecanoate	NaC12	Anionic	8, 10, 35
Sodium tetradecanoate	NaC14	Anionic	8
Sodium perfluorooctanoate	NaPF8	Anionic	26
Sodium perfluoroheptanoate	NaPF7	Anionic	26
Sodium perfluoropentanoate	NaPF5	Anionic	26
Octadecylphosphonic acid	ODP	Anionic	27
Sodium dodecylsulphate	NaDS	Anionic	38
Sodium dioctyl sulfosuccinate	DOSS	Anionic	33
Decyltrimethylammonium bromide	C10Br	Cationic	10 and 16
Tetradecyltrimethylammonium bromide	C14Br	Cationic	16
Hexadecyltrimethylammonium bromide	C16Br	Cationic	16 and 34

magnetically stirred for 48 h at 20 °C. For the modification of the HNT inner surface the suspensions could be taken under vacuum overnight in order to favor the encapsulation of the surfactant into the cavity.[26] The fizzing of the suspensions indicates the evaporation of water from the HNT lumen and replacement with the surfactant solution. Successively, the dispersions were centrifuged to recover the functionalized nanomaterials, which were washed with water several times to remove the unbound surfactant. Concerning the anionic surfactants, the washing procedure was carried out until the rinsed water reached a surface tension close to the value of pure water (72 mN m^{-1}). This result proves that eventually free surfactant is no longer present in the centrifuged materials, which contain only functionalized HNTs with a very low surface chemical potential in the aqueous phase because of surfactant adsorption onto the inner surface of the HNTs. The encapsulated surfactant is not released within at least one month. On the other hand, for alkyltrimethylammonium bromides the efficiency of the washing procedure was estimated by checking the presence of bromide in the rinsing water through $AgNO_3$ addition. In particular, the functionalized nanomaterials were rinsed with water until no formation of AgBr precipitate was observed. Finally, the obtained powder was dried at *ca.* 80 °C and stored in a desiccator at room temperature.

Modification of the HNT lumen with octadecylphosphonic acid (ODP) was carried out in EtOH–H_2O solution.[27] Further modification of ODP–HNTs by *N*-(2-aminoethyl)-3-aminopropyltrimethoxysilane (AEAPS) was performed by mixing AEAPS and the dried ODP in dry toluene.

6.2.2 Characterization

Surfactant–HNTs composite materials were extensively characterized by means of several techniques (Table 6.2), which provided a rather complete picture on the properties and perspectives of the functionalized nanomaterials.

Table 6.2 Experimental techniques for the characterization of surfactant–HNTs hybrid materials.

Technique	Property/insight	References
Thermogravimetry (TG)	HNT functionalization efficiency, surfactant loading, thermal stability	8, 10, 16, 26, 33
Fourier transform infrared spectroscopy (FTIR)	Specific interactions	8, 10, 16, 27
X-ray diffraction (XRD)	HNT multilayer wall spacing, surfactant intercalation	27 and 34
X-ray photoelectron spectroscopy (XPS)	Surface chemical composition and surface bonds in the modified HNTs	27
Nuclear magnetic resonance (NMR)	Surface bonds in the modified HNTs	27
Dispersive X-ray spectrometry (EDX)	Local elemental analysis	16 and 38
Scanning electron microscopy (SEM)	Morphology and size distribution	8, 10, 16, 26, 33, 38
Transmission electron microscopy (TEM)	Cavity loading	16, 27, 33, 38
Atomic force microscopy (AFM)	Roughness	35
Water contact angle	External/internal hydrophobization	16
Dynamic light scattering (DLS)	Average translational diffusion dynamic, aggregation in solvent media	8, 10, 16, 26
ζ-Potential	Particle charge	8, 10, 16, 26
Turbidimetry	Sedimentation kinetics	8 and 10

6.2.2.1 *Thermogravimetry*

Thermogravimetry (TG) is an established technique to investigate the thermal behavior of hybrid materials, such as polymeric nanocomposites[28–30] and functionalized nanoparticles.[8,10] TG measurements were performed by heating the sample at a constant scanning rate within a temperature range. These experiments provided the dependence of the mass loss on temperature (TG curve). Generally, the analysis of TG curves enables the quantitative composition of composite materials to be determines using several mathematical approaches. Therefore, TG experiments are appropriate to estimate the functionalization efficiency of nanoparticle surfaces. Concerning the modified HNTs, the amount of surfactant adsorbed onto the nanoclay surface was estimated by using the rule of mixture once the TG curves of the pure components were known. In particular, the surfactant weight percent (C_S) in the hybrid materials was determined by focusing on the weight percentage of degraded matter at 600 °C (MD_{600}), which can be expressed for each measured sample as

$$MD_{600} = 100 - (RM_{600} - ML_{150}) \tag{6.1}$$

where RM_{600} is the weight percentage of residual matter at 600 °C while ML_{150} is the weight percentage of the water content calculated as the mass loss to 150 °C. Assuming that the surfactant–HNTs interactions in the hybrid do not alter the MD_{600} of each component, the weight percentage of degraded matter at 600 °C for the modified HNTs (MD_{600SH}) is given by

$$MD_{600SH} = MD_{600H} + C_S \times 10^{-2} (MD_{600S} - MD_{600H}) \tag{6.2}$$

where MD_{600H} and MD_{600S} are the weight percentages of degraded matter at 600 °C for HNTs and surfactant, respectively.

The combination of eqn (6.1) and (6.2) represents a successful protocol for the estimation of the degree of functionalization of the HNTs. The blank method[31] was used to correct the error in the mass determination due to the reduction of the buoyancy force on increasing temperature. Specifically, a TG run with an empty pan was preliminarily performed in the same experimental conditions used for the samples. Then, the obtained blank curve was subtracted from those of the measured samples. It should be noted that even very low surfactant loadings ($C_S < 1$ wt%) can be determined using this approach.

Furthermore, TG experiments are proper tools to investigate the solubilization capacity of modified HNTs toward hydrocarbons by using the thermogravimetric volatilization method. The first-order differentiation curves of mass loss to temperature (DTG) were used to monitor the volatilization rates, which enables calculation of the amount of hydrocarbons solubilized by the functionalized HNTs. Additionally, the adsorption of the hydrocarbons onto the inner surface was evidenced by the temperature at the maximum volatilization rate (T_m). According to the Gibbs–Thomson effect, a shift of T_m to a value lower than that of pure hydrocarbons is expected in a confined environment such as the HNT lumen. As pristine HNTs possess hydrophilic surfaces, T_m reduction indirectly proved the formation of a hydrophobic pocket in the functionalized nanoparticles.

6.2.2.2 Spectroscopy

Spectroscopy is fundamental to study the chemical modification of nanoparticle surfaces as well as the adsorption ability in aqueous media. Studies on surfactant–HNTs hybrids were based on several spectroscopic techniques (FTIR, NMR, XPS and XRD), adopted to provide evidence of the nature of interactions between the two components. The successful functionalization of HNT surfaces was evidenced by comparing spectra of the composites with those of the corresponding pristine components. In particular, the surfactant–HNTs interactions induced variations (shifting, splitting, *etc.*) of the characteristic vibration bands explored through FTIR. Isotopic chemical shifts evidenced by NMR spectra were attributed to surface bonds in the modified HNTs. Changes in the HNT surface chemical composition were analyzed by XPS spectra. XRD data provided insights into the HNT multilayer wall spacing, which can be affected by the intercalation of the surfactant into the nanoclay interlayer. Furthermore, UV-vis spectroscopy enabled the study of the solubilization ability of functionalized HNTs toward hydrophobic compounds in aqueous media. In order to determine the adsorption isotherms, different amounts of surfactant–HNTs hybrids were added to a fixed quantity of water saturated with hydrocarbons. The dispersions were stirred and equilibrated for one week. The concentration of the solubilized hydrocarbons was determined by measuring the absorption spectra. Moreover, the kinetics of hydrocarbon release from modified HNTs were studied. UV-vis absorption spectra were collected over time providing the release profiles. Both the adsorption isotherms and the release kinetics were influenced by the chemical modification of HNT surfaces, which possess tunable hydrophobic/hydrophilic characteristics because of the surfactant loading.

6.2.2.3 *Microscopy*

Morphological characterization of nanoparticles is generally carried out by means of several microscopic techniques, such as SEM, TEM and AFM. Concerning the surfactant–HNTs hybrid nanomaterials, SEM gives insights on the shape as well as the size polydispersity. SEM coupled with EDX enabled determination of the local elemental analysis along the surfaces of modified HNTs. Because of the better resolution, TEM images were helpful to estimate the effect of the functionalization on the size of HNTs. Changes in the dimensions of the inner and outer diameters were correlated to the selectivity of the HNT modification. AFM was used to analyze the roughness of the modified HNTs.

6.2.2.4 *Water Contact Angle*

Water contact angle measurements are good tools to investigate the hydrophobicity/hydrophilicity of a solid surface. These experiments were performed on surfactant–HNT hybrids using the sessile drop method. Data analysis was focused on the initial water contact angle (θ_i), which corresponds to a contact angle value after just the deposition of the water droplet on the solid surface. θ_i values of the hybrids were useful to estimate the hydrophobicity/hydrophilicity of the external surface of the HNTs.

6.2.2.5 *DLS, ζ-Potential and Turbidimetry*

Dynamic light scattering (DLS), ζ-potential and turbidity measurements are commonly used to explore the mechanisms influencing the aqueous colloidal stability.

DLS data were used to determine the dynamic behavior of modified HNTs in water. In detail, the field-time autocorrelation functions were well described by a mono-exponential decay, which provided the rate (Γ) of the diffusive mode. For the translational motion, the apparent diffusion coefficient (D_t) can be expressed as

$$D_t = \Gamma/q^2 \tag{6.3}$$

where q is the scattering vector given by $4\pi n\lambda^{-1}\sin(\theta/2)$, where n is the water refractive index, λ the wavelength (632.8 nm) and θ the scattering angle (173°).

D_t results were helpful to investigate aggregation processes between modified HNTs as well as the balance between the repulsive and attractive interactions.

ζ-Potential measurements allowed us to determine the surface electrical potential of modified HNTs. For this purpose, a diluted aqueous dispersion of surfactant–HNTs hybrids was transferred into the special cuvette and an electric field was applied across the suspension causing the movement of the charged nanoparticles toward the electrode with opposite charge. The ζ-potential of the particles was determined by the Smoluchowski formula using particle speed measured with DLS. Pristine HNTs possess a net negative charge. ζ-Potential values of the hybrids evidenced inversions of the charge and/or variations of the ζ-potential induced by the functionalization of the HNT surfaces. These results were useful to investigate both the selectivity of the surfactant adsorption onto HNT surfaces and the aqueous colloidal stability of functionalized nanoclays interpreted in terms of the DLVO theory.

Turbidimetry enabled the study of the sedimentation kinetics of functionalized HNTs. Turbidity experiments on aqueous dispersions of surfactant–HNTs composites were performed using a UV-vis spectrometer at a wavelength of 600 nm, where no absorption bands were detected. The sedimentation was strongly affected by the selective surfactant interactions with the HNT surfaces. In addition, turbidity measurements under variable ionic strength conditions were performed with the aim to investigate the crucial role of the electrostatic forces in controlling the aqueous colloidal stability of the hybrid materials.

6.3 Results and Discussion

In this paragraph, we review the characterization of surfactant–HNTs hybrids. Aqueous colloidal stability and solubilization capacity are crucial to evaluate the effectiveness of the hybrids as inorganic tubular micelles and their applicability in water media to encapsulate apolar compounds.

6.3.1 Evidencing the Hybrid Formation

As a general feature, the modification of HNTs may generate exfoliation[32] inducing the loss of the hollow cylinder. SEM images of the hybrids evidenced that regardless of the nature of the interactions between the nanotubes and surfactants, the cylindrical morphology is preserved (Figure 6.2).

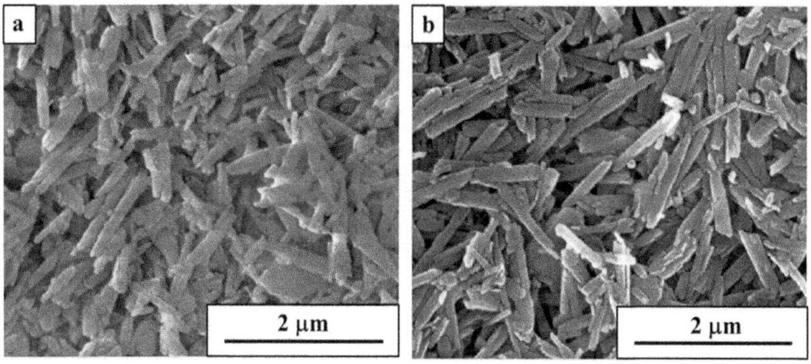

Figure 6.2 SEM micrographs of NaPFC5–HNTs (a) and NaC12–HNTs (b). Adapted with permission from G. Cavallaro, G. Lazzara, S. Milioto, F. Parisi, and V. Sanzillo, *ACS Appl. Mater. Interfaces*, 2014, **6**, 606–612.[8] Copyright (2014) American Chemical Society and reprinted from *J. Colloid Interface Sci.*, **417**, G. Cavallaro, G. Lazzara, S. Milioto, G. Palmisano, and F. Parisi, Halloysite nanotube with fluorinated lumen: Non-foaming nanocontainer for storage and controlled release of oxygen in aqueous media, 66–71.[26] Copyright 2014 with permission from Elsevier.

TEM micrographs showed the hollow nanotubular structure of pristine HNTs while in the case of the HNTs filled with anionic surfactants, the hollow was no longer evident as a consequence of the higher electron density generated by the surfactants.[33]

It was experimentally evidenced that the loaded amount of surfactant depends on both the hydrophobicity of the chains as well as the nature of the polar head (Table 6.3). Based on geometric considerations, the maximum loading value of the cavity size is *ca.* 10% in volume. As far as the sodium alkanoate surfactant is concerned, assuming the presence of a monolayer in the lumen and the head group surface area, from geometrical considerations, a loading value of 1.2 wt% was predicted; that means that above such a value more hydrophobic surfactants form complex structures. In the case of cationic surfactants, the amount of surfactant adsorbed onto the nanotube external surface never reached the predicted value for a monolayer configuration.

FTIR spectra of hybrid materials highlighted the interactions between HNTs and surfactants, as the characteristic bands of both the surfactants and HNTs were influenced. It was intriguing[10] to note that the stretching of the carboxylate group of NaC12 was split into two peaks (1541.4 and 1578.1 cm^{-1}) due to its interactions with the positive charged sites of the alumina groups; consequently, the negative charge delocalization on the carboxylate group was hindered. Moreover, FTIR as well as NMR data proved the covalent binding of ODP to the lumen of HNTs through bidentate and tridentate P–O–Al linkages.[27] The successful modification of the HNT lumen with ODP was confirmed by XPS measurements. Namely, the wall thickness of the pristine HNTs (*ca.* 20 nm) exceeds the penetration depth of XPS (10 nm) and

Table 6.3 Characteristics of surfactant–HNTs hybrid materials.

Materials	C_S/wt%	ζ/ζ_0
NaC11–HNTs	1.4	1.46[a]
NaC12–HNTs	8.0	1.30[a]
NaC14–HNTs	6.6	1.75[a]
NaPFC5–HNTs	0.29	1.29[b]
NaPFC7–HNTs	0.62	1.38[b]
NaPFC8–HNTs	0.86	1.52[b]
C10Br–HNTs	0.89	0.90[c]
C14Br–HNTs	1.49	0.40[c]
C16Br–HNTs	1.95	−0.80[c]

[a]From ref. 8 (ζ_0 = −19 mV).
[b]From ref. 26 (ζ_0 = −21 mV).
[c]From ref. 16 (ζ_0 = −30 mV).

therefore, it was difficult to disclose the chemical composition of the inner lumen surface where the surfactant bonding occurred. Nevertheless, being that the nanotubes were oriented randomly, the inner surface was partially exposed to X-rays, and therefore, the presence of specific atoms (P and C) evidenced the functionalization of the HNTs. XRD spectra evidenced that the HNT multilayer wall spacing is not altered by surface functionalization with either ODP[27] or C16Br,[34] indicating that no intercalation of the surfactants into the nanoclay interlayer walls occurred.

The selective functionalization of HNT surfaces influenced the wettability properties of the hybrids. In particular, the presence of cationic surfactants (C10Br, C14Br and C16Br) generated an increase of θ_i values according to the enhancement of the surface hydrophobicity. Based on AFM micrographs, the surface roughness is not altered by the presence of alkyltrimethylammonium bromides.[35] Therefore, the θ_i increase is due only to the presence of hydrophobic chains on the HNT external surface. On the other hand, literature[25] reports that the functionalization of the HNT cavity with anionic polymers did not affect the hydrophilicity of the HNTs.

ζ-Potential data revealed that changes in the modification of HNTs are sensitive to variations in the charge of the plane. The simultaneous presence of positive and negative charges at the inner and external surfaces, respectively, renders the ζ-potential value sensitive to some parameters including pH.[23] The adsorption of surfactants onto HNTs (Table 6.3) generated strong ζ-potential variations. The ζ-potential values of alkyltrimethylammonium bromide–HNTs hybrids agree with the neutralization of negative sites of the external surface from the polar head charge side. Interestingly, an inversion of the charge was observed for C16Br–HNTs. Sodium alkanoates and sodium perfluoroalkanoates included into the lumen cancel the positive charges, generating an increase in the total negative charge of the HNTs. The same result was detected when HNTs were functionalized using sodium hexametaphosphate.[36]

According to the ζ-potential results, the diffusion dynamics of nanohybrids were affected by the selective modification of HNT surfaces. The nanostructures containing anionic surfactants showed average translational diffusion coefficients (D_t) close to those reported for pristine HNTs in water, which is consistent with the penetration of the surfactants into the hollow cylinder of HNTs. The situation was rather different when cationic surfactants were involved. Accordingly, the much larger D_t values than pristine HNTs reflect the tendency to form large aggregates, enhanced by hydrophobic forces exercised between the surfactant chains on the HNTs outer surfaces.

6.3.2 Colloidal Behavior of the Hybrid in Solvent Media

The sedimentation of the clay nanotubes assumes a key role in understanding their colloidal stability and, consequently, their potential applications. Experiments done in 1952 on HNT particle-size distribution revealed anomalous sedimentation behavior.[37] In particular, a loose lattice structure in the suspension was invoked to interpret the formation of a clear layer and a stable concentrated bottom phase. Very recently, such a phenomenon was quantitatively explored and interpreted. Figure 6.3 shows the two phases, namely a bottom milk-like phase and an upper transparent phase, obtained after equilibration of HNT dispersion in water. The lower phase represents the sedimentation volume and it has a specific concentration depending on the HNTs characteristics. It can be defined as a critical concentration above which the dispersion is stable because the highest packing of the nanotubes is approached and shorter distances between the nanoparticles are hindered

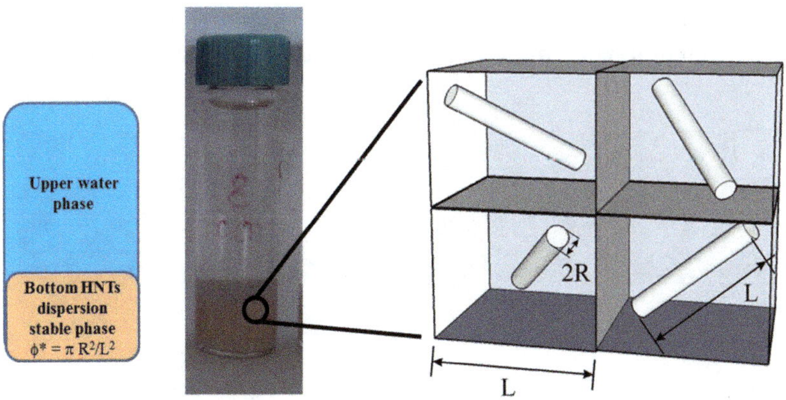

Figure 6.3 Optical image of the sedimentation volume in a glass vial for NaPFC8–HNTs and a sketch representation of the simple cubic model used for the determination of the overlapping concentration generating the nanotube entanglement. Adapted from *J. Colloid Interface Sci.*, **417**, G. Cavallaro, G. Lazzara, S. Milioto, G. Palmisano, and F. Parisi, Halloysite nanotube with fluorinated lumen: Non-foaming nanocontainer for starage and controlled release of oxygen in aqueous media, 66–71.[26] Copyright 2014 with permission from Elsevier.

by the excluded volume and/or the electrostatic repulsions. A simple cubic model considering a nanotube contact distance given by the average length of the nanotubes (see Figure 6.3) nicely describes the experimental data. Assuming that rotation is much quicker than translation, the overlapping concentration that generates the HNT entanglement (ϕ^*) is given by

$$\phi^* = \pi R^2/L^2 \tag{6.4}$$

where R and L are the external radius and the length of the nanotubes.

For pristine HNTs, the sedimentation volume is mainly driven by hard-cylinder interactions while hybrid surfactant–HNTs present stronger electrostatic repulsions in agreement with the ζ-potential alteration. Such a strong repulsion generates a reduction of the overlapping concentration that generates a stable milky phase at a concentration lower than that required for pristine HNTs.

TEM micrographs of HNTs compared to those of sodium dodecyl sulphate–HNTs systems confirms these hypotheses (Figure 6.4). In particular, they indicate the presence of a large number of aggregated nanotubes in the raw HNTs; however, most of the aggregates have been well dispersed by adding the surfactant.[38] The aggregate density of the hybrid decreases and the average distance between nanoparticles increases in agreement with the stronger electrostatic repulsion expected between the modified HNTs.

Another approach in studying the stability of nanotube dispersion in water takes into account the low concentration regimes much lower than the overlapping concentration. These results are relevant in any application requiring dilute dispersions such as drug delivery.

DLS and ζ-potential findings predict a better colloidal stability of the hybrid anionic surfactant–HNT materials compared to the pristine nanotubes due

Figure 6.4 TEM images of HNTs (a) and NaDS–HNTs (b). Adapted with permission from ref. 38 with kind permission from Springer Science and Business Media.

to the low tendency to aggregate controlled by the increased electrostatic repulsions between the particles. The opposite effect on the colloidal stability of HNTs is expected by functionalization with cationic surfactants because of the hydrophobic forces that favor the formation of clusters between the hybrid nanomaterials. On the contrary, the hydrophobization of the HNT external surface enhances the affinity toward nonpolar solvents, as evidenced by partition experiments.[16] For example, the chloroform–water partition constant of the C16Br–HNTs composite is 20 times larger than that of pure HNTs.[16]

As observed in Figure 6.5, the turbidity value of the aqueous dispersions after equilibration follows the order NaC12–HNTs > HNTs > C10Br–HNTs, highlighting that the cationic surfactant enhances the sedimentation whereas the alkanoate is an efficient stabilizing agent for the dispersion of

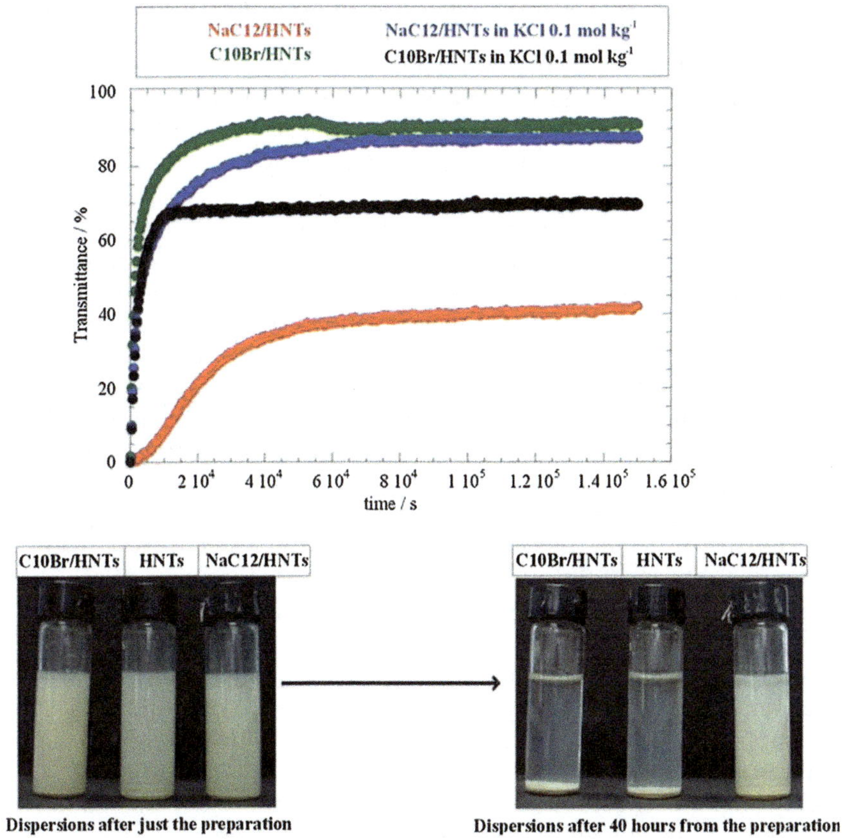

Figure 6.5 Photographs and transmittance as a function of time for HNTs and functionalized HNT dispersions in water and in a water + KCl mixture. The concentration of the dispersions was 1 wt%. Adapted with permission from G. Cavallaro, G. Lazzara, and S. Milioto, *J. Phys. Chem. C*, 2012, **116**, 21932–21938.[10] Copyright (2012) American Chemical Society.

HNTs. In accordance with ζ-potential data, the presence of KCl strongly destabilizes the NaC12 modified HNTs because of the salt screening effect while C10Br–HNTs sedimentation is slightly influenced by the ionic strength.

6.3.3 Solubilization Ability

Filling the HNT nanocavity is strategic for the removal of substances from a certain matrix, like decontamination processes, and for any application where the controlled release of the payload should occur, such as in drug delivery or films with active protection ability. In both cases, the loading efficiency is crucial and research was devoted to maximize it. The most common and efficient loading protocol consists of the use of a vacuum, which favors solvent evaporation within the HNT lumen due to the Gibbs–Thomson confinement effect and therefore generates an accumulation of the solute within the cavity. This strategy was successfully reported for pristine HNT loading with several active small molecules.[1,39] As an example, end-capped HNTs were proposed as smart containers for controlled release of a corrosion inhibitor.[40] Oil entrapment experiments in surfactant–HNTs hybrids are reported from aqueous media and, in the case of volatile oils, from the vapor saturated gas phase. In other words, the hydrophobized HNTs were employed as inorganic micelles to capture hydrophobic compounds.

The most common experiments are related to the solubilization/release of hydrophobic compounds in aqueous dispersion. HNT lumen modification through specific reaction with ODP generated an efficient hydrophobic cavity responsible for an enhanced ferrocene incorporation. A detailed investigation of the release profile of ferrocene as a model molecule showed a two-step release profile reflecting desorption of the outside surface and from the nanotube lumen.[27] Sodium alkanoate–HNTs hybrids exhibited a higher capturing ability toward toluene from aqueous solution compared to HNTs. The surfactant tail length played a role. Accordingly, the NaC14–HNTs with the largest surfactant loading exhibited the strongest solubilizing power as a consequence of the formation of complex structures within the lumen.[8]

Comparing the solubilization power of such inorganic micelles with that of sodium alkanoate micelles in water showed that inorganic micelles are much more efficient than the conventional ones as concentrations far below the critical micellar concentration (cmc) endowed oil capturing. This result is relevant in the light of any oil solubilization within applications as smaller amounts of surfactant than required for forming conventional micelles are required to form the hydrophobic pockets.

A synergistic effect on dodecane-in-water emulsion stabilization was observed for HNTs loaded with dioctyl sulfosuccinate sodium salt (DOSS) surfactant.[33] Interestingly, HNTs exercised a stabilization effect on oil droplets because of the rigid and protective interfacial film formed at the oil–water interface. Consequently, the DOSS loaded HNTs generated stable emulsions with a droplet size controlled by the DOSS/HNTs ratio (Figure 6.6).

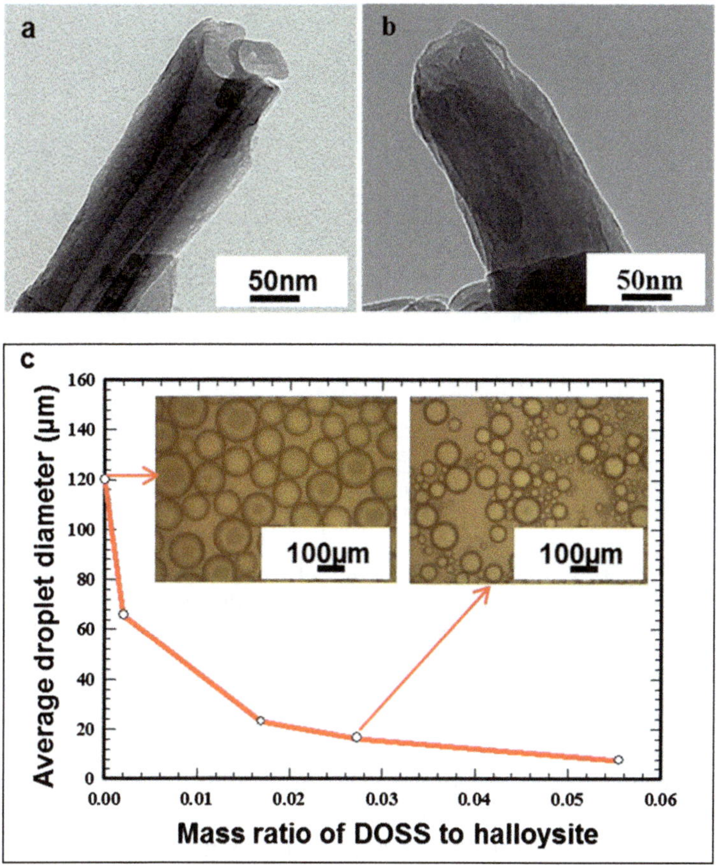

Figure 6.6 TEM images of HNTs (a) and DOSS–HNTs (b). Panel (c) shows the effect of DOSS loading and release from HNTs on the average droplet size of dodecane-in-water emulsions. The inset shows optical microscopy images of dodecane-in-water emulsions stabilized by HNT alone and DOSS-loaded HNT at 2.83 wt% DOSS loading. Adapted with permission from O. Owoseni, E. Nyankson, Y. Zhang, S. J. Adams, J. He, G. L. McPherson, A. Bose, R. B. Gupta, and V. T. John, *Langmuir*, 2014, **30**, 13533–13541.[33] Copyright (2014) American Chemical Society.

Incorporation/release of oil vapors into/from modified HNTs are reported.[8] The thermodynamic property which properly evidences the affinity of gaseous hydrocarbons toward the hybrid composite is the adsorption constant, calculated as

$$K_{\mathrm{ads}} = Q_{\mathrm{ads}}/P_{\mathrm{vap}} \tag{6.5}$$

where P_{vap} is the vapor pressure and Q_{ads} is the adsorbed amount of oil per adsorbent mass. The K_{ads} for all the investigated surfactants–HNTs hybrids

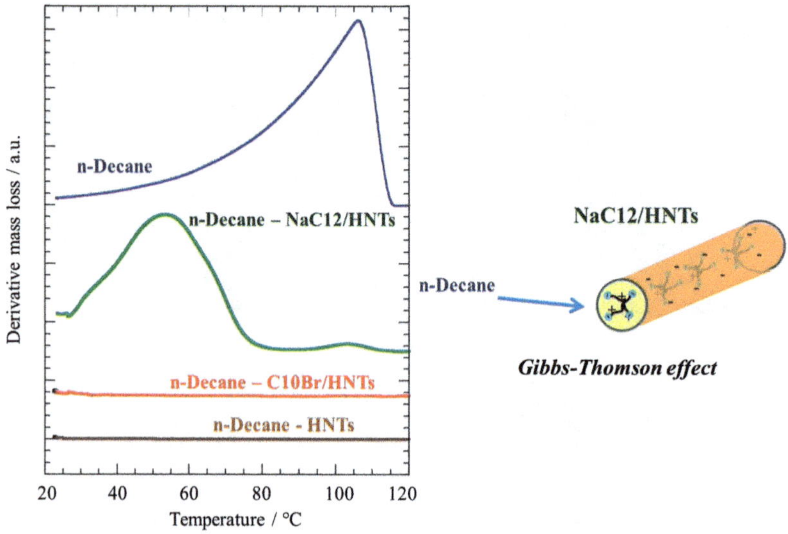

Figure 6.7 Volatilization rate as a function of temperature for *n*-decane and HNT based materials after equilibration with *n*-decane, and a schematic illustration of the Gibbs–Thomson effect occurring for NaC12–HNTs. Adapted with permission from G. Cavallaro, G. Lazzara, and S. Milioto, *J. Phys. Chem. C*, 2012, **116**, 21932–21938.[10] Copyright (2012) American Chemical Society.

are nearly independent of temperature and exhibit a larger affinity toward *n*-decane than pristine HNTs. Despite the amount of the surfactant as well as the alkyl chain lengths being different, the anhydrous hybrids evidenced a comparable adsorption ability, likely because the thermodynamic properties reveal the interactions exercised between the molecule and its environment in the initial and final states. Concerning the oil vapor release from the HNT cavity, vaporization rate measurements evidenced the depletion of the oil boiling point (Figure 6.7), as well predicted by the Gibbs–Thomson equation for confined solvents in a nanosized cavity.[10]

The perspective of gas solubilization as oxygen in aqueous media were exploited using hybrid materials based on HNTs where the lumen was modified with perfluorinated anionic surfactants. It should be noted that, contrary to conventional micellar solutions, nanomaterials based on fluorinated surfactants and HNTs do not generate foam during oxygen bubbling as well as upon stirring. Concerning the gas desaturation curves (Figure 6.8), both HNTs and NaPFC8–HNTs are efficient in keeping the supersaturation state for a longer time compared to water; nevertheless, after 6 h the HNT dispersion released half of the O_2 amount while the hybrid nanotubes showed an oxygen loss of only 10%. It is likely that NaPFC8–HNTs behave like an O_2 reservoir, releasing the gas in water over time in contrast to the desaturation of the aqueous dispersion.

Finally, the capability to remove an oil film at the water–air interface using surfactant–HNT systems was proved by time-resolved surface tension

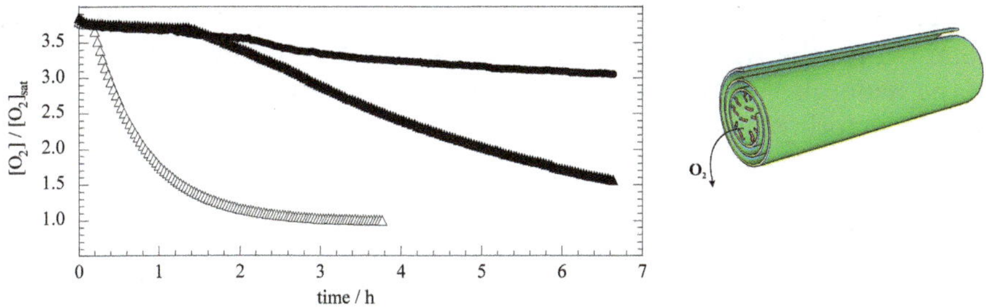

Figure 6.8 Oxygen concentration as a function of time in water (Δ), in aqueous dispersions of HNTs (▲) and NaPFC8–HNTs (●). The oxygen reservoir ability of the fluorinated lumen is sketched. Adapted from *J. Colloid Interface Sci.*, **417**, G. Cavallaro, G. Lazzara, S. Milioto, G. Palmisano, and F. Parisi, Halloysite nanotube with fluorinated lumen: Non-foaming nanocontainer for storage and controlled release of oxygen in aqueous media, 66–71.[26] Copyright 2014 with permission from Elsevier.

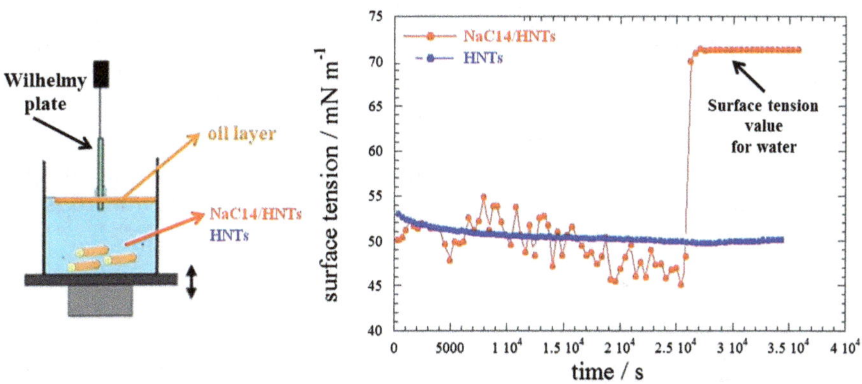

Figure 6.9 Surface tension as a function of time for aqueous HNTs and NaC14 dispersions in the presence of an *n*-decane film. A schematic illustration of the surface tension experiments is reported. Adapted with permission from G. Cavallaro, G. Lazzara, S. Milioto, F. Parisi, and V. Sanzillo, *ACS Appl. Mater. Interfaces*, 2014, **6**, 606–612.[8] Copyright (2014) American Chemical Society.

measurements. In this case, the jump of surface tension to the pure water value indicates oil removal from the interface. The pristine HNTs were revealed to be unsuitable for oil incorporation while NaC14–HNTs were able to clean the surface of liquid *n*-decane (Figure 6.9). The efficiency is dependent on the alkyl chain length of the surfactant, with NaC11–HNTs being inappropriate for the purpose.

Functionalized HNTs with anionic and cationic surfactants can be used for inducing a robust, differential adhesion response of cells on coated surfaces.[35] To this aim, HNTs were immobilized in a polyurethane matrix and a coated surface with nanoscale roughness was obtained. Although the roughness was not altered by functionalization, its cancer cell capturing ability was enhanced by the increase of the negative charge of the HNTs, which was obtained for the NaC12–HNTs system. Conversely, diminishing the charge of the nanotubes with a cationic surfactant abolished cancer cell adhesion while promoting leukocyte adhesion.[35]

6.4 Conclusions

The selective surface modification of HNTs was successfully controlled by electrostatic interactions. The lumen was loaded with anionic surfactants forming either a monolayer or more complex ordered structures. These nanohybrids showed an enhancement of the nanoparticle net charge that controlled the stability of the dispersed nanotubes in aqueous media. The outer HNT surface modification by cationic surfactants generated nanotubes with a better affinity toward apolar solvents. On the contrary, HNTs can be dispersed in apolar solvents by modifying the external surface with cationic surfactants as a result of the hydrophobization of the nanotube shell. Hybrids based on HNTs and perfluoroalkylated anionic surfactants can be used as oxygen nanocontainers in aqueous media. Adsorption of sodium alkanoates onto the inner surface of HNTs enables fabrication of inorganic tubular micelles able to encapsulate aliphatic and aromatic hydrocarbons.

The functionalization of HNTs with surfactants represents an efficient route to obtain composite nanomaterials at low cost, which are biocompatible and easy to separate. The inorganic micelles are suitable for industrial applications over a large scale.

Acknowledgements

The work was financially supported by the University of Palermo, PRIN 2010–2011 (prot. 2010329WPF), FIRB 2012 (prot. RBFR12ETL5).

References

1. Y. Lvov and E. Abdullayev, *Prog. Polym. Sci.*, 2013, **38**, 1690–1719.
2. Y. M. Lvov, D. G. Shchukin, H. Mohwald and R. R. Price, *ACS Nano*, 2008, **2**, 814–820.
3. E. Joussein, S. Petit, G. J. Churchman, B. Theng, D. Righi and B. Delvaux, *Clay Miner.*, 2005, **40**, 383–426.
4. P. Pasbakhsh, G. J. Churchman and J. L. Keeling, *Appl. Clay Sci.*, 2013, **74**, 47–57.
5. S. Barrientos-Ramírez, E. V. Ramos-Fernández, J. Silvestre-Albero, A. Sepúlveda-Escribano, M. M. Pastor-Blas and A. González-Montiel, *Microporous Mesoporous Mater.*, 2009, **120**, 132–140.
6. R. Wang, G. Jiang, Y. Ding, Y. Wang, X. Sun, X. Wang and W. Chen, *ACS Appl. Mater. Interfaces*, 2011, **3**, 4154–4158.

7. G. S. Machado, K. A. D. de Freitas Castro, F. Wypych and S. Nakagaki, *J. Mol. Catal. A: Chem.*, 2008, **283**, 99–107.
8. G. Cavallaro, G. Lazzara, S. Milioto, F. Parisi and V. Sanzillo, *ACS Appl. Mater. Interfaces*, 2014, **6**, 606–612.
9. P. Luo, J. Zhang, B. Zhang, J. Wang, Y. Zhao and J. Liu, *Ind. Eng. Chem. Res.*, 2011, **50**, 10246–10252.
10. G. Cavallaro, G. Lazzara and S. Milioto, *J. Phys. Chem. C*, 2012, **116**, 21932–21938.
11. F. Dong, J. Wang, Y. Wang and S. Ren, *J. Mater. Chem.*, 2012, **22**, 11093–11100.
12. G. Cavallaro, G. Lazzara and S. Milioto, *Polym. Degrad. Stab.*, 2013, **98**, 2529–2536.
13. G. Cavallaro, G. Lazzara, S. Konnova, R. Fakhrullin and Y. Lvov, *Green Mater.*, 2014, **2**, 232–242.
14. E. Abdullayev, K. Sakakibara, K. Okamoto, W. Wei, K. Ariga and Y. Lvov, *ACS Appl. Mater. Interfaces*, 2011, **3**, 4040–4046.
15. W. Wei, R. Minullina, E. Abdullayev, R. Fakhrullin, D. Mills and Y. Lvov, *RSC Adv.*, 2014, **4**, 488–494.
16. G. Cavallaro, G. Lazzara, S. Milioto and F. Parisi, *Langmuir*, 2015, **31**, 7472–7478.
17. A. Joshi, E. Abdullayev, A. Vasiliev, O. Volkova and Y. Lvov, *Langmuir*, 2013, **29**, 7439–7448.
18. E. Abdullayev and Y. Lvov, *J. Mater. Chem.*, 2010, **20**, 6681–6687.
19. D. G. Shchukin and H. Möhwald, *Small*, 2007, **3**, 926–943.
20. F. Arcudi, G. Cavallaro, G. Lazzara, M. Massaro, S. Milioto, R. Noto and S. Riela, *J. Phys. Chem. C*, 2014, **118**, 15095–15101.
21. C. Chao, J. Liu, J. Wang, Y. Zhang, B. Zhang, Y. Zhang, X. Xiang and R. Chen, *ACS Appl. Mater. Interfaces*, 2013, **5**, 10559–10564.
22. G. Cavallaro, G. Lazzara, M. Massaro, S. Milioto, R. Noto, F. Parisi and S. Riela, *J. Phys. Chem. C*, 2015, **119**, 8944–8951.
23. V. Vergaro, E. Abdullayev, Y. M. Lvov, A. Zeitoun, R. Cingolani, R. Rinaldi and S. Leporatti, *Biomacromolecules*, 2010, **11**, 820–826.
24. Z. Luo, H. Song, X. Feng, M. Run, H. Cui, L. Wu, J. Gao and Z. Wang, *Langmuir*, 2013, **29**, 12358–12366.
25. Y. Zhao, G. Cavallaro and Y. Lvov, *J. Colloid Interface Sci.*, 2015, **440**, 68–77.
26. G. Cavallaro, G. Lazzara, S. Milioto, G. Palmisano and F. Parisi, *J. Colloid Interface Sci.*, 2014, **417**, 66–71.
27. W. O. Yah, A. Takahara and Y. M. Lvov, *J. Am. Chem. Soc.*, 2011, **134**, 1853–1859.
28. I. Blanco, L. Abate, F. A. Bottino and P. Bottino, *Polym. Degrad. Stab.*, 2014, **102**, 132–137.
29. G. Cavallaro, G. Lazzara and S. Milioto, *Langmuir*, 2011, **27**, 1158–1167.
30. M. Du, B. Guo and D. Jia, *Eur. Polym. J.*, 2006, **42**, 1362–1369.
31. S. Vyazovkin, K. Chrissafis, M. L. Di Lorenzo, N. Koga, M. Pijolat, B. Roduit, N. Sbirrazzuoli and J. J. Suñol, *Thermochim. Acta*, 2014, **590**, 1–23.

32. P. Sidheswaran, A. N. Bhat and P. Ganguli, *Clays Clay Miner.*, 1990, **38**, 29–32.
33. O. Owoseni, E. Nyankson, Y. Zhang, S. J. Adams, J. He, G. L. McPherson, A. Bose, R. B. Gupta and V. T. John, *Langmuir*, 2014, **30**, 13533–13541.
34. Y. Xi, M. Mallavarapu and R. Naidu, *Appl. Clay Sci.*, 2010, **48**, 92–96.
35. M. J. Mitchell, C. A. Castellanos and M. R. King, *Biomaterials*, 2015, **56**, 179–186.
36. N. Houta, G.-L. Lecomte-Nana, N. Tessier-Doyen and C. Peyratout, *J. Colloid Interface Sci.*, 2014, **425**, 67–74.
37. I. McDowall and W. Vose, *Nature*, 1952, **170**, 368.
38. H. Lun, J. Ouyang and H. Yang, *Phys. Chem. Miner.*, 2014, **41**, 281–288.
39. D. Fix, D. V. Andreeva, Y. M. Lvov, D. G. Shchukin and H. Möhwald, *Adv. Funct. Mater.*, 2009, **19**, 1720–1727.
40. Y. Lvov and E. Abdullayev, US 8507056 B2, 2013 Patent.

CHAPTER 7

Halloysite and Related Mesoporous Carriers for Advanced Catalysis and Drug Delivery

KATSUHIKO ARIGA*[a], HIDEKI ABE[a], QINGMIN JI[a,b] AND YURI M. LVOV[c,d]

[a]World Premier International (WPI) Research Center for Materials Nanoarchitectonics (MANA), National Institute for Materials Science (NIMS), 1-1 Namiki, Tsukuba 305-0044, Japan; [b]Herbert Gleiter Institute of Nanoscience, Nanjing University of Science and Technology, China; [c]Institute of Micromanufacturing, Louisiana Tech University, 911 Hergot Avenue, Ruston, Louisiana 71272, United States; [d]Kazan Federal University, Kreml Uramı 18, Kazan, Republic of Tatarstan, 420008, Russian Federation
*E-mail: ARIGA.Katsuhiko@nims.go.jp

7.1 Introduction

For various demands in current societies – including energy creation, optimization of energy consumption, minimization of waste materials, rapid and comfortable information exchanges, and smart materials for biomedical applications – fine control of nanostructures in target materials becomes an important key, because optimization of nanostructures often results in high

RSC Smart Materials No. 22
Functional Polymer Composites with Nanoclays
Edited by Yuri Lvov, Baochun Guo and Rawil F. Fakhrullin

Published by the Royal Society of Chemistry, www.rsc.org

efficiency and specificity in energy and materials conversion systems. Various microfabrication techniques, such as photolithography and electron-beam lithography, have been developed toward these targets. However, most of these technologies remain in micrometer-scale precision for structural control. Instead, for more precise structure control, bottom-up approaches have especially been paid much attention to control nanoscale structures on the basis of the self-organizing capabilities of the materials themselves.[1]

Recently, a more advanced concept, nanoarchitectonics, has been proposed to totally unify these scientific and technological flows.[2] This terminology, nanoarchitectonics, was first proposed and defined by Masakazu Aono at the *1st International Symposium on Nanoarchitectonics Using Suprainteractions* in 2000 in Tsukuba, Japan. With nanoarchitectonic strategies, functional materials with controlled nanostructures are architected through manipulation of atoms, molecules, and assemblies on the basis of the synergetic combination of chemical nanofabrication, self-organization, and field-controlled organization. The most important key would be the fundamental difference between nanoscopic-scale phenomena and those for the macroscopic-scale/micro-scale worlds. In the nanoscale dimension, perturbations from surrounding moieties or components, such as thermal fluctuation, unexpected mutual interactions, and statistic and quantum effects, lead to a requirement for concerted harmonization of the various interactions and effects for control of the organization of materials and stimulating spontaneous processes. The combined actions enable functional materials to be architected rather than simply being an assembly of techniques. Use of the nanoarchitectonics concept has been initiated in various kinds of research fields including nanomaterial development,[3] nanostructure fabrication,[4] application-oriented science including sensing and biological/biomedical applications.[5]

Among countless numbers of materials candidates, structurally-defined inorganic materials can be nice hosts for development of functional material structures according to nanoarchitectonics (Figure 7.1). For example, assembling materials on the surface of flat substrates and colloidal particles by layer-by-layer (LbL) adsorption is a fine versatile methodology to architect functional structures at external surfaces of inorganic host structures.[6] On the other hand, decoration with functional groups on the inside of mesoporous materials and confinement of functional units within mesopore channels is also a nice strategy to architect functional systems at the interior of inorganic structures.[7] More advanced nanoarchitectonics would be selective modification on the interior and exterior surfaces of inorganic hosts. For such a possibility, we briefly introduce recent research efforts on functionalization of the clay nanotubes, halloysites, in this chapter.

Halloysite clay is a natural tubular aluminosilicate that is chemically identical to kaolin.[8] Halloysite tubes are made through rolling typically 10–15 aluminosilicate layers up into cylindrical tubular structures. The external surface of the halloysite nanotubes is formed from silica, and its innermost is made from alumina, resulting in strong negative zeta-potential on the external surface and positive potential in the innermost tube at ambient pH

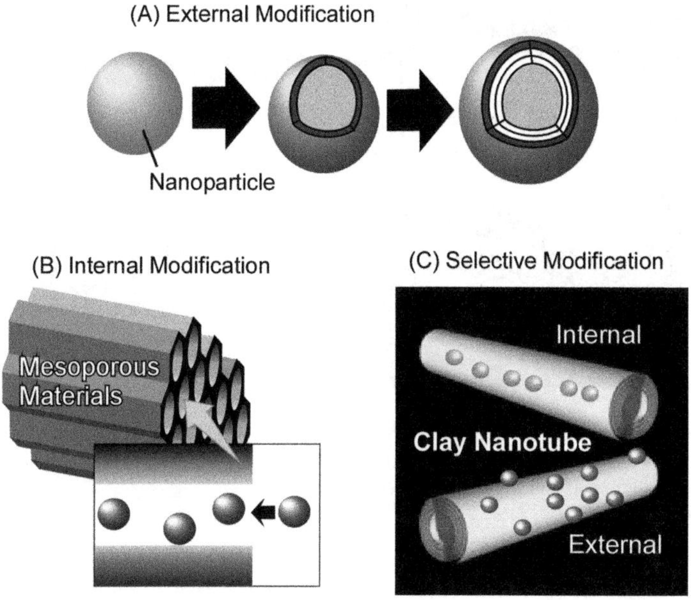

Figure 7.1 The nanoarchitectonics approach for functional materials using structurally-defined inorganic hosts: (A) external modification; (B) internal modification; (C) selective modification.

conditions in aqueous media. Their diameter ranges from 40–70 nm with an inner lumen diameter of 10–15 nm and their typical length becomes 1500 ± 500 nm. Undoubtedly, the halloysites are attractive inorganic hosts for further nanoarchitectonics with functional units. In this chapter, we mainly focus on introduction of research examples on functionalization of the halloysite nanotubes for catalysis and antibacterial activities. In addition, several examples of hierarchic silica nanostructures for applications such as drug delivery are also briefly introduced.

7.2 Halloysite: Catalyst Carrier

Because the halloysite clay nanotubes have asymmetric structural features at the tube interior and exterior, catalysts can be selectively immobilized on either the interior or exterior of the nanotubes. This characteristic in catalyst immobilization on a halloysite carrier provides us certain insights into materials designs for optimized catalytic performances, as seen in the following example.[9] In this example, as catalysts for removal of toxic nitrogen monoxide (NO) and carbon monoxide (CO) (with reaction scheme $NO + CO = 1/2N_2 + CO_2$ and/or $NO + CO = 1/2N_2O + 1/2CO + 1/2CO_2$), Cu–Ni nanoparticles (NPs) were selectively immobilized at the interior (AlO_x side) and the exterior (SiO_x side) of the halloysite clay nanotube materials (Figure 7.2). The Cu–Ni NPs immobilized at the interior of halloysite exhibited excellent catalytic activity in the

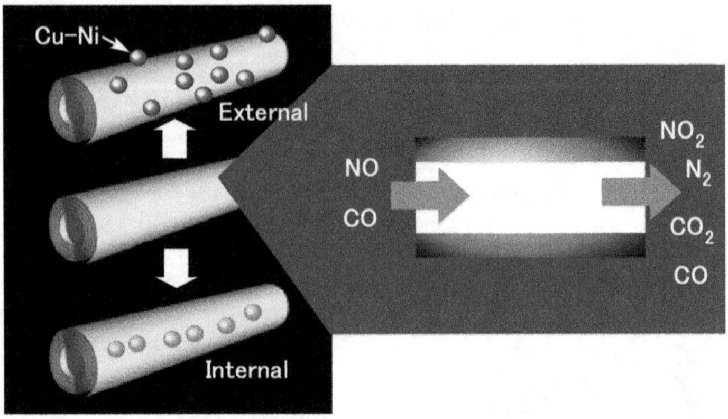

Figure 7.2 Removal of toxic nitrogen monoxide (NO) and carbon monoxide (CO) with Cu-Ni nanoparticles immobilized at the interior and exterior of the halloysite clay nanotubes.

purification of simulated exhaust comprising an equimolar mixture of NO and CO, while those at the exterior of the clay nanotubes were catalytically much less active, probably because of particle agglomeration.

Basically, Cu–Ni NPs were deposited on the surface of the halloysite nanotubes through co-reduction of $CuCl_2$ and $NiCl_2$ in aqueous ethanol. In the presence of sodium citrate, the Cu–Ni NPs were selectively immobilized on the AlO_x interior. Because the AlO_x interior of the halloysite clay nanotube has positive charges in aqueous solution in the range of pH = 6–7, negatively charged citrate ions coordinated to Cu^{2+} and Ni^{2+} are preferably delivered to the interior AlO_x surface. A reduction process with hydrazine at 80 °C finally converted these cations into bimetal NPs. On the other hand, the Cu–Ni NPs were formed only at the SiO_x exteriors of the halloysite clay nanotubes when citrate was not used in the synthetic process.

The prepared Cu–Ni NPs were extensively characterized (Figure 7.3). Transmission electron microscopy (TEM) images confirmed the localization of the Cu–Ni NPs of size 2–3 nm at a defined place (the interior or exterior of the halloysite nanotubes). Elemental mapping by energy-dispersive spectroscopy (EDS) in conjunction with TEM revealed intense EDS signals from Cu and Ni at the nanotube interior for the interior-immobilized Cu–Ni NPs. In contrast to the former case, elemental mapping revealed all the elements, including Ni and Cu, are distributed throughout the sample of the exterior-immobilized one. Powder X-ray diffraction profiles for both the samples showed peaks at diffraction angles of 43.5° and 50.6°, corresponding to the (111) and (200) reflections for the Cu–Ni NP. In addition, inductively-coupled plasma mass-spectroscopy confirmed that both the Cu–Ni NPs have similar chemical compositions.

X-ray photoemission spectra (XPS) for the exterior Cu–Ni NPs exhibited an intense $Cu2p_{3/2}$ peak at 932 eV, deconvoluted to two peaks centered at 932.1 and 934.0 eV, corresponding to the photoemissions from metallic Cu(0) and

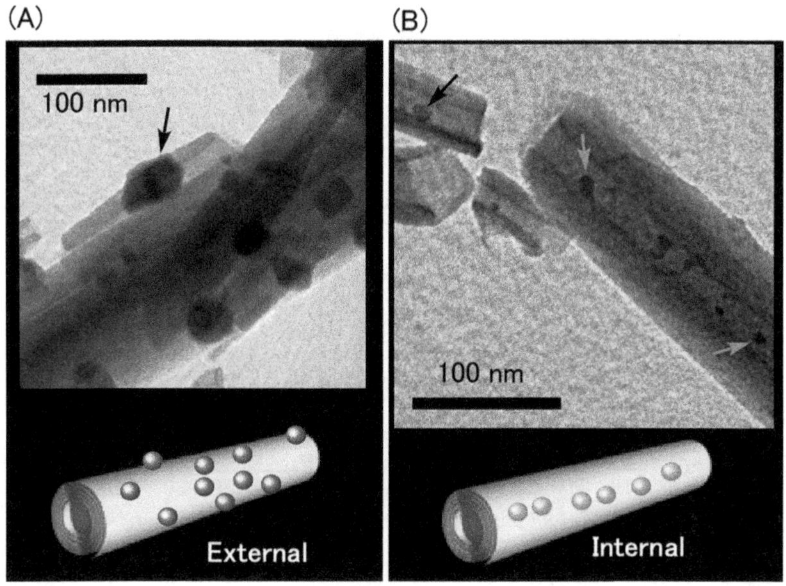

Figure 7.3 Localization of the Cu–Ni nanoparticles confirmed by transmission electron microscopy (TEM) images: (A) external surface; (B) internal surface.

divalent Cu^{2+}, respectively. The obserbed peaks at 941.9 and 962.0 eV also indicate the presence of Cu^{2+}-containing surface phases. Because the Cu^{2+} peak has a lower intensity than that for Cu(0), the surface of the Cu–Ni NPs primarily consists of metallic Cu. An intense peak at 855.5 eV in the Ni2*p* region is assigned to metallic Ni(0). The presence of two relatively intense shake-up peaks at 860.29 and 863.68 eV indicates partial oxidation of the surface Ni to Ni^{2+}. In contrast to these XPS characteristics for the exterior Cu–Ni NPs, neither Cu nor Ni peaks were observed for the interior Cu–Ni NPs because the profiling depth of the XPS nm is actually smaller than the thickness of the tube wall of halloysite. The latter fact confirms complete encapsulation of the Cu–Ni NPs in the interior of the halloysite clay nanotubes.

In order to investigate the catalytic capability of the immobilized Cu–Ni NPs, the purification of simulated automobile exhaust, an equimolar mixture of NO and CO gases, at elevated temperatures was examined in a circulating-gas reactor ($NO + CO = 1/2N_2 + CO_2$ and/or $NO + CO = 1/2N_2O + 1/2CO + 1/2CO_2$). Figure 7.4 demonstrates time-course profiles of conversion with the interior-type and exterior-type Cu–Ni NP catalysts at reaction temperatures ranging from 275 to 400 °C. The interior-type Cu–Ni catalyst showed high NO conversion activity at 325 °C. Under this condition, 50% NO conversion was achieved at 50 min after exposure of the Cu–Ni catalysts to the simulated exhaust gas mixture. In contrast, the exterior-type Cu–Ni catalysts exhibited only 10% reaction of NO at 300 °C. Furthermore, increasing the temperature to 400 °C even decreased the reaction conversion. The

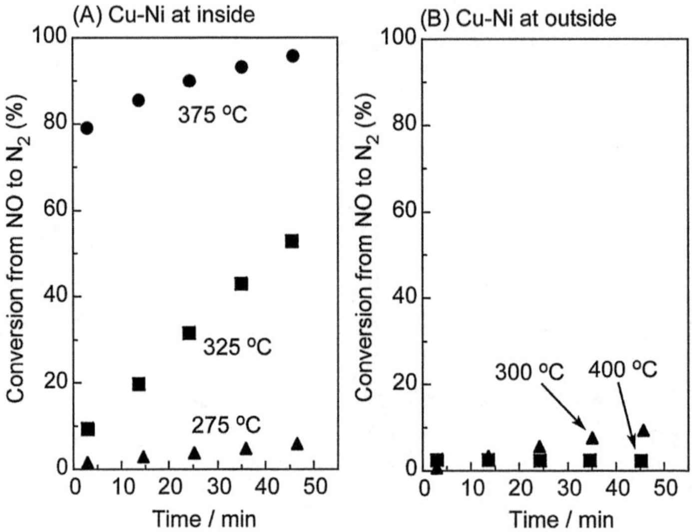

Figure 7.4 Time-course profiles of NO conversion with (A) the interior-type and (B) exterior-type Cu-Ni nanoparticle catalysts at reaction temperatures ranging from 275 to 400 °C.

latter results suggest particle agglomeration at the exterior surface, resulting in lowered catalytic capability at high temperatures. Immobilization of Cu–Ni NPs in the interior space is advantageous to avoid unfavorable agglomerations of catalytic NPs.

Detailed mechanisms of the enhanced NO conversion performance of the interior-immobilized Cu–Ni NPs were investigated by *in situ* Fourier-transform infrared (FT-IR) spectroscopy. The CO-stretching peak at the interior Cu–Ni NP is more intense than that for the exterior-immobilized Cu–Ni NPs. More deep analyses on the CO-stretching region actually demonstrate that CO molecules are predominantly adsorbed to the Cu sites over the Cu–Ni NPs. The peak position of the CO band on the Cu(0) site of the interior Cu–Ni NP locates at a higher frequency compared with that on the exterior one, probably due to weakened CO adsorption at the alloy surface. Formation of strong bonds between the Cu–Ni NP and the AlO_x interior of halloysite would deplete the density-of-states at the NP surfaces, leading to weaker CO adsorption. This situation enables the adsorbed CO molecules to migrate over the surface of the NPs and provide greater opportunity to react with NO molecules and/or oxygen atoms. As a result, NO conversion can be efficiently enhanced at the Cu–Ni NPs immobilized at the interior of the halloysite clay nanotubes.

The above-mentioned example clearly demonstrated advantages of the halloysite clay nanotubes. Platinum group free catalysts, Cu–Ni alloy NPs, can be selectively immobilized at the AlO_x interior of the halloysite clay nanotubes through ligand-assisted co-reduction of the Cu and Ni metal cations. The enhanced NO conversion performance was attained because of the specific tubular structure of the host materials. The Cu–Ni NPs strongly immobilized at the AlO_x interior are protected against particle agglomeration even at elevated temperatures. A similar strategy can be applied to other catalytic systems that require harsh conditions such as desulfurization of petroleum, methane steam reformation and/or photocatalytic water splitting.

7.3 Halloysite: Carrier for Antibacterials

Halloysite clay nanotubes can be used as carriers for various properties. In this section, use of the halloysite nanotubes for antibacterial functions is exemplified. Among various additives in paint products, Ag-based composite particles are especially powerful components as antimicrobial additives to prevent bacterial corrosion. However, direct addition of the Ag components to the paints often has disadvantages such as unfavorable color changes upon decomposition of Ag components. In addition, emission of plasmons from Ag upon UV light absorption may cause decomposition of polymers in paints. Therefore, addition of encapsulated Ag components is more desirable. Halloysite clay nanotubes would be an ideal carrier of Ag components for antibacterial activities.[10]

As illustrated in Figure 7.5, Ag nanorods can be selectively formed within the inner lumen of halloysite clay nanotubes through a series of simple procedures including loading of silver acetate into the nanotube lumen under reduced pressure, washing, separation, and heating without any harsh conditions. Silver nanorods were selectively synthesized within the halloysite inner lumen *via* loading of silver acetate from aqueous solution. In this method, the aqueous suspension of the halloysite clay nanotubes containing silver acetate was first out-gassed under vacuum. Next, loading of silver acetate was carried out by pulling and breaking the vacuum. After the composite materials were washed and dried, these materials were heated at

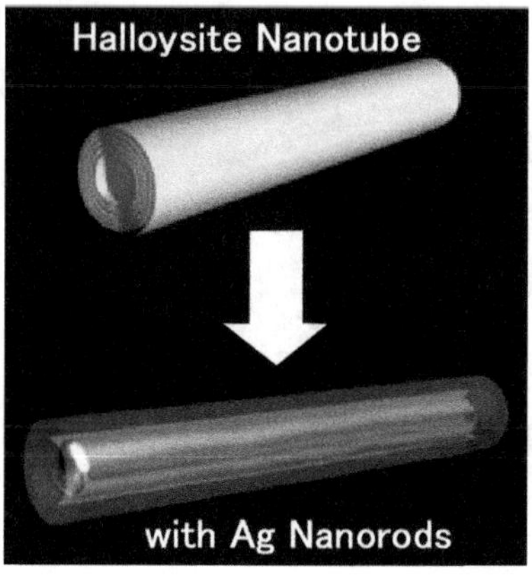

Figure 7.5 Formation of Ag nanorods selectively within the inner lumen of halloysite clay nanotubes.

300 °C for 1.5 hours to reduce silver ions to produce Ag nanorods within the clay nanotubes.

The Ag nanorods were formed according to the following two processes: (i) decomposition of silver acetate into small Ag nanocrystals within halloysite channels and (ii) conversion of the Ag crystals to grow into the Ag nanorods. Observation by electron microscopy of the resulting materials confirmed unbroken Ag nanorods with diameters corresponding to the lumen diameter of *ca.* 15 nm and a length of several hundred nanometers, which extended along the tube lumen. These processes resulted in effective encapsulation of the Ag metal core into the halloysite tubes that can protect Ag nanorods from the effects of the external environment, oxygen, light, *etc.*

The antibacterial performance of the formed encapsulated Ag nanorods was tested as a functional demonstration (Figure 7.6). Many additives, such as titanium dioxide, silica, clay, mica, and latex NPs, are included in recent paint products. Encapsulation of Ag nanorods into the halloysite preserves the colors of the paint main body and exhibits an extended lifetime of the antimicrobial effect.

After one-week exposure to *Escherichia coli* and *Staphylococcus aureus* bacterial cultures, polymer-based coatings with and without halloysite-Ag nanocomposites indicated fast bacterial growth around the pure paint without halloysite-Ag additives and significant inhibition of bacterial growth around that with halloysite-Ag additives. Probably, Ag^+ ions slowly leak from the tubes, resulting in efficient inhibition of bacterial growth. The metal plate coated with the halloysite-Ag paint nanocomposite kept the metal bar in its original state, while metal coated with pure paint began to have many cracks on it due to rapid deformation of the plate. A similar fabrication of metallic

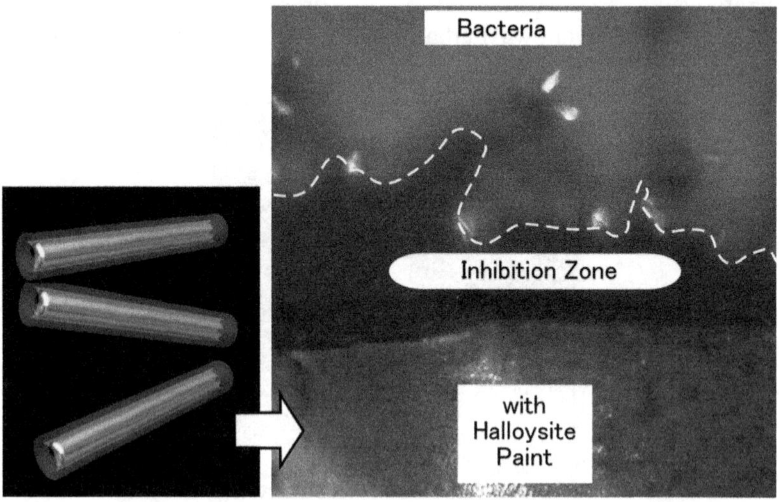

Figure 7.6 Antibacterial performance of Ag nanorods encapsulated within halloysite clay nanotubes.

nanorods within halloysite ceramic nanocomposites would create dispersible antimicrobial additives in plastic composites, nanoelectronic and optical materials, some of which have to be used under biocompatible environments.

7.4 Other Nanostructured Carriers for Drug Delivery

Not limited to halloysite materials, various nanostructures of silica-related stuffs have been used for drug carriers. In the remaining parts of this chapter, several examples of silica nanostructures for delivery of drugs and related materials are briefly explained. Spontaneously formed silica nanostructures – three-dimensional hollow silica capsules, which have shell structures made by assembly of two-dimensional silica nanosheets – the so called flake–shell capsule, were used for delivery of various drugs, including anticancer reagents (Figure 7.7).[11] The flake–shell capsules were spontaneously formed under hydrothermal conditions (*ca.* 75 °C, 20 h) in the presence of $NaBH_4$. Silica flakes with nanosheet morphology appear around a silica NP that gradually dissolves, followed by formation of hollow silica spheres through precipitation and self-assembly of nanosheets at the expense of the particle cores. The spherical hollow capsules were confirmed to have a diameter of 550–570 nm with a shell thickness of 60–80 nm.

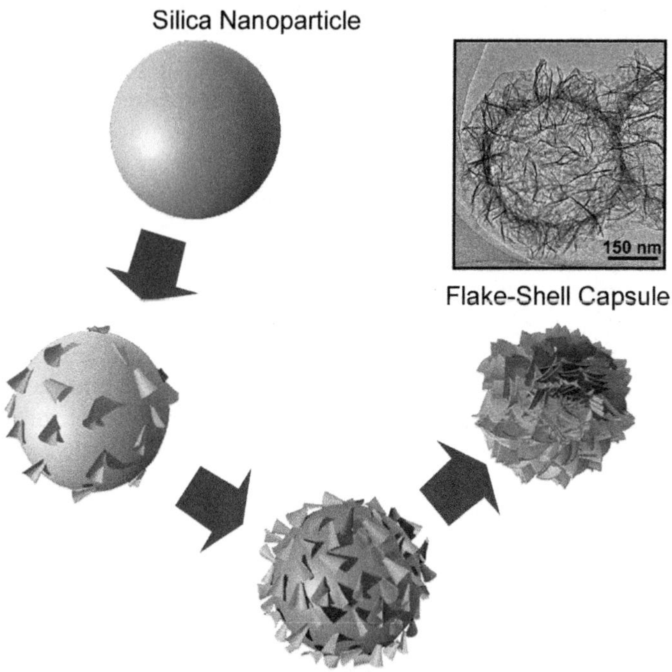

Figure 7.7 Spontaneous formation of silica nanostructures – three-dimensional hollow silica capsules with a shell made by assembly of two-dimensional silica nanosheets – the so called flake–shell capsule.

More intentional structural control of the flake–shell capsules can be made through regulation of the pH conditions of the surrounding media, because the silica flakes prepared under hydrothermal conditions have a silanol-rich surface. Although exposing the flake–shell capsules to strongly acidic conditions (pH 3) resulted in assembly of silica flakes upon destruction of the capsule structure, probably due to unfavourable protonation and/or hydrolysis of silica, immersing the flake–shell capsules in either weakly acidic solution (pH 5) or basic solution (pH 9) induced certain morphological changes without causing collapse of the capsule structures. For example, *ca.* 30% shrinkage of the capsule diameter was observed after treatment at pH 5 with the capsule structure itself kept intact. Treatment at pH 5 makes the sheet network denser because of shrinkage of the shell. On the other hand, the capsule diameter did not show significant changes at pH 9 treatment but flakes fused together to induce an aggregation of the shell capsules. Such structural modifications can be made for both macroscopic morphology and the nanoscopic pore size. Although the as-synthesized capsules possess rather broad pore size distributions, the pH 5 treatment resulted in a decrease of capsule diameter with a narrower pore size distribution. Similarly, a sharper distribution of the pore size was achieved upon treatment under weakly basic conditions. If the capsules have smooth continuous silica structures, such flexible structural changes cannot be fulfilled. Based on a hierarchic structure of 2D-in-3D, the flake–shell capsules show changes of three-dimensional structures (capsule diameter) and one-dimensional motifs (pore size) with two-dimensional (sheets) materials assembly.

This structural flexibility of flake–shell silica capsules is reflected in a dynamic function, regulation of the delivery of doxorubicin (anticancer drug) and estradiol (steroid hormone). Under selected conditions, encapsulation efficiency of doxorubicin to the flake–shell capsules was 95% while that for conventional mesoporous silica capsules remained at 23%. The flake–shell capsules with loosely assembled flake structures contain unavoidable amounts of macro-holes in the shell, leading to enhanced encapsulation of larger quantities of doxorubicin. Interestingly, the release speed of doxorubicin from the flake–shell capsules is not as fast as expected from their loose structures, probably due to hydrogen bonding formation between doxorubicin molecules and silica surfaces. Sustained release, with approximately 13% drug release at 12 h, was actually observed. The flexible structural nature and surface functionality result in enhanced encapsulation and sustained release of the anticancer drug, doxorubicin.

More precise regulation of drug release from the flake–shell capsules was made for release of the hydrophobic drug, estradiol. After introduction of estradiol to the flake–shell capsules, structural changes on the capsule shell were made at pH 5 or pH 9. The untreated capsules showed rapid estradiol release in ethanol, while the corresponding release speed from the flake–shell capsules treated under alkaline conditions became slower. The observed difference originates from differences in their pore sizes. In contrast, the release profile of estradiol from flake–shell capsules treated under weakly acidic conditions has a two-step nature where the slow release process interestingly obeys zero-order kinetics, enabling a constant release rate over rather long periods. By selecting conditions, it is possible to realize prolonged stable drug release at a constant rate. Such optimized drug release comes from the structural flexibility and adjustability of the hierarchic nature of the fabricated materials, flake–shell capsules.

The prepared flake–shell capsule structures would be nice media for immobilization of biomaterials such as enzymes and DNA to make bioactive hierarchic structures. This expectation was realized through immobilization of three naturally abundant enzymes with different molecular weights, sizes, and isoelectric points, *i.e.*, lysosyme, lipase, and chymotrypsin (Figure 7.8).[12] We demonstrated that the soft shell composed of a silica flake network is flexible to embed different enzymes efficiently and has potential for use in various bio-applications. Silica materials have intrinsic advantages for biomaterial immobilization, including structural stability, biocompatibility, and a wide capability for surface functionalization. For example, enzyme loading and activity would be controlled through modification of the silica surface with appropriate functional groups such as amine and dextran groups. With the exception of lysosyme as a cell substrate, the apparent catalytic activities of the immobilized enzymes were retained. Moreover, slow degradation under physiological conditions and a low cytotoxicity of the fabricated materials were confirmed. Coating of the capsules with a polyethyleneimine coating promotes loading of CpG oligodeoxynucleotides. Strategies for immobilization of biomaterials to flake–shell capsules would create huge potential to immobilize bio-functions at defined places with controlled morphology because of their hierarchic structural nature, which could be a useful tool kit for adjustable systems.

Figure 7.8 Imobilization of enzymes onto silica flake–shell capsules.

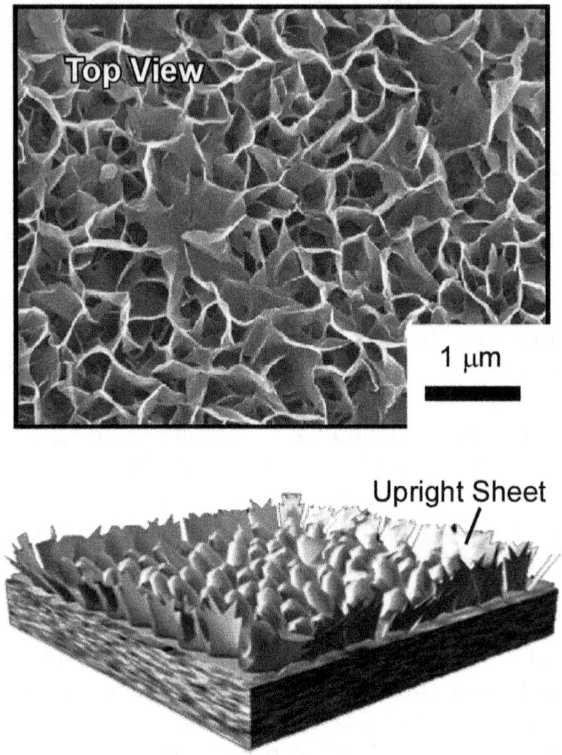

Figure 7.9 A surface with upright silica nanosheets.

As another silica-based hierarchic structure, silica upright sheets were utilized for reverse gene transfection (Figure 7.9).[13] According to a similar reaction at the silica surface, transformation of the regrown silica layer into a network structure created cross-linked upright nanosheets with a thickness of several nanometers. Finally, the formed silica nanosheets formed a porous-film-like surface with pore sizes of several hundred nanometers and a sharp edge with a thickness of around 5 nm. Accommodation of DNA strands was enhanced at the upright sheet. In addition, the *in vitro* reverse transfection capability of the silica upright sheet structures was examined with green fluorescent protein reporter genes using the human embryonic kidney mammalian cell line. A typical spread morphology was observed for the cells adsorbed on the dense upright sheet silica film. The fabricated systems enhance DNA binding and facilitate gene delivery to cells without using recombinant proteins.

Higher hierarchic silica nanostructures were assembled by layer-by-layer adsorption of mesoporous hollow silica capsules that were synthesized by a multi-step sol–gel and carbonization process using a pseudohexagonal prismatic zeolite (Figure 7.10).[14] The synthesized hierarchic silica nanostructured films accommodated liquid guest molecules inside their large empty core

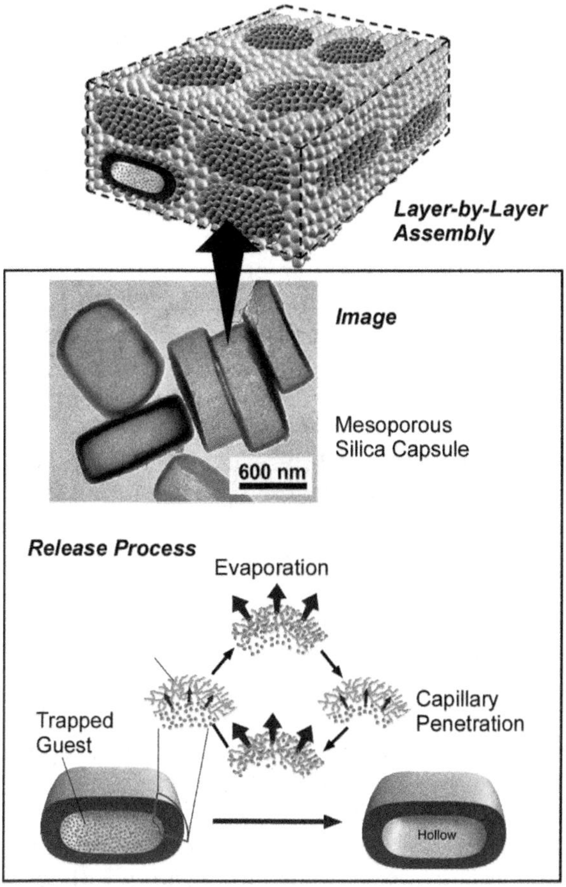

Figure 7.10 Layer-by-layer adsorption of mesoporous hollow silica capsules that exhibited unusual release profiles of these guest molecules with automatic on and off behaviors without any external stimuli.

and exhibited unusual release profiles of these guest molecules, with automatic on and off behaviors without any external stimuli. The mechanisms of the observed unusual behavior are explained by several non-equilibrium processes. The observed stimulus-free controlled on–off release profiles of drug molecules would be useful for drug delivery within certain intervals without any external inputs.

7.5 Summary

For more precise structure control, bottom-up approaches have especially been paid much attention to control nanoscale structures on the basis of the organizing capabilities of the materials themselves. A more advanced concept, nanoarchitectonics, has recently been proposed to promote these nanotechnology flows. With nanoarchitectonic strategies, functional materials with controlled nanostructures are architected through manipulation of atoms, molecules, and assemblies on the basis of the synergetic combination of chemical nanofabrication, self-organization, and field-controlled organization. Among countless numbers of materials candidates, structurally-defined inorganic materials can be nice hosts for development of functional materials structures according to nanoarchitectonics. For selective modification on interior and exterior surfaces of an inorganic host, halloysite clay nanotubes are highlighted in this chapter.

Here, we mainly focus on introduction of research examples on functionalization of halloysite nanotubes for catalysis and antibacterial activities. In the first example, as catalysts for removal of toxic nitrogen monoxide (NO) and carbon monoxide (CO) with reaction scheme ($NO + CO = 1/2N_2 + CO_2$ and/or $NO + CO = 1/2N_2O + 1/2CO + 1/2CO_2$), Cu–Ni NPs were selectively immobilized at the interior (AlO_x side) and the exterior (SiO_x side) of halloysite clay nanotube materials. The Cu–Ni NPs immobilized at the interior of the halloysite nanotubes exhibited excellent catalytic activities in the purification of simulated exhaust comprising an equimolar mixture of NO and CO, while those at the exterior of the clay nanotubes were catalytically much less active, probably because of particle agglomeration. The Cu–Ni NPs strongly immobilized at the AlO_x interior are protected against particle agglomeration even at elevated temperatures. A similar strategy can be applied to other catalytic systems that require harsh conditions, such as desulfurization of petroleum, methane steam reformation and/or photocatalytic water splitting.

In the second example, Ag nanorods were selectively formed within the inner lumen of halloysite clay nanotubes through a series of simple procedures, including loading of silver acetate into the nanotube lumen under reduced pressure, washing, separation, and heating without any harsh conditions. Encapsulation of Ag nanorods into the halloysite preserved the colors of the paint main body and exhibited an extended lifetime of the antimicrobial effect. A similar fabrication of metallic nanorods within halloysite ceramic nanocomposites would create dispersible antimicrobial additives in plastic composites, nanoelectronic and optical materials, some of which have to be used under biocompatible environments.

In addition, several examples were given of hierarchic silica nanostructures for applications such as drug delivery. Spontaneously formed silica nanostructures – three-dimensional hollow silica capsules with shell structures made by assembly of two-dimensional silica nanosheets – the so called flake–shell capsule, were used for delivery of various drugs including anticancer reagents. As another silica-based hierarchic structure, silica upright sheets were utilized for reverse gene transfection. Furthermore, higher hierarchic silica nanostructures were assembled by layer-by-layer adsorption of mesoporous hollow silica capsules. The observed stimulus-free controlled on–off release profiles of drug molecules would be useful for drug delivery within certain intervals without any external inputs.

We hope that the examples introduced in this chapter successfully stimulate readers to consider new strategies for functional nanocomposites. Wise nanoarchitectonics using structurally-defined inorganic host materials such as halloysite clay nanotubes would create various kinds of functional materials for many practical uses including environmental catalysis and biomedical applications.

Acknowledgements

We thank World Premier International Research Center Initiative (WPI Initiative), MEXT, Japan. YL thanks the Russian Government Program of Competitive Growth of Kazan Federal University for support.

References

1. K. Ariga, J. P. Hill, M. V. Lee, A. Vinu, R. Charvet and S. Acharya, *Sci. Technol. Adv. Mater.*, 2008, **9**, 014109.
2. (a) M. Aono, *Sci. Technol. Adv. Mater.*, 2011, **12**, 040301; (b) K. Ariga, M. Li, G. J. Richards and J. P. Hill, *J. Nanosci. Nanotechnol.*, 2011, **11**, 1; (c) K. Ariga, Q. Ji, J. P. Hill, Y. Bando and M. Aono, *NPG Asia Mater.*, 2012, **4**, e17; (d) K. Ariga, Q. Ji, W. Nakanishi, J. P. Hill and M. Aono, *Mater. Horiz.*, 2015, **2**, 406; (e) K. Ariga, Y. Yamauchi and M. Aono, *APL Mater.*, 2015, **3**, 061001; (f) M. Aono and K. Ariga, *Adv. Mater.*, 2016, **28**, 989.
3. (a) M. Ramanathan, L. K. Shrestha, T. Mori, Q. Ji, J. P. Hill and K. Ariga, *Phys. Chem. Chem. Phys.*, 2013, **15**, 10580; (b) L. K. Shrestha, Q. Ji, T. Mori, K. Miyazawa, Y. Yamauchi, J. P. Hill and K. Ariga, *Chem.–Asian J.*, 2013, **8**, 1662; (c) M. Ramanathan, L. K. Shrestha, T. Mori, Q. Ji, J. P. Hill and K. Ariga, *Phys. Chem. Chem. Phys.*, 2013, **15**, 10580.
4. (a) K. Ariga, T. Mori and J. P. Hill, *Langmuir*, 2013, **29**, 8459; (b) R. Rajendran, L. K. Shrestha, K. Minami, M. Subramanian, R. Jayavel and K. Ariga, *J. Mater. Chem. A*, 2014, **2**, 18480; (c) K. Ariga, J. Li, J. Fei, Q. Ji, J. P. Hill, *Adv. Mater.*, 2016, **28**, 1251.
5. (a) K. Ariga, S. Ishihara, H. Abe, M. Li and J. P. Hill, *J. Mater. Chem.*, 2012, **22**, 2369; (b) K. Ariga, Q. Ji, M. J. McShane, Y. M. Lvov, A. Vinu and J. P. Hill, *Chem. Mater.*, 2012, **24**, 728; (c) K. Ariga, Q. Ji, T. Mori, M. Naito, Y. Yamauchi, H. Abe and J. P. Hill, *Chem. Soc. Rev.*, 2013, **42**, 6322; (d) W. Nakanishi, K. Minami, L. K. Shrestha, Q. Ji, J. P. Hill and K. Ariga, *Nano Today*, 2014, **9**, 378; (e) K. Ariga, Y. Yamauchi, Q. Ji, Y. Yonamine and J. P. Hill, *APL Mater.*, 2014, **2**, 030701; (f) S. Ishihara, J. Labuta, W. Van Rossom, D. Ishikawa, K. Minami, J. P. Hill and K. Ariga, *Phys. Chem. Chem. Phys.*, 2014, **16**, 9713; (g) H. Abe, J. Liu and K. Ariga, *Mater. Today*, 2016, **19**, 12.
6. K. Ariga, Y. Yamauchi, G. Rydzek, Q. Ji, Y. Yonamine, K. C.-W. Wu and J. P. Hill, *Chem. Lett.*, 2014, **43**, 36.

7. (a) K. Ariga, A. Vinu, Y. Yamauchi, Q. Ji and J. P. Hill, *Bull. Chem. Soc. Jpn.*, 2012, **85**, 1; (b) V. Malgras, Q. Ji, Y. Kamachi, T. Mori, F.-K. Shieh, K. C.-W. Wu, K. Ariga and Y. Yamauchi, *Bull. Chem. Soc. Jpn.*, 2015, **886**, 1171.
8. (a) Y. Lvov and E. Abdullayev, *Prog. Polym. Sci.*, 2013, **38**, 1690; (b) Y. M. Lvov, D. G. Shchukin, H. Mohwald and R. R. Price, *ACS Nano*, 2008, **2**, 814; (c) E. Abdullayev, R. Price, D. Shchukin and Y. Lvov, *ACS Appl. Mater. Interfaces*, 2009, **1**, 1437; (d) W. O. Yah, A. Takahara and Y. M. Lvov, *J. Am. Chem. Soc.*, 2012, **134**, 1853; (e) M. R. Dzamukova, E. A. Naumenko, Y. M. Lvov and R. F. Fakhrullin, *Sci. Rep.*, 2015, **5**, 10560; (f) Y. Fu, D. Zhao, P. Yao, W. Wang, L. Zhang and Y. Lvov, *ACS Appl. Mater. Interfaces*, 2015, **7**, 8156; (g) J. Xue, Y. Niu, M. Gong, R. Shi, D. Chen, L. Zhang and Y. Lvov, *ACS Nano*, 2015, **9**, 1600.
9. N. M. Sanchez-Ballester, G. V. Ramesh, T. Tanabe, E. Koudelkova, J. Liu, L. K. Shrestha, Y. Lvov, J. P. Hill, K. Ariga and H. Abe, *J. Mater. Chem. A*, 2015, **3**, 6614.
10. E. Abdullayev, K. Sakakibara, K. Okamoto, W. Wei, K. Ariga and Y. Lvov, *ACS Appl. Mater. Interfaces*, 2011, **3**, 4040.
11. (a) Q. Ji, C. Guo, X. Yu, C. J. Ochs, J. P. Hill, F. Caruso, H. Nakazawa and K. Ariga, *Small*, 2012, **8**, 2345; (b) Q. Ji, S. Ishihara, T. G. Terentyeva, K. Deguchi, S. Ohki, M. Tansho, T. Shimizu, J. P. Hill and K. Ariga, *Chem.–Asian J.*, 2015, **10**, 1379.
12. T. G. Terentyeva, A. Matras, W. Van Rossom, J. P. Hill, Q. Ji and K. Ariga, *J. Mater. Chem. B*, 2013, **1**, 3248.
13. (a) Q. Ji, T. Yamazaki, N. Hanagata, M. V. Lee, J. P. Hill and K. Ariga, *Chem. Commun.*, 2012, **48**, 8496; (b) N.-C. Huang, Q. Ji, K. Ariga and S.-H. Hsu, *NPG Asia Mater.*, 2015, **7**, e184.
14. (a) Q. Ji, M. Miyahara, J. P. Hill, S. Acharya, A. Vinu, S. B. Yoon, J.-S. Yu, K. Sakamoto and K. Ariga, *J. Am. Chem. Soc.*, 2008, **130**, 2376; (b) Q. Ji, S. Acharya, J. P. Hill, A. Vinu, S. B. Yoon, J.-S. Yu, K. Sakamoto and K. Ariga, *Adv. Funct. Mater.*, 2009, **19**, 1792.

CHAPTER 8

Application of Clay Materials as Nanocontainers for Self-Healing Coatings

DMITRY G. SHCHUKIN*[a] AND ELENA SHCHUKINA[a]

[a]Stephenson Institute for Renewable Energy, University of Liverpool, Chadwick Building, Peach Street, L69 7ZF, Liverpool, UK
*E-mail: d.shchukin@liverpool.ac.uk

8.1 Introduction

Total annual losses directly related to metal corrosion are tremendous and estimated worldwide at around $200 billion. Even in developed countries, material performance failure costs related to corrosion are 3–4% of GDP.[1] Prevention of corrosion is one of the most challenging problems facing industrial society and includes various scientific, technical and engineering issues. There are different approaches to, at least, control corrosion to some extent. The most frequently used methods are application of different organic, ceramic and metallic protective coatings,[2] the use of corrosion-resistant alloys[3] and composites.[4,5]

Contemporary protective coatings have a quite complex structure, closely related to the nature of the substrate and to demands of the corrosion resistance. There are some common elements present in the majority of the coatings. A typical protective coating has (i) pretreatment layer or conversion

RSC Smart Materials No. 22
Functional Polymer Composites with Nanoclays
Edited by Yuri Lvov, Baochun Guo and Rawil F. Fakhrullin

Published by the Royal Society of Chemistry, www.rsc.org

layer, (ii) primer layer and (iii) topcoat layer. The first two layers have corrosion protection performance while the topcoat is mostly esthetic. Quite often, a layer of electrochemically active metal like Zn, Cd is deposited on the top of the steel to improve corrosion protection.[6] The most effective conversion layers used to date are layers based on the compounds of hexavalent chromium.[7] A tight solid layer of $Cr_2O_3 \cdot nH_2O$ is formed on the substrate surface due to the strong oxidation activity of Cr(VI), which has both excellent barrier and self-healing properties.[8] The latter results from excess chromate ions incorporated into the conversion layer and only loosely bonded with it, which provide gradual release of chromate into the defects during coating exploitation.

However, the strong oxidizing ability of chromates is responsible for their extremely high environmental toxicity and can cause various cruel diseases such as cancer or mutagenic damage.[9] For these reasons, chromates have been banned for use in many corrosion protection applications since the beginning of the 21st century,[10] and almost all main industry stopped the utilization of chromates for corrosion protection due to this legislative ban.

This situation led to the intensive search for environmentally friendly chromate alternatives. Various solutions were proposed. First of all, compounds with properties similar to chromates were studied as components of conversion layers; these are transition metals in high oxidation states like titanium,[11] vanadium,[12] zirconium and niobium,[13] manganese,[14] cobalt,[15] molybdenum,[16] tungsten[17] and cerium.[18,19] Other approaches are electrochemical chromate-free oxidation treatment (anodizing),[20] diverse organic–inorganic composite pretreatments[21] and application of organic inhibitors directly on the metal surface.[22,23] However, an equivalent substitution for chromates has still not been found and most comparative studies of chromate-free corrosion protection solutions led, unfortunately, to the unambiguous conclusion that chromates provide outstandingly better corrosion resistance than any other coatings known until now.

In chromate-free coating systems, the contributions of primer and topcoat layers increase the protective efficiency of the whole coating system. These layers (especially topcoat) should facilitate good barrier properties, leading to the safe isolation of the metal substrate from corrosive factors and from charge transfer occurring as a consequence of electrochemical corrosion processes. A huge variety of organic paint systems was developed for this purpose.[24] A typical organic coating provides a physical barrier preventing access of corrosive agents and can eliminate corrosion as long as the integrity of the coating remains. These organic layers cannot withstand various damage, like scratches, stone chipping effects, appearing during their service life. Being broken, the organic coating cannot effectively resist the penetration of corrosives into the defect area.

The first evident solution to provide healing activity to the organic coatings is to impart the coating matrix with an "active" anticorrosion functionality by direct introduction of corrosion inhibitors in the pretreatment, primer or topcoat layers of the protective coating system.[25] This, contrary to expectations, leads to a significant reduction of the corrosion protection performance of the coatings due to specific physicochemical properties of anticorrosive agents and their interaction with other components of the coating formulation. A high efficiency of inhibitor action can only be attained if a defect site lies within certain "proper" limits. Very low solubility leads to a lack of inhibitor in the defect area and, consequently, to weak feedback activity. In the opposite case of too high solubility, substrate can be protected only for a relatively short time after which the inhibitor will be rapidly leached out from the coating. Another drawback, which can appear due to high solubility, is the osmotic pressure that leads to blistering and, finally, to delamination of the coating from the substrate active surface. The osmotic pressure can stimulate water to be transported through the coating, which acts as a semipermeable membrane, causing the destruction of the passive coating matrix.[26,27] Additionally, inhibitor freely dispersed inside the active matrix is often subjected to spontaneous leakage from the surface, reducing the inhibitor content during aging.[28]

The mostly negative experiences with the development of corrosion protection coatings by direct addition of inhibitor resulted in new ideas for more complex coatings with self-healing ability, where coating multifunctionality is expressed not only across the coating layers but also within every layer. The next generation of protective coatings should be much thinner (thus reducing CO_2 emission and the quantity of raw components per 1 m^2 of the coating) and should possess self-healing effects in the scratched (damaged) areas. This will require the development of components of the coatings that react to external or internal impacts (pH, humidity changes or distortion of the coating integrity, *etc.*) without inhibitor deactivation and, if possible, are able to combine the self-healing property with other functionalities (*e.g.*, sensor, self-cleaning).

8.2 General Concept of Nanocontainer-Based Self-Healing Coatings

The self-healing activity of the materials is based on their feedback action. Feedback is often a part of naturally occurring or constructed systems. In a feedback system, cause–effect relationships between system variables form a closed-loop wherein the output affects the input.[29] In this way, a feedback system can respond to its environment (like local pH changes during the corrosion process). The input can also be an external signal applied to the system (like light or mechanical force). The output is the restored functionality of the coating.

Mimicking the natural concept of feedback active systems (like blood clothing[30]) in the field of synthetic coatings opens up a number of opportunities for the development of "smart" coatings with stimuli-responsive behavior. These coatings are of great scientific and technological importance, as they can be applied in various fields such as medicine, biotechnology or materials science. Their main passive function is to protect the underlying substrate from contact with the environment. In addition to their barrier properties, they also possess active properties due to their responsive, active function.

A very successful approach to impart feedback functionality to a coating is incorporation of encapsulated corrosion inhibitor in the coating matrix (Figure 8.1).[31,32] Capsules or nanocontainers can isolate corrosion inhibitor from the coating matrix, terminate spontaneous leakage of the inhibitor and, at the same time, provide controlled release of the inhibitor directly into the corroded area on demand. The nanocontainers can have inorganic, organic or composite origin and be applied for encapsulation of drugs, vitamins, quantum dots, bactericides, oils, biopolymers, corrosion inhibitors, low molecular weight compounds and many other active molecules. The most important task is to develop nanocontainers with good compatibility with the matrix components. The nanocontainers should also be of a size less than 300–400 nm; nanocontainers of a larger size can damage the integrity of the coating matrix, forming large hollow cavities which reduce the passive protective properties of the coating.

The most interesting property of the nanocontainers is their ability of controlled release/upload. The controlled release/upload is, in general, dependent on the interaction between the material of the shell and the environment surrounding the container. The shell of the containers should be sensitive to external impact or changes in the local environment; moreover, it should

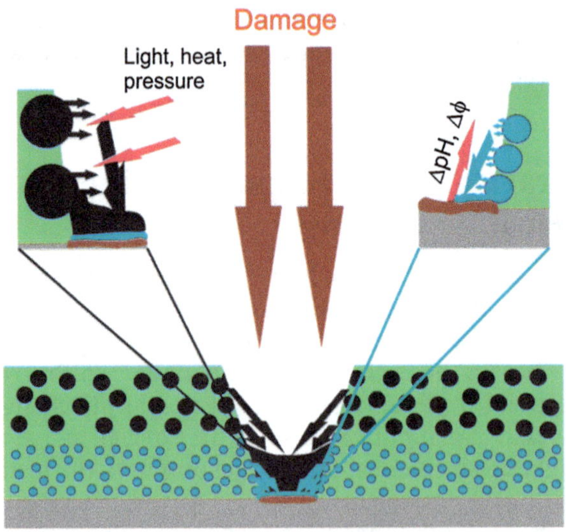

Figure 8.1 Schematic representation of nanocapsule-based self-healing coatings. Reprinted from ref. 31 with permission from Wiley-VCH Verlag. Copyright © 2007 Wiley-VCH Verlag GmbH & Co. KGaA, Weinheim.

selectively react only to one or two triggers changing its permeability while the others should keep the shell intact. The following triggers are utilized for opening/closing of the nanocontainer shell: local pH changes, temperature changes, electromagnetic irradiation, mechanical pressure (also ultrasonic treatment), humidity, electric (electrochemical) potential, ionic strength and dielectric permeability of the solvent.

Mechanical impact is the first external stimulus utilized for capsule-based self-healing materials and coatings. Adopting mechanical stress as a trigger to recover the mechanical stability by autonomic crack healing was demonstrated for the first time for microcapsule-based systems in 2001 by the group of S. R. White.[33] They encapsulated healing agent (a monomer) into polymer microcapsules embedded in an epoxy matrix containing a catalyst. Upon crack formation the embedded microcapsules are ruptured and the healing agent is released in the crack due to capillary forces and comes into contact with the embedded catalyst. The resulting polymerization of the healing agent leads to the crack healing and recovery of the barrier properties of the coating. Simultaneous incorporation of two types of capsules containing different active agents[34] or micrometer size capsules filled with reactive epoxy resin or oil[35,36] were also reported for successful crack healing. Encapsulation of moisture reactive materials, such as metal oxide precursors,[37] or linseed oil[38] in polymer microcapsules could adapt response of self-healing coatings to combined external stimuli. The active material is first released due to crack propagation and forms an impermeable layer only after oxidation or hydrolysis by atmospheric oxygen or humidity.

Utilization of pH shift occurring on the substrate surface during the corrosion process as a stimulus for corrosion inhibitor release is the second most popular way to design anticorrosive coatings with a smart self-healing response. That is because the corrosion process leads to a local pH decrease in anodic areas and a local pH increase in cathodic ones. Therefore, the nanocontainer shell should be sensitive to both acidic and alkaline shifts of the local pH environment in order to release inhibitor for corrosion termination. Polyelectrolyte capsules, hydrogels and porous materials with bio-inspired nanovalves with weak acidic or basic functional groups in the shell demonstrate reversible and (or) irreversible changes of the shell permeability in a wide pH range (*e.g.*, at low $pH < 4$ or high $pH > 9$).[39,40] For opening by electromagnetic irradiation, the container shell should have sensitive components like metal (silver) nanoparticles for IR light,[41] dyes for visible light[42] and semiconductors (TiO_2 nanoparticles) for UV light.[43] Ultrasonic treatment can also be applied to irreversibly open containers of any size.[44] Other triggers for capsule opening were demonstrated, presumably for hydrogels (temperature, electric potential, high ionic strength) and polyelectrolyte capsules (temperature, high ionic strength, dielectric permeability of the solvent).[45,46] These triggers are usually involved in feedback active systems of specific functionality (*e.g.*, coatings with electrochemically reversible permeability).[47]

The current level of the development of nanocontainer-based self-healing coatings has a sufficient number of demonstrations of highly-efficient corrosion protection coatings at the laboratory scale. It is time now to move the research on these materials from the laboratory to the industrial level. However, there are two main difficulties here – the availability of valid results of industry-accepted corrosion protection tests (mostly salt-spray tests) and the cost of the nanocontainers. The industry wants to be sure that a new generation of paints will be reliable and cost effective. The first drawback can be overcome with increased investments of time and resources into the characterization of self-healing coatings on a commercial level while the solution of the second problem requires the search for cheap nanocontainer hosts which can be available in multiton-scale quantities. Layered clay materials and halloysites can be perfect candidates for these hosts – they are much cheaper compared to other types of nanocontainers and can be directly mined from mineral deposits in large-scale quantities. Moreover, the interest in using inorganic systems is that their mechanical and thermal stability allow their utilization in different coating layers (pretreatment, primer, topcoat) subjected to high mechanical loads or significant thermal stresses.

8.3 Self-Healing Coatings Based on Ion-Exchange Nanoclays

Increasing the complexity in terms of inhibitor release from simple desorption to pH-triggered release of inhibitors, another mechanism was suggested for smart release based on release by ion-exchange.[48] There are two types of ion-exchange release of the entrapped inhibitor from nanoclays: the less studied cation-exchange release and the more developed anion-exchange release.

Bentonite is a cation-exchanger, consisting of stacks of negatively charged aluminosilicate sheets between which inhibiting cations like Ca^{2+} and Ce^{3+} are intercalated.[49,50] Ce-loaded bentonites dispersed in polyester resin layers and applied to galvanized steel substrates display active protection avoiding coating decomposition. The performance of different pigments was measured by the rate of coating delamination from the cut edge of samples exposed to salt-spray testing. Ce-loaded bentonites were also combined with Ce-exchanged Shieldex and dispersed in polyvinyl butyral coatings applied to hot dip galvanized steel showing the decrease of the coating delamination.

The protective performance of Na-montmorillonite (Na-MMT) intercalated with zinc cations (Zn-MMT), benzimidazole (BIAMMT) and the mixture of these two inhibitors (Zn-MMT + BIAMMT) is described as an example of another natural clay material for smart corrosion coatings.[51] The modified clay particles were added to the epoxy ester coating. Results showed that the interlayer distance of the clay particles was increased after Na^{+} cations exchange with inhibitors. The highest corrosion resistance and the lowest adhesion loss were observed when the mixture of Zn-MMT and BIA-MMT clay nanocontainers was used. The crack filling behavior of the transparent, multilayer coating consisting of a polysiloxane film on top of a thin montmorillonite interlayer was monitored using confocal and SEM microscopy.[52] Healing of the coating was carried out under high and low humidity conditions at 50 °C. Filling of cracks in the polymeric topcoat occurred primarily within the first two hours of exposure to moisture saturated air and resulted in a good restoration of the surface flatness. Beyond the 2 h healing period, further expansion of the clay layers resulted in a "pushing-up" effect of the swollen clay in the proximity of the damaged coating with potential restoration of the barrier properties of the coating. The overall healing process consists of a partial expansion of the clay layer, physically restoring the coating surface profile, followed by further expansion of the clay layer, producing a pushing-up effect up until saturation of the montmorillonite is reached. A similar self-healing mechanism was observed for montmorillonite-hydrogel layers.[53,54]

However, the release of the inhibitor in the case of cation-exchanger clays can only be triggered by metal cations available in the surroundings, which may not be directly associated with corrosion processes, leading to inhibitor waste.

The anion-exchanging clays are more effective in retaining the corrosion inhibitor inside the coating without spontaneous leakage. An additional possibility of the anion-exchange clays is the entrapment of aggressive anionic species such as chlorides that initiate corrosion processes. Hydrotalcite-based conversion films demonstrated good corrosion protection, and some research groups have been trying to improve the interaction between these conversion films and organic coatings.[55] The hydrotalcites are typical anion-exchange materials consisting of stacks of positively charged mixed hydroxide layers separated by anions and water molecules. The most attractive property of hydrotalcites is the ability to modify the nature of the films by varying the layer composition and intercalation of other anions, as well as the excellent adsorption capability imparted by the laminar structure which also provides a large surface area. An environmentally friendly corrosion inhibitor, phytic acid, was successfully added to the hydrotalcite film by adsorption and ion-exchange.[56] The metal substrate used in this study was AZ31 Mg alloy. Compared with the original hydrotalcite film, the corrosion behavior of the hydrotalcite film with intercalated phytic acid is much better. The resistance slightly increased and was notably higher than that of the original hydrotalcite film after 12 h immersion. Micro-cracks appeared after two days, which is much later than for the original hydrotalcite film, due to the sustained release of phytic acid. The same group later studied the performance of molybdate intercalated into hydrotalcite particles for magnesium protection.[57] The molybdate intercalated hydrotalcite coating, characterized by interlocking plate-like nanostructures, ion-exchange and a self-healing ability, has the potential to be a "smart" coating capable of responding to stimuli from the environment, which was proved by SEM cross-sectional views. By means of competitive adsorption, protective deposition and oxidation reactions, the released MoO_4^{2-} ions acted as a good anodic inhibitor to protect Mg alloys from corrosion attacks.

Layered double hydroxides (LDHs) are potential artificial inorganic hosts which can be used in industrial self-healing coatings. They can be expressed by the general formula $[M_{1-x}^{2+}M_x^{3+}(OH)_2]A_{x/n}^{n-}\cdot mH_2O$, where the cations M^{2+} (Mg^{2+}, Zn^{2+}, Fe^{2+}, Co^{2+}, Cu^{2+} and others) and M^{3+} (Al^{3+}, Cr^{3+}, Fe^{3+}, Ga^{3+} and others) occupy the octahedral holes in a brucite-like layer and the anion A^{n-} is located in the hydrated interlayer galleries.[58] The ability to vary the LDH composition during synthesis over a wide range allows materials with a wide variety of properties to be prepared. Figure 8.2 demonstrates SEM and TEM images of LDHs intercalated with vanadates. Here, LDHs are polycrystalline particles with a plate-like morphology and dimensions ranging 200–400 nm diameter and lateral size 20–40 nm. Moreover, both dimensions and morphologies were found to be retained upon anion-exchange. X-ray diffraction results show that these materials are crystalline, and the peak positions at low 2Θ angles allow correlation between the gallery height, size and orientation of different anions in the intergallery space.

In some studies, the LDHs were produced *in situ* on the top of metallic substrates as protective films.[59,60] The applied methodology explores the weakness of the native aluminium oxide layer in the zones of intermetallic phases promoting differentiated growth of island-like LDHs on active intermetallics. The loading of corrosion inhibitor into the intergallery space of LDH by ion-exchange reaction creates nanostructured container islands deposited on the corrosion activity zones. After one month of immersion in 0.05 NaCl, the plates coated with inhibitor-loaded LDH films barely showed any signs of corrosion (one–two minor pits) while the reference metal surface was completely destroyed. Another approach is to use these anionic clays as containers for corrosion inhibitors and incorporate them into the organic

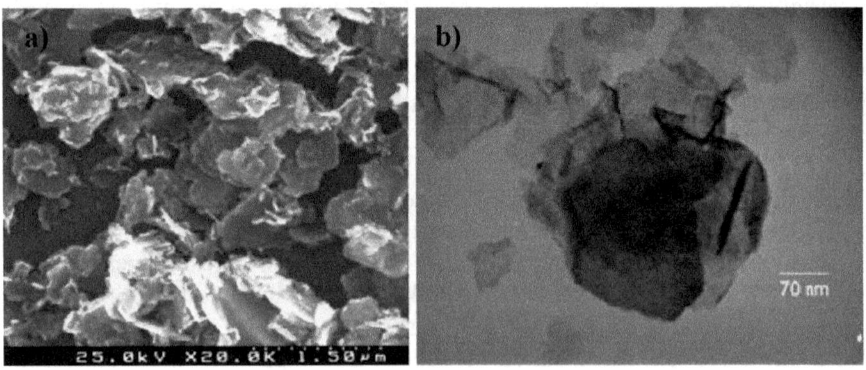

Figure 8.2 SEM (a) and TEM (b) images of vanadate-loaded LDHs. Reprinted from *Electrochimica Acta*, **82**, M. L. Zheludkevich, J. Tedim and M. G. S. Ferreira, "Smart" coatings for active corrosion protection based on multi-functional micro and nanocontainers, 314–323, Copyright 2012 with permission from Elsevier.[48]

coatings. In this case, not only release of the corrosion inhibitors but also trapping of the corrosive agents (Cl^-, SO_4^{2-}) are efficient tools to increase corrosion protection performance. Regarding this topic, several works can be found in the literature demonstrating loading of LDHs with anionic inhibitors.[61,62] Williams and McMurray prepared LDHs with different organic species (benzotriazole, ethyl xantate and oxalate) by rehydration of commercial [$Mg_6Al_2(OH)_{16}\cdot CO_3\cdot 4H_2O$]. The resulting layered systems were inserted into a poly(vinylbutyral) coating, prepared by a bar cast on the top of AA2024-T3 aluminium alloy. In another report, Kendig and Hon prepared LDHs intercalated with 2,5-dimercapto-1,3,4-thiadiazolate in a similar way and studied the inhibiting properties of this anion with respect to the oxygen reduction reaction on copper.[63] Zn–Al and Mg–Al LDHs loaded with quinaldate and 2-mercaptobenzothiazolate were synthesized *via* anion-exchange reaction.[64] Spectrophotometric measurements demonstrated that the release of organic anions from these LDHs into the bulk solution is triggered by the presence of chloride anions, evidencing the anion-exchange nature of the release trigger. The coating barrier properties were found to be the best for LDH-doped primer, evidencing the anticorrosion effect of LDHs. Poznyak *et al.*[65] demonstrated the active corrosion protection rendered by coatings typically used in the aeronautical industry to aluminium alloy 2024-T3, when LDHs were intercalated with 2-mercaptobenzothiazole and quinaldic acid. The anion-exchange of corrosion inhibitors by chloride anions was investigated using atomic absorption and UV-visible spectroscopy.

An additional improvement of the self-healing activity can be achieved by using simultaneously inhibitor-loaded LDHs and cerium molybdate nanocontainers.[66] Both of these nanocontainers were loaded with mercaptobenzothiazole (4 wt%) and added into epoxy primers, either individually or in 1 : 1 combination. The thickness of the coatings varied between 2 and 2.2 μm. The evolution of the self-healing activity was studied by the scanning vibrating electrode technique (SVET) or scanning ion-selective electrode technique (SIET) with a defect (0.2 mm size) created in the coating prior to immersion into 0.05 M NaCl solution.

For the blank coating, corrosion activity was detected immediately after immersion in the aggressive solution. After 1 h of immersion, the anodic current density measured over the defect was above 70 μA cm^{-2} and increased continuously with time. The first, but weak, signs of corrosion activity for the coating modified with mercaptobenzothiazole-loaded LDH were detected after 4–5 h of immersion. From this time, and up to one day of immersion, there was very little activity, revealing a strong inhibition effect when compared to the blank coating. However, after two days of immersion the corrosion activity increased, reaching 100 μA cm^{-2} for single mercaptobenzothiazole-loaded LDH nanocontainers, which is two times lower than the activity of the reference sample without nanocontainers. Different behaviour was observed for mercaptobenzothiazole-loaded cerium molybdate nanocontainers, where evident corrosion activity was observed at the early stages of immersion (3–5 h). After this peak activity, the current densities decreased attaining 5 μA cm^{-2} at 20 h of immersion. This trend remained for more than three days of immersion. An increase of the corrosion activity was observed after four days of immersion and coating deterioration was detected after one week.

When the two types of containers are mixed, the localized electrochemical measurements reveal a kind of "mixed behaviour" (Figure 8.3). Corrosion at early stages is delayed, being negligible. The activity over the defect became evident between 18 and 24 h, reaching a maximum value. The behaviour is very similar to that observed for the singe LDH nanocontainers – inhibition at earlier immersion times and increase of corrosion activity after approximately 18 h of immersion. Increased activity in the defect creates the conditions for activating cerium molybdate containers and the anodic currents decrease, being negligible up to three days of immersion. Cerium molybdate containers have slow kinetics of inhibition if compared with LDH containers. The addition of a mixture of nanocontainers, filled with 2- mercaptobenzothiazole, results in a synergistic inhibition effect that combines early corrosion protection and longer term inhibition. This is related to the different activation mechanism of the containers. Such behaviour clearly demonstrates the potential of nanocontainer mixtures in the design of smart coatings with self-healing abilities for prolonged coating lifetime. A similar synergetic effect was reported for only LDH nanocontainers but loaded with different inhibitors: vanadate, phosphate and 2-mercaptobenzothiazole.[67] The combination of LDH nanocontainers loaded with vanadate and phosphate anions, and vanadate and mercaptobenzothiazole anions conferred a significant improvement in corrosion protection on AA2024 aluminium alloy immersed in NaCl solution. The inhibition effect of the LDH mixture is significantly higher than that offered by the individual systems and is attributed to stabilization of the native oxide film on the alloy surface. To achieve this effect, an inhibitor working at low pH should be placed in the coating system as close to the metal-coating interface as possible. The novel concept of the combination of clay nanocontainers was extended to aluminium alloy galvanically coupled to carbon fiber reinforced plastic.[68] The anionic layered double hydroxide and cationic bentonite were used as functional nanocontainers for 1,2,3-benzotriazole and Ce^{3+} inhibitors, respectively. The possible mechanism is the targeted inhibition actions for spatially separated cathodes (Ce^{3+}) and anodes (benzotriazole).

Recently, the combination of Layer-by-Layer (LbL) assembly with LDH nanoparticles was demonstrated for humidity-triggered self-healing films with oxygen barrier performance.[69] Hybrid films were fabricated *via* LbL assembly of LDH nanoplatelets and poly(sodium styrene-4-sulfonate) (PSS) followed by subsequent permeation of poly(vinyl alcohol) (PVA) (Figure 8.4).

After being treated by humidity for 75 h, the $(LDH–PSS)_{20}$–PVA–PET film recovered its original oxygen barrier property, demonstrating an excellent humidity-triggered self-healing capability. In contrast, both the pure PET and $(LDH–PSS)_{20}$–PET film did not show self-healing behaviour. It is proposed that the hydrogen bond network formed in the $(LDH–PSS)_{20}$–PVA–PET

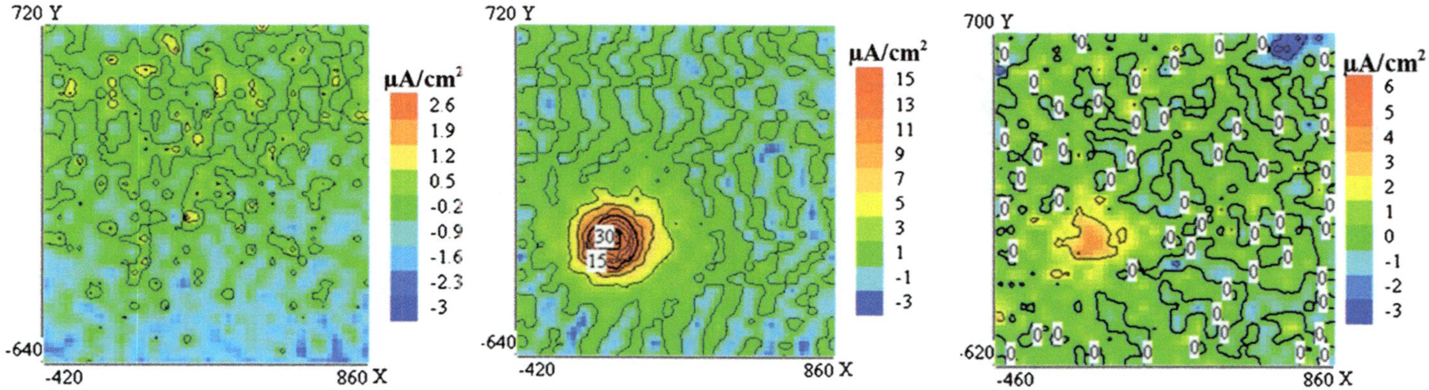

Figure 8.3 Coating modified with a mixture of mercaptobenzothiazole-loaded LDH and cerium molybdate nanocontainers. Current density maps obtained after different immersion times in 0.05 M NaCl: from left to right: after 3 h, after 18 h, after 40 h. Adapted from *Electrochimica Acta*, **60**, M. F. Montemor *et al.*, Evaluation of self-healing ability in protective coatings modified with combinations of layered double hydroxides and cerium molibdate nanocontainers filled with corrosion inhibitors, 31–40, Copyright 2012 with permission from Elsevier.[66]

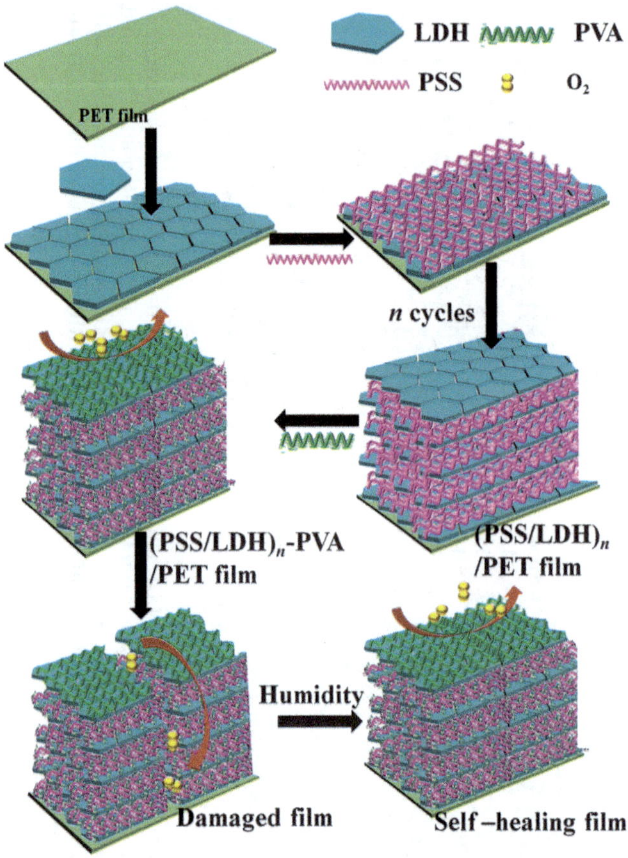

Figure 8.4 Schematic illustration of the fabrication of (LDH–PSS)n–PVA film on a polyester (PET) substrate serving as an oxygen barrier film with self-healing abilities. Reproduced from ref. 69 with permission from the Royal Society of Chemistry.

film with the assistance of water vapour hinders the diffusion of oxygen and maintains the durability of the barrier film. In addition, the self-healing capability of (LDH–PSS)$_{20}$–PVA–PET film can be repeated for several cycles by flexing and humidity treatment alternately. Corrosion inhibitors mercaptobenzothiazole and cerium(III) nitrate were later incorporated into polyelectrolyte–LDH multilayers.[70] The structural and compositional data showed that layered double hydroxides behaved as a dynamic system during the LbL process, exchanging part of the anionic inhibitor previously intercalated inside with species available in cerium- and polyelectrolyte-containing aqueous solutions. The release studies performed by UV-vis spectrophotometry indicated that the main triggering conditions for release of mercaptobenzothiazole from LDHs changed from anionic exchange to pH triggering when LDH nanoplatelets were covered with polyelectrolytes.

8.4 Self-Healing Coatings Based on Halloysites

Layered mineral clays, described above, can only load corrosion inhibitor between their layers in a small, 0.5–2 nm sized lumen. This results in a low loading capacity of the clays, which is usually 5–8 wt%. One needs to increase the loading capacity of nanocontainers in order to prolong the service lifetime of self-healing materials. The more corrosion inhibitor loaded into the containers, the larger the corrosion defect that can be healed. Inner halloysite lumen can provide additional space and increase loading capacity for corrosion inhibitors up to 20 wt%.[71] Additional selective etching of the alumina inside halloysite lumen with sulfuric acid can increase lumen capacity by two to three times and corrosion inhibitor loading up to four times compared to the untreated halloysites.[72] This loading capacity is close to the loading capacity of polymer nanocontainers. Besides corrosion inhibitors, halloysite can be a cargo for biocides and exploited in antifouling coatings.[73,74]

The typical procedure of the loading of halloysite nanotubes is as follows.[75] Halloysites are mixed with a solvent possessing high solubility for the desired corrosion inhibitor and a low temperature boiling point (*e.g.*, acetone, ethanol). Then, the vial containing the solution is placed in a desiccator under vacuum, which deaerates the halloysite lumen. Due to the rapid evaporation of the solvent, the inhibitor concentration increases thus improving the loading efficiency. After completing the loading procedure, the dispersion is mixed with a magnetic stirrer. The vacuum treatment is followed by washing and centrifugation. This procedure can be repeated several times. In the final stage, halloysites are removed from the centrifuge tube and dried.

Embedding of the inhibitor-loaded halloysites into the coating requires intensive mixing of the dried halloysites with the coating formulation using high-speed stirrers, UltraTurrex or ultrasound. It is a very important step to avoid the aggregation of the halloysites in the coating formulation. Formation of any aggregated nanocontainers will make defects in the coating integrity thus reducing coating barrier properties and corrosion protection performance. The halloysite should be homogeneously distributed on the coated area to protect every part of the metal (Figure 8.5).

The self-healing properties of benzotriazole and 8-hydroxyquinoline-loaded halloysite nanotubes (Atlas mining, USA) were studied in zirconia–silica sol–gel coatings deposited on the surface of aluminium alloy A2024 by periodic measurements of SVET current density profiles.[75] Migration of the corroding centre along the defect appeared for aluminium plates coated with pure sol–gel film. This is not the case for hybrid films containing halloysite nanocontainers. The time propagation of the extreme current densities reveals the development of the corrosion process. The maximum current density continuously increases with time for the pure sol–gel coating and the sol–gel film with empty halloysite tubes.

SVET observations on aluminium plates coated with sol–gel films containing inhibitor-loaded halloysite revealed significantly different results. In the

Figure 8.5 Distribution of inhibitor-loaded halloysite nanotubes inside a sol–gel coating. Reprinted from ref. 75 with permission from WILEY-VCH Verlag. Copyright © 2009 WILEY-VCH Verlag GmbH & Co. KGaA, Weinheim.

case of benzotriazole-loaded halloysites, the maximal anodic current was reduced down to the noise level within 4.5 hours (Figure 8.6). The inhibiting effect is also obvious for 8-hydroxyquinoline-loaded nanotubes. However, the maximal current density did not remain constant at the noise level in this case.

Another corrosion inhibitor, mercaptobenzothiazole, was loaded into halloysites nanotubes.[76] To prevent undesirable leakage of the loaded inhibitor from the halloysite interior at neutral pH, the outer surface of the mercaptobenzothiazole-loaded halloysite nanotubes was modified by deposition of alternating polyelectrolyte multilayers poly(allylamine hydrochloride)–poly (styrene sulfonate), PAH/PSS. The polyelectrolyte shell also provides release of the encapsulated inhibitor, controlled by pH changes in the environment surrounding the halloysite nanotubes.

Approximately 10% of the loaded material is released from halloysite nanotubes without polyelectrolyte shells after six days of storage in neutral water (Figure 8.7). On the contrary, the polyelectrolyte shell terminates the spontaneous release of the 2-mercaptobenzothiazole at neutral pH, blocking the openings of the halloysite nanotubes. A polyelectrolyte shell made of weak polycations (polyallylamine) and strong polyanions (polystyrene sulfonate) can be opened by increasing the pH value because of the formation of defects between the polyelectrolyte multilayers. The release of the mercaptobenzothiazole from halloysite nanotubes with polyelectrolyte shells is triggered at pH 10, similar to the pH value of the localized corrosion process. Visual observation of the surface of the samples after two weeks of immersion in the corrosion medium demonstrated that the non-coated aluminium sample immersed in pure aqueous sodium chloride solution had many pits on the surface and is completely corroded. Addition of mercaptobenzothiazole in solution decreases the number of pits and reduces the degree of

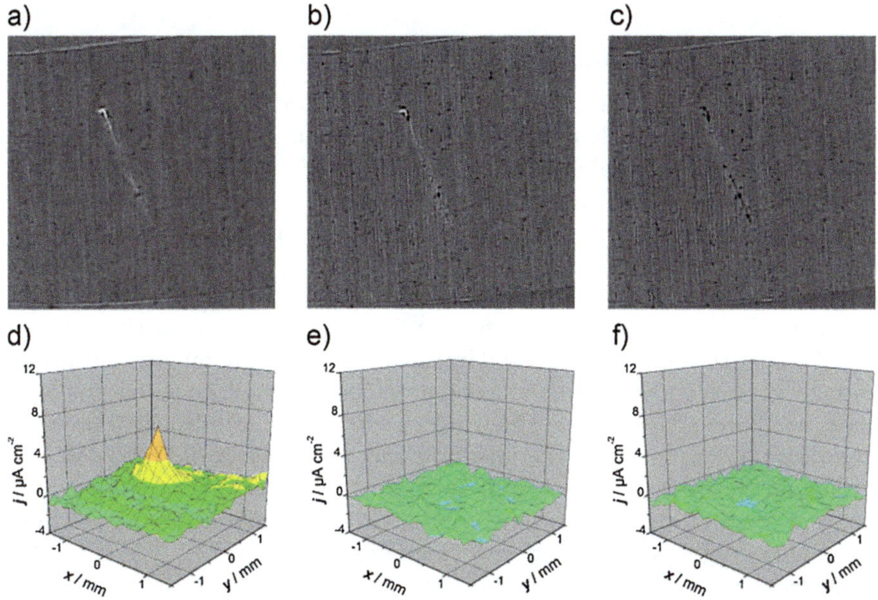

Figure 8.6 (a–c) CCD images and (d–f) current density maps of Al coated with a sol–gel layer doped with benzotriazole-loaded halloysite nanotubes immersed in 0.1 M NaCl after 0, 4.5 and 10 h, respectively. Reprinted from ref. 75 with permission from WILEY-VCH Verlag. Copyright © 2009 WILEY-VCH Verlag GmbH & Co. KGaA, Weinheim.

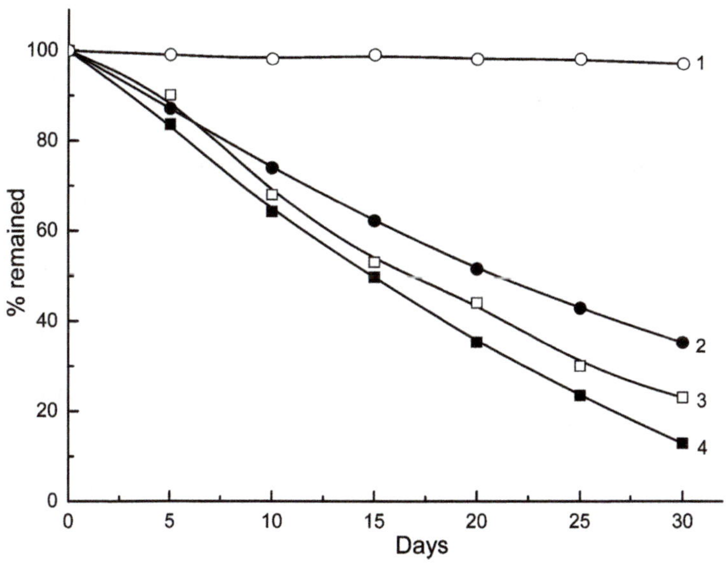

Figure 8.7 Release of the mercaptobenzothiazole in water solution at pH 6.5 from (1) halloysite nanotubes with a PAH/PSS shell and (3) halloysite nanotubes without a PAH/PSS shell; in water solution at pH 10 from (2) halloysite nanotubes with a PAH/PSS shell and (4) halloysite nanotubes without a shell. Reprinted with permission from D. G. Shchukin, S. V. Lamaka, K. A. Yasakau, M. L. Zheludkevich, M. G. S. Ferreira, H. Mohwald, *J. Phys. Chem. C*, 2008, **112**, 958.[76] Copyright (2008) American Chemical Society.

corrosion attack. However, the sample is still corroded after immersion in 0.3% NaCl solution, while the sample coated with a thin sol–gel film doped by mercaptobenzothiazole-loaded halloysites and immersed in 3% NaCl solution shows the intact surface.

Halloysite clay nanotubes loaded with corrosion inhibitors benzotriazole, mercaptobenzimidazole and mercaptobenzothiazole were used as additives in self-healing composite paint coating of copper.[77,78] These inhibitors form protective films on the metal surface and mitigate corrosion. The release rate of benzotriazole was controlled by the formation of metal–benzotriazole stoppers at tube endings. This method provides high halloysite loading efficiency and a long release time which gives additional possibilities for process optimization. Formation of the insoluble metal–benzotriazole complex was studied for Cu(II), Fe(II), Fe(III) and Co(II) ions. The best release control was achieved for the Cu–benzotriazole complex. The benzotriazole release rate decreased at higher concentrations of copper.

Corrosion tests were performed on 110 copper alloy strips. One strip was coated with pure acrylic and polyurethane paints, whereas three others were coated with paint composites containing 10 wt% halloysite loaded with the inhibitors. Coatings were artificially scratched and exposed to salt water containing 0.5 M of NaCl for six months. The corrosion process was monitored by measuring the concentration of Cu(II) ions released (Figure 8.8a). Extensive release of metal ions from the blank sample indicates a severe corrosion process taking place on the metal. A green patina observed underneath the paint also indicates severe corrosion (Figure 8.8b). Adding halloysites loaded with corrosion inhibitors significantly reduced corrosion in all the composite coatings. Some corrosion was developed within the first 15 days, which is evident from the rise of the Cu(II) ions in the solution, but then it was suppressed with the release of inhibitors in the coating defects. The efficiency of the halloysite lumen loading ascended in the order of benzotriazole < mercaptobenzothiazole < mercaptobenzimidazole.

The same corrosion inhibitors were used for protection of ASTM steel plates.[79] The inhibitor-loaded nanotubes were admixed at 5–10 wt% with oil-based alkyd paint providing sustained agent release and corrosion healing in the coating defects. The slow 20–30 h release of the inhibitors at defect points caused a remarkable improvement of the coating anticorrosion efficiency. A further time expansion of anticorrosion agent release was achieved by the formation of release stoppers at nanotube ends with urea–formaldehyde copolymer and copper-inhibitor complexation. Corrosion inhibitor-loaded halloysite samples were exposed to 0.5 mL of urea–formaldehyde prepolymer solution for 10 min. Then, sulfuric acid, which acts as a cross-linking agent, was added dropwise. The best protection was found using mercaptobenzimidazole and benzotriazole inhibitors. The results from undercoat corrosion indicated much less rust formation in cases of stopper formation compared to those in uncapped halloysites loaded with benzotriazole.

Despite the large number of papers devoted to nanocontainer-based self-healing coatings, most of them use lab-scale analytical methods for

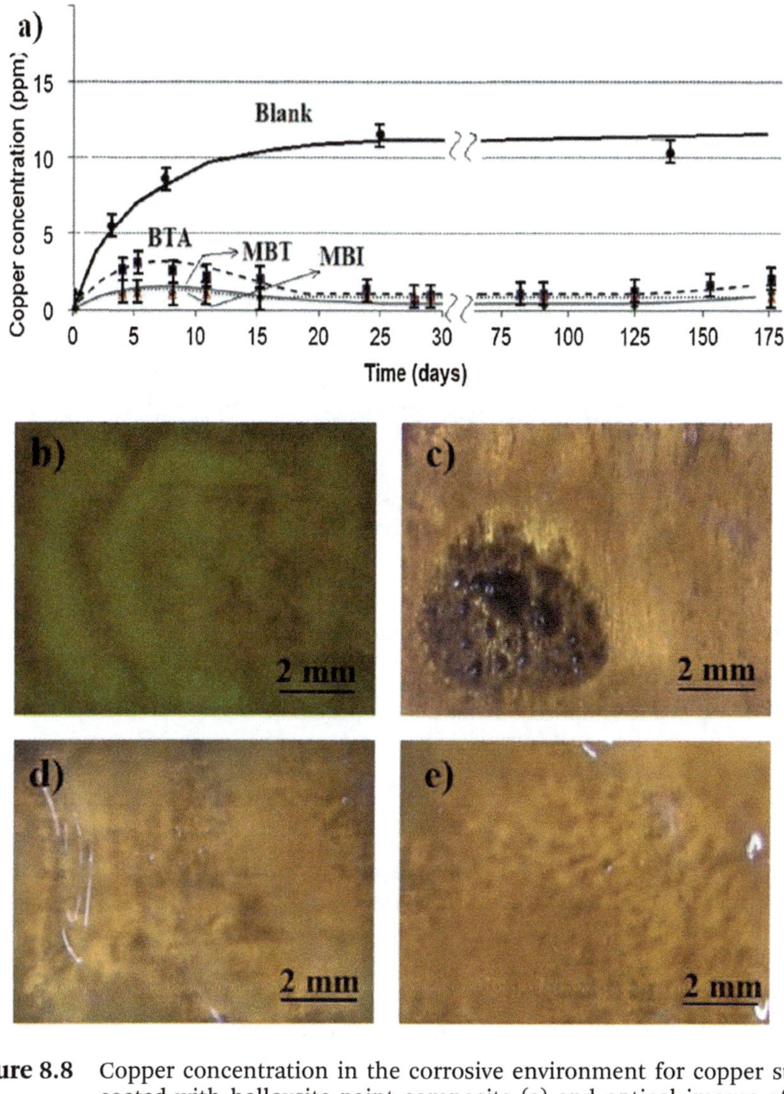

Figure 8.8 Copper concentration in the corrosive environment for copper strips coated with halloysite paint composite (a) and optical images of the strips through transparent polyurethane film; (b) blank coating and protective halloysite coating loaded with (c) benzotriazole (BTA), (d) mercaptobenzimidazole (MBI) and (e) mercaptobenzothiazole (MBT). Reprinted with permission from D. G. Shchukin, S. V. Lamaka, K. A. Yasakau, M. L. Zheludkevich, M. G. S. Ferreira, H. Mohwald, *J. Phys. Chem. C*, 2008, **112**, 958.[76] Copyright (2008) American Chemical Society.

characterization of their self-healing performance. Typical methods are electrochemical impedance spectroscopy, polarization, SVET and various adapted electrochemical techniques. However, the current numbers of reported successful laboratory studies call for the transfer of results from the laboratory to industry for further commercialization. To check halloysite

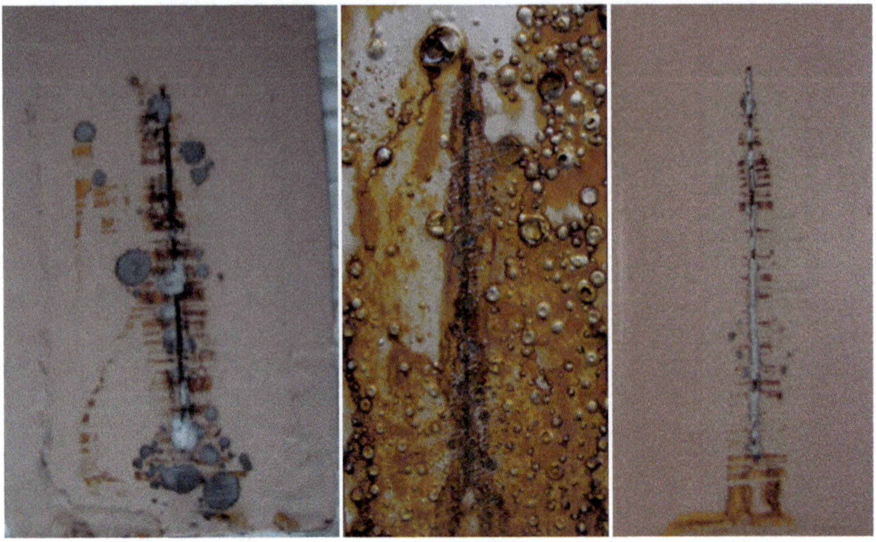

Figure 8.9 Neutral salt-spray test results for pure polyepoxy coating (A, 1000 h), polyepoxy coating directly loaded with Korantin SMK corrosion inhibitor (B, 500 h) and polyepoxy coating in the presence of Korantin SMK-loaded halloysite nanotubes (C, 1000 h).

efficiency, we conducted anticorrosion performance tests using the industrial neutral salt-spray test (ISO 9227 standard, 5 wt% NaCl, 35 °C).

As a benchmark, we used a standard commercial polyepoxy coating. We added corrosion inhibitor Korantin SMK, which is an alkylphosphoric ester produced by BASF with a chain length of alkyls in the ester group ranging from C6 to C10, into the coating in free form (1 wt%) and in the same amount but encapsulated into halloysite nanotubes. As one can see in Figure 8.9, addition of free corrosion inhibitor into the organic coating drastically reduces corrosion protection performance even after 500 h of the neutral salt-spray test. On the contrary, encapsulated inhibitor – with controlled and sustained release ability – increased corrosion protection by five times. This is clear evidence on an industrial level, employing a widely spread industrial test, that halloysite nanotubes loaded with industrial inhibitor can develop a new, revolutional generation of self-healing anticorrosion coatings.

8.5 Conclusions and Outlook

In the present chapter, we concentrated on the application of clay materials in self-healing anticorrosion coatings. This topic is of high importance because of the modern trend to save raw material and increase the security of metal structures. We showed in our review a number of promising candidates for application as nanocontainers with controlled release of inhibitor in self-healing coatings. As for any other material, there are pros and cons in application of clays as nanocontainers. They have a loading capacity lower than that of polymer nanocontainers, and they do not have large versatility in the mechanisms triggering and controlling the release of corrosion inhibitors. However, they are mechanically and thermally robust (important for coating formation), cheap and available in thousands of ton amounts (attractive for the coating industry).

One has to be aware that there are also other parameters that finally decide if clay materials can be marketed as additional pigments for self-healing coatings. There are technical issues like adhesion, optical appearance, environmental safety and recyclability. Additional knowledge is required about the behavior of clay-based self-healing coatings in industrial tests on a large scale. On the other hand, the progress in recent years is fast and sufficiently encouraging that one can be confident of finding solutions for the next steps for application of clay materials as smart nanocontainers.

Acknowledgements

We acknowledge ERC Consolidator grant ENERCAPSULE for financial support.

References

1. G. H. Koch, M. P. H. Brongers, N. G. Thompson, Y. P. Virmani and J. H. Payer, *Suppl. Mater. Perform.*, 2002, **42**, 2.
2. G. Pritchard, *Anti-Corrosion Polymers: PEEK, PEKK and Other Polyaryls*, ChemTec Publishing, Toronto, 1995.
3. Institute of Materials (Great Britain), *Corrosion Resistant Alloys for Oil and Gas Production: Guidance on General Requirements and Test Methods for H2S Service*, Maney Publishing, Leeds, 2nd edn, 2002.
4. M. C. Gupta and A. P. Gupta, *Polymer Composite*, New Age International (P) Limited, New Delhi, 2005.
5. V. P. McConnell, *Reinf. Plast.*, 2005, **49**, 20.
6. T. Brock, M. Groteklaes and P. Mischke, *Lehrbuch der Lacktechnologie*, Vincentz Verlag, Hannover, 3rd edn, 2009.
7. M. W. Kendig and R. G. Buchheit, *Corrosion*, 2003, **59**, 379–400.
8. L. Xia and R. L. McCreery, *J. Electrochem. Soc.*, 1998, **145**, 3083.
9. R. L. Twite and G. P. Bierwagen, *Prog. Org. Coat.*, 1998, **33**, 91.
10. Directive 2000/53/EC of the European Parliament and the council of 18.09.2000 on end-life vehicles, *Off. J. Eur. Communities*, L269/34–L269/42.
11. Y. T. Tsai, K. H. Hou, C. Y. Bai, J. L. Lee and M. D. Ger, *Thin Solid Films*, 2010, **518**, 7541.
12. Z. L. Zou, N. Li, D. Y. Li, H. P. Liu and S. L. Mu, *J. Alloys Compd.*, 2011, **509**, 503.

13. H. Ardelean, I. Frateur and P. Marcus, *Corros. Sci.*, 2008, **50**, 1907.
14. A. E. Hughes, J. D. Gorman, T. G. Harvey, A. Galassi and G. McAdam, *Corrosion*, 2006, **62**, 773.
15. A. E. Hughes, J. Gorman, T. G. Harvey, D. McCulloch and S. K. Toh, *Surf. Interface Anal.*, 2004, **36**, 1585.
16. A. A. Magalhaes, I. C. Margarit and O. R. Mattos, *J. Electroanal. Chem.*, 2004, **572**, 433.
17. Z. Li, C. Dai, Y. Liu and J. Zhu, *Electroplat. Pollut. Control*, 2007, **1**, 16.
18. C. S. Lin and W. J. Li, *Mater. Trans.*, 2006, **47**, 1020.
19. Y. Kobayashi and Y. Fujiwara, *Electrochem. Solid-State Lett.*, 2006, **9**, B15.
20. O. O. Knudsen, B. S. Tanem, A. Bjorgum, J. Mardalen and M. Hallenstvet, *Corros. Sci.*, 2004, **46**, 2081.
21. W. E. G. Hansal, S. Hansal, M. Polzler, A. Kornherr, G. Zifferer and G. E. Nauer, *Surf. Coat. Technol.*, 2006, **200**, 3056.
22. L. A. Hernandez-Alvarado, L. S. Hernandez, J. M. Miranda and O. Dominguez, *Anti-Corros. Methods Mater.*, 2009, **56**, 114.
23. E. J. Dufek and D. A. Buttry, *Electrochem. Solid-State Lett.*, 2008, **11**, C9.
24. A. S. Khanna, *High-performance organic coatings*, Woodhead, Cambridge, 2008.
25. E. W. Brooman, *Met. Finish.*, 2002, **100**, 48.
26. D. G. Shchukin and H. Mohwald, *Small*, 2007, **3**, 926.
27. D. Raps, T. Hack, J. Wehr, M. L. Zheludkevich, A. C. Bastos, M. G. S. Ferreira and O. Nuyken, *Corros. Sci.*, 2009, **51**, 1012.
28. A. J. Vreugdenhi and M. E. Woods, *Prog. Org. Coat.*, 2005, **53**, 119.
29. J. J. DiStefano, A. R. Stubberud and I. J. Williams, *Schaum's outline of theory and problems of feedback and control systems : continuous (analog) and discrete (digital)*, McGraw-Hill, New York, 1995.
30. K. Pelis, *Ann. Sci.*, 1997, **54**, 331.
31. D. G. Shchukin and H. Mohwald, *Small*, 2007, **3**, 926.
32. D. G. Shchukin and H. Mohwald, *Chem. Commun.*, 2011, **47**, 8730.
33. S. R. White, N. R. Sottos, P. H. Geubelle, J. S. Moore, M. R. Kessler, S. R. Sriram, E. N. Brown and S. Viswanathan, *Nature*, 2001, **409**, 794.
34. M. W. Keller, S. R. White and N. R. Sottos, *Adv. Funct. Mater.*, 2007, **17**, 2399.
35. M. M. Caruso, B. J. Blaiszik, S. R. White, N. R. Sottos and J. S. Moore, *Adv. Funct. Mater.*, 2008, **18**, 1898.
36. X. He and X. Shi, *Prog. Org. Coat.*, 2009, **65**, 37.
37. H. Liu, E. Gnade and K. J. Balkus, Preparation of a Delivery System for Smart Coatings by Electrostatic Deposition, *Smart Coatings III*, American Chemical Society, 2010, pp. 31–44.
38. C. Suryanarayana, K. C. Rao and D. Kumar, *Prog. Org. Coat.*, 2008, **63**, 72.
39. D. G. Shchukin and G. B. Sukhorukov, *Adv. Mater.*, 2004, **16**, 671.
40. Z. Zheng, X. Huang, M. Schenderlein, H. Mohwald, G. K. Xu and D. G. Shchukin, *Nanoscale*, 2015, **7**, 2409.
41. A. M. Javier, P. de Pino, M. F. Bedard, D. Ho, A. G. Skirtach, G. B. Sukhorukov, C. Planck and W. J. Parak, *Langmuir*, 2008, **24**, 12517.

42. X. Tao, J. B. Li and H. Mohwald, *Chem.–Eur. J.*, 2004, **10**, 3397.
43. D. G. Shchukin and D. V. Sviridov, *J. Photochem. Photobiol., C*, 2006, **7**, 23.
44. D. G. Shchukin, D. A. Gorin and H. Mohwald, *Langmuir*, 2006, **22**, 7400.
45. A. A. Antipov and G. B. Sukhorukov, *Adv. Colloid Interface Sci.*, 2004, **111**, 49.
46. K. Köhler, D. G. Shchukin, H. Mohwald and G. B. Sukhorukov, *J. Phys. Chem. B*, 2005, **109**, 18250.
47. D. S. Shchukin, K. Kohler and H. Mohwald, *J. Am. Chem. Soc.*, 2006, **128**, 4560.
48. M. L. Zheludkevich, J. Tedim and M. G. S. Ferreira, *Electrochim. Acta*, 2012, **82**, 314.
49. S. Bohm, H. N. McMurray, S. M. Powell and D. A. Worsley, *Mater. Corros.*, 2001, **52**, 896.
50. H. N. McMurray, D. Williams, G. Williams and D. Worsley, *Corros. Eng., Sci. Technol.*, 2003, **38**, 112.
51. A. Ghazi, E. Ghasemi, M. Mahdavian, B. Ramezanzadeh and M. Rostami, *Corros. Sci.*, 2015, **94**, 207.
52. F. Micciché, H. Fischer, R. Varley and S. van der Zwaag, *Surf. Coat. Technol.*, 2008, **202**, 3346.
53. X. Qi, Y. Guan, G. Chen, B. Zhang, J. Ren, F. Peng and R. Sunab, *RSC Adv.*, 2015, **5**, 41006.
54. G. Gao, G. Du, Y. Sun and J. Fu, *ACS Appl. Mater. Interfaces*, 2015, **7**, 5029.
55. R. B. Leggat, S. A. Taylor and S. R. Taylor, *Colloids Surf., A*, 2002, **210**, 83.
56. J. Chen, Y. Song, D. Shan and E.-H. Han, *Corros. Sci.*, 2013, **74**, 130.
57. R.-C. Zeng, Z.-G. Liu, F. Zhang, S. Q. Li, H.-Z. Cui and E.-H. Han, *J. Mater. Chem. A*, 2014, **2**, 13049.
58. G. R. Williams and D. O'Hare, *J. Mater. Chem.*, 2006, **16**, 3065.
59. R. G. Buchheit, S. B. Mamidipally, P. Schmutz and H. Guan, *Corrosion*, 2002, **58**, 3.
60. J. Tedim, M. L. Zheludkevich, A. N. Salak, A. Lisenkov and M. G. S. Ferreira, *J. Mater. Chem.*, 2011, **21**, 15464.
61. P. V. Mahajanarn and R. G. Buchheit, *Corrosion*, 2008, **64**, 230.
62. G. Williams and H. N. McMurray, *Electrochem. Solid-State Lett.*, 2004, **7**, B13.
63. M. H. Kendig and M. Hon, *Electrochem. Solid-State Lett.*, 2005, **8**, B10.
64. M. L. Zheludkevich, S. K. Poznyak, L. M. Rodrigues, D. Raps, T. Hack, L. F. Dick, T. Nunes and M. G. S. Ferreira, *Corros. Sci.*, 2010, **52**, 602.
65. S. K. Poznyak, J. Tedim, L. M. Rodrigues, A. N. Salak, M. L. Zheludkevich, L. F. P. Dick and M. G. S. Ferreira, *ACS Appl. Mater. Interfaces*, 2009, **1**, 2353.
66. M. F. Montemor, D. V. Snihirova, M. G. Taryba, S. V. Lamaka, I. A. Kartsonakis, A. C. Balaskas, G. C. Kordas, J. Tedim, A. Kuznetsova, M. L. Zheludkevich and M. G. S. Ferreira, *Electrochim. Acta*, 2012, **60**, 31.
67. J. Tedim, S. K. Poznyak, A. Kuznetsova, D. Raps, T. Hack, M. L. Zheludkevich and M. G. S. Ferreira, *ACS Appl. Mater. Interfaces*, 2010, **2**, 1528.

68. M. Serdechnova, S. Kallip, M. G. S. Ferreira and M. L. Zheludkevich, *Electrochem. Commun.*, 2014, **41**, 51.
69. Y. Dou, A. Zhou, T. Pan, J. Han, M. Wei, D. G. Evans and X. Duan, *Chem. Commun.*, 2014, **50**, 7136.
70. J. Carneiro, A. F. Caetano, A. Kuznetsova, F. Maia, A. N. Salak, J. Tedim, N. Scharnagl, M. L. Zheludkevich and M. G. S. Ferreira, *RSC Adv.*, 2015, **5**, 39916.
71. R. Price, B. Gaber and Y. Lvov, *J. Microencapsulation*, 2001, **18**, 713.
72. E. Abdullayev, A. Joshi, W. B. Wei, Y. F. Zhao and Y. Lvov, *ACS Nano*, 2012, **6**, 7216.
73. Y. Lvov and R. Price, Halloysite Nanotubules, A Novel Substrate for the Controlled Delivery of Bioactive Molecules, in *Bio-Inorganic Hybrid Nanomaterials*, ed. E. Ruiz-Hitzky, K. Ariga and Y. Lvov, Wiley, London, Berlin, 2007, ch. 12, pp. 419–441.
74. Y. M. Lvov, D. G. Shchukin, H. Mohwald and R. R. Price, *ACS Nano*, 2008, **2**, 814.
75. D. Fix, D. V. Andreeva, Y. M. Lvov, D. G. Shchukin and H. Mohwald, *Adv. Funct. Mater.*, 2009, **19**, 1720.
76. D. G. Shchukin, S. V. Lamaka, K. A. Yasakau, M. L. Zheludkevich, M. G. S. Ferreira and H. Mohwald, *J. Phys. Chem. C*, 2008, **112**, 958.
77. E. Abdullayev, R. Price, D. Shchukin and Y. Lvov, *ACS Appl. Mater. Interfaces*, 2009, **1**, 1437.
78. E. Abdullayev, V. Abbasov, A. Tursunbayeva, V. Portnov, H. Ibrahimov, G. Mukhtarova and Y. Lvov, *ACS Appl. Mater. Interfaces*, 2013, **5**, 4464.
79. A. Joshi, E. Abdullayev, A. Vasiliev, O. Volkova and Y. Lvov, *Langmuir*, 2013, **29**, 7439.

CHAPTER 9

Flame Retardant Polymer–Halloysite Nanocomposites

ZHENGHAI TANG[a] AND BAOCHUN GUO*[a]

[a]Department of Polymer Materials and Engineering, South China University of Technology, Guangzhou 510540, China
*E-mail: psbcguo@scut.edu.cn

9.1 Introduction

The incorporation of nanofillers into polymers has drawn great attention in past decades. Nanofillers not only can improve the mechanical performance of polymers, but also may impart them with functional properties based on the nanofillers used. Most polymer materials are flammable; it is necessary to decrease the flammability by adding flame retardant compounds to fulfill their practical applications. In particular, the use of halogenated flame retardant additives has been greatly restricted due to environmental regulation in recent years. Nanofillers are highly attractive as alternative flame retardant additives because they are environmentally friendly and can simultaneously improve both the physical properties and flammability properties of the polymers. Since Blumstein and Ries first reported the improved thermal stability of poly(methyl methacrylate)–montmorillonite (MMT) nanocomposites,[1] studies concerning the thermal stability and flame-retardant properties of nanofiller-reinforced polymer nanocomposites have inspired great scientific interest.[2] It has been reported that the integration of nanofillers such as MMT,[3] magnesium hydroxide,[4] layered double hydroxides,[5] carbon

RSC Smart Materials No. 22
Functional Polymer Composites with Nanoclays
Edited by Yuri Lvov, Baochun Guo and Rawil F. Fakhrullin

Published by the Royal Society of Chemistry, www.rsc.org

nanotubes (CNTs),[6] and polyhedral silsesquioxanes[7] in polymeric matrices can significantly improve their fire performance.

Recently, halloysite nanotubes (HNTs), a type of naturally occurring aluminosilicate with nanotubular structures, have drawn numerous attention due to their wide applications in various fields such as polymer reinforcement, controlled release, and nanotemplating.[8–10] The structure and composition of HNTs have been introduced in Chapter 5. HNTs show a three-stage decomposition within the temperature range between 35 and 800 °C. The first thermal decomposition stage occurring below 200 °C is due to the loss of physically adsorbed water on the HNT surface. The second stage between 200 and 500 °C corresponds to release of adsorbed water on the surface and interlayers of the HNTs. In this stage, the weight loss is about 10 wt%, with a product of metahalloysite [$Al_2Si_2O_5(OH)_4$]. The final weight loss above 500 °C is ascribed to the dehydroxylation of the HNT interlayer surface to form $Al_2S_2O_5$.[11] Similar to other inorganic nanoparticles, HNTs can be used as an effective halogen-free flame retardant in polymers through the formation of a thermally stable layer on the sample surface. Up to now, HNTs have been added as a flame retardant in various polymer matrices.[12–14] In this chapter, the flame retardancy of polymer–HNTs and the underlying role of HNTs during combustion were first introduced. Then, the factors related to the dispersion of HNTs and modification of HNTs that affect the flame retardant properties of the polymer–HNTs nanocomposites were discussed. After that, the synergistic effects of the combination of HNTs with other flame retardants (an intumescent flame retardant system and other nanofillers) on the flame retardancy performance of the nanocomposites were reviewed. Finally, the flame retardancy of other clays (MMT, vermiculite, sepiolite, kaolinite) towards polymers and the comparison between HNTs and other clay were briefly illustrated.

9.2 Use of HNTs as a Flame Retardant in Polymers

HNTs have been incorporated as flame retardants into various polymer matrices such as low density polyethylene (LDPE),[12] polypropylene (PP),[13,14] acrylonitrile–butadiene–styrene (ABS),[15] polyamide,[16] nitrile rubber (NBR),[17] and so on. HNTs were demonstrated to be very efficient in improving the flame retardant properties of the polymers without any sacrifice of the mechanical properties. Due to the tubular shape and few hydroxyl groups located on the HNT surface, HNTs can be readily dispersed in polymers through direct melt mixing.

Marney *et al.* showed that HNTs had a positive effect on the fire performance of polyamide 6 (PA6).[16] A cone calorimeter (CC) test showed that the heat release rate (HRR), peak of heat release rate (PHRR), and total heat released (THR) of the nanocomposites were consistently decreased, and the residual char was gradually increased with increased HNT loading. The CC test enables quantitative assessment of material flammability through oxygen consumption calorimetry based on the principle that there is a constant relationship between the mass of oxygen consumed and the amount of heat released during combustion of a material. It can provide a variety of fire properties such as HRR, PHRR, THR, time to ignition (TTI), mass loss rate (MLR), specific extinction area (SEA), total smoke release (TSR), and residual char to determine how a material may or may not be appropriate for a particular fire risk scenario.[18,19] The HRR curves for neat PA6 and its nanocomposite with low HNT content (5 wt%) exhibited a narrow plateau followed by a sharp peak (Figure 9.1). The narrow plateau suggested that a thin skin was formed on the sample surface prior to ignition. This thin skin eventually broken down due to the release of the trapped volatile products and the material ignited; consequently, HRR increased until it reached a peak. When HNT loading was higher than 10 wt%, a flat HRR curve featured the char forming systems. In this case, the char layer did not break down during combustion, instead it grew and became more tough with increasing HNT loadings, and formed an insulating layer between the melted polymer and flame zone, which inhibited the heat transfer into the sample to slow down its decomposition rate and acted as a mass barrier for the in-diffusion of oxygen and the out-diffusion of volatile degradation products. Apart from the protective coating formation, HNTs underwent dehydration of physisorbed water and dehydroxylation. The endothermic decomposition of HNTs absorbed part of the external heat to keep the surrounding polymer cooler and hence delay the thermal decomposition. However, HNTs accelerated the rate of smoke production of PA6

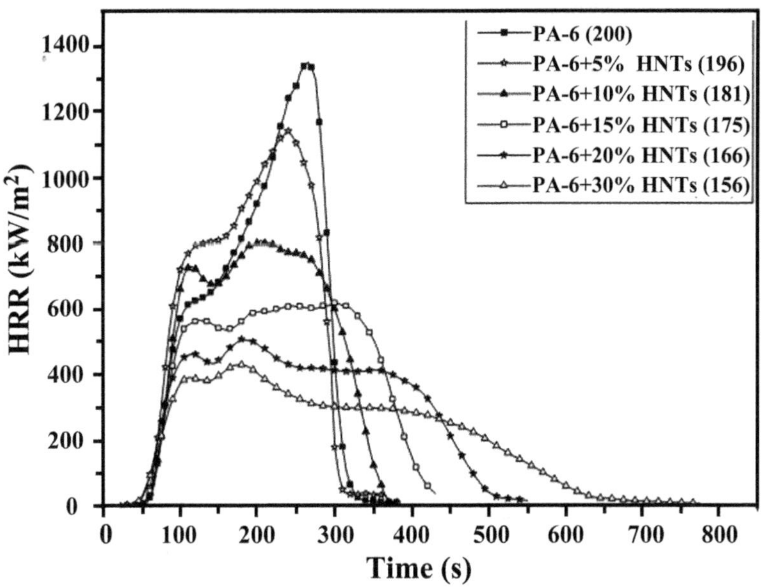

Figure 9.1 Heat release rate as a function of time for PA6 and its nanocomposites. (Reproduced from Polymer Degradation and Stability, 93, D. C. O. Marney *et al.*, The suitability of halloysite nanotubes as a fire retardant for nylon 6, 1971–1978, Copyright (2008) with permission from Elsevier.[16])

during the burning process, which was related to an increase in unburned higher molecular weight hydrocarbons associating with secondary pyrolysis reactions undergone by the evolved gases. Moreover, the limiting oxygen index (LOI) was gradually increased with increasing HNT content, and it was increased from 22% for neat PA6 to 30% for the nanocomposite with 30 wt% HNTs.[16] In a study by Lecouvet *et al.*, the flammable behavior of polyamide 12 (PA12)–HNTs nanocomposites was similar to that observed in PA6–HNTs nanocomposites. It was proposed that the reduction in the HRR was attributed to physical processes occurring in the condensed phase instead of in the gas phase because the ratio of HRR to MLR was almost constant for all the samples.[20]

Liu *et al.* found that HNTs could effectively delay the degradation process and improve the thermal stability of poly(vinyl chloride) (PVC), especially in the first-stage degradation related to release of hydrogen chloride, which might be because HNTs trapped the $Cl^{\bullet}$ radicals during the initial degradation stage. The smoke production rate and the total volume of smoke obtained from CC measurement were significantly reduced with the addition of HNTs. During the degradation of PVC, hydrogen chloride was first released and conjugated double bonds yielded; these conjugated bonds eventually cyclized to form aromatic compounds, which gave lots of smoke. The addition of HNTs into PVC reduced the production of hydrogen chloride in the first-stage degradation of PVC, thus decreasing the formation of aromatic compounds and suppressing smoke generation. In addition, the HRR curves of PVC–HNTs nanocomposites exhibited multi-peaks and the intensities of these peaks were much lower in comparison with virgin PVC, which was because the thermal insulation properties and barrier effects of HNTs blocked the heat and hindered the diffusion of the volatile decomposition products.[21] In another study, Shi *et al.* mentioned that HNTs could act as antimicrobial additives and a flame retardant in PVC. The obtained PVC–HNTs nanocomposites exhibited excellent antimicrobial ability and non-flammability properties.[22]

Du *et al.* demonstrated that incorporation of HNTs in PP improved the thermal stability and flammability of PP. Compared with neat PP, the temperature at maximum weight loss for the PP nanocomposites with 10 phr HNTs was increased by 10 °C. Such enhancement could be comparable to that achieved in PP–layered silicate nanocomposites. Because the barrier effect of the tubular HNTs is inferior to that of layered silicates, the barrier effect is not the only reason for the remarkably improved thermal stability of PP–HNTs nanocomposites. Given that HNTs have typical dimensions of 2–40 μm length, 10–50 nm in outer diameter and 5–20 nm in internal diameter, and the volume percentage of the lumen of HNTs may be as high as 25%, the authors proposed that the lumen of the HNTs played an important role in improving the thermal stability of the nanocomposites. In the initial degradation stage, the degradation products of PP could be considerably entrapped into the lumens of HNTs, resulting in effective delay in mass transport and significantly increased thermal stability. The entrapment mechanism is proposed in Figure 9.2.[13] Along with this study, Lecouvet *et al.* reported that the

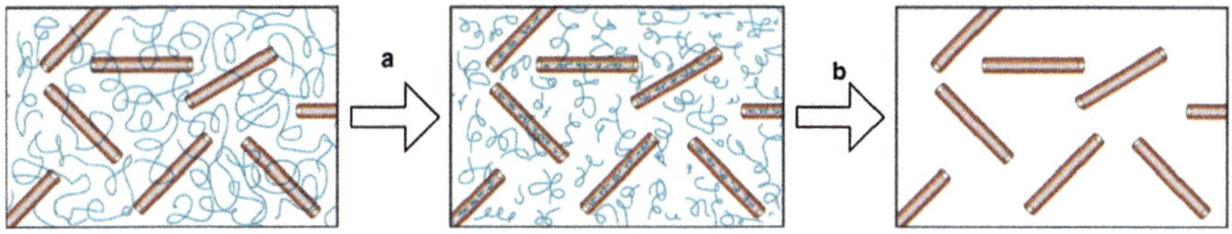

Figure 9.2 Scheme of entrapment mechanism of thermal decomposition products of PP–HNTs nanocomposites: (a) entrapping of decomposition products by HNTs lumens and (b) escaping of decomposition products from the lumens. (Reproduced from European Polymer Journal, 42, M. Du, B. Guo and D. Jia, Thermal Stability and flame retardent effect of halloysite nanotubes on poly(propylene), 1362–1369, Copyright (2006) with permission from Elsevier.[13])

activation energies of degradation kinetics for PP–HNTs nanocomposite were higher than the neat PP in both air and nitrogen atmospheres, which also suggested an increase of the thermal stability of PP in the presence of HNTs.[23] Although the thermal stability of PP was improved by adding HNTs, the nanocomposites exhibited inferior resistance to thermal oxidation aging, which might be due to the acidic sites and volume of entrapped oxygen in the lumen of HNTs and cavities in the nanocomposites.[24] Furthermore, a CC test showed that the time for the ignition of the nanocomposites was decreased with the incorporation of HNTs. This behavior was because the viscosity of the condensed phase was increased by adding HNTs during the burning process, which lowered convection and decreased the heat conductivity, and thus more heat was accumulated at the sample surface and the ignition temperature of the material reached faster. In addition, unlike neat PP where incident radiation tended to be absorbed inside the PP matrix perpendicularly to the sample surface, the nanocomposites might absorb and/or scatter incident radiation mainly near the sample surface as HNTs could scatter isotropically incident radiation, thereby a shorter time was required to reach the ignition temperature.[14]

Jia *et al.* reported that the introduction of HNTs into linear LDPE endowed the nanocomposites with excellent flame retardant properties, especially when the HNT content was higher than 30 wt%. The authors suggested that HNTs decomposed and released crystal water in the temperature range of 430–570 °C, at which the maximum thermal degradation temperatures of LDPE occurred. The released water vapor might dilute and cool flammable gases. Meanwhile, the barrier effects created by HNTs also slowed down the release of flammable gases.[12]

Adding HNTs into cellulose fiber-reinforced vinyl ester (VER) resin also gave rise to improvements in the thermal stability and fire resisting properties. Compared with neat VER, the maximum thermal degradation temperature of the nanocomposites with 5 wt% HNTs was increased by 30 °C and the fire velocity was decreased by nearly half. The improved thermal stability and fire retardancy were due to the multi-effects brought by HNTs as follows.

(1) The interfacial adhesion between cellulose fibers and VER was improved by adding HNTs. (2) The char formation of HNTs acted as a barrier that prevented heat transfer and mass transport. (3) The hollow tubular structure of HNTs enabled the entrapment of degradation products inside the lumens. (4) The iron oxides in HNTs trapped radicals during the degradation process to enhance the thermal stability.[25]

In ethylene propylene diene monomer (EPDM), the thermal stability was gradually enhanced by increasing HNT loading from 0 to 100 phr. When HNT loading was lower than 15 phr, the samples failed to pass a UL-94 test, but the dripping and total ignition time were remarkably decreased. While when HNT content was increased to 100 phr, the nanocomposites could pass the V-0 rating. The improved flame retardancy was related to the uniform dispersion of HNTs, interfacial and inter-tubular interactions between EPDM and HNTs as well as the formation of zig-zag structures, which increased char residue and inhibited diffusion of the volatile decomposition products.[26]

Nakamura *et al.* prepared a bio-based material comprising jute fabric and soy protein concentrate (SPC), and investigated the effects of HNTs on the flammability properties. In virgin SPC, the formed char was fragile and badly damaged. However, the char was more intact and had more carbon content in SPC–HNTs nanocomposites. Micro-combustion calorimetry (MCC) tests showed that the flammability and total heat release were decreased by addition of HNTs. It was worthy of note that the ignition time of the nanocomposite measured by CC test was shorter than neat SPC, while the peak pyrolysis temperature of the nanocomposites measured by MCC was unchanged. This disparity could be explained by the differences in sample size and method of heating, which had a significant effect on heat transfer.[27]

9.3 Factors Affecting the Flame Retardancy of Polymer–HNTs Nanocomposites

9.3.1 Dispersion of HNTs

The uniform dispersion of nanofillers in a polymer matrix is a critical factor to obtain a good flame retardant performance as it plays a key role in enhancing the structural stability of the char.[2,28] The dispersion of HNTs is highly related to the processing methods for the preparation of the nanocomposites. Although HNTs can be readily dispersed in some polymers by direct melt mixing, as mentioned above, other processing approaches such as master batch, water-assisted extrusion, and solution mixing, have been developed to further improve the dispersion of HNTs.

Lecouvet *et al.* mixed PA12 and HNTs together in a twin-screw extruder equipped with a water injection system and a degassing zone (water-assisted extrusion technique) to prepare PA12–HNTs nanocomposites. The water-assisted extrusion process is shown in Figure 9.3a. HNTs and polymer are

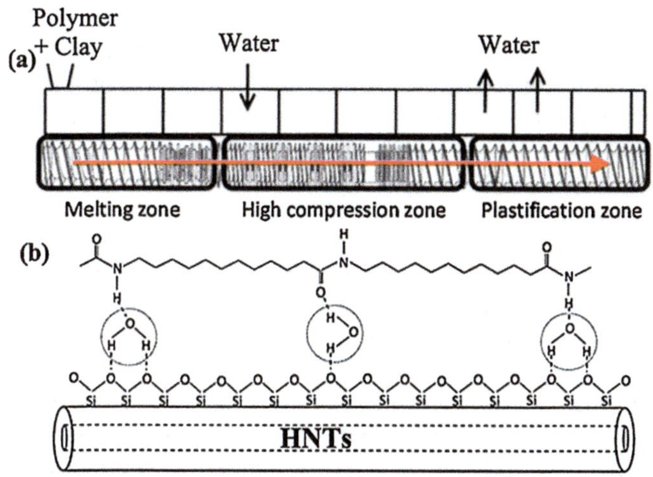

Figure 9.3 (a) Scheme of the water-assisted extrusion process of polymer–HNT nanocomposites: melting zone, high compression zone, and plastification zone. (Reproduced from Polymer, 52, B. Lecouvet, M. Sclavons, S. Bourbigot, J. Devaux, C. Bailly, Water-assisted extrusion as a novel processing route to prepare polypropylene/halloysite nanotube nanocomposites: Structure and properties, 4284–4295, Copyright (2011) with permission from Elsevier.[14]); (b) schematic illustration of the hydrogen bonds between water-solvated PA12 chains and HNTs. (Reproduced from ref. 20 with permission from John Wiley & Sons. Copyright © 2013 John Wiley & Sons, Ltd.)

introduced simultaneously and melt mixed together in the first melting zone, and water is injected in the high compression zone followed by water degassing in the terminal plasticization zone. In the high compression zone, water and PA12 were miscible and formed a single molten phase due to high pressure and high temperature. The melt viscosity of the blend was decreased, and the PA12 polymer chains could readily diffuse into the HNTs. Moreover, the one phase system had higher polarity than the neat PA12, which enhanced the affinity between PA chains and HNTs *via* the formation of additional hydrogen bonds, as illustrated in Figure 9.3b. The water-assisted extrusion facilitated the uniform dispersion of HNTs throughout the polymer matrix and strengthened the interfacial adhesion, which enabled the formation of a stronger inorganic-rich layer with enhanced barrier properties, but also the release of a higher amount of water during combustion.[20] The same group also demonstrated that the water-assisted strategy was efficient in improving the dispersion of HNTs in poly(lactic acid) (PLA) and the flame retardancy.[29]

PA12–HNTs nanocomposites were produced by batch internal mixer or twin-screw mini-compounder. The delay in the thermal degradation was observed in both PA12–HNTs nanocomposites prepared by these two different compounding methods. At a given HNT content, the nanocomposites prepared by twin-screw mini-compounding showed inferior thermal stability in comparison with the nanocomposites prepared by a batch internal mixer. This was because twin-screw mini-compounding produced an alignment of HNTs in the nanocomposites, which gave rise to relatively bad barrier properties.[30]

Fernández *et al.* prepared PA6–HNTs nanocomposites by diluting the master batch with a combination of an extrusion-compounding process and injection molding. A UL94 test showed that the flame resistance of the nanocomposites was improved by the addition of HNTs; especially, the flame duration was consistently decreased with HNT content. This process method produced nanocomposites with uniformly dispersed HNTs in industry scale size.[31]

Liu *et al.* found that poly(vinyl alcohol) (PVA)–HNTs nanocomposites prepared by the coagulation method had superior thermal stability than those prepared by solution casting. This was explained by the fact that the coagulation method could yield composites with better dispersion, while HNTs tended to aggregate during the evaporation of solvent in the drying process.[32]

9.3.2 Modification of HNTs

Due to the great polarity discrepancy between HNTs and polyolefins, it is essential to modify HNTs to enhance the compatibility between HNTs and polymers, especially for non-polar polymers.

In a study by Du *et al.*, the surface treatment of HNTs with silane made HNTs organophilic and enabled the preparation of PP–HNTs nanocomposites. Compared with composites filled with unmodified HNTs, the temperature at maximum weight loss for the PP nanocomposites containing 10 phr silane modified-HNTs was increased by 27 °C. The compatibility between HNTs and PP was improved and HNTs were more evenly dispersed in PP matrices upon modification of HNTs, thus leading to higher randomness of lumen ends, which enabled the entrapment of the volatile products inside the lumens more effectively.[13]

Lecouvet *et al.* prepared PP–HNT nanocomposites by different mixing strategies with or without the use of a PP-*graft*-maleic anhydride (PP-*g*-MA) compatibilizer. It was found that a uniform dispersion of HNTs was only achieved when the compatibilizer and water-assisted extrusion were combined together. On one hand, water injected at high pressure into the extruder remained liquid and created an aqueous suspension of HNTs during the melt blending step, which promoted the dispersion of HNTs. Moreover, the injected water in the high compression zone could hydrolyze anhydride functions of PP-*g*-MA into carboxylic diacid functions, leading to the formation of interfacial hydrogen bonds with the siloxane groups and aluminol groups at the surface of HNTs, which further enabled the homogeneous dispersion of HNTs into molten polymer. The nanocomposites obtained in this way exhibited the highest thermal stability and lowest flammability. Compared with pure PP, the PHRR of the nanocomposites with 16 wt% HNTs was decreased from 620 to 220 kW m^{-2}, which was because the uniform dispersed HNTs migrated to the surface of the condensed phase and formed a rigid and stable inorganic residue that acted as a barrier to both heat and mass transport during combustion.[14]

As mentioned above, adding HNTs into LDPE improved the flame retardancy of LDPE. Furthermore, the introduction of interfacial modifier, maleic anhydride, methyl methacrylate and butyl acrylate grafted-PE, into LDPE–HNTs nanocomposites could further improve the flame retardancy and thermal stability of the nanocomposites. As shown in Figure 9.4, the PHRR, average HRR, and mass loss rate of the LDPE–HNTs–graft nanocomposites were lower than those of the LDPE–HNTs nanocomposite. It was argued that the graft copolymer improved the dispersion of HNTs and strengthened the barrier layers formed during combustion, and consequently further retarded the release of flammable gases and the degradation of LDPE.[12]

Rybinski *et al.* treated HNTs with sulfuric acid to remove organic carbon and aluminosilicate acids from its surface. The activated HNTs were then reacted with chlorosulfonated polyethylene to yield nucleus-sheath-HNTs. Both activated HNTs and nucleus-sheath-HNTs could remarkably decrease the thermal decomposition rate of styrene–butadiene rubber (SBR) and NBR. The residue after thermal decomposition in NBR nanocomposites was improved when compared to unfilled-NBR, while that in SBR was almost unchanged. This indicated that HNTs increased the capability of NBR to form rings and to carbonize, which facilitated the formation of a thermally stable barrier during combustion, whereas HNTs failed to improve the carbonization processes of SBR. Compared with activated HNTs, nucleus-sheath-HNT filled rubbers exhibited better flammable retardant properties with a higher ignition temperature, lower PHRR and oxygen consumption. This was because nucleus-sheath-HNTs released hydrogen chloride, which impeded radical reactions taking place during combustion.[33]

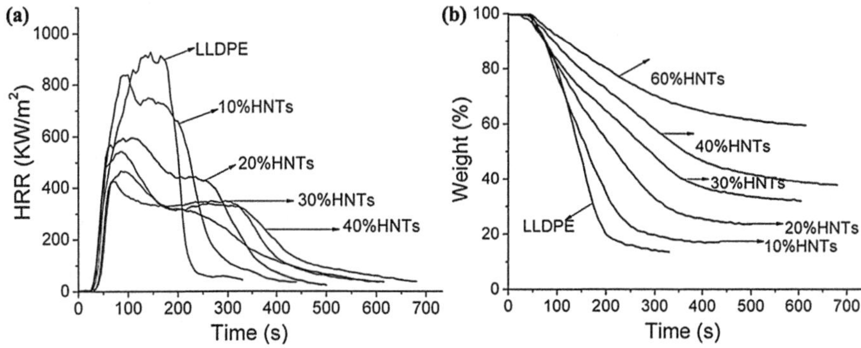

Figure 9.4 (a) HRR and (b) mass loss rate curves for neat LDPE, LDPE nanocomposites with 40 wt% HNTs and grafted HNTs. (Reproduced from ref. 12 by permission of Taylor & Francis Ltd.)

Tham *et al.* used *N*,*N*′-ethyl-enebis(stearamide) (EBS) to modify HNTs and prepared PLA–HNTs nanocomposites. The amide groups in EBS could interact with hydroxyl groups on HNTs and with carbonyl groups in PLA, thus improving the dispersion of HNTs and enhancing interfacial adhesion between PLA and HNTs. Compared with unmodified HNTs, the PLA–modified HNTs nanocomposites exhibited better flame retardancy. This was because the improved dispersion of HNTs and enhanced interfacial adhesion facilitated the formation of a char barrier. In addition, the amide group in EBS enabled the trapping of free radicals generated during combustion, thus slowing down the thermal decomposition rate.[34]

Fu *et al.* utilized chitosan to modify HNTs and then added the chitosan-modified HNTs into waterborne polyurethane. The amino groups in the modified HNTs could react with polyurethane prepolymer, resulting in uniform dispersion of HNTs and improved interfacial interaction between HNTs and the polyurethane matrix. The well dispersed HNTs acted as thermal insulators and formed chars on the polyurethane surface, which hindered the decomposition of nanocomposites. In addition, the cross-linking density of nanocomposites was increased as the modified HNTs served as the cross-linking sites, which was also helpful to improve the thermal stability.[35]

When replacing the –Cl chain ends of polyethersulfone (PES) with –OH groups, the dispersion of HNTs was significantly improved in the PES terminated with –OH (PES-OH), which benefited from the covalent linkage between PES-OH chains and HNTs during extrusion. In PES-OH–HNTs nanocomposites, the flame retardancy was greatly improved in terms of the remarkable reduction in PHRR and THR. The mechanism of action of HNTs for the fire retardancy of PES was because an inorganic-rich layer was formed at the sample surface during combustion, which reinforced the char structure to delay crack formation. Besides, HNTs acted as an aluminosilicate "skeleton frame" that supported the expanded carbonaceous char and reduced the size of pores in the intumescent layer. The enhanced structural integrity of the insulating layer acted as a physical barrier to both heat and mass transport, thus slowing down its decomposition rate and escape of volatiles.[36]

Jing *et al.* selectively modified the inner lumen of HNTs with octadecylphosphonic acid (ODP) to impart the inner lumen with hydrophobicity. Then bisphenol-A bis(diphenyl phosphate) (BDP), a widely used non-halogenated flame retardant, was loaded into the hydrophobized HNT lumens, resulting in a high loading and controlled release of BDP. The preparation schematic is displayed in Figure 9.5(a). In Figure 9.5(b), a linear 70 h lasting release of BDP from the nanotubes was observed for ODP-modified HNTs, while a much faster release of BDP was completed within 3 h for the unmodified HNTs. In addition, the loading of BDP in modified HNTs was threefold over the unmodified one, indicating a high loading capacity after modification of HNTs. The use of HNTs as nanocontainers by loading with flame retardant might have promising applications for low-flammable smart materials.[37]

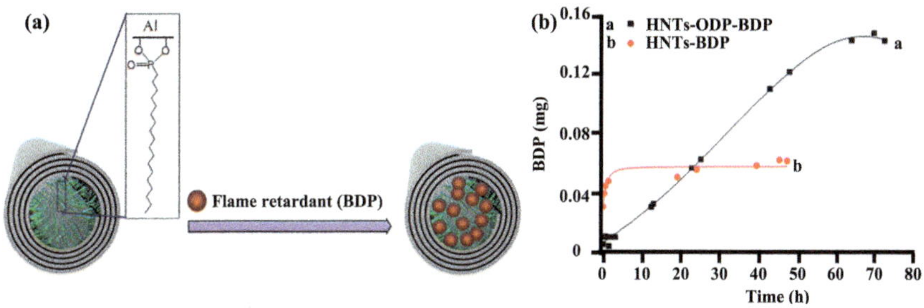

Figure 9.5 (a) Schematic representation of loading BDP into ODP-modified HNTs, (b) BDP release curves of time plots from BDP loaded HNTs and ODP-modified HNTs in THF at room temperature. (Reproduced with permission from ref. 37. Copyright 2013 The Chemical Society of Japan.)

9.3.3 Hierarchical Structure in the Nanocomposites

Lecouvet *et al.* demonstrated that the fire retardancy of PA–HNTs nanocomposites could be substantially improved by coating highly loaded PA12–HNTs nanocomposite films on the surface of virgin PA substrate. This hierarchical structure enabled the rapid build-up of a uniform inorganic-rich layer on the sample surface, which consequently limited heat and mass transfer and improved flame retardancy. The flame retardancy of a hierarchical nanocomposite coated with a 30 wt% HNT-loaded PA nanocomposite film (only 3.1 wt% HNTs in the hierarchical nanocomposite) was compared to that of a "homogeneous" nanocomposite with 16 wt% HNTs. In the conventional homogeneous nanocomposites, an ablative reassembling of HNTs at the sample surface and the migration of HNTs from the bulk towards the external surface, propelled by rising gas bubbles of degradation products and the convection flow in the melt, yielded an inorganic-rich residue char during combustion. However, the burst of gas bubbles at the sample surface might propel the accumulated HNTs outward from the bursting area and form island-like floccules instead of a continuous inorganic structure. As the protective floccules increased progressively, a substantial amount of polymer was decomposed to form a protective layer. In contrast, the hybrid materials with a highly loaded nanocomposite film rapidly formed a continuous and strong layer consisting of a net-like structure of aluminosilicate nanotubes, which efficiently impeded the formation of cracks in the char. Furthermore, the addition of a small amount of CNTs into PA substrate could alleviate the bubbling process by increasing the melt viscosity of the condensed phase. The combination of CNTs and HNTs in the nanocomposite film could further improve the fire retardancy, which was because CNTs with high aspect ratio could form a jammed network in the char layer. This hybrid materials by covering highly loaded nanocomposite film onto the sample surface are highly effective in increasing flame retardancy with fewer part of nanofillers.[38]

9.4 Synergistic Effects of HNTs and Other Flame Retardants

Although addition of HNTs can significantly improve the flame retardant properties in terms of the decreased HRR and PHRR evaluated by CC test, the nanocomposites usually failed some important fire tests, especially those with samples in vertical positions such as UL-94 and LOI.[39] For example, the addition of HNTs into PA6 led to a decrease in the UL-94 rating because HNTs enabled a more effective build-up and retention of heat during the first exposure to the flame, thereby resulting in sustained burring after the second exposure to the flame.[16] In addition, compared with neat polymer, the time to ignition is almost shorter[14,40,41] and the total heat release is slightly reduced[42,43] when nanofillers are added. The protective layer only slows down the burning process without decreasing the total amount of combustible materials.[44] In view of these limitations, nanofillers should preferably be regarded as a part of flame retardant formulations, instead of stand-alone flame retardants.

9.4.1 Synergistic Effects with Intumescent Flame Retardants

Intumescent flame retardants (IFR) are widely used as halogen-free additives due to their advantages of being environmentally friendly, low smoke and producing low toxicity during burning.[45,46] The mechanism of IFR is due to the formation of a multiple foam cellular char on the polymer surface, which reduces the mass and heat transfer between the flame and polymer, thus inhibiting the combustion. However, the IFR and the formed char are not thermally stable, resulting in a low flame-retarded efficiency.[47] The recent trend in polymer fire retardancy is to combine conventional flame retardants with nanoparticles to achieve synergistic enhancements along with satisfactory thermal and mechanical properties.[48–50] The use of IFR in combination with HNTs to produce synergistic flame retardancy in polymers has been widely studied.

Phenyl phosphonic acid (PPA) was intercalated into HNTs (PPA-HNTs) by covalent bonds between PPA and hydroxyl groups on the HNTs, and then PPA-HNTs were incorporated into PA6. The HRR curve of the nanocomposite with 10 wt% HNTs displayed a two-stage combustion process, which had a similar profile to virgin PA6. However, the HRR curve of the nanocomposite with 10 wt% PPA-HNTs was steadily decreased after an initial increase. This suggested that the char layer of the nanocomposites did not break down during combustion; instead, it formed a continuous insulating layer. SEM images indicated that the surface of the char residue of the nanocomposites with 10 wt% HNTs after a cone calorimeter test was relatively smooth, implying a dense and continuous char layer composed of carbon, aluminium, and silicon (Figure 9.6a and b), while the char layer of the nanocomposites with 10 wt% PPA-HNTs showed a highly porous structure, which had little carbon and was predominantly made up of silicon, aluminium, and phosphorus

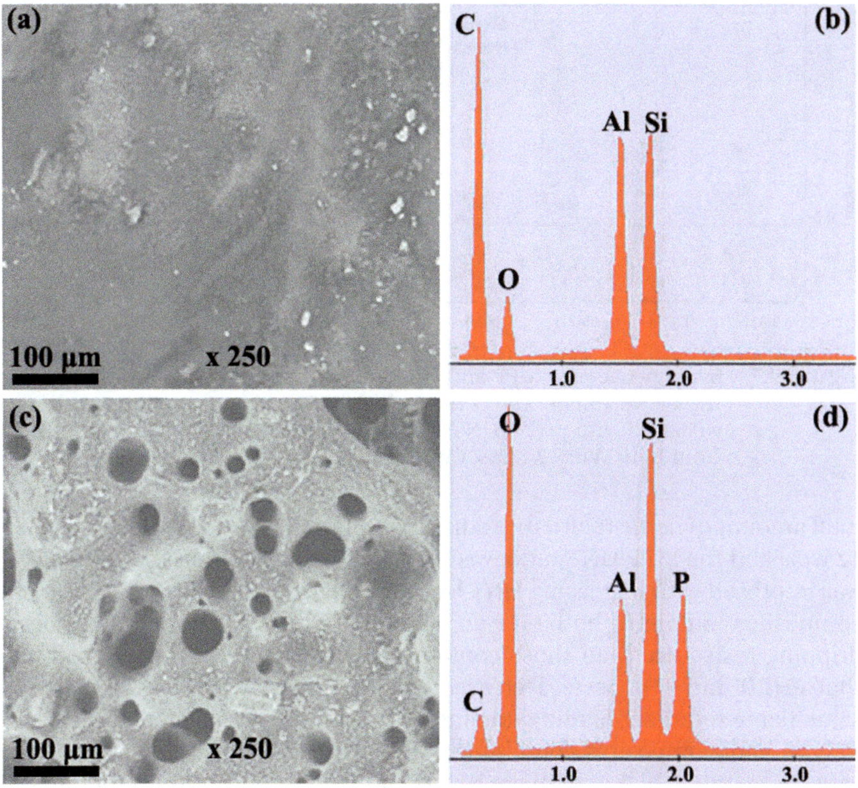

Figure 9.6 (a) SEM image and (b) elemental analysis of the char residue from the nanocomposite with 10 wt% HNTs, (c) SEM image and (d) elemental analysis of the char residue from the nanocomposite with 10 wt% PPA-HNTs. The char was obtained after cone calorimeter testing. (Reproduced from ref. 51 with permission from John Wiley & Sons. Copyright © 2012 John Wiley and Sons, Ltd.)

(Figure 9.6c and d). The porosity was caused by escaping gases during the foaming-highly viscous stage, which provided good thermal insulation between the heat source and the polymer. However, the amount of smoke production in the nanocomposites with PPA-HNTs was higher. The PPA-HNTs reacted with PA6 to initially accelerate the burning, which was followed by the formation of a strong char layer that retarded the escape of combustible molecules. Due to the tougher char layer, volatile products could not escape and therefore reacted with other small chain pyrolysis products to form longer chain condensed/aromatic carbon or char. When sufficient pressure was built up beneath the insulative layer, the relatively incombustible soot type compounds escaped by “blowing” holes through the char layer, which contributed to the large amount of smoke evolved during the decomposition.[51]

In a study by Sun *et al.*, melamine polyphosphate (MPP) and HNTs were combined and introduced into PA6. To maintain the spinnability of PA6, the

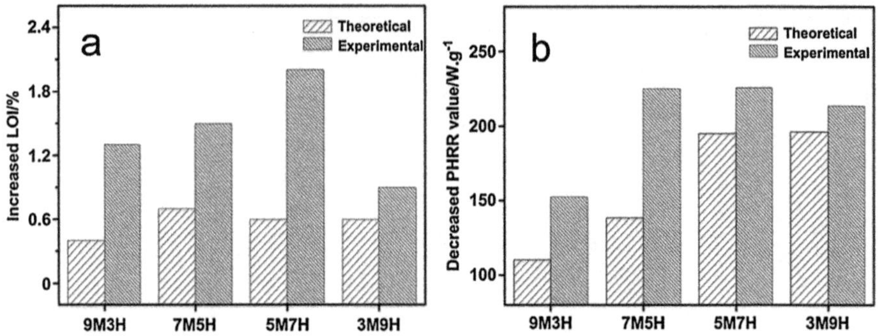

Figure 9.7 Synergy between MPP and HNTs in (a) increasing LOI and (b) decreasing PHRR values. *x*M*y*H means the PA6 nanocomposites containing *x* wt% MPP and *y* wt% HNTs. (Reproduced from ref. 52 with permission from John Wiley & Sons. Copyright ©2014 John Wiley & Sons, Ltd.)

total amount of flame retardant additive was fixed at a relatively low content of 12 wt% and the MPP/HNT ratio was varied. It was found that the spinnability was improved with increasing HNT loading. LOI tests indicated that the nanocomposites containing both MPP and HNT had a higher LOI value and a better dripping resistance than those containing MPP alone. MCC results showed that PHRR and THR were decreased with increased HNT/MPP ratio. At the same flame retardant loading, the increment in LOI and decrement in PHRR for the nanocomposites containing MPP and HNTs together (experimental value) were much higher than the sum of individual contributions for the composites containing sole MPP or HNTs (theoretical value). Figure 9.7 illustrates the synergy between MPP and HNTs in increasing the LOI and decreasing the PHRR values. For example, the addition of 7 wt% MPP or 5 wt% HNTs alone had a PHRR reduction of 111 W g^{-1} and 27 W g^{-1}, respectively. However, when combined together, the reduction of PHRR was 226 W g^{-1}, which was much higher than the sum of the individual contributions. The synergy between MPP and HNTs in improving the flame resistance was because HNTs interacted with MPP by chemical interactions to form a compact and integrated network-structured residue layer, rather than the thin and soft char layer for the sample with MPP or HNTs alone. The char layer not only efficiently hindered oxygen and heat spearing into the sample surface but also prevented the release of flammable volatiles.[52]

Zhao *et al.* used HNTs as a synergistic agent to substitute part of ammonium polyphosphate (APP) to enhance the flame retardancy of LDPE. Table 9.1 summarizes the flame retardancy parameters of neat LDPE and its nanocomposites with different amounts of HNTs and APP, where PE–A*x*–H*y* represent PE with *x* wt% APP and *y* wt% HNTs. The LOI was increased from 17 for neat LDPE to 29 for PE/A30, and the LOI further increased to 36.5 for PE–A28–H2. In addition, CC results showed that an obvious decrease in the PHRR, THR, and TSR were observed for PE–A30, and these parameters were further decreased by increasing the HNT content to 2 wt%. However, the

Table 9.1 Flame retardant properties of neat LDPE and LDPE with varied amounts of IFR and HNTs.

Samples	LOI (%)	UL-94	PHRR (kW m^{-2})	THR (MJ m^{-2})	Residue (wt%)	PSPR ($\times 10^{-2}$ m^2 s^{-1})	TSP (m^2)
PE	17.0 ± 0.5	N.R.	2271	337	0	12.9	20.5
PE–A30	29.0 ± 0.3	V-0	210	191	15.7	2.7	19.0
PE–A29.5–H0.5	32.7 ± 0.4	V-0	209	212	15.1	2.0	12.6
PE–A29–H1	33.1 ± 0.3	V-0	182	194	19.8	1.5	6.5
PE–A28–H2	36.5 ± 0.2	V-0	162	85	60.6	1.4	3.6
PE–A27–H3	35.0 ± 0.5	V-0	219	213	20.4	5.1	22.7

flame retardant properties were decreased when HNT loading increased to 3 wt%. After combustion, neat LDPE was completely burnt up and the char residue for PE–A30 was 15.7. The char residue gradually increased with the addition of HNTs, which reached a maximum value of 60.6 wt% for PE–A28–H2.[53]

Lecouvet *et al.* reported that an appropriate substitution of IFR (ammonium polyphosphate coated with a carbonization agent) by HNTs (1.5–3.0 wt%) greatly improved the flame retardancy of PP with reduced HRR and longer burning time. The synergistic flame retardant effect between HNTs and IFR was due to both physical and chemical mechanisms occurring in the condensed phase during combustion. HNTs not only speeded up the expansion of the carbonizing material through a nucleating action exerted on foaming, but also contributed to the mechanical reinforcement of the intumescent char through the formation of an aluminosilicate "skeleton frame" for the phosphocarbonaceous structure, enabling the formation of a more homogeneous, denser, and tougher char. HNTs could react with IFR upon heating to yield aluminophosphates; these chemical species enhanced the thermal stability of the char at high temperatures. Moreover, the aluminophosphate complexes formed a "top-bottom pillared" structure, which displayed higher mechanical strength at high temperature, and hence was able to better withstand the internal pressure arising from the degradation products against cracking in the char.[54]

HNTs can absorb both cationic and anionic compounds because they have negative silanols on the outer surface and positive aluminols on the inner lumen surface.[55] Dong *et al.* utilized the adsorption–precipitation method to synthesize ferrum ammonium phosphate-loaded HNTs (FAP-HNTs) by adding ammonium nitrogen, phosphate, and ferrum sulfate into the aqueous dispersion of HNTs. The synergistic effects arising from FAP-HNTs and Exolit OP 1230 (OP) on the flame retardancy of epoxy (EP) were studied. According to the MCC test, with 20 wt% flame retardants, FAP, FAP-HNT, and FAP-HNT–OP (weight ratio of FAP : HNTs is 1 : 1) led to 26, 26, and 37% reductions in PHRR, respectively, in comparison with the neat EP. The EP–FAP-HNTs nanocomposite with 20 wt% FAP-HNTs failed to pass the UL-94 V-0 rating, while composites with 20 wt% FAP-HNTs–OP could pass the UL-94 V-0 rating. The char of EP–FAP-HNTs–OP had a higher carbon content and lower oxygen content, indicating better char-forming performance. A combination of FAP-HNTs and OP had better flame retardant properties, which were attributed to the synergistic effect between FAP-HNTs and OP. During combustion, FAP ($FeNH_4PO_4{\cdot}12H_2O$) decomposed with loss of crystal water, structural water and ammonia, and a barrier layer formed by FAP-HNTs and OP slowed down the heat and mass transfer between the gas and condensed phases and shielded the underlying material from further burning. Moreover, FAP-HNTs acted as a catalyst and a gas source to form an expanded charred layer.[56]

Attia *et al.* investigated the synergy between HNTs and an IFR system consisting of ammonium polyphosphate, melamine polyphosphate, and pentaerythritol as a fire retardant system for ABS. CC measurement showed that neat ABS resin burned fast after ignition with a sharp PHRR. The sole incorporation of HNTs into ABS improved the thermal stability due to the barrier properties of the HNT char. With increasing HNT content, the PHRR and THR gradually decreased, and 30 wt% HNTs gave rise to a 38% reduction in PHRR and 28.7% decrease in THR. However, the char barrier formed by the combustion of ABS–HNTs nanocomposites was not strong enough to protect the melted polymer from contacting the fire and to prevent the release of volatile products, as indicated by the present of cracks in the char layer (Figure 9.8b). The combination of HNTs and IFR yielded better fire retardant properties, and the ratio between IFR and HNTs had a significant effect on the fire-retardancy. When the total HNT–IFR loading was 30 wt%, the optimum ratio of HNTs and IFR was 1 : 1, at which the largest decrease of 56.2% in the PHRR was obtained. The synergic effect was explained by the formation of an acidic catalyst due to the chemical reaction between phosphoric acid released from IFR and HNTs, which improved the acid source, leading to an enhancement in the oxidative dehydration cross-linking charring process and improving the char strength. The char of ABS–HNTs–IFR remained intact after combustion (Figure 9.8c).[15]

The synergistic effects of aluminium diethylphosphinate, phosphorus-based flame retardants and nanofillers with different geometries on the fire

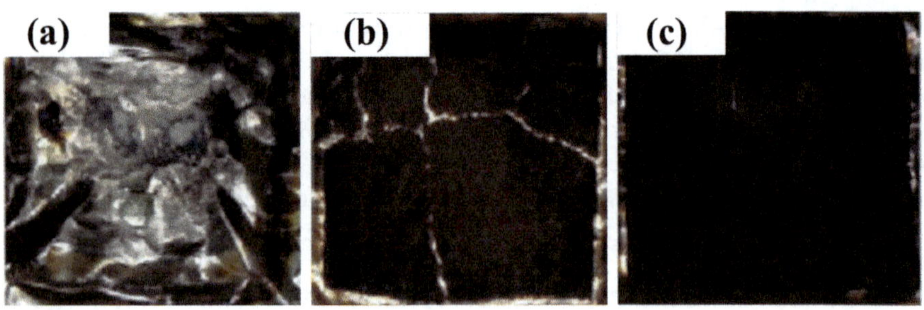

Figure 9.8 Digital photos of residues for (a) neat ABS resin, (b) ABS–HNTs nanocomposites with 30 wt% HNTs and (c) ABS–HNTs–IFR composites with 15 wt% HNTs and 15 wt% IFR. (Reproduced from ref. 15 with permission from John Wiley & Sons. © 2013 Society of Chemical Industry.)

retardancy of PLA were investigated. It was found that the spherical (0-D) silica inhibited intumescence and resulted in a smoother char surface. Rod-like (1-D) HNTs did not significantly affect the intumescence and char thickness, but could reduce the number and size of macro-pores present in the intumescent char. The plate-like (2-D) clay could rapidly migrate and accumulate on the exposed surface, which hindered intumescence and facilitated the formation of a much thicker residue reinforced by the clay platelets. This eventually helped establish an effective barrier to heat and mass transfer as combustion proceeded.[57]

9.4.2 Synergistic Effects with Other Nanofillers

The combination of HNTs with other nanofillers also exhibited synergistic effects in enhancing the flame retardancy of polymers. In a study by Li *et al.*, HNTs were covalently grafted onto graphene oxide (GO) to create a HNTs–GO hybrid filler, which was then added into PA6 for fire retardant applications. The fire retardant properties of PA6 were improved with the addition of HNTs–GO when compared with the composites with either HNTs, or GO, or a physical mixture of HNTs and GO. The co-existence of HNTs and GO improved the network structure that hindered the movement of polymer chains and thus improved flame retardancy. Besides, the combination of the stable silica layer created by HNTs and the barrier effect of GO gave a synergistic barrier effect.[58]

Adding a hybrid filler of HNTs and phthalocyanine (zinc phtalocyanine or chloroaluminophthalocyanine) into NBR led to improvements in the thermal and mechanical properties as well as a significant reduction in flammability expressed with PHRR and THR. This was because the phthalocyanine pigments not only facilitated the dispersion of HNTs in the polymer matrix, but also themselves considerably reduced the flammability of the polymer. In addition, the colored composites with hybrid filler exhibited an increase in heat capacity, which increased the fire safety by promoting self-extinguishing processes.[59]

Rybinski *et al.* used three halogen-free flame retardant compounds including antimony trioxide, magnesium hydroxide, and melamine cyanurate together with HNTs, and their effects on the thermal stability and flame retardancy of NBR were studied. Only addition of HNTs could improve the thermal stability of NBR due to the barrier effect of HNTs and limited amplitude of thermal vibration caused by the restricted mobility of segments. The combination of HNTs with the three flame retardant compounds could further decrease the thermal decomposition rate and endowed the nanocomposites with self-extinguishing properties. The combination of antimony trioxide and HNTs improved the thermal stability of NBR due to the intrinsic high thermal stability of antimony trioxide. But excess antimony trioxide decreased the barrier properties of HNTs, causing a decrease in flame retardant properties. In the case of magnesium hydroxide and HNTs, 15 phr magnesium hydroxide and 5 phr HNTs resulted in a comparable flammability with 50 phr magnesium hydroxide. The synergy was because the water vapor released from the decomposition of magnesium hydroxide decreased the energetic balance and diluted the volatile products. Moreover, HNTs with magnesium oxide and carbonization products formed a protective isolating layer on the sample surface during combustion. In the combination of melamine cyanurate and HNTs, the decomposition of melamine cyanurate could release non-flammable ammonia and formed condensation products such as melam, melem, and melon, which impeded the transport of mass and heat between the solid and gaseous phases. Besides, the decomposition process was endothermic.[17]

9.5 Comparison Between HNTs and Other Clays

9.5.1 Other Flame Retardant Clays in Polymers

Besides HNTs, other natural clays such as MMT, vermiculite, sepiolite, and kaolinite have also been added as flame retardants into polymers.

In the 1980s, PA6–MMT nanocomposites were first synthesized by Toyota Central Research and Development Laboratories who found that a small amount of clay could greatly improve the thermal stability and mechanical properties.[60] Due to the success of flame retarded clay nanocomposites in PA6, polymer–MMT nanocomposites were expended rapidly to other polymer matrices such as PP,[61] PS,[62] PE,[63] rubber,[64] and so on.[40] The improved flame properties of the polymers are due to multi-effects brought by the addition of MMT. Zanetti *et al.* reported that the incorporation of 2–5 wt% MMT into poly(ethylene-*co*-vinyl acetate) (EVA) led to a 70–80% reduction in the HRR.[65] The layered clay in a polymer matrix enabled the formation of a protective inorganic-rich layer at the sample surface during the combustion process, which insulated the polymer from the heat source and prevented the escape of volatile gases from the polymer combustion.[40] It was stressed that the charring process might be accelerated by the protonic sites on the silicate layers by the Hoffman decomposition of the organic modifier within MMT.[65,66] In addition, MMT could also play a catalytic role in the promotion of a char-forming reaction because the confined and superheated environment caused by MMT trapped decomposition products, making additional carbonization pathways available.[67] Some studies also reported that the presence of iron in the silicate fillers could act as a flame retardant and improve the thermal stability of nanocomposites. It appeared that iron or other paramagnetic sites within the clay could trap radicals during the degradation process.[68] There are some excellent literature reports concerning the flame retardancy of polymer–MMT nanocomposites and the underlying mechanism for the improved fire properties.[2,40,45,69]

To achieve satisfactory flame retardant properties of polymer–MMT nanocomposites, the uniform dispersion of MMT is a prerequisite. Due to the hydrophilic nature of MMT and strong interaction among clay sheets, organic modification of MMT is necessary to improve the dispersion of MMT in the polymer matrix. For example, quaternary ammonium chloride modified-MMT was well dispersed in EVA with an exfoliated/intercalated structure, while raw MMT tended to aggregate in EVA and formed a micro-composite. Compared with virgin EVA, the PHRR was decreased by 25% for the EVA–MMT micro-composite and by 50% for the EVA–modified-MMT nanocomposite.[42] It should be noted that the organic intercalating agents, on one hand, improved the dispersion of clay in the polymer matrix to improve the flame retardancy, but on the other hand, reduced the ignition time due to the inherent low thermal stability and high flammability.[70] It was mentioned that the products of thermal decomposition of organic modifiers could accelerate the polymer matrix degradation by the volatilization of the organic modifier of acidic sites occurring on the MMT layer.[71]

Vermiculite ($(MgFeAl)_3(Al,Si)_4O_{10}(OH)_2{\cdot}4H_2O$) is a natural hydrated magnesium aluminium silicate clay mineral and is known for its inherent flame-retardant properties, which can withstand a flame up to about 1200 °C.[72,73] Bulk vermiculite can be exfoliated when heated at an elevated temperature (*e.g.*, >200 °C) through loss of water in the sheet interlayers, leading to the self-intumescing properties of vermiculite. With its heat-resistant characteristics, vermiculite was demonstrated to be an effective flame retardant material for many polymers such as siloxane[74] and PP.[75] Cheong *et al.* prepared a fire retardant, flexible polymer–vermiculite film by photo-initiated polymerization of poly(ethylene glycol)$_n$ dimethacrylate with micronized vermiculite powders. A fire endurance test showed that the composite film (20 μm thick) containing 70 wt% vermiculite could withstand actual fire for more than a minute.[76] In a study by Chen *et al.*, organ-modified-vermiculite (OVMT) was prepared by ion exchange with octadecyltrimethyl ammonium bromide. The OVMT showed high synergistic effects in improving the flame retardancy of a PP–IFR system. LOI and UL-94 tests revealed that low loading of OVMT improved the flame retardancy and retarded the dripping of PP. With 1 wt% OVMT, the LOI value increased from 30.7% for PP–IFR to 32.9%, which was due to the advantageous combination of thermal and oxygen insulation of OVMT.[77]

Sepiolite is a kind of fibrous hydrated magnesium silicate having the ideal formula $Si_{12}Mg_8O_{30}(OH)_4(H_2O)_4{\cdot}8H_2O$, with two tetrahedral silica sheets linked by a central sheet of octahedral magnesia.[78] Sepiolite has been incorporated into many polymers to improve their flame retardant properties.[79–81] For example, with the addition of 5 wt% sepiolite into EVA, the LOI value of the EVA composites reached 24.3%, contrasting with a value of 18.2% for the virgin EVA, and the UL-94 classification reached V-1. This was due to a restriction on the mobility of polymer chains and mass transport barrier imposed by sepiolite.[81] Zhan *et al.* reported that the addition of sepiolite into polyamide 6,6 (PA66)–aluminium diethylphosphinate (AlPi) composites gave rise to a synergistic effect in improving the flame retardancy. Compared with PA66–AlPi composites with 10 wt% AlPi, the LOI of the composites with 9.5 wt% AlPi and 0.5 wt% sepiolite increased from 30.5 to 32.9%, and the UL-94 flammability rating changed from No rating to V-0 rating. It was proposed that the sepiolite in the PA66–AlPi composites accelerated the catalytic degradation reaction of AlPi to form silicon phosphate and magnesium phosphate, which enhanced the thermal stability and char yield of PA66.[80] The synergistic effects between sepiolite and other flame retardants such as zinc borate,[82] magnesium hydroxide,[83] and IFR[84] were also disclosed.

Kaolinite is also a naturally occurring alumino-silicate clay with a similar chemical composition to HNTs, but differs by having a predominantly plate-like structure, rather than a hollow tubular structure. Tang *et al.* exfoliated kaolinite by intercalating dimethyl sulfoxide and potassium acetate followed by ultrasonic irradiation, and then incorporated kaolinite or exfoliated-kaolinite into PP containing aluminium hydroxide (AH) and APP. Compared with PP–APP–AH, addition of 1 phr kaolinite or exfoliated-kaolinite led to 22% and 27% decrease in the PHRR, respectively, and 10% and 25% decrease in the peak of smoke production rate, respectively. Both kaolinite and exfoliated-kaolinite had positive effects in improving the flame retardancy of PP, and exfoliated kaolinite was more effective. Char analysis suggested that the more uniform dispersion of exfoliated-kaolinite in the PP matrix promoted char formation and resulted in a more compact char layer during combustion, which restricted the diffusion of oxygen and volatile gases.[85]

9.5.2 Comparing HNTs with Other Nanofillers in Flame Retardant Polymers

Rybinski *et al.* made a comparison of the effects of MMT and HNTs on the flame retardancy of SBR. Compared with HNT-filled SBR, the onset degradation temperature of MMT-filled SBR was much lower, which was due to the degradation of organic modifiers in MMT. Both HNTs and MMT led to a great decrease in the thermal decomposition rate, the maximum heat release rate, and total heat release, and also led to an increase in the oxygen index, combustion time, and ignition temperature of SBR. In SBR–MMT nanocomposites, the char was significantly higher than that in SBR–HNTs nanocomposites. The layered MMT appeared to enhance the performance of the char layer, which acted as a heat insulator and mass transport barrier. Also, the "labyrinth effect" of layered MMT might also improve the flame retardancy. It needs to be pointed out that only the MMT with exfoliated or intercalated status could improve the flame retardancy. In HNT-filled SBR nanocomposites, 1D HNTs have a lower aspect ratio and specific surface area than 2D MMT, thus the barrier effects created by HNTs are expected to be inferior to those created by layered MMT. However, the entrapment of degradation products into lumens of HNTs also could contribute to the delay in mass transport and remarkably increased flame retardancy of the nanocomposites.[86] In another study, the same authors reported that the NBR nanocomposite filled with CNF exhibited a lower thermal decomposition rate in comparison with that containing HNTs. This was because CNF was a radical scavenger to inhibit the free radical reactions during decomposition. However, the residue after thermal decomposition in HNT-filled NBR was much higher as CNF completely burned over 400–450 °C.[87]

Vahabi *et al.* made a systematic comparison of the effects of three inorganic fillers including needle-like sepiolite, tubular HNTs, and lamellar organ-modified MMT (OMMT) and also in combination with APP on the flame retardancy of a PP–PA6 blend. When only nanofillers were incorporated into the blend, OMMT could significantly decrease the PHRR compared to sepiolite or HNTs due to its better barrier effects. When the nanofillers were combined with APP, the sepiolite–APP composition led to a better fire performance. The sepiolite fibers could reinforce the char to form a very compact charred protective structure, which in turn enabled more phosphorus to act in the condensed phase. While, when compared to OMMT and sepiolite, time to ignition was greatly increased in HNT-filled PP–PA6 blends, regardless of the inclusion of APP, which was because HNTs could enhance the thermal diffusivity of the sample and thereby delay ignition.[88]

PVA-based aerogel nanocomposites with different inorganic nanofillers of silica, HNTs, MMT, and LAPONITE® were prepared *via* a freeze-drying method. Among these four nanocomposites, the PVA–HNTs aerogel exhibited the highest onset decomposition temperature and the lowest char residue (Figure 9.9a). CC tests showed that the HRR, THR, and TSR of the four aerogel nanocomposites were much lower than those for neat aerogel (Figure 9.9b). In particular, the layered systems PVA–MMT and PVA–laponite had comparatively lower flammability due to the enhanced barrier effects associated with layered nanofillers. Whereas PVA–HNTs exhibited the highest PHRR and THR, probably because of its low gas barrier and high thermal conductivity. Time for the ignition of the four aerogel composites was reduced with the addition of fillers. This undesirable phenomenon was explained by the clay catalyzing the degradation of the polymer, or the temperature on the

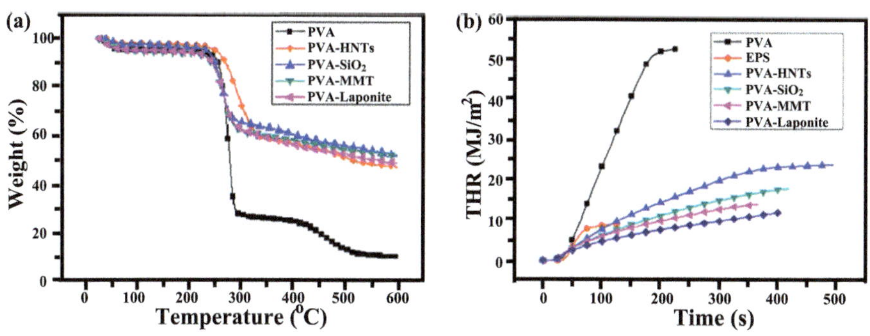

Figure 9.9 (a) Weight loss *versus* temperature and (b) THR *versus* time for neat PVA aerogel and the four aerogel nanocomposites. (Reproduced with permission from H. B. Chen, Y. Z. Wang and D. A. Schiraldi, ACS Appl. Mater. Inter., 2014, 6, 6790.[89] Copyright (2014) American Chemical Society.)

nanocomposite surface drastically increasing as the thermal diffusion was blocked by the increased melt viscosity.[89]

9.6 Conclusion

HNTs show promising applications as flame retardants in polymer–HNTs nanocomposites. With the inclusion of HNTs into polymers, the flame retardant properties of the resulting nanocomposites are greatly increased. Especially, fire parameters such as HRR, PHRR, and THR evaluated by the CC test are remarkably decreased with increasing HNT loading. The improved thermal stability and flame retardancy of polymer–HNTs nanocomposites are due to the multi-effects brought by HNTs. (1) HNTs migrate to the surface of the condensed phase and form an insulating char layer between the melted polymer and flame zone, which acts as a barrier to limit the heat transfer into the sample and to inhibit the in-diffusion of oxygen and the out-diffusion of volatile degradation products during combustion. (2) HNTs undergo dehydration of physisorbed water and dehydroxylation at high temperature. The endothermic decomposition of HNTs can absorb part of the external heat to keep the surrounding polymer cooler and hence delay the thermal decomposition. Besides, the released water vapor can dilute and cool flammable gases. (3) Due to the hollow tubular structure of HNTs, the degradation products can be considerably entrapped into the lumens of HNTs, resulting in an effective delay in mass transport and significantly improved thermal stability. (4) The iron oxides in HNTs can trap radicals generated during the process of degradation, thus slowing down the thermal decomposition rate. An undesired phenomenon is that the time for the ignition of the sample is usually reduced with the incorporation of HNTs, which is caused by the decreased convection and increased scatter of incident radiation near the sample surface.

The uniform dispersion of HNTs in the polymer matrix is a critical factor to produce polymer–HNTs nanocomposites with better fire properties. Due to the tubular shape and low hydroxyl group density on the HNT surface, HNTs can be readily dispersed in many polymer matrices by direct melt mixing. Other processing approaches such as master batch, water-assisted extrusion, and solution mixing, have shown great advantages in improving the dispersion of HNTs. In particular, for the non-polar polymers, surface modifications of HNTs is necessary to enhance the compatibility between the HNTs and polymer matrices. The improved dispersion of HNTs can enhance the structural stability of the formed char, thus promoting the barrier effects created by HNTs. The structural design of the materials is also of benefit to the improved flame retardant properties. Covering the surface of the polymer substrate with highly loaded nanocomposite films facilitates the rapid formation of a strong layer at a lower HNT loading. Increasing the melt viscosity of the condensed phase can also improve the quality of the protective barrier through alleviating the bubbling process.

Although HNTs can significantly improve the flame retardancy parameters such as HRR and PHRR evaluated by CC, the nanocomposites usually fail some fire tests such as UL-94 and LOI. The combination of HNTs with IFR gives rise to significant synergistic effects in improving the fire properties of the polymers. Generally, HNTs can reinforce the intumescent char and react with IFR upon heating to yield a thermally stable char, which can better withstand the internal pressure arising from the degradation products against cracking in the char. In addition, the combination of HNTs with other nanofillers can improve their mutual dispersion and network structure in the nanocomposites, and hence improve the flame retardancy.

Similar to HNTs, other natural clays such as layered MMT, kaolinite, vermiculite, and fibrous sepiolite have also been incorporated as flame retardants into polymers. These clays exert a similar role as that for HNTs during combustion. Because the 1D HNTs have a lower aspect ratio and specific surface area when compared to 2D clays, the barrier effects created by HNTs are expected to be inferior to those by layered clays. These layered clays tend to severely aggregate in polymer matrices due to strong interaction among clay sheets. Thus the elaborate modification of layered clays is particularly important for their implementation.

Acknowledgements

The authors thank the National Natural Science Foundation (Grant Numbers 50603005, 51222301 and U1462116), China Postdoctoral Science Foundation (2015M570709), and Fundamental Research Funds for the Central Universities for financial support.

References

1. A. Blumstein and H. E. Ries, *J. Polym. Sci., Part B: Polym. Lett.*, 1965, **3**, 927–931.
2. S. Bourbigot and S. Duquesne, *J. Mater. Chem.*, 2007, **17**, 2283–2300.
3. J. Zhu, A. B. Morgan, F. J. Lamelas and C. A. Wilkie, *Chem. Mater.*, 2001, **13**, 3774.
4. S. Bourbigot, M. Le Bras, R. Leeuwendal, K. K. Shen and D. Schubert, *Polym. Degrad. Stab.*, 1999, **64**, 419.
5. C. M. Becker, A. D. Gabbardo, F. Wypych and S. C. Amico, *Composites, Part A*, 2011, **42**, 196.
6. T. Kashiwagi, F. M. Du, K. I. Winey, K. A. Groth, J. R. Shields, S. P. Bellayer, H. Kim and J. F. Douglas, *Polymer*, 2005, **46**, 471.
7. D. M. Fox, M. Novy, K. Brown, M. Zammarano, R. H. Harris, M. Murariu, E. D. Mccarthy, J. E. Seppala and J. W. Gilman, *Polym. Degrad. Stab.*, 2014, **106**, 54.
8. M. L. Du, B. C. Guo and D. M. Jia, *Polym. Int.*, 2010, **59**, 574.
9. Y. Lvov and E. Abdullayev, *Prog. Polym. Sci.*, 2013, **38**, 1690.

10. Y. M. Lvov, D. G. Shchukin, H. Mohwald and R. R. Price, *ACS Nano*, 2008, **2**, 814.
11. E. Joussein, S. Petit, J. Churchman, B. Theng, D. Righi and B. Delvaux, *Clay Miner.*, 2005, **40**, 383.
12. Z. X. Jia, Y. F. Luo, B. C. Guo, B. T. Yang, M. L. Du and D. M. Jia, *Polym.–Plast. Technol. Eng.*, 2009, **48**, 607.
13. M. L. Du, B. C. Guo and D. M. Jia, *Eur. Polym. J.*, 2006, **42**, 1362.
14. B. Lecouvet, M. Sclavons, S. Bourbigot, J. Devaux and C. Bailly, *Polymer*, 2011, **52**, 4284.
15. N. F. Attia, M. A. Hassan, M. A. Nour and K. E. Geckeler, *Polym. Int.*, 2014, **63**, 1168.
16. D. C. O. Marney, L. J. Russell, D. Y. Wu, T. Nguyen, D. Cramm, N. Rigopoulos, N. Wright and M. Greaves, *Polym. Degrad. Stab.*, 2008, **93**, 1971.
17. P. Rybinski and G. Janowska, *Thermochim. Acta*, 2013, **557**, 24.
18. A. Mack, *Flame Retardants, Halogenated*, John Wiley & Sons, Inc., 2004.
19. M. Alexandre and P. Dubois, *Mater. Sci. Eng., R*, 2000, **28**, 1.
20. B. Lecouvet, M. Sclavons, S. Bourbigot and C. Bailly, *Polym. Adv. Technol.*, 2014, **25**, 137.
21. C. Liu, Y. F. Luo, Z. X. Jia, B. C. Zhong, S. Q. Li, B. C. Guo and D. M. Jia, *Int. J. Polym. Mater.*, 2013, **62**, 128.
22. H. Shi, X. Q. Liu and Y. F. Zhang, *Iran. Polym. J.*, 2014, **23**, 297.
23. B. Lecouvet, S. Bourbigot, M. Sclavons and C. Bailly, *Polym. Degrad. Stab.*, 2012, **97**, 1745.
24. M. L. Du, B. H. Guo, M. X. Liu and D. M. Jia, *Polym. Compos.*, 2007, **15**, 321.
25. A. M. Alhuthali and I. M. Low, *J. Reinf. Plast. Compos.*, 2013, **32**, 233.
26. H. Ismail, P. Pasbakhsh, M. N. A. Fauzi and A. Abu Bakar, *Polym. Test.*, 2008, **27**, 841.
27. R. Nakamura, A. N. Netravali, A. B. Morgan, M. R. Nyden and J. W. Gilman, *Fire Mater.*, 2013, **37**, 75.
28. D. R. Paul and L. M. Robeson, *Polymer*, 2008, **49**, 3187.
29. G. Stoclet, M. Sclavons, B. Lecouvet, J. Devaux, P. Van Velthem, A. Boborodea, S. Bourbigot and N. Sallem-Idrissi, *RSC Adv.*, 2014, **4**, 57553.
30. B. Lecouvet, J. G. Gutierrez, M. Sclavons and C. Bailly, *Polym. Degrad. Stab.*, 2011, **96**, 226–235.
31. A. Fernandez, M. Muniesa and J. Gonzalez, *J. Mech. Eng.*, 2013, **59**, 183.
32. M. Liu, B. Guo, M. Du and D. Jia, *Appl. Phys. A*, 2007, **88**, 391.
33. P. Rybinski, G. Janowska, M. Jozwiak and A. Pajak, *J. Therm. Anal. Calorim.*, 2012, **107**, 1243.
34. W. L. Tham, B. T. Poh, Z. A. M. Ishak and W. S. Chow, *J. Therm. Anal. Calorim.*, 2014, **118**, 1639.
35. H. Q. Fu, Y. Wang, W. F. Chen and J. Xiao, *Appl. Surf. Sci.*, 2015, **346**, 372.
36. B. Lecouvet, M. Sclavons, S. Bourbigot and C. Bailly, *Polym. Degrad. Stab.*, 2013, **98**, 1993.
37. H. Jing, Y. Higaki, W. Ma, H. Wu, W. O. Yah, H. Otsuka, Y. M. Lvov and A. Takahara, *Chem. Lett.*, 2013, **42**, 121.

38. B. Lecouvet, M. Sclavons, S. Bourbigot and C. Bailly, *J. Fire Sci.*, 2014, **32**, 145.
39. M. Bartholmai and B. Schartel, *Polym. Adv. Technol.*, 2004, **15**, 355.
40. P. Kiliaris and C. D. Papaspyrides, *Prog. Polym. Sci.*, 2010, **35**, 902.
41. A. Dasari, Z. Z. Yu, Y. W. Mai and S. Liu, *Nanotechnology*, 2007, 18.
42. S. Duquesne, C. Jama, M. Le Bras, R. Delobel, P. Recourt and J. M. Gloaguen, *Compos. Sci. Technol.*, 2003, **63**, 1141.
43. T. Kashiwagi, R. H. Harris, X. Zhang, R. M. Briber, B. H. Cipriano, S. R. Raghavan, W. H. Awad and J. R. Shields, *Polymer*, 2004, **45**, 881.
44. B. Schartel, M. Bartholmai and U. Knoll, *Polym. Adv. Technol.*, 2006, **17**, 772.
45. C. I. Idumah, A. Hassan and A. C. Affam, *Rev. Chem. Eng.*, 2015, **31**, 149.
46. E. A. Weil and S. V. Levchik, *J. Fire Sci.*, 2008, **26**, 5.
47. B. Schartel and T. R. Hull, *Fire Mater.*, 2007, **31**, 327.
48. H. Demir, E. Arkis, D. Balkose and S. Ulku, *Polym. Degrad. Stab.*, 2005, **89**, 478.
49. R. J. Song, Z. Wang, X. Y. Meng, B. Y. Zhang and T. Tang, *J. Appl. Polym. Sci.*, 2007, **106**, 3488.
50. Y. T. Li, B. Li, J. F. Dai, H. Jia and S. L. Gao, *Polym. Degrad. Stab.*, 2008, **93**, 9.
51. D. C. O. Marney, W. Yang, L. J. Russell, S. Z. Shen, T. Nguyen, Q. Yuan, R. Varley and S. Li, *Polym. Adv. Technol.*, 2012, **23**, 1564.
52. J. Sun, X. Y. Gu, M. Coquelle, S. Bourbigot, S. Duquesne, M. Casetta and S. Zhang, *Polym. Adv. Technol.*, 2014, **25**, 1552.
53. J. Zhao, C. L. Deng, S. L. Du, L. Chen, C. Deng and Y. Z. Wang, *J. Appl. Polym. Sci.*, 2014, 131.
54. B. Lecouvet, M. Sclavons, C. Bailly and S. Bourbigot, *Polym. Degrad. Stab.*, 2013, **98**, 2268.
55. M. X. Liu, Z. X. Jia, D. M. Jia and C. R. Zhou, *Prog. Polym. Sci.*, 2014, **39**, 1498.
56. Y. M. Dong, B. Lisco, H. Wu, J. H. Koo and M. Krifa, *J. Appl. Polym. Sci.*, 2015, 132.
57. N. A. Isitman, M. Dogan, E. Bayramli and C. Kaynak, *Polym. Degrad. Stab.*, 2012, **97**, 1285.
58. L. L. Li, S. H. Chen, W. J. Ma, Y. H. Cheng, Y. P. Tao, T. Z. Wu, W. P. Chen, Z. Zhou and M. F. Zhu, *eXPRESS Polym. Lett.*, 2014, **8**, 450.
59. P. Rybinski, A. Pajak, G. Janowska and M. Jozwiak, *J. Appl. Polym. Sci.*, 2015, 132.
60. A. Okada, Y. Fukushima, M. Kawasumi, S. Inagaki, A. Usuki, S. Sugiyama, T. Kurauch and O. Kamigaito, Composite material and process for manufacturing same: US, US 4739007A[P]. 1988.
61. A. Valera-Zaragoza, E. Ramirez-Vargas, F. J. Medellin-Rodriguez and B. M. Huerta-Martinez, *Polym. Degrad. Stab.*, 2006, **91**, 1319.
62. S. Bourbigot, J. W. Gilman and C. A. Wilkie, *Polym. Degrad. Stab.*, 2004, **84**, 483.
63. Y. Peneva, E. Tashev and L. Minkova, *Eur. Polym. J.*, 2006, **42**, 2228.

64. H. F. Zhang, Y. Q. Wang, Y. P. Wu, L. Q. Zhang and J. Yang, *J. Appl. Polym. Sci.*, 2005, **97**, 844.
65. M. Zanetti, T. Kashiwagi, L. Falqui and G. Camino, *Chem. Mater.*, 2002, **14**, 881.
66. H. L. Qin, S. M. Zhang, C. G. Zhao, G. J. Hu and M. S. Yang, *Polymer*, 2005, **46**, 8386.
67. H. L. Qin, Q. S. Su, S. M. Zhang, B. Zhao and M. S. Yang, *Polymer*, 2003, **44**, 7533.
68. J. Zhu, F. M. Uhl, A. B. Morgan and C. A. Wilkie, *Chem. Mater.*, 2001, **13**, 4649.
69. A. Dasari, Z. Z. Yu, G. P. Cai and Y. W. Mai, *Prog. Polym. Sci.*, 2013, **38**, 1357.
70. H. Y. Ma, P. A. Song and Z. P. Fang, *Sci. China: Chem.*, 2011, **54**, 302.
71. F. Laoutid, L. Bonnaud, M. Alexandre, J. M. Lopez-Cuesta and P. Dubois, *Mater. Sci. Eng., R*, 2009, **63**, 100.
72. S. A. Suvorov and V. V. Skurikhin, *Refract. Ind. Ceram.*, 2003, **44**, 186.
73. M. A. Osman, M. Ploetze and U. W. Suter, *J. Mater. Chem.*, 2003, **13**, 2359.
74. J. E. Connell, E. Metcalfe and M. J. K. Thomas, *Polym. Int.*, 2000, **49**, 1092.
75. S. C. Tjong, Y. Z. Meng and A. S. Hay, *Chem. Mater.*, 2002, **14**, 44.
76. J. Y. Cheong, J. Ahn, M. Seo and Y. S. Nam, *RSC Adv.*, 2015, **5**, 61768.
77. S. H. Chen, B. Wang, J. Kang, J. Y. Chen, J. G. Gai, L. Yang and Y. Cao, *J. Macromol. Sci., Part B*, 2013, **52**, 1212.
78. A. Nohales, L. Solar, I. Porcar, C. I. Vallo and C. M. Gomez, *Eur. Polym. J.*, 2006, **42**, 3093.
79. A. Zotti, A. Borriello, M. Ricciardi, V. Antonucci, M. Giordano and M. Zarrelli, *Composites, Part B*, 2015, **73**, 139.
80. Z. S. Zhan, M. J. Xu and B. Li, *Polym. Degrad. Stab.*, 2015, **117**, 66.
81. H. C. Bidsorkhi, M. Soheilmoghaddam, R. H. Pour, H. Adelnia and Z. Mohamad, *Polym. Test.*, 2014, **37**, 117.
82. M. He, W. C. Cao, L. J. Wang and C. A. Wilkie, *Polym. Adv. Technol.*, 2013, **24**, 1081.
83. R. Gul, A. Islam, T. Yasin and S. Mir, *J. Appl. Polym. Sci.*, 2011, **121**, 2772.
84. Y. Liu, J. Zhao, C. L. Deng, L. Chen, D. Y. Wang and Y. Z. Wang, *Ind. Eng. Chem. Res.*, 2011, **50**, 2047.
85. W. F. Tang, X. Y. Gu, Y. Jiang, J. R. Zhao, W. J. Ma, P. Jiang and S. Zhang, *J. Appl. Polym. Sci.*, 2015, 132.
86. P. Rybinski, G. Janowska, M. Jozwiak and A. Pajak, *J. Therm. Anal. Calorim.*, 2012, **109**, 561.
87. P. Rybinski and G. Janowska, *Thermochim. Acta*, 2012, **549**, 6.
88. H. Vahabi, R. Sonnier, B. Otazaghine, G. Le Saout and J. M. Lopez-Cuesta, *Polimery*, 2013, **58**, 350.
89. H. B. Chen, Y. Z. Wang and D. A. Schiraldi, *ACS Appl. Mater. Interfaces.*, 2014, **6**, 6790.

CHAPTER 10

Polymer–Halloysite Composite Membranes for Ultrafiltration and Proton Exchange Applications

YATAO ZHANG[a], JINGTAO WANG[a], JUNYONG ZHU[a], YAHUA LIU [a] AND JINDUN LIU*[a]

[a]School of Chemical Engineering and Energy, Zhengzhou University, Science Road 100, Zhengzhou 450001, P. R. China
*E-mail: liujindun@zzu.edu.cn

10.1 Polymer–Halloysite Composite Membranes for Ultrafiltration Applications

10.1.1 Introduction

Halloysite nanotubes (HNTs) are a kind of hydrophilic and low-cost clay. Their basic structure and properties have been discussed in Chapter 5, the following references are also suggested for more information.[1–3] Their special structure and physiochemical properties make them promising candidates as nanoscale fillers/carriers or functionalized nanomaterials. For instance, drugs can be loaded in HNTs for controlled/sustained release due to their

RSC Smart Materials No. 22
Functional Polymer Composites with Nanoclays
Edited by Yuri Lvov, Baochun Guo and Rawil F. Fakhrullin

Published by the Royal Society of Chemistry, www.rsc.org

cellular nontoxicity, environmental friendliness, biocompatibility and the lumen space.[4] In addition, during past decades, HNT-derived composites have been well-known for the adsorption of contaminants or pollutants due to their hollow tubular structure, high surface area and high adsorption properties,[5] and HNTs can also be used to improve the fire resistance of polymers and reduce thermal expansion properties.[1] Besides, HNTs have emerged as promising additives/carriers for the fabrication of mixed matrix membranes (MMMs) to optimize the performance of organic–inorganic hybrid membranes, including mechanical strength, thermal stability, and permeation flux and antifouling properties.[6–10]

Available water resources have registered a sharp decline with the continuous growth of the global population. The demand for water and wastewater treatment has driven a series of technologies for water purification. As an alternative candidate for wastewater treatment, membrane separation is regarded to be environmentally friendly, easy to control and operate, as well as requiring low initial investment and energy consumption. Herein, ultrafiltration (UF), one of the most active separation and purification fields, has been widely applied in the applications of water treatment, food processing, the pharmaceutical industry, oil–water separation, membrane bioreactors, and so forth.[11–15] Polymer-based UF membranes have the following excellent advantages: superior pH and chlorine tolerance, high mechanical strength and good thermal resistance.[16,17] But, unfortunately, the intrinsic hydrophobicity of polymeric materials, mainly ascribed to their repeated ether and sulfone linkages alternating between aromatic rings, plays a significant role in them suffering membrane organic fouling. In other words, the hydrophobic membrane surface is vulnerable to foulant adhesion and contamination, thus notably limiting its further application in control of water systems. The priority is to introduce hydrophilic nanomaterial and optimize the morphology and performance (roughness, charge properties, *etc.*) of the membranes without affecting their processability. A great deal of modifications including blending,[18] coating,[19] graft polymerization,[20] layer-by-layer (LBL) assembly,[21] interfacial polymerization (IP)[22] have been brought to the forefront in view of improving antifouling properties of polymeric membranes. Another major impediment to the wider application of membranes is biofouling. Membrane biofouling induced by microorganism adherence and growth is a major conditioning factor in the membrane separation process, giving rise to a distinct reduction of system productivity and a sharp increase of energy consumption. In addition, once a biofilm matures, extracellular polymeric substances (EPS) involving polysaccharides, proteins, glycoproteins, lipoproteins and other biomacromolecules in turn provide protection from antibacterial agents, rendering microorganisms embedded in biofilms more difficult to kill.[23] Hence, it is essential to prevent cell adhesion to the membrane surface for long-term stability and service life of the membranes.

It was acknowledged that membranes functionalized with neutral and smooth hydrophilic surfaces exhibited a low propensity for bacterial adhesion. Nevertheless, even if the majority of bacteria fail to achieve adhesion to the membrane surface, the rest of the microorganisms, which can grow and reproduce rapidly, will remain a risk of severe biofouling. Hence, for a long-term antimicrobial effect, the design of a UF membrane functionalized with antibacterial properties is of great importance, and it has been considered as the effective and long-lasting approach to reduce membrane biofouling. As previously mentioned, HNTs can be used as hydrophilic carriers of various antibacterial agents, and then introduced into the membrane matrix for effective antibacterial activities. In this section, polymeric membranes functionalized with modified HNTs for UF applications are focused on and thoroughly reviewed. Additionally, the main topics mainly target how to immobilize different antibacterial agents (organic, inorganic and other antimicrobial agents) onto HNT surfaces.

10.1.2 Special Structure and Properties of HNTs for Functionalization with Antibacterial Agents

Figure 10.1(a) and (b) present TEM images of HNTs and it is found that HNTs typically have a tubular structure and different sizes (external diameter of 20–50 nm and inner diameter of 15–30 nm) with a unique hollow space, which could facilitate the permeability of hybrid membranes in comparison with other nanoparticles (NPs), such as SiO_2, TiO_2, graphene oxide (GO), *etc.* A broad peak around 7.3 Å, corresponding to the first order (001) basal reflections, is due to the less ordered multilayer structure and the high intensity peak around 4.4 Å is the distinguishing feature of halloysite (shown in Figure 10.1(c)). The FT-IR spectrum of HNTs is shown in Figure 10.1(d). The absorption peaks at 3688 cm^{-1} and 3634 cm^{-1} are the stretch vibration of the Al–OH bonds and the peaks at around 1036 cm^{-1} are ascribed to the Al–OH blending vibration. A strong absorption peak at 1030 cm^{-1} is assigned to indicative stretching vibrations of Si–O–Si groups.[6–8] From the above analysis, the existence of certain hydroxyl groups on the surface of HNTs makes them easily modified with functionalized polymer brushes, providing a convenient route for the introduction of antibacterial agents. Additionally, the strong electronegativity of oxygen atoms makes the outer surface negatively charged. The lower density of hydroxyl groups on the outermost surface along with the charged property endow HNTs with good dispersion in nonpolar polymers, facilitating the construction of HNTs–polymer nanocomposites.

10.1.3 Application of HNT-Derived Nanomaterials for Antibacterial UF Membranes

As previously mentioned, HNTs are capable of enhancing the mechanical strength, thermal stability and separation properties of HNT-containing polymer membranes. A great deal of studies show an improved permeation

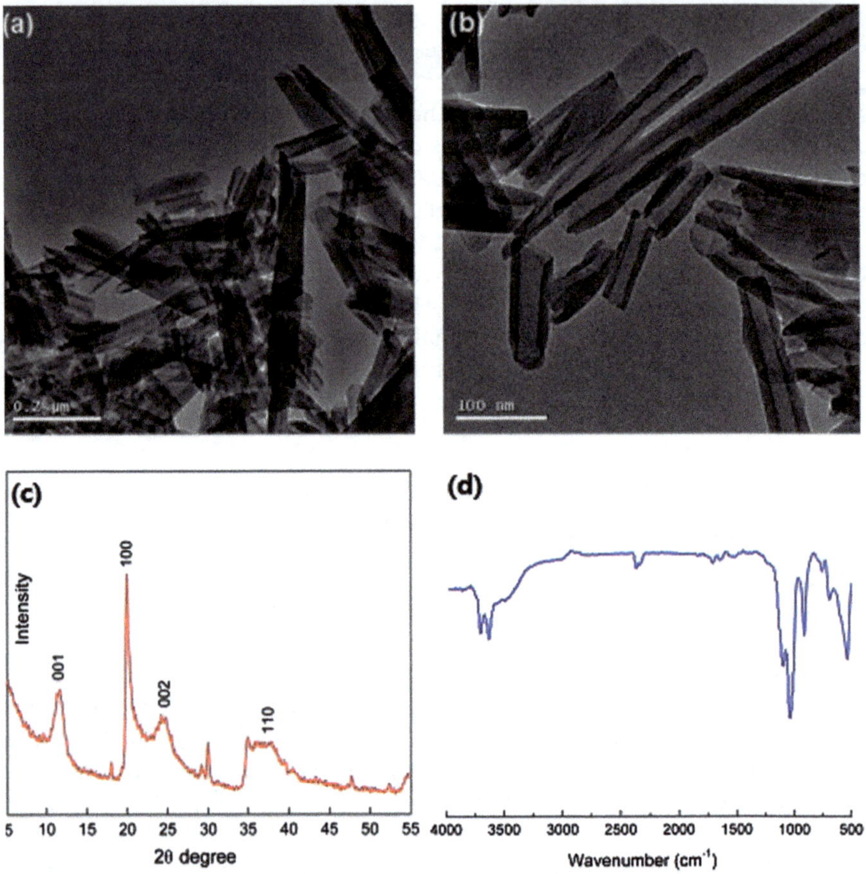

Figure 10.1 (a) and (b) TEM images, (c) XRD pattern and (d) FT-IR curve of HNTs.

flux of HNTs–polymer membranes compared with control membranes, including UF,[24] nanofiltration (NF),[10] forward osmosis (FO)[25] and reverse osmosis (RO).[26]

The possible explanation is as follows: (1) HNTs, as a hydrophilic clay, can enhance the hydrophilicity of hybrid membranes; (2) HNTs can also undermine the polymer chain packing to a certain extent and increase the free volume between polymer chains, rendering interface voids around HNTs, and the hollow inner of HNTs can provide short pathways for permeation of water molecules[10,27] (shown in Figure 10.2). Herein, a kind of NP, the carbon nanotube (CNT), also contains a similar hollow and tubular structure. However, the hydrophobicity of the CNT greatly limited its development in fabrication of UF membranes. Although a hydrophilic CNT could be realized through acidification using concentrated acid, which is quite harmful to the environment. In addition to this, HNTs are superior in mechanical strength and thermal stability compared with CNTs. More importantly, the low-cost HNTs without complex pre-treatment provide many advantages in

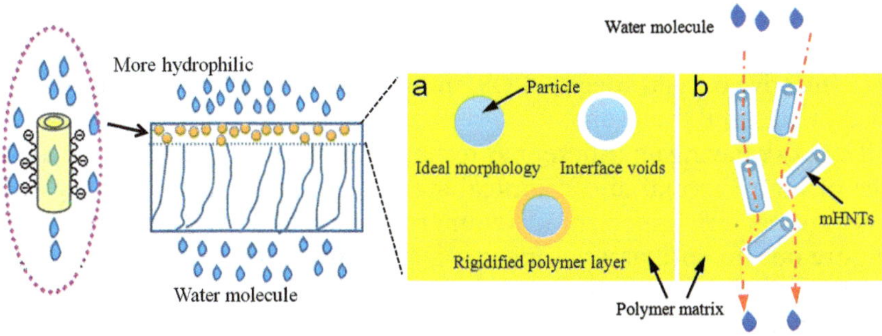

Figure 10.2 Schematic diagram of increased water flux of the membranes ascribed to HNTs. Reprinted from *Journal of Membrane Science*, **465**, J. Zhu *et al.*, Preparation and characterization of negatively charged PES nanofiltration membrane by blending with halloysite nanotubes grafted with poly (sodium 4-styrenesulfonate) *via* surface-initiated ATRP, Copyright 2014 with permission from Elsevier (ref. 11) and reprinted with permission from J. Zhu, Y. Zhang, M. Tian and J. Liu, *ACS Sustainable Chem. Eng.*, 2015, **3**, 690. Copyright (2015) American Chemical Society (ref. 27).

the preparation of UF membranes. Due to the big challenge of membrane biofouling, more and more researchers are focusing on the incorporation of antibacterial agents on/into the membrane surface/matrix to reduce/control surface biofouling. The incorporation of antibacterial inorganic NPs into polymer solution has been performed to fabricate hybrid membranes through phase inversion.[28] Although it has a positive effect on the anti-biofouling properties of polymeric membranes, inevitable leaching of NPs is bound to pose an adverse impact on long-term antibacterial activity. Therefore, HNTs are regarded as an ideal carrier of antibacterial agents, not only enhancing the overall performance as reported, but also effectively restraining NP leaching during phase inversion.

Considerable research has been reported on modification methods for the introduction of organic, inorganic or bio-enzymatic antibacterial agents, which endow polymer-based membranes with highly efficient antibacterial performance.[29–31] Herein, HNTs functionalized with different types of antibacterial agents and the application of modified HNTs for the construction of antibacterial UF membranes are highlighted in the following sections.

10.1.3.1 HNTs Functionalized with Inorganic NPs

Inorganic antibacterial agents are typically prepared from antibacterial constituents including silver, copper, titanium oxide, zinc oxide, graphene, CNTs, *etc.*[32–37] However, the exact mechanisms of NP toxicity against bacteria are not fully understood. It has been reported that it conforms to a contact-action mode: NPs can adhere to the bacterial cell surface *via* electrostatic interaction and thus the integrity of the bacterial cell suffers from disruption. Furthermore, the toxicity of NPs against bacteria derives from oxidative stress (reactive oxygen species, ROS) induced by free radicals, which are produced by NPs and their ions.

Silver (Ag), having an inherent antimicrobial property, presents the advantages of high thermal and chemical stability, good catalytic activity and a broad antibacterial spectrum.[38] Numerous studies have been carried out to figure out its antibacterial mechanism against model bacteria including *Escherichia coli* (*E. coli*, Gram-negative) and *Staphylococcus aureus* (*S. aureus*, Gram-positive). Despite the perfect antibacterial properties of silver ions, it is a big challenge to make silver ions exist stably in a polymer matrix without adding carriers, and an inevitable loss of silver ions during practical application will greatly impair their antimicrobial activity.

Hence, Zhang *et al.*[24] synthesized modified HNTs loaded with silver NPs (Ag NPs) and incorporated modified HNTs into polyethersulfone (PES) membrane to investigate the antibacterial activity of hybrid membranes. Specifically, poly(4-vinylpyridine) (P4VP) was firstly grafted onto the HNT surface *via in situ* polymerization and silver ions were immobilized on P4VP *via* complex reaction. HNTs–P4VP–Ag NPs were obtained *via in situ* reduction using $NaBH_4$ as the reducing agent (Figure 10.3). The hybrid membranes containing modified HNTs loaded with Ag NPs were prepared *via* the

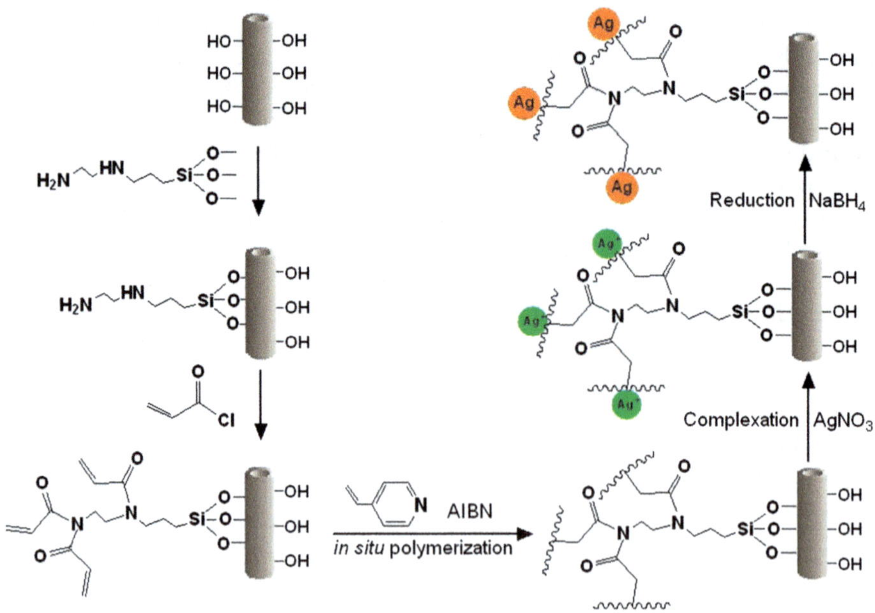

Figure 10.3 The synthesis route of HNTs–P4VP–Ag NPs. Reprinted with permission from J. Zhang, Y. Zhang, Y. Chen, L. Du, B. Zhang, H. Zhang, J. Liu and K. Wang, *Ind. Eng. Chem. Res.*, 2012, **51**, 3081. Copyright (2012) American Chemical Society (ref. 24).

"blending-phase inversion" method. The permeation flux of hybrid membranes was distinctly superior to pristine membrane. Especially when the modified HNT content was 3%, the water flux of the membrane reached as high as 396.5 L m^{-2} h^{-1}. In addition, the rejection slightly decreased with an increase of modified HNTs, whereas they had a little influence on the structure and morphology of hybrid membranes. The antibacterial rates of the hybrid membranes against *E. coli* and *S. aureus* were *ca.* 99.9% and 99.8%, which showed an excellent antibacterial activity. However, due to the complicated synthesis process, Zhang *et al.*[39] proposed a facile and effective method to immobilize Ag NPs onto the surface of HNTs. The synthesis procedure is as follows: modified HNTs functionalized with silane coupling agent (KH-792) were firstly complexed with silver ions, which were then reduced into Ag NPs with about 5 nm diameters using $NaBH_4$ as reducing agent. The results showed Ag NPs–HNTs composites manifested good antibacterial performance against both *E. coli* and *S. aureus*. On that basis, Chen *et al.*[40] introduced chitosan (CS) onto the surface of HNTs, which was then complexed with silver ions. HNTs–CS@Ag composites were developed for biofouling control *via in situ* reduction. The hybrid membranes containing HNTs–CS@Ag composites prepared by phase inversion exhibited both good antifouling and anti-biofouling properties due to their hydrophilic and antibacterial surface. In addition to Ag NPs, copper-based NPs have gained growing attention as effective biocidal agents for biofouling control due to their excellent antibacterial properties, low toxicity and low cost. Hence, PES UF membrane containing HNTs loaded with copper ions (Cu^{2+}-HNTs–PES) were prepared *via* the blending-phase inversion method.[41] Herein, Cu^{2+}-HNTs are regarded as an antibacterial agent, synthesized by chemical modification of HNTs with a silane coupling agent, and then complexed with Cu^{2+}. The hybrid membranes exhibited a good antibacterial effect with a higher water flux and enhanced mechanical strength compared with pristine PES membranes. However, the concentration of Cu^{2+} would inevitably decrease during membrane filtration due to the depletion of Cu^{2+}. This may lead to a relatively short antibacterial effect for UF membranes. Duan *et al.*[42] reduced Cu^{2+} into Cu NPs and loaded on HNTs using $NaBH_4$ as the reducing agent to decrease the depletion rate of Cu^{2+}. The results showed the hydrophilicity of hybrid membranes was distinctly enhanced and their surface became smoother compared with virgin membranes. Importantly, the antibacterial test indicated that the hybrid membranes showed good antibacterial activity against *E. coli* with a high bacteriostasis rate of 94.5%. Yu *et al.*[43] devised a sandwich-like antibacterial agent (Ag–HNTs–reduced graphene oxide (rGO)) through the direct growth of Ag NPs on the surface of graphene-based HNT nanosheets, which are shown in Figure 10.4.

The Ag–HNTs–rGO composites markedly increased antibacterial ability against *E. coli* and *S. aureus* compared with individual Ag NPs due to the synergistic effect of rGO and Ag, rendering them promising for the fabrication of antibacterial membranes. Zhao *et al.*[44] designed an antibacterial and antifouling PES UF membrane by embedding as-synthesized Ag–HNTs–rGO

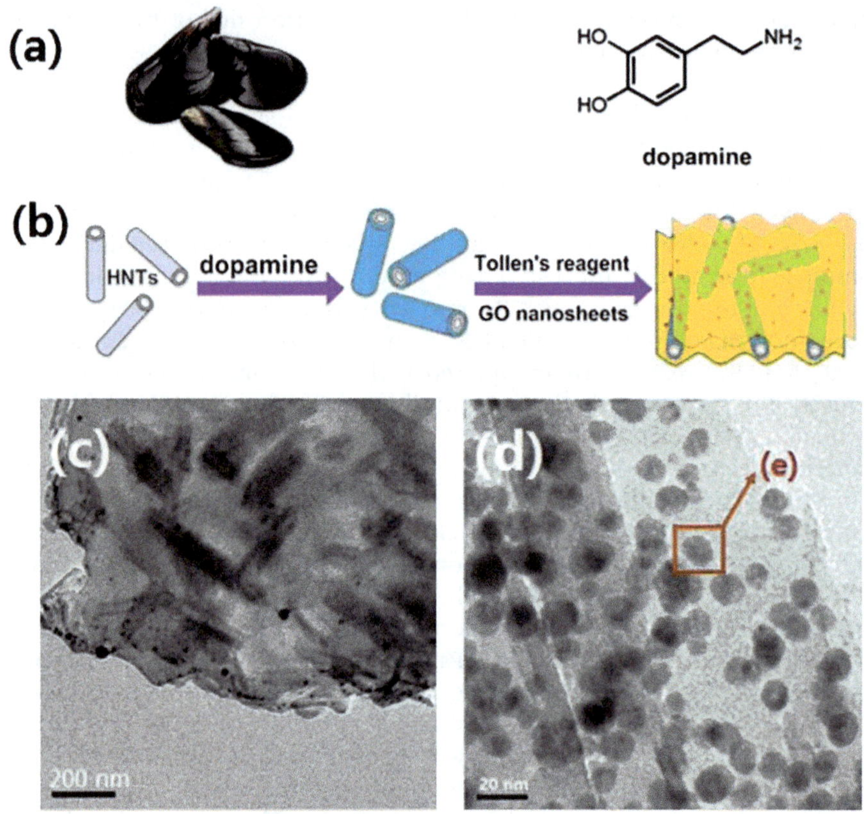

Figure 10.4 (a) Bioadhesive proteins of marine organisms (mussels) and chemical structure of dopamine, (b) synthesis process of sandwich-like Ag–HNTs–rGO, (c) and (d) TEM images of sandwich-like nanomaterials at different magnifications. Reprinted with permission from ref. 43. Copyright (2014) Nature Publishing Group.

into PES solution through phase inversion. As seen from Figure 10.5a, the flux decline of the hybrid membrane was quite a bit lower than that of the pristine membrane, and chemical cleaning could effectively recover its initial permeation flux, indicating the improved antifouling performance of the hybrid membrane. The incorporated Ag NPs were evenly attached onto the rGO support with an average size of 10 nm, which ensured its good antibacterial performance: the hybrid membrane had an ideal bacteriostasis rate against *E. coli* even after six months storage (Figure 10.5b).

In addition, the water flux of the membranes is distinctly superior to that of hybrid membranes containing graphene-based nanomaterials. The high flux can be explained by an improved surface hydrophilicity and the formation of nanoscale flow channels ascribed to HNTs between neighboring GO sheets. In conclusion, HNT-based inorganic NPs have the advantages of good antibacterial properties, high hydrophilicity, thermal stability and mechanical

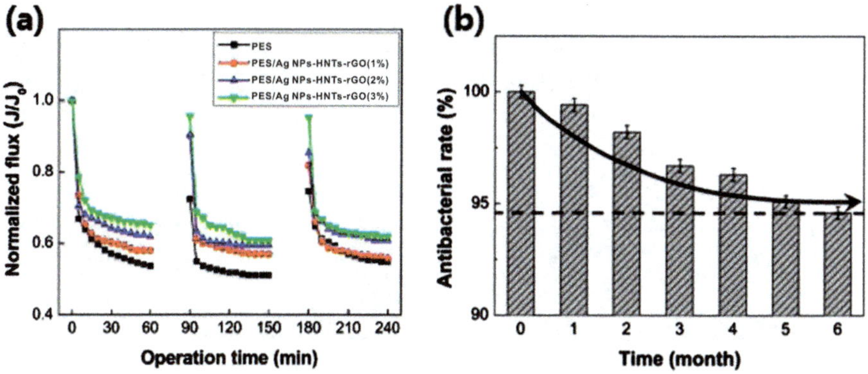

Figure 10.5 (a) Normalized flux profiles of the tested membranes during filtration of BSA solution at 25 °C and 0.1 MPa. (b) Long-lasting antibacterial rate of the hybrid membranes with 3 wt% of AgNPs–HNTs–rGO against *E. coli*. Reprinted from ref. 44 with permission from The Royal Society of Chemistry.

strength. However, there are still thorny issues existing in the design of UF membranes: (a) the aggregation phenomenon is hard to avoid if the NP content is too high; (b) most NPs are buried within the polymer matrix through blending-phase inversion, which makes it hard to maximize the antibacterial efficiency of the NPs; (c) gradual leaching of antibacterial agents is difficult to avoid (NPs or ions) over the long-term operation, especially during the membrane filtration process when the hydraulic shear force is applied, which may lead to a relatively short lifespan of the antibacterial efficiency and the detached ions may result in secondary pollution to the permeate.

10.1.3.2 HNTs Functionalized with Organic Antibacterial Agents (N-halamine)

In recent decades, antibacterial polymers and their nanostructures have gained a lot of attention in human and animal health care due to their intrinsic antibacterial groups. However, the limited number of specific studies were focused on the improvement of antibacterial membranes modified with polymeric nanostructures. Herein, *N*-halamine materials have received intensive interest as promising organic antibacterial agents due to their unique properties: antibacterial efficacy, stability in aqueous solution and in dry storage, low toxicity, low-expense and so on.[45] Duan *et al.*[46] synthesized *N*-halamine grafted HNTs (*N*-halamine@HNTs), which were applied as antibacterial agents to fabricate PES UF hybrid membranes. The results indicated that the hydrophilicity of the membranes was greatly improved with the addition of *N*-halamine@HNTs. AFM results indicated that the hybrid had lower surface roughness than the pure PES membrane, and the water flux of the hybrid membrane could rise up to 248.3 L m^{-2} h^{-1} when the

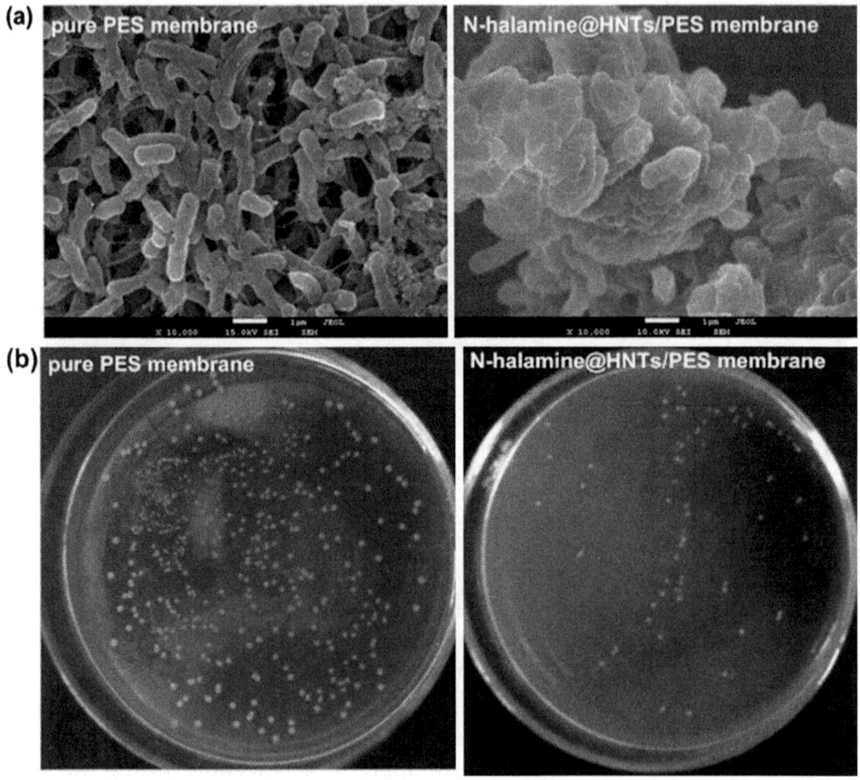

Figure 10.6 (a) SEM images of *E. coli* attached to pure PES membrane and PES membrane containing 3% *N*-halamine@HNTs; (b) measurement of antibacterial activity of the membranes by the bacteriostasis rate. Reprinted from ref. 46 with permission from The Royal Society of Chemistry.

content of *N*-halamine@HNTs was 1.0 wt%. In addition, the antibacterial test indicated that the hybrid membranes exhibited a good antibacterial activity against *E. coli* (Figure 10.6). Although organically functionalized HNTs are beneficial for the fabrication of antibacterial UF membranes, the synthesis of HNT-based NPs is relatively complex and contains hazardous solvents which are unfriendly to the environment. Therefore, the optimization of the preparation process is of great importance for sustainability and cost reduction.

10.1.3.3 HNTs Functionalized with Bio-Enzymatic Antibacterial Agents (Lysozyme)

Alternatively, enzymes are environmentally friendly natural products which possess the advantages of high activity, selectivity and specificity, and most of the products after their catalysis are nontoxic and safe.[47] Their anti-biofilm

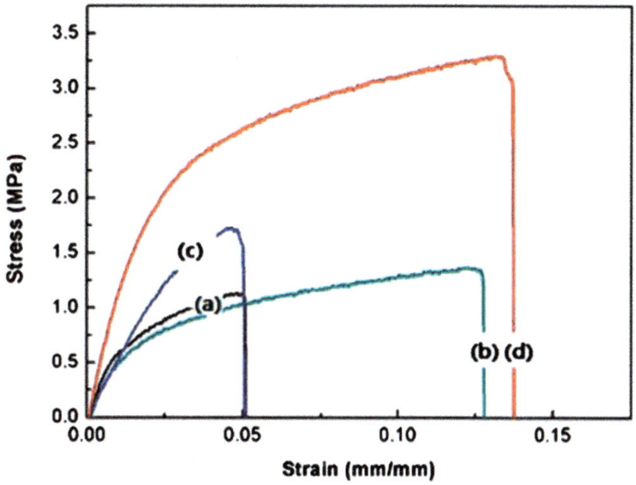

Figure 10.7 Stress–strain curves of (a) PES membrane, (b–d) PES–HNTs-Ly membranes with different HNTs-Ly contents (1 wt%, 2 wt% and 3 wt%, respectively). Reprinted from ref. 50 with permission from The Royal Society of Chemistry.

behavior generally includes destabilization of the physical integrity of the biofilm, degradation of biofilm compounds, and bacterial cell lysis, which eventually prevent biofilm formation or promote biofilm detachment. Herein, one enzyme, lysozyme (Ly), acts as a nonspecific innate immunity molecule against the invasion of bacterial pathogens and catalyzes cell wall lysis, rendering it promising for antibacterial application.[48,49] Generally, free enzymes exhibit low stability and durability, and an immobilization technique is needed to improve the enzyme properties. Zhao *et al.*[50] devised a method to immobilize Ly onto the surface of HNTs (HNTs-Ly) *via* a covalent binding reaction, and then incorporated HNTs-Ly into a PES matrix to prepare hybrid antibacterial UF membranes through phase inversion. Briefly, the introduction of carboxyl groups onto the surface of the HNTs *via* a two-step chemical reaction was carried out, and the immobilization of Ly on HNTs-COOH was carried out through electrostatic and hydrogen-bonding interactions.

It was found that the surface hydrophilicity and water flux of the hybrid membranes significantly improved after the addition of HNTs-Ly, as well as the mechanical strength (Figure 10.7). Particularly, water flux could achieve a highest value of 400 L m^{-2} h^{-1} when HNTs-Ly content was 3.0 wt%, and the hybrid membrane's rejection for PVA 30 000–70 000 remained as high as 99.6%. Moreover, the hybrid membrane showed good antibacterial activity against *E. coli* with a high bacteriostasis rate of 63%, which exhibited the potential to reduce fouling in water treatment. In conclusion, functionalization of HNTs with enzymes using immobilization techniques is an effective strategy to improve the enzyme features, *e.g.*, stability, activity, specificity and selectivity, reduction of inhibition. Although the addition of immobilized enzymes into the bulk phase in a membrane system exhibits an obvious antifouling property, more concern will turn to the membrane surface, which is easily contaminated by bacterial strains. However, to a certain extent, the high cost of bio-enzymes will limit their progress in the development of antibacterial membranes.

10.2 Polymer–Halloysite Composite Membranes for Proton Exchange Applications

10.2.1 Introduction

Proton-conducting materials with high conductivity are essential both in biological contexts and in materials for alternative energy, such as fuel cells.[51–54] Fuel cells are energy conversion devices converting chemical energy directly into electrical energy, which have been considered as a promising technology with great potential to alleviate the ever-growing demands for power and a decrease in pollution.[55–57] Fuel cells possess many advantages – such as high energy density, quiet operation, no requirement for a fuel reforming process and low emission of pollutants – and are potential power sources for application in transportation and in stationary and portable electronics. The proton exchange membrane fuel cells (PEMFCs), as one kind of fuel cell, are attracting increasing attention all over the world by virtue of their high energy density, fast start-up and easy fuel handling merits. The proton exchange membrane (PEM) is the core component of PEMFCs and plays a critical role in the development of PEMFCs; it is no exaggeration to say that improvement in the properties of the membrane would significantly improve the performance of PEMFCs.[58] The PEM functions as an electrolyte for transporting protons from the anode to the cathode as well as blocking the passage of electrons and fuel between the electrodes. To be applied in PEMFCs, the desired PEM should meet the following requirements: (i) high proton conductivity; (ii) low fuel permeability; (iii) good thermal and mechanical stabilities; (iv) sufficient water uptake and moderate swelling; (v) a competitive low-cost and sufficiently long-term durability. And among them, high proton conductivity is the foremost parameter for PEM, which directly affects the operational fuel cell voltage and current output of a fuel cell.

Currently, the perfluorosulfonic acid membranes, represented by Nafion, are the most commonly used PEMs due to their high proton conductivity as well as exceptional chemical and mechanical stability.[59,60] However, Nafion suffers from some drawbacks, such as high fuel crossover, high cost and considerable conductivity loss at high temperature, which restrict its practical application. In past decades, significant efforts have been dedicated to the design and fabrication of alternative PEMs. Among the various candidates developed, organic–inorganic hybridization to prepare composite membranes has become a hot topic among researchers, which can combine the advantages from the inorganic fillers and organic polymer matrix.

To date, spherical, sheet and tubular fillers have been utilized as inorganic fillers to facilitate the proton conduction of composite membranes. Among them, tubular fillers display the most significant effects due to their high aspect ratios, which allow the formation of long-range uninterrupted pathways for proton transport.[61] HNTs have been employed as additives to fabricate composite PEMs for enhancement of proton conductivity. HNTs possess a large specific surface area, and excellent mechanical and thermal stability. Sufficient Si–OH and Al–OH groups on the surface endow HNTs with good hydrophilicity as well as facile modification abilities. Matusik and Wścisło[62] found that the edge aluminol Al–OH and/or silanol Si–OH groups may be protonated or deprotonated by the competitive absorption of protons, which would inhibit the absorption of metal ions. These superior properties afford HNTs with potential applications in PEMs. In this section, two types of composite PEMs embedded with raw HNTs and surface-modified HNTs are focused on, and the properties and performances of these membranes are described one after another and compared in the following sections.

10.2.2 HNT-Incorporated Composite PEMs

Generally, the structure has great influence on the properties. Thereby, the structural peculiarity of Nafion is briefly described, which endows it with a high proton transfer capability and excellent structural stability. Within Nafion, the amphiphilic feature between the hydrophobic backbone and the hydrophilic pendant side chain drives the sulfonic acid groups to assemble into 4.0 nm clusters as inverted micelles arranged on a lattice, which are interconnected by pores or channels (about 1.0 nm).[63,64] These clusters and channels are well-established, which work as fast proton transfer pathways and afford Nafion high proton conductivity (0.1 S cm^{-1}) and low activation energy (0.092 eV).[65] Zhang *et al.* synthesized sulfonated poly(ether ether ketone) (SPEEK) through a facile sulfonation reaction to graft sulfonic acid groups into poly(ether ether ketone) (PEEK) chains.[66] Similar to Nafion, SPEEK membrane also displays nanophase separation between the hydrophobic backbone and hydrophilic side chain and has been considered to be one of the promising alternatives for Nafion. However, owing to its weak amphiphilic features, SPEEK membrane exhibits a narrower ion-channel size (0.5–1.0 nm), more "dead ends" and phase discontinuity compared with Nafion.[67] Such morphologies result in a higher energy barrier for proton transfer and lower proton conductivity than those of Nafion. The proton-conducting one-dimensional nanostructures, HNTs, were incorporated into SPEEK to interconnect the ionic channels and hence form a long-range uninterrupted trajectory for proton transport. The representative morphologies of HNTs and the resultant membranes are displayed in Figure 10.8. As can be seen, HNTs exhibit a well-defined tubular structure with a diameter of about 60 nm and a length of about 400–750 nm. From the SEM of

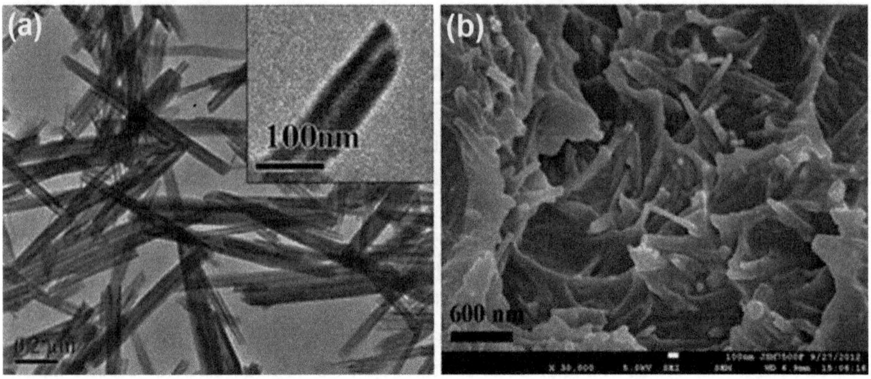

Figure 10.8 (a) TEM of HNTs, (b) SEM image of the cross-section of SPEEK–HNT-15%. Reprinted from *International Journal of Hydrogen Energy*, **39**, H. Zhang *et al.*, Enhancement of proton conductivity of polymer electrolyte membrane enabled by sulfonated nanotubes, Copyright 2014 with permission from Elsevier (ref. 66).

SPEEK–HNTs-15%, it can be seen that the nanotubes keep their intact tubular structure during membrane preparation, creating the possibility of building a continuous pathway for proton transfer by embedding these nanotubes into the polymer matrix. And it also shows that HNTs display favorable compatibility with the SPEEK matrix owing to hydrogen-bonding interactions and thus disperse homogeneously within the polymer matrix.

Theoretically, the protons are transferred inside the PEM through two mechanisms: (i) a vehicle mechanism, in which the protons diffuse together with H_2O molecules by forming different hydronium ions, *e.g.*, H_3O^+, $H_5O_2^+$ and $H_9O_4^+$; (ii) the Grotthuss mechanism, in which the protons jump *via* adjacent proton hopping sites.[68,69] Generally, the proton transfer through the membrane is a complex process involving both of the mechanisms. Zhang *et al.* found that the incorporation of HNTs reduced the channel size.[70] Due to the relatively lower dissociation degree and hydrophilicity of –(Al/Si)–OH than those of $-SO_3H$, the incorporation of HNTs decreases the concentration of sulfonic acid groups and hence the water uptake and ion exchange capacity (IEC) of HNT-filled membranes. Compared with the proton conductivity of the SPEEK control membrane (0.011 S cm^{-1} at 25 °C and 100% RH), the incorporation of HNTs increases the proton conductivity of HNT-filled membranes when the HNT content is below 15%. For phase-separated membranes, the reduced channel size, IEC, and water uptake ought to reduce proton carriers and suppress the diffusion of hydronium ions across the membrane, thus causing the reduction of vehicle-type proton conductivity. Therefore, an increment of the proton conductivities of the SPEEK–HNTs composite membranes reasonably arises from Grotthuss-type transfer. The hydrogen-bonding interactions at the SPEEK–HNTs interface will affect SPEEK chain packing and thus induce the ion-channel arranging along the HNT surface. In this manner, these channels become more continuous compared with those in the SPEEK control membrane, and the Grotthuss-type transfer is enhanced. This is different from the results of the incorporation of CNTs (another one-dimensional inorganic material) into a SPEEK matrix. Tripathi *et al.* found that the introduction of CNTs into SPEEK resulted in a decrease of proton conductivity of composite membranes due to a deficiency of proton conductive groups on the surface of CNTs.[71] Nevertheless, when HNT content is up to 20%, the proton conductivity of SPEEK–HNTs membrane is reduced, which is attributed to the aggregation of HNTs at high content. The results of PEMFC performances exhibit that the incorporation of HNTs affords elevated maximum current densities and power densities to HNT-filled membranes compared to SPEEK control membrane.

In addition to phase-separated membranes, HNTs have also been incorporated into other polyelectrolytes. Bai *et al.* selected CS as a polymer matrix, considering its excellent fuel-barrier properties and low cost.[72] CS is a linear polysaccharide composed of randomly distributed β-(1–4)-linked D-glucosamine and *N*-acetyl-D-glucosamine. During membrane preparation, the hydrophilic groups (*e.g.*, $-NH_2$ and –OH) generate strong intramolecular and intermolecular hydrogen-bonding interactions, which drive CS chains to form numerous crystalline regions and cross-linked networks. These networks have small free volume cavities with a diameter of about 0.56 nm, which is smaller than the size of methanol–methanol or methanol–water associations. Therefore, CS-based membranes exhibit high methanol-barrier properties. In this research, a series of hybrid membranes are prepared by embedding HNTs into a CS matrix *via* the solution casting method with a filler content of 3%, 6%, 9%, 12% and 15%. The microstructure of the nanohybrid membranes is representatively presented in Figure 10.9. The surface images of the nanohybrid membranes suggest that the membrane surface is dense and uniform without appreciable defects. The cross-section image reveals that HNTs maintain an intact tubular structure without destruction. However, HNTs display poor dispersion behavior within the CS matrix, and obvious nanotube agglomerations and voids along the HNTs are observed. And the number of agglomerations and voids increases with increasing HNT content. Such observations are probably ascribed to the weak interfacial compatibility between HNTs and CS, which increases the surface tension of HNTs and thus drives the formation of nanotube agglomerations.

The results of thermogravimetric analysis and differential scanning calorimetry, and a stress–strain test indicated that hybrid membranes with an appropriate nanotube content possess improved thermal and mechanical stabilities over those of the CS control membrane. During the preparation of hybrid membranes, the presence of nanotubes would interfere with the chain packing and motion of the CS phase *via* the interfacial interactions,

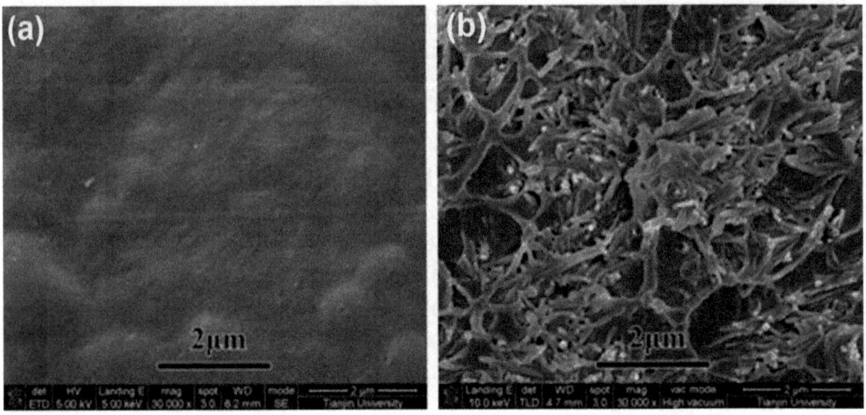

Figure 10.9 (a) The surface image, and (b) cross-section image of CS–HNT-9. Reprinted from *Journal of Membrane Science*, **454**, H. Bai *et al.*, Enhanced proton conduction of chitosan membrane enabled by halloysite nanotubes bearing sulfonate polyelectrolyte brushes. Copyright 2014 with permission from Elsevier (ref. 72).

such as hydrogen-bonding interactions. In such a way, the degradation and stretching of the polymer back bone is inhibited, *i.e.* the structural stability of the hybrid membranes is reinforced. Unfortunately, according to the results of proton conductivity measurement, HNT-filled membranes exhibit slightly reduced proton transfer compared with the CS control membrane. The reason might be the following aspects: (i) the decreased water uptake of the hybrid membrane, resulting from the weaker water adsorption capacity of HNTs than that of the CS matrix, (ii) the decreased IEC, resulting from the lower concentration of available protons of HNTs than the groups of the CS matrix, (iii) the tortuous and lengthened transfer paths for hydronium ions and protons in the presence of nanotubes, (iv) the inhibited chain mobility of the CS matrix in hybrid membranes, which will increase the transfer resistance of hydronium ions and protons. Accordingly, with the incorporation of HNTs, proton transfer *via* both vehicle and Grotthuss mechanisms is reduced.

10.2.3 Modified-HNT-Incorporated Composite PEMs

The incorporation of raw HNTs often accompanies the phenomenon of agglomeration of nanotubes within the composite membrane at high content and limitation of the enhancement of proton conductivity. To improve the compatibility between the polymer matrix and nanotubes and the proton conductivity of the hybrid membrane, surface modification of HNTs is proposed. One facile and efficient approach to address this issue is to import

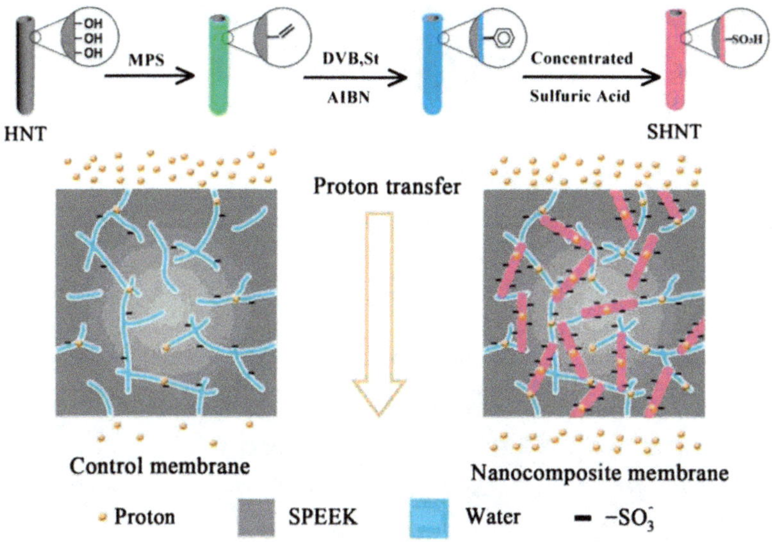

Scheme 10.1 Synthesis process of sulfonated halloysite nanotubes and the proton transfer behaviors in SPEEK control and SPEEK–SHNT nanocomposite membranes. Reprinted from *International Journal of Hydrogen Energy*, **39**, H. Zhang *et al.*, Enhancement of proton conductivity of polymer electrolyte membrane enabled by sulfonated nanotubes, Copyright 2014 with permission from Elsevier (ref. 66).

conductive groups onto the surface of inorganic fillers, similar to the method utilized in previous research.[73,74]

Zhang *et al.* designed and synthesized sulfonated halloysite nanotubes (SHNTs) by importing sulfonate polyelectrolyte brushes onto the nanotube surface *via* distillation–precipitation polymerization.[66] Furthermore, they synthesized two forms of sulfonated HNTs by adjusting the time of sulfonation to 2 h or 4 h, which were designated S2HNTs and S4HNTs, respectively. Then these SHNTs were incorporated into a SPEEK matrix to prepare hybrid membranes. For comparison, a SPEEK control and HNT-filled membranes were prepared. The synthesis process of the sulfonated HNTs and the proton transfer behavior through the hybrid membrane are illustrated in Scheme 10.1.

The microstructures of the SHNTs and the hybrid membrane are shown in Figure 10.10. It can be seen that S2HNTs and S4HNTs dispersed uniformly within the SPEEK matrix. Such an improvement is contributed by the introduction of the sulfonated polymeric layer, which improves the interfacial compatibility between the inorganic fillers and organic polymer.

SHNT-filled membranes exhibit better thermal and mechanical stabilities than those of both the SPEEK control and HNT-filled membranes under

Figure 10.10 TEM of S2HNTs (a) and S4HNTs (b), and image of SPEEK–S2HNT-7.5 (c) and SPEEK–S4HNT-7.5 (d). Reprinted from *International Journal of Hydrogen Energy*, **39**, H. Zhang *et al.*, Enhancement of proton conductivity of polymer electrolyte membrane enabled by sulfonated nanotubes. Copyright 2014 with permission from Elsevier (ref. 66).

identical conditions, and the properties can be further improved with an increase of nanotube content. As the key parameter of PEM, the proton conductivities also confirm the superiority of SHNT-filled membranes, and the proton conductivities of the as-prepared membranes are shown in Figure 10.11. The results reveal that SPEEK–S2HNTs-7.5 and SPEEK–S4HNTs-10 obtain highest conductivities of 0.0189 and 0.0245 S cm^{-1}, which are 24% and 61%, respectively, higher than that of the SPEEK control membrane. These findings are probably due to the presence of sulfonated nanotubes within the SPEEK matrix, which interconnect the sulfonic-aggregated ionic channels, especially the dead ends, and then donate more continuous ionic networks *via* the conducting –SO_3H groups on the nanotube surface. These networks then act as proton pathways and allow efficient proton transfer

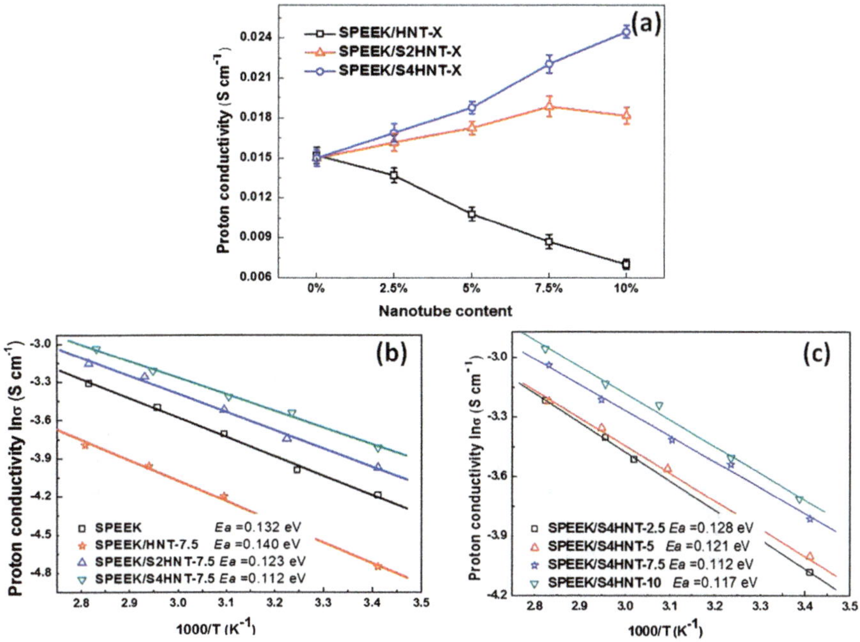

Figure 10.11 (a) The proton conductivity of the as-prepared membranes at 25 °C and 100% RH, (b) and (c) the temperature-dependent conductivity of the SPEEK control and nanocomposite membranes at 100% RH. Reprinted from *International Journal of Hydrogen Energy*, **39**, H. Zhang *et al.*, Enhancement of proton conductivity of polymer electrolyte membrane enabled by sulfonated nanotubes, Copyright 2014 with permission from Elsevier (ref. 66).

with low resistance, as illustrated in Scheme 10.2. The activation energy (Figure 10.11), which reflects the transfer barrier properties of protons through the membranes, implies the co-existence of Grotthuss-type and vehicle-type proton transfer in the as-prepared membranes. Meanwhile, the higher sulfonic acid loading of S4HNTs affords S4HNT-filled membranes better properties than those of S2HNT-filled membranes.

Among the various kinds of alternative inorganic–organic PEMs, acid–base composite membranes have proved to be a novel and effective approach for high proton conductivity and attract extensive attention due to their unique manner of transport. Within these composites, proton donor (acid group) and acceptor (base group) are closely linked, and protons can shuttle directly between donor and acceptor *via* the Grotthuss mechanism without water. This transfer manner is a common motif in biological proton transport. For example, protons are transported along the carboxylic acid–Schiff base pair through the proton pump of bacteriorhodopsin within 80 microseconds.[75] These composite membranes are typically fabricated by incorporating

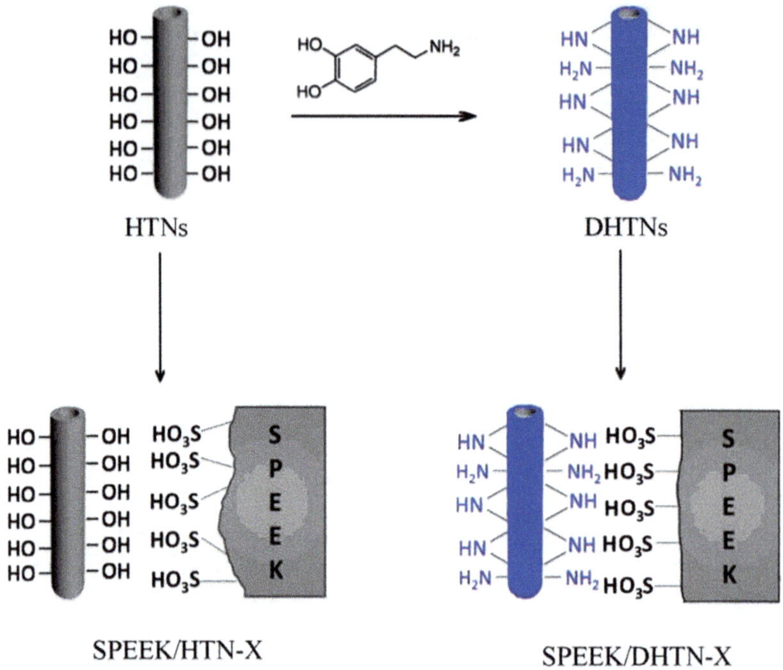

Scheme 10.2 Preparation of dopamine-modified halloysite nanotubes and the interfacial microstructure between the SPEEK bulk and nanotube. Reprinted from ref. 70 with permission from John Wiley and Sons. Copyright © 2013 WILEY-VCH Berlag GmbH & Co. KGaA, Weinheim.

acidic fillers into basic polymers or incorporating basic fillers into acidic polymers.[75–78]

Zhang *et al.* designed and synthesized dopamine-modified halloysite nanotubes (DHNTs) bearing $-NH_2$ and –NH– groups by directly immersing natural HNTs into dopamine aqueous solution under mild conditions.[70] Then the DHNTs were incorporated into a SPEEK matrix to prepare composite membranes. The preparation of DHNTs and the interfacial microstructure between the SPEEK bulk and nanotube are illustrated in Scheme 10.2.

When the HNT content is higher than 25%, obvious nanotube aggregation is observed in raw HNT-filled membranes due to their high surface tension, which in turn causes numerous voids at the HNTs–HNTs interface and SPEEK–HNTs interface. After surface modification, the interfacial compatibility between DHNTs and SPEEK is improved and the surface tension of DHNTs is effectively reduced (Figure 10.12) due to the stronger electrostatic interaction. DHNTs therefore display better dispersion, without aggregation and voids. In such a way, DHNTs may donate much more surface-induced pathways for proton transfer.

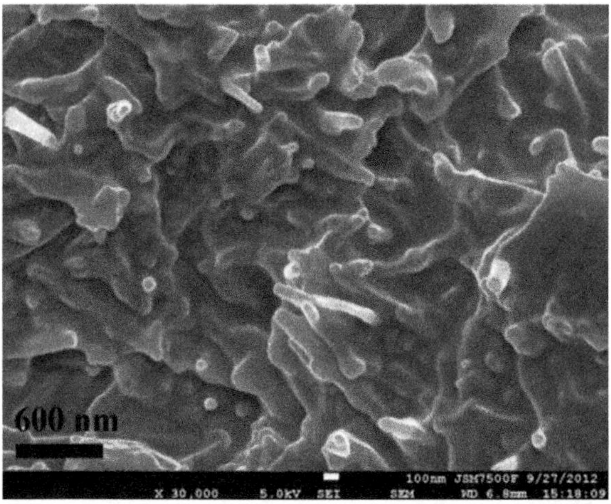

Figure 10.12 An image of the cross-section of SPEEK–DHNT-15%. Reprinted from ref. 70 with permission from John Wiley and Sons. Copyright © 2013 WILEY-VCH Berlag GmbH & Co. KGaA, Weinheim.

The presence of DHNTs generates strong electrostatic interaction with the SPEEK chains, yielding numerous ordered acid–base pairs (*e.g.*, $–S–O^{-}\cdots{}^{+}H–HN–$, $–S–O^{-}\cdots{}^{+}H–N–$) at the SPEEK–DHNTs interface. Such acid–base pairs could significantly facilitate the proton transfer through the membrane. Accordingly, DHNT-filled membranes exhibit surprisingly high proton conductivities compared with HNT-filled and SPEEK control membranes (Figure 10.13). On the other hand, the strong interfacial interaction can induce SPEEK chain packing more effectively, which is beneficial for the improvement of the structural stability and the construction of continuous ion-channels and proton transfer pathways (Scheme 10.1). Moreover, with the content of DHNTs increasing from 5% to 30%, the proton conductivity continuously elevates from 0.012 to 0.018 S cm^{-1} without conductivity decline, which is due to the better dispersion of DHNTs. SPEEK–DHNTs-30 achieves the highest proton conductivity of 0.018 S cm^{-1}. Although it is lower than that of Nafion (about 0.1 S cm^{-1}), it is still high enough (>0.01 S cm^{-1}) to serve as the PEM for fuel cell application. The enhanced conductivity confers higher H_2/O_2 PEMFC performances on the SPEEK–DHNTs hybrid membrane than those of the SPEEK–HNTs hybrid and SPEEK control membranes.

Bai *et al.* synthesized HNTs bearing sulfonate polyelectrolyte brushes (SHNTs) *via* distillation–precipitation polymerization and then embedded them into a CS matrix to fabricate nanohybrid membranes.[72] Similar to that in the last case, the strong electrostatic attractions between $–SO_3H$ of SHNTs and $–NH_2$ of CS generate acid–base pairs along the CS–SGO interface region, which work as facile proton-hopping sites and construct continuous proton transfer pathways. The synthesis process of SHNTs and the

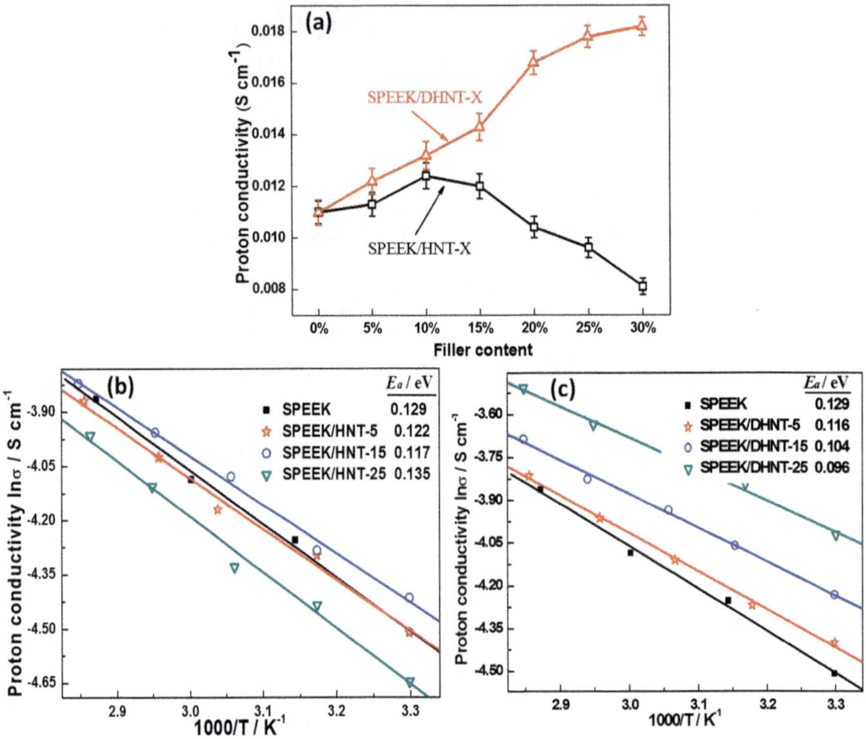

Figure 10.13 (a) The proton conductivity of the as-prepared membranes under 25 °C and 100% RH, (b) and (c) the temperature-dependent conductivity of the SPEEK control and nanocomposite membranes at 100% RH. Reprinted from ref. 70 with permission from John Wiley and Sons. Copyright © 2013 WILEY-VCH Berlag GmbH & Co. KGaA, Weinheim.

promotion of proton transfer within the hybrid membrane are illustrated in Scheme 10.3.

The images of the surface and cross-section of CS–SHNTs-9 are displayed in Figure 10.14. It can be seen that SHNTs display much better dispersion within the CS matrix than that of raw HNTs. The presence of sulfonate polyelectrolyte brushes should be responsible for this observation, which generates strong and extensive electrostatic attractions between the $-SO_3H$ groups of SHNTs and the $-NH_2$ groups of CS. The attractions efficiently reduce the surface tension of SHNTs and improve the interfacial compatibility, endowing SHNTs with uniform dispersion. Meanwhile, the attractions may drive the long brushes to insert into CS chains with the aid of brush motion. In such a way, SHNTs are wholly coated by CS chains and no bare nanotube can be observed on the cross-section of SHNT-filled membranes. However, a high amount of nanotubes would increase the viscosity

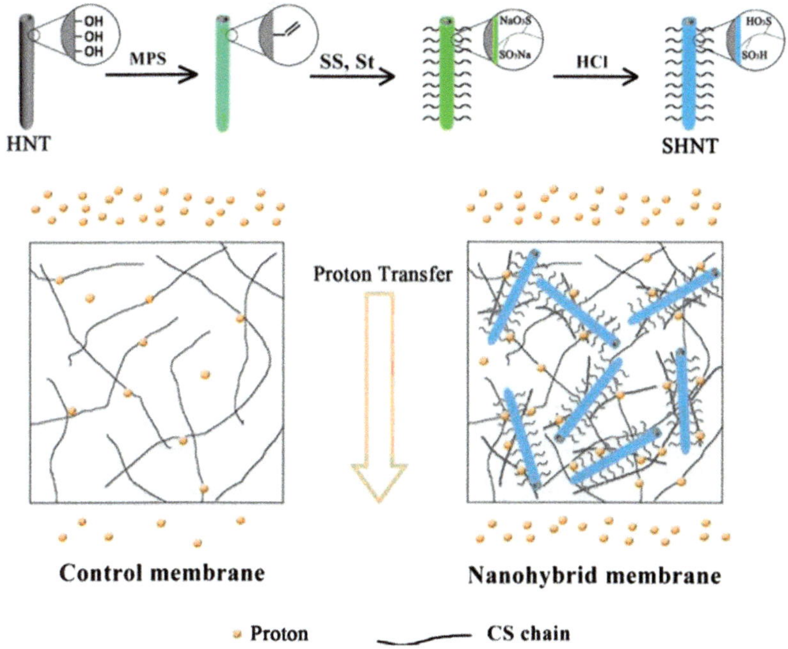

Scheme 10.3 Synthesis process of SHNTs and the microstructure of the CS–SHNT-X hybrid membranes. Reprinted from *Journal of Membrane Science*, **454**, H. Bai *et al.*, Enhanced proton conduction of chitosan membrane enabled by halloysite nanotubes bearing sulfonate polyelectrolyte brushes. Copyright 2014 with permission from Elsevier (ref. 72).

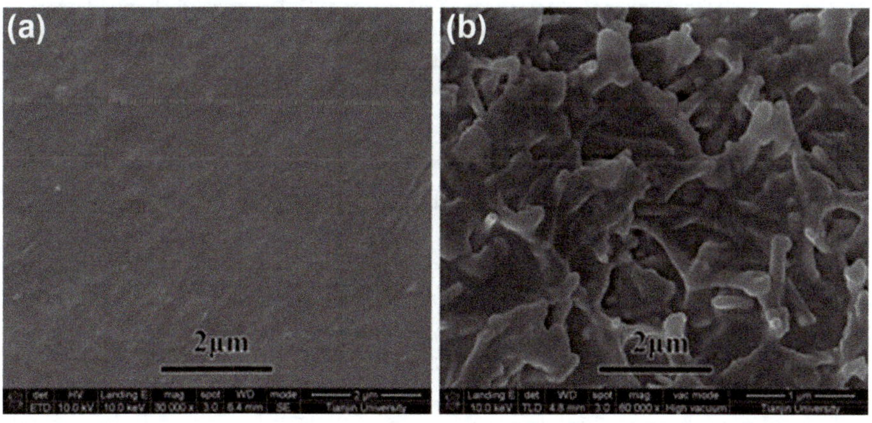

Figure 10.14 Images of the surface (a) and cross-section (b) of CS-SHNT-9. Reprinted from *Journal of Membrane Science*, **454**, H. Bai *et al.*, Enhanced proton conduction of chitosan membrane enabled by halloysite nanotubes bearing sulfonate polyelectrolyte brushes, Copyright 2014 with permission from Elsevier (ref. 72).

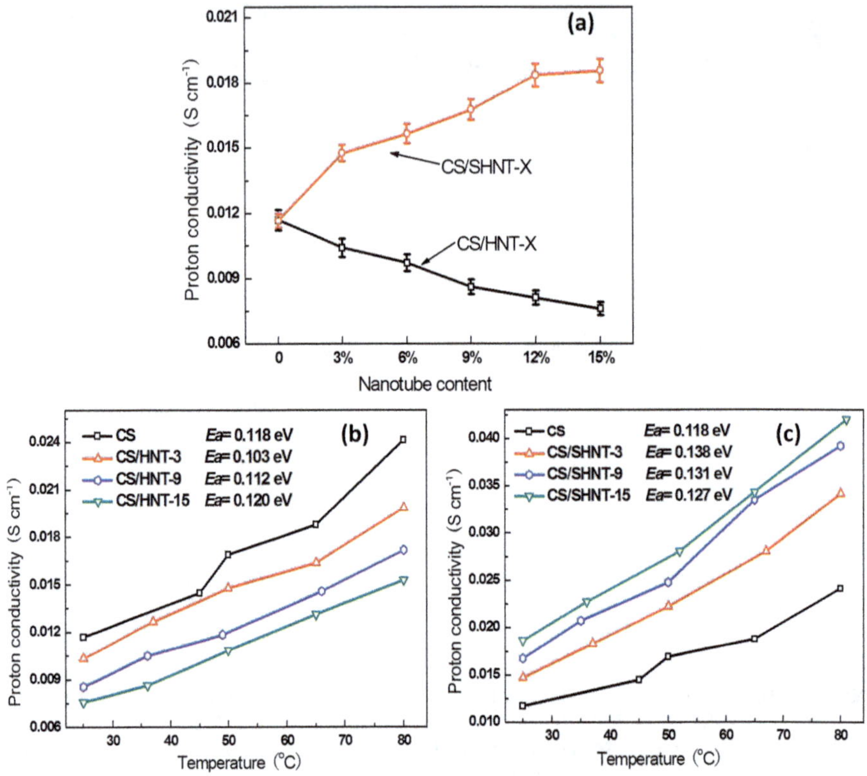

Figure 10.15 (a) The proton conductivity of the as-prepared membranes under 25 °C and 100% RH, (b) and (c) the temperature-dependent conductivity of the CS control and hybrid membranes at 100% RH. Reprinted from *Journal of Membrane Science*, **454**, H. Bai *et al.*, Enhanced proton conduction of chitosan membrane enabled by halloysite nanotubes bearing sulfonate polyelectrolyte brushes, Copyright 2014 with permission from Elsevier (ref. 72).

of the membrane casting solution and affect the mixing of nanotubes with the polymer matrix.

Proton conductivity measurement reveals that SHNT-filled membranes obtain enhanced conductivities compared with HNT-filled and CS control membranes (Figure 10.15). Meanwhile, increasing SHNT content can further elevate the conductivity, and CS–SHNTs-15 achieves the highest conductivity of 0.0186 S cm^{-1} with an enhancement of about 60% when compared to the CS control membrane. Since incorporation of SHNTs leads to a reduction of water uptake and swelling of the hybrid membrane, which would result in reduced vehicle-type transfer, Grotthuss-type transfer should contribute to the enhancement of conductivities.

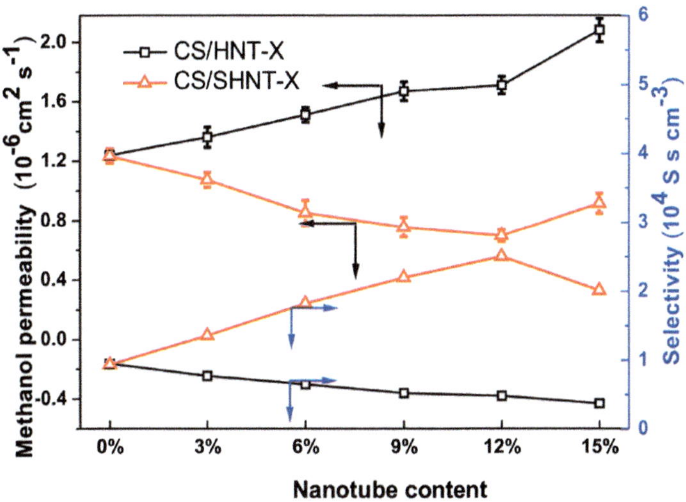

Figure 10.16 Methanol permeability and selectivity of the CS control and nanohybrid membranes. Reprinted from *Journal of Membrane Science*, **454**, H. Bai *et al.*, Enhanced proton conduction of chitosan membrane enabled by halloysite nanotubes bearing sulfonate polyelectrolyte brushes, Copyright 2014 with permission from Elsevier (ref. 72).

The proton transfer mechanisms within the SHNT-filled membranes are shown in Scheme 10.4. Thereinto, SHNTs play at least three roles for enhancing the conductivity: (1) generating acid–base pairs at the SHNTs–CS interface, which work as low-barrier proton-hoping sites; (2) creating continuous pathways along the nanotubes due to their high aspect ratios; and (3) inserting into the CS chains and then forming wider pathways with the aid of the long brushes on SHNTs. In such a way, SHNTs promote the Grotthuss-type transfer of hybrid membrane in an efficient way. Additionally, the strong attractive interactions from SHNTs suppress the chain motion of CS and make the stretching difficult, conferring improved thermal and mechanical stabilities on CS–SHNTs hybrid membranes.

The results of methanol crossover (Figure 10.16) show that embedding 3% SHNTs into the CS matrix reduce the methanol permeability from 1.24×10^{-6} to 1.08×10^{-6} cm^2 s^{-1}, and increasing the SHNT content to 12% reduces the permeability continuously to 0.70×10^{-6} cm^2 s^{-1}. Such phenomena should be attributed to the well-dispersed SHNTs, which on the one hand reduce the chain mobility and free volume, and on the other hand induce tortuous paths for methanol diffusion. Both of these two facts will increase the diffusion resistance for methanol and thus suppress the methanol crossover.

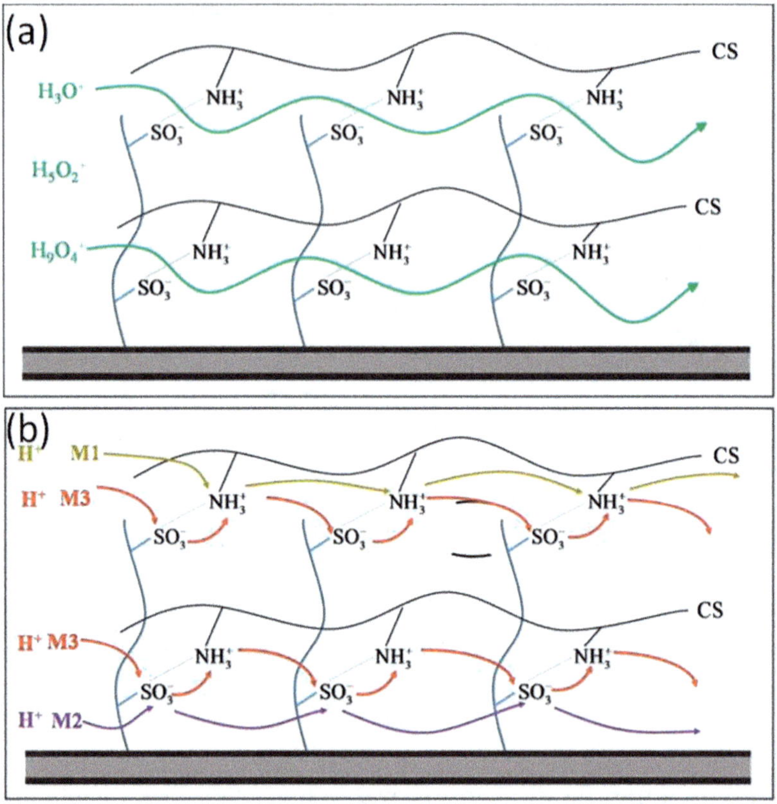

Scheme 10.4 The proton transfer mechanisms in CS–SHNT-X: (a) vehicle-type transfer; and (b) Grotthuss-type transfer. Reprinted from *Journal of Membrane Science*, **454**, H. Bai *et al.*, Enhanced proton conduction of chitosan membrane enabled by halloysite nanotubes bearing sulfonate polyelectrolyte brushes. Copyright 2014 with permission from Elsevier (ref. 72).

10.3 Conclusions

In this Chapter, the general structure and main properties of HNTs are mainly summarized, and an overview of very recent research toward HNTs–polymer for UF and proton exchange applications is provided. Due to the interesting characteristics of HNTs – nanoscale lumen, high aspect ratio, different internal and external surface chemistry, as well as good mechanical strength and thermal stability – HNT-based nanocomposites have gained a growing interest as multi-functional additives for the fabrication of UF membranes as well as PEMs.

The introduction of natural HNTs in the polymer matrix can substantially improve surface hydrophilicity, mechanical properties, water permeability and the antifouling performance of polymer-based composite membranes for UF membranes. As for PEMs, although the proton conductivity values of the HNT-embedded composite membrane are lower than that of Nafion (about 0.1 S cm^{-1}), the efficient promotion of sulfonated nanotubes to proton transfer of proton-conducting materials may offer a promising strategy in the pursuit of PEM with high proton conductivity and to obtain PEMFC with high performance for diverse relevant energy and environmental fields.

However, HNTs are prone to aggregation in the casting solution, forming an uneven dispersion which restricts their further application. Additionally, the compatibility and adhesion strength between the HNTs and polymer are poor, giving rise to a concomitant defect interface which would deteriorate the membrane separation selectivity and mechanical properties of the composite membranes, as well as their antibacterial properties. Therefore, future studies should focus on how to modify HNTs so as to improve their dispersion in solvent and enhance the interface compatibility between HNTs and polymer.

The modification of natural HNTs could not only endow HNT-based nanomaterials with good antibacterial activity, but improve interface compatibility to a great extent, rendering hybrid membranes qualified for UF and PEM applications.

Acknowledgements

This work was supported by National Natural Science Foundation of China (No. 21476215, 21576244, and 21376225).

References

1. M. Du, B. Guo and D. Jia, *Polym. Int.*, 2010, **59**, 574.
2. Y. Lvov and E. Abdullayev, *Prog. Polym. Sci.*, 2013, **38**, 1690.
3. J. Zhu, Y. Wang, J. Liu and Y. Zhang, *Ind. Eng. Chem. Res.*, 2014, **53**, 13711.
4. Y. M. Lvov, D. G. Shchukin, H. Mohwald and R. R. Price, *ACS Nano*, 2008, **2**, 814.
5. P. Luo, Y. Zhao, B. Zhang, J. Liu, Y. Yang and J. Liu, *Water Res.*, 2010, **44**, 1489.
6. J. Zhu, N. Guo, Y. Zhang, L. Yu and J. Liu, *J. Membr. Sci.*, 2014, **465**, 91.
7. L. Yu, Y. Zhang, H. Zhang and J. Liu, *Desalination*, 2015, **359**, 176.
8. H. Yu, Y. Zhang, X. Sun, J. Liu and H. Zhang, *Chem. Eng. J.*, 2014, **237**, 322.
9. Z. Wang, H. Wang, J. Liu and Y. Zhang, *Desalination*, 2014, **344**, 313.
10. Y. Wang, J. Zhu, G. Dong, Y. Zhang, N. Guo and J. Liu, *Sep. Purif. Technol.*, 2015, **150**, 243.
11. J. Zhu, N. Guo, Y. Zhang, L. Yu and J. Liu, *J. Membr. Sci.*, 2014, **465**, 91.
12. A. W. Mohammad, C. Y. Ng, Y. P. Lim and G. H. Ng, *Food Bioprocess Technol.*, 2012, **5**, 1143.
13. M. Safarpour, A. Khataee and V. Vatanpour, *Sep. Purif. Technol.*, 2015, **140**, 32.
14. W. Zhang, Z. Shi, F. Zhang, X. Liu, J. Jin and L. Jiang, *Adv. Mater.*, 2013, **25**, 2071.

15. S. Liang, G. Qi, K. Xiao, J. Sun, E. P. Giannelis, X. Huang and M. Elimelech, *J. Membr. Sci.*, 2014, **463**, 94.
16. J. Zhu, M. Tian, Y. Zhang, H. Zhang and J. Liu, *Chem. Eng. J.*, 2015, **265**, 184.
17. J. Yin and B. Deng, *J. Membr. Sci.*, 2015, **479**, 256.
18. L. Yu, Y. Zhang, Y. Wang, H. Zhang and J. Liu, *J. Hazard. Mater.*, 2015, **287**, 373.
19. D. Rana and T. Matsuura, *Chem. Rev.*, 2010, **110**, 2448.
20. Y. Chang, C.-Y. Ko, Y.-J. Shih, D. Quémener, A. Deratani, T.-C. Wei, D.-M. Wang and J.-Y. Lai, *J. Membr. Sci.*, 2009, **345**, 160.
21. T. Ishigami, K. Amano, A. Fujii, Y. Ohmukai, E. Kamio, T. Maruyama and H. Matsuyama, *Sep. Purif. Technol.*, 2012, **99**, 1.
22. Q.-F. An, W.-D. Sun, Q. Zhao, Y.-L. Ji and C.-J. Gao, *J. Membr. Sci.*, 2013, **431**, 171.
23. J. Wingender, T. R. Neu and H.-C. Flemming, *What are Bacterial Extracellular Polymeric Substances?*, Springer-Verlag, Berlin Heidelberg, 1999.
24. J. Zhang, Y. Zhang, Y. Chen, L. Du, B. Zhang, H. Zhang, J. Liu and K. Wang, *Ind. Eng. Chem. Res.*, 2012, **51**, 3081.
25. M. Ghanbari, D. Emadzadeh, W. J. Lau, S. O. Lai, T. Matsuura and A. F. Ismail, *Desalination*, 2015, **358**, 33.
26. M. Ghanbari, D. Emadzadeh, W. J. Lau, T. Matsuura and A. F. Ismail, *RSC Adv.*, 2015, **5**, 21268.
27. J. Zhu, Y. Zhang, M. Tian and J. Liu, *ACS Sustainable Chem. Eng.*, 2015, **3**, 690.
28. H. Basri, A. F. Ismail, M. Aziz, K. Nagai, T. Matsuura, M. S. Abdullah and B. C. Ng, *Desalination*, 2010, **261**, 264.
29. A. Nguyen, L. Zou and C. Priest, *J. Membr. Sci.*, 2014, **454**, 264.
30. G. Xueli, W. Haizeng, W. Jian, H. Xing and G. Congjie, *J. Membr. Sci.*, 2013, **445**, 146.
31. J. H. Kim, D. C. Choi, K. M. Yeon, S. R. Kim and C. H. Lee, *Environ. Sci. Technol.*, 2011, **45**, 1601.
32. X. Li, R. Pang, J. Li, X. Sun, J. Shen, W. Han and L. Wang, *Desalination*, 2013, **324**, 48.
33. M. J. Hajipour, K. M. Fromm, A. A. Ashkarran, D. Jimenez de Aberasturi, I. R. de Larramendi, T. Rojo, V. Serpooshan, W. J. Parak and M. Mahmoudi, *Trends Biotechnol.*, 2012, **30**, 499.
34. C. Yang, J. Mamouni, Y. Tang and L. Yang, *Langmuir*, 2010, **26**, 16013.
35. J. Dang, Y. Zhang, Z. Du, H. Zhang and J. Liu, *Water Sci. Technol.*, 2012, **66**, 799.
36. F. Perreault, A. Fonseca de Faria and M. Elimelech, *Chem. Soc. Rev.*, 2015, **44**, 5861.
37. L. Karimi, M. E. Yazdanshenas, R. Khajavi, A. Rashidi and M. Mirjalili, *Cellulose*, 2014, **21**, 3813.
38. S. Chernousova and M. Epple, *Angew Chem., Int. Ed. Engl.*, 2013, **52**, 1636.
39. Y. Zhang, Y. Chen, H. Zhang, B. Zhang and J. Liu, *J. Inorg. Biochem.*, 2013, **118**, 59.

40. Y. Chen, Y. Zhang, H. Zhang, J. Liu and C. Song, *Chem. Eng. J.*, 2013, **228**, 12.
41. Y. Chen, Y. Zhang, J. Liu, H. Zhang and K. Wang, *Chem. Eng. J.*, 2012, **210**, 298.
42. L. Duan, Q. Zhao, J. Liu and Y. Zhang, *Environ. Sci.: Water Res. Technol.*, 2015, **1**, 874.
43. L. Yu, Y. Zhang, B. Zhang and J. Liu, *Sci. Rep.*, 2014, **4**, 4551.
44. Q. Zhao, J. Hou, J. Shen, J. Liu and Y. Zhang, *J. Mater. Chem. A*, 2015, **3**, 18696.
45. Z. Cao and Y. Sun, *ACS Appl. Mater. Interfaces*, 2009, **1**, 494.
46. L. Duan, W. Huang and Y. Zhang, *RSC Adv.*, 2015, **5**, 6666.
47. M. Khan, S. Danielsen, K. Johansen, L. Lorenz, S. Nelson and A. Camper, *Biofouling*, 2014, **30**, 153.
48. A. Maartens, P. Swart and E. P. Jacobs, *J. Membr. Sci.*, 1996, **119**, 9.
49. J. Chen, L. Wang and Z. Zhu, *Desalination*, 1992, **86**, 301.
50. Q. Zhao, C. Liu, J. Liu and Y. Zhang, *RSC Adv.*, 2015, **5**, 38646.
51. K. D. Kreuer, *Chem. Mat.*, 1996, **8**, 610.
52. C. A. Wraight, *Biochim. Biophys. Acta*, 2006, **1757**, 886.
53. K. D. Kreuer, S. J. Paddison, E. Spohr and M. Schuster, *Chem. Rev.*, 2004, **104**, 4637.
54. D. Marx, *Chem. Phys. Chem.*, 2006, **7**, 1848.
55. M. Winter and R. J. Brodd, *Chem. Rev.*, 2004, **104**, 4245.
56. M. A. Hickner, H. Ghassemi, Y. S. Kim, B. R. Einsla and J. E. McGrath, *Chem. Rev.*, 2004, **104**, 4587.
57. T. Ous and C. Arcoumanis, *J. Power Sources*, 2013, **240**, 558.
58. H. Zhang and P. Shen, *Chem. Rev.*, 2012, **112**, 2780.
59. B. G. Choi, Y. S. Huh, Y. C. Park, D. H. Jung, W. H. Hong and H. S. Park, *Carbon*, 2012, **50**, 5395.
60. H. Liao, K. Zhang, G. Xiao and D. Yan, *J. Membr. Sci.*, 2013, **447**, 43.
61. B. P. Tripathi and V. K. Shahi, *Prog. Polym. Sci.*, 2011, **36**, 945.
62. J. Matusik and A. Wścisło, *Appl. Clay Sci.*, 2014, **100**, 50.
63. H.-L. Lin, T. L. Yu, C. H. Huang and T.-L. Lin, *J. Polym. Sci., Part B: Polym. Phys.*, 2005, **43**, 3044.
64. T. Schultz, S. Zhou and K. Sundmacher, *Chem. Eng. Technol.*, 2001, **24**, 1223.
65. Y.-L. Liu, *Polym. Chem.*, 2012, **3**, 1373.
66. H. Zhang, C. Ma, J. Wang, X. Wang, H. Bai and J. Liu, *Int. J. Hydrogen Energy*, 2014, **39**, 974.
67. M. L. Einsla, Y. S. Kim, M. Hawley, H.-S. Lee, J. E. McGrath, B. Liu, M. D. Guiver and B. S. Pivover, *Chem. Mat.*, 2008, **20**, 5636.
68. J. Wang, Z. Zhang, X. Yue, L. Nie, G. He, H. Wu and Z. Jiang, *J. Math. Chem. A*, 2013, **1**, 2267.
69. K. D. Kreuer, *Chem. Mat.*, 1996, **8**, 610.
70. H. Zhang, T. Zhang, J. Wang, F. Pei, Y. He and J. Liu, *Fuel Cells*, 2013, **13**, 1155.
71. B. P. Tripathi, M. Schieda, V. K. Shahi and S. P. Nunes, *J. Power Sources*, 2011, **196**, 911.

72. H. Bai, H. Zhang, Y. He, J. Liu, B. Zhang and J. Wang, *J. Membr. Sci.*, 2014, **454**, 220.
73. M. L. D. Vona, Z. Ahmed, S. Bellitto, A. Lenci, E. Traversa and S. Licoccia, *J. Membr. Sci.*, 2007, **296**, 156.
74. B. P. Tripathi and V. K. Shahi, *Prog. Polym. Sci.*, 2011, **36**, 945.
75. Y. He, J. Wang, H. Zhang, T. Zhang, B. Zhang, S. Cao and J. Liu, *J. Math. Chem. A*, 2014, **2**, 9548.
76. Y. Liu, J. Wang, H. Zhang, C. Ma, J. Liu, S. Cao and X. Zhang, *J. Power Sources*, 2014, **269**, 898.
77. J. Wang, Y. He, L. Zhao, Y. Li, S. Cao, B. Zhang and H. Zhang, *J. Membr. Sci.*, 2015, **482**, 1.
78. J. Wang, H. Bai, H. Zhang, L. Zhao, H. Chen and Y. Li, *Electrochim. Acta*, 2015, **152**, 443.

CHAPTER 11

Rubber Functionalized with Halloysite Loaded with Antioxidants and Antibacterials

LIQUN ZHANG*[a,b,c] AND JIAJIA XUE[a,b,c]

[a]State Key Laboratory of Organic-Inorganic Composites, Beijing University of Chemical Technology, Beijing, 100029, China; [b]Beijing Laboratory of Biomedical Materials, Beijing University of Chemical Technology, Beijing, 100029, China; [c]Key Laboratory of Beijing City for Preparation and Processing of Novel Polymer Materials, Beijing University of Chemical Technology, Beijing, 100029, China
*E-mail: zhanglq@mail.buct.edu.cn

11.1 Introduction

Nanocomposites are defined as composite materials with the filler component having at least one dimension in the nanoscale.[1] Rubber nanocomposite is a cross-linked elastomeric polymer with dispersed nanoparticles in the rubber matrix.[2] The nanoparticles contribute reinforcement and functional features to the pristine rubber. Halloysite (HNTs) is an environmentally-friendly natural nanomaterials.[3] Its structure and properties have been described in detail in Chapter 5. Loading the hollow tubes' lumens with chemicals, including bioactive molecules (self-healing, anticorrosion, antimicrobial agents, drugs, *etc.*),[4] and compounding the loaded tubes into the

RSC Smart Materials No. 22
Functional Polymer Composites with Nanoclays
Edited by Yuri Lvov, Baochun Guo and Rawil F. Fakhrullin

Published by the Royal Society of Chemistry, www.rsc.org

rubber matrix allows a sustained release of the loaded substances, providing these nanocomposites with new smart properties. Aging and bacterial colonization are two main problems that affect the long-term usage of rubber. Compounding the HNTs loaded with antioxidants or antibacterials into rubber matrix allow for material formulations with sustained release which could be tuned by the tube end-stoppers or the clogging of polymeric matrix. The fabrication and performance of various HNTs–rubber composites and those with loaded HNTs for sustained release of antioxidants or antibacterials will be reviewed in this chapter.

11.2 Rubber–Halloysite Nanocomposites

Various kinds of rubber–HNTs nanocomposite systems have been fabricated and investigated. The rubber–HNTs nanocomposites have been reported to possess improved mechanical, dynamic mechanical, thermal and other functional properties endowed by the addition of pure HNTs or functional agent-loaded HNTs.[5] Different processing techniques, including melt mixing, latex blending and solution blending, are generally adopted to fabricate rubber–HNTs nanocomposites depending on the target properties of the rubber matrix. Due to the relatively low density of surface hydroxyl groups, the inter-tubular interactions for HNTs are relatively weak, resulting in higher dispersity of HNTs in various rubber matrices.[6] Better dispersion of HNTs in the rubber matrix can be achieved by surface functionalization of HNTs with organic modifiers. Strong interfacial interaction between rubber chains and HNTs could be obtained *via* interfacial covalent bonds, hydrogen bonds or multiple interactions, which improves the dispersion of HNTs and leads to much enhanced overall performances.[7]

According to the targeted applications, the rubber–HNTs nanocomposites can be mainly classified into two aspects: industrial products and biomedical products. HNTs derived from natural resources are environmentally friendly and cost-effective. Such promising advantages promote their application as nanofillers for rubbers in developing commercial industrial rubber products. Alternatively, because of the good biocompatibility of HNTs, the rubber –HNTs nanocomposites are also promising in biomedical applications by incorporating HNTs with biomedical elastomers like polyurethane (PU) and silicone rubber. The properties of rubber–HNTs nanocomposites are mainly affected by two important factors: the dispersion of HNTs in the rubber matrix and the compatibility between HNTs and the rubber matrix.

11.2.1 Rubber–HNTs Nanocomposites for Engineering Applications

Diverse rubber–HNTs nanocomposites have been prepared with many kinds of rubber matrices[7] such as natural rubber (NR), styrene–butadiene rubber (SBR), ethylene propylene diene monomer (EPDM), nitrile rubber (NBR), polybutadiene rubber (BR), fluoroelastomers (FKM), PU rubber and some other thermoplastic elastomers.

NR is widely used in both industrial and biomedical areas due to its high green strength and strain-induced crystallization behavior.[8] A green composite based on natural HNTs and NR was prepared, and the ultimate properties of the composite were affected by both the type of NR and the compounding method. HNTs were added to two types of NR (Standard Malaysian Rubber (SMR) L and epoxidized natural rubber (ENR) 50).[9] The addition of HNTs caused a decrement in fatigue life for both nanocomposites at any extension ratio. Addition of more HNTs caused a decrement of stress for HNTs–SMR L nanocomposites, and *vice versa* for HNTs–ENR 50 nanocomposites at any strain. Two different compounding methods, *i.e.*, mechanical and solution mixing, were used in preparation of HNTs–NR nanocomposites with various contents of HNTs.[10] Nanocomposites prepared by the solution blending method exhibited higher tensile strength, tensile modulus and fatigue life than those achieved by the mechanical compounding method, because of the better dispersion of HNTs resulting from the solution blending method.

In order to improve the rubber–filler interaction and the dispersion of fillers, HNTs were surface modified by different methods. HNTs were modified with silane coupling agent bis (triethoxysilylpropyl)-tetrasulphide and then compounded with the NR matrix by mechanical mixing on a laboratory-sized two-roll mill.[11] It was observed that the reinforcing efficiency of HNTs was superior to commercial silica coupled with the same amount of silane coupling agent. The tensile strength of the modified HNTs–NR composite was 15 MPa while that of the modified silica–NR composite was 12 MPa. HNTs were also surface modified by plasma polymerization to change their surface polarity and chemistry and then used in a NR–BR blend in the presence of carbon black.[12] The surface modification resulted in a higher bound rubber content and lower Payne effect, indicating better filler–polymer interaction.

SBR is a widely used synthetic rubber for various applications such as pneumatic tires. SBR and 40 phr of HNTs were compounded with rubber additives with a two-roll mill, and methacrylic acid (MAA) was used to improve the performance of the nanocomposites by direct blending.[13] The mechanical properties, especially the modulus and hardness, were significantly enhanced by the incorporation of HNTs and the increase in MAA content. The tensile strength of the nanocomposites could be improved to 14.0 MPa, which was attributed to the strong interfacial bonding between HNTs and the rubber matrix. To further improve the performance of the SBR–HNTs nanocomposites, sorbic acid (SA) was added to the nanocomposite by direct blending as an *in situ* modifier.[14] The tensile strength of the nanocomposite increased from 4.1 MPa to 18.6 MPa by increasing the SA content from 0 phr to 15 phr. The significantly improved mechanical properties were correlated with the largely improved interfacial bonding between the HNTs and rubber and the enhanced dispersion of the HNTs. Interfacial modification was also realized using functional thiol ionic liquids as interfacial modifiers for SBR–HNTs composites.[15] Substantiated hydrogen bonding between the HNTs and ionic liquids largely facilitated the dispersion of HNTs in the rubber matrix. The grafting reactivity of HNTs onto SBR chains was mainly ascribed to the thiol–ene reaction. The mechanical properties of SBR–HNTs vulcanizates were largely improved and better than those of the 3-mercaptopropyl trimethoxysilane modified system. In addition, other fillers were also combined with HNTs and then compounded in SBR to study the performance of the nanocomposites. A hybrid colloid of tannic acid functionalized graphene and HNTs was co-coagulated with SBR latex to prepare elastomeric composites.[16] The hybrid fillers dispersed homogeneously in the SBR matrix. The well-developed hybrid filler network and the enhanced interfacial interaction were considered to be responsible for the significant synergetic effects in the reinforcement of the rubber matrix.

EPDM is an essentially saturated polyolefin rubber which has attracted much attention for outdoor applications. To improve the properties of EPDM, EPDM–HNTs nanocomposites were prepared with HNTs up to 100 phr.[17] The dispersion of HNTs in the EPDM matrix was homogeneous and there were also interfacial interactions between HNTs and EPDM. These were the main reasons for the simultaneous improvement in tensile strength, stiffness and ductility of EPDM–HNTs nanocomposites, even though, due to the fact that EPDM did not bear any polar groups in its backbone, EPDM and HNTs were incompatible. To improve the compatibility, EPDM grafted with maleic anhydride (MAH-*g*-EPDM) was prepared by peroxide-initiated melt grafting using a HAAKE internal mixer.[18] Due to the creation of an interphase which helped to make stronger interfacial interactions between EPDM and HNT, the tensile properties of the composite enhanced as well as the swelling resistance. HNTs were also modified by γ-methacryloxypropyl trimethoxysilane to improve their dispersion in EPDM.[19] The EPDM–HNTs nanocomposites were prepared by mixing 0, 5, 10 and 30 phr of modified HNTs with EPDM on a two-roll mill. The tensile strength and tensile modulus of the nanocomposites were higher than those for the control sample, although the elongation at break decreased a little after modification of the HNTs. The tensile strength increased from 1.3 MPa for neat EPDM rubber to 4.81 MPa for the nanocomposites loaded with 30 phr modified HNTs. To study the effect of HNTs on the properties of EPDM composites reinforced with commercial fillers like silica or calcium carbonate ($CaCO_3$), partial silica or $CaCO_3$ were replaced by HNTs.[20] A higher tensile strength and E_b of EPDM–silica–HNTs were realized compared with EPDM–silica nanocomposites and, as a result, HNTs can be used as a co-filler with silica to achieve better dispersion of silica inside the EPDM. Similarly, the tensile properties of EPDM–$CaCO_3$–HNTs composites continuously increased with an increase of HNT loading from 0 to 30 phr because of the better dispersion of HNTs inside the EPDM matrix. In other work, microcellular elastomeric EPDM–HNTs nanocomposite foam was prepared when the foaming was processed in supercritical CO_2.[21] Such a nanocomposite foam can potentially be used in a variety of industrial applications involving gaskets and seals for automotives and electrical enclosures.

PUs are the only class of polymers that display thermoplastic, elastomeric and thermoset behavior depending on their chemical and morphological makeup. PU elastomer is widely used in industrial and biomedical products. Nanocomposites comprised of millable polyurethane (MPU) and very low loadings of HNTs have been prepared (Figure 11.1).[6] The composites exhibited high tensile strength and elongation at break, extending the applications of HNT-based nanocomposites in hot melt adhesives. Hybrid nanofillers, composed of acid-treated multi-walled carbon nanotubes (a-CNTs) and silane-treated HNTs, were incorporated in PU elastomers by a simple solution-casting method.[22] A synergistic reinforcement effect of CNTs and HNTs within the hybrid remarkably improve the strength and toughness of the PU composite.

Other rubber–HNTs composites were also fabricated. The incorporation of HNTs clearly improved the thermal stability and reduced the flammability and fire hazard of the NBR elastomers.[23,24] NBR–carbon black–HNTs hybrid composites showed an increasing trend of tensile strength with the increase of HNT content, which was mainly influenced by the NBR–carbon

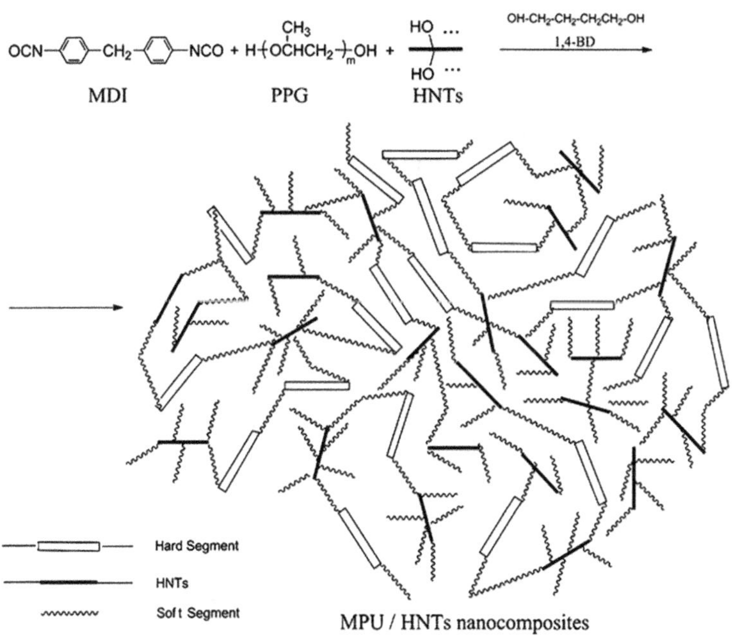

Figure 11.1 Reaction scheme for the synthesis of elastomeric gum polyurethane nanocomposites. Reproduced from ref. 6 with permission from the Royal Society of Chemistry.

black–HNTs interaction and the intercalation of rubber and carbon black into the tubules.[25] By incorporating HNTs in the fluoroelastomers, the mechanical, dynamic mechanical and thermal properties of fluoroelastomers were improved.[26]

11.2.2 Elastomer–HNTs Nanocomposites for Biomedical Applications

Elastomers for biomedical applications can be tracked back to the time when the rubber industry began to develop. Various kinds of elastomers have been used as clinical products. Elastomeric nanocomposites with adjustable properties to satisfy clinical needs have received a lot of attention recently.[27] The specific requirements of bioelastomers include biocompatibility, mechanical properties, biodegradability, manufacturability, sterilizability and so on.[28] HNTs have been verified to possess low cytotoxicity and good biocompatibility. It has been demonstrated that HNTs improve biological cell proliferation. In addition, a study with a mouse primary peritoneal macrophage model demonstrated the anti-inflammatory activity of HNTs. Such activity increased with lumen volume and space, and decreased with average pore size.[29] Bionanocomposites consisting of HNTs have been reported.[30,31] The potential benefits of using nanotubular halloysite in bionanocomposites are not only to maintain the initial mechanical compliance and slow down the degradation rate *in vivo*, but also to achieve an ideal degradation profile in terms of mechanical strength, which is a reasonable match to the healing profile of injured tissue.[32] Poly (glycerol sebacate) (PGS) is a promising elastomer for use in soft tissue engineering. PGS–HNTs nanocomposites were synthesized and investigated.[32] The good compliance, enhanced stretchability, stable mechanical behavior over an extended period and reduced degradation rates make the composites promising candidates for application in soft tissue engineering.

Although rubber–HNTs composites have been widely studied, functionalized halloysite as the delivery vehicle in rubber composite systems was rarely reported. For industrial and biomedical rubber products, antioxidative and antibacterial effects are highly desired. The application of HNTs as a nanocarrier of antioxidants and antibacterials for rubber composites will be reviewed in the following section.

11.3 Antioxidative Rubber Composites

11.3.1 Aging of Elastomers and Their Antioxidative Progress

In the industrial field, aging of rubber is usually caused by high temperature, oxygen, oil, acids, alkali or other active chemicals. Thermal-oxidative aging is one of the most common types of rubber aging, leading to the deterioration of the properties of the rubber.[33] Thus, antioxidant is an essential additive for rubber products to prolong their lifetime. However, conventional antioxidants are mainly small molecules which are easy to volatilize at high temperature during polymer processing and to bloom to the surface of rubber products during long-term service. This results in the formation of microdefects inside rubber composites, the loss of antioxidative activity and environmental pollution. Therefore, adoption of antioxidants with higher molecular weight or immobilization of antioxidant onto the surface of nanofillers are proposed and have been under research to overcome the above problems.[34]

Immobilization of antioxidant onto the surface of a nanofiller is a useful way to restrain the migration and extractability of the antioxidants and thus maintains the antioxidative ability of the rubber composite. Different kinds of silica functionalized with antioxidant were synthesized through the reaction between precipitated silica and antioxidant coupling agent.[34,35] After incorporating the above modified silica into SBR, the composites showed greatly improved mechanical properties while the stability of the composites in thermo-oxidative aging and damp heat aging was also improved. Antioxidant 2-mercaptobenzimidazole (MB) was chemically bonded onto the surface of silica particles *via* chlorosilane coupling agent to obtain silica-supported antioxidant SiO_2-s-MB.[36,37] The SiO_2-s-MB particles were homogeneously dispersed in the SBR matrix. The antioxidative efficiency of SBR–SiO_2-s-MB was superior to that of SBR–MB. Similarly, the nanosilica functionalized with antioxidant, 2-*tert*-butyl-6-(3-*tert*-butyl-2-hydroxy-5-methylbenzyl)-4-methyphenyl acrylate (GM), was prepared using antioxidant coupling agent and then compounded with NR.[38] The NR composite filled with GM functionalized nanosilica exhibited excellent extraction resistance. The corresponding aging coefficient and oxidation induction time increased compared with NR–silica mixed with GM.

In the biomedical field, elastomer materials containing small molecule antioxidants are frequently used in medical devices like catheters, angioplasty balloons, stents, *etc.* To improve the anti-aging properties of the medical device.[39] For example, PU is a very broad class of material widely used in the medical field.[40] Irganox 3114® polymorphs, a phenolic antioxidant (Irganox 3114®), were used as an additive in a medical grade PU.[41] Moreover, for the regeneration of tissue, free radicals are present to a great extent in all types of trauma.[42] The process of wound healing can be hampered by the presence of free radicals, therefore, scavenging free radicals can prevent further damage to cells surrounding the wound and consequently increase the rate of the healing process.[43] Hence, antioxidants with radical scavenging ability are inevitably needed in elastomers which come into contact with the wound. The antioxidant added in the biomedical elastomer could also adjust the biodegradation rate of the polymer *in vivo* by capturing oxygen radicals that would otherwise cause polymer chain scission and cross-linking.[44] However, the small molecular antioxidant undergoes recrystallization into a more stable polymorphic form after a period of time at a temperature near the physiological temperature. The blooming of the small molecular antioxidant onto the surface of products will increase the inflammatory risks and promote bacterial adhesion which are harmful to the patients.[45] The blooming and recrystallization of antioxidants will reduce the biocompatibility of the material. Therefore, anti-blooming aging resistant elastomer composites with sustained antioxidant release are highly desired.

11.3.2 Halloysite Loaded with Antioxidants

Halloysite mostly displays a hollow tubular morphology formed of rolling layers, which could be employed for the delivery of antioxidants. One interesting thing is that it was reported that HNT themselves showed anti-inflammatory and antioxidant abilities.[46] Halloysite surfaces favored the stabilization of radical species (R•) and their oxidation to RO•. Antioxidant-loaded HNTs showed a sustained release profile. Resveratrol is a polyphenolic antioxidant but its usage is limited because of its low solubility in water.[47] Resveratrol-loaded HNTs showed slow 48 hour release with a constant release rate with time. Additional functionalization for controlling the release rate was realized using layer-by-layer polyelectrolyte multilayer coating onto the tube external surface and a further decrease of the release rate was demonstrated. A natural antioxidant, quercetin, was loaded in halloysite nanotubes.[48] The high surface energy of the halloysite resulted in strong bonding of the additive to the surface. Thyme essential oil, an active antioxidant agent, was encapsulated into HNT using a vacuum process to sustain the release rate and to solidify the thyme oil (TO) from a liquid state.[49] Moreover, the TO-loaded HNT capsules (TO–HNTs capsules) were coated with Eudragit EPO polymer to avoid burst release and to prolong the release time. TO–HNTs capsules exhibited excellent antioxidant activity and a sustained TO release profile. Based on these results, halloysite clay nanotubes are proposed as novel natural antioxidant carriers.

11.3.3 Rubber Functionalized with Antioxidant-Loaded Halloysite

As discussed above, the sustained release of antioxidant is needed to realize a sustained and on demand antioxidative ability for a rubber system, regardless of whether it is for application in the industrial or the biomedical field. To date, only a few works directly adopt antioxidant-loaded HNTs as nanofillers in rubber–HNTs composites for sustained antioxidant function.

11.3.3.1 Highly Aging-Resistant SBR Elastomers with Antioxidant-Loaded Halloysites

A novel aging-resistant SBR composite was prepared using antioxidant *N*-isopropyl-*N*′-phenyl-*p*-phenylenediamine (4010NA)-loaded HNTs as fillers for SBR.[50] The loading procedure of antioxidant into HNTs and surface

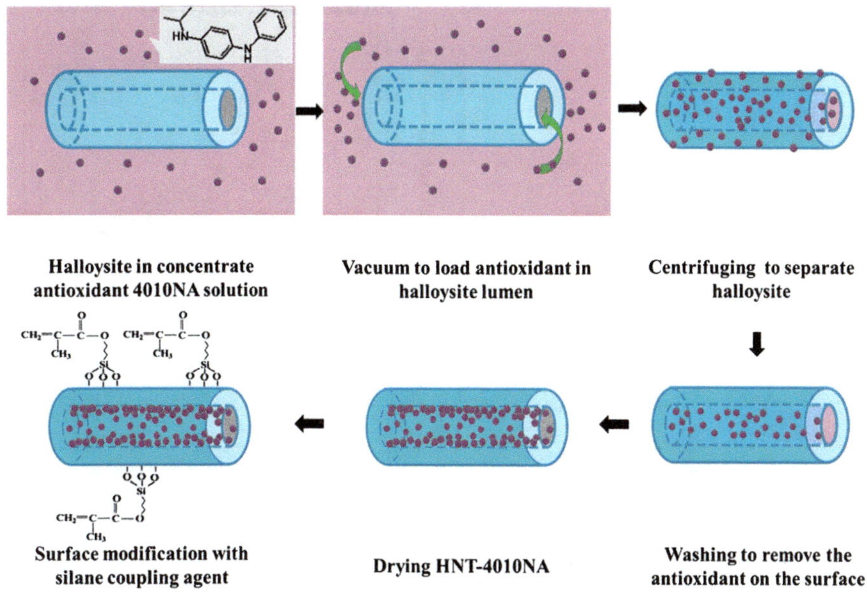

Figure 11.2 Loading procedure of antioxidant into halloysite and surface modification with silane coupling agent. Reproduced from ref. 6 with permission from the Royal Society of Chemistry and reprinted with permission from Y. Fu, D. Zhao, P. Yao, W. Wang, L. Zhang and Y. Lvov, *ACS Appl. Mater. Inter.*, 2015, 7, 8156 (ref. 50). Copyright (2015) American Chemical Society.

modification with silane coupling agent are shown in Figure 11.2. HNTs were mixed thoroughly with antioxidant in acetone. Then the mixture was placed in a vacuum chamber for 30 min, then at atmospheric pressure for 15 min. This vacuum cycle was repeated three times. During this process, air inside the tubes' lumen was removed and replaced by antioxidant solution. Then the HNTs were washed with deionized water to remove the antioxidant attached on the surface or intercalated into the multilayer walls. The antioxidant-loading efficiency was analyzed by TGA, and the maximum loading was 8.1 wt%. Moreover, to improve the dispersion of HNTs in the SBR matrix, the antioxidant-loaded HNTs were further dispersed in ionized water and then modified by silane coupling agent KH570. During loading, modification and compounding with the rubber matrix, HNTs did not change their tubular structure and no chemical reaction between the antioxidant and HNT was identified.

The antioxidant release kinetics in both water and cyclohexane were analyzed by ultraviolet visible spectrophotometry. The initial burst release was acceptable to prolong the aging-induction of the rubber. The rate of antioxidant release in cyclohexane was much faster than in water (3 and 6 h, respectively), attributed to its higher solubility in cyclohexane (Figure 11.3a).

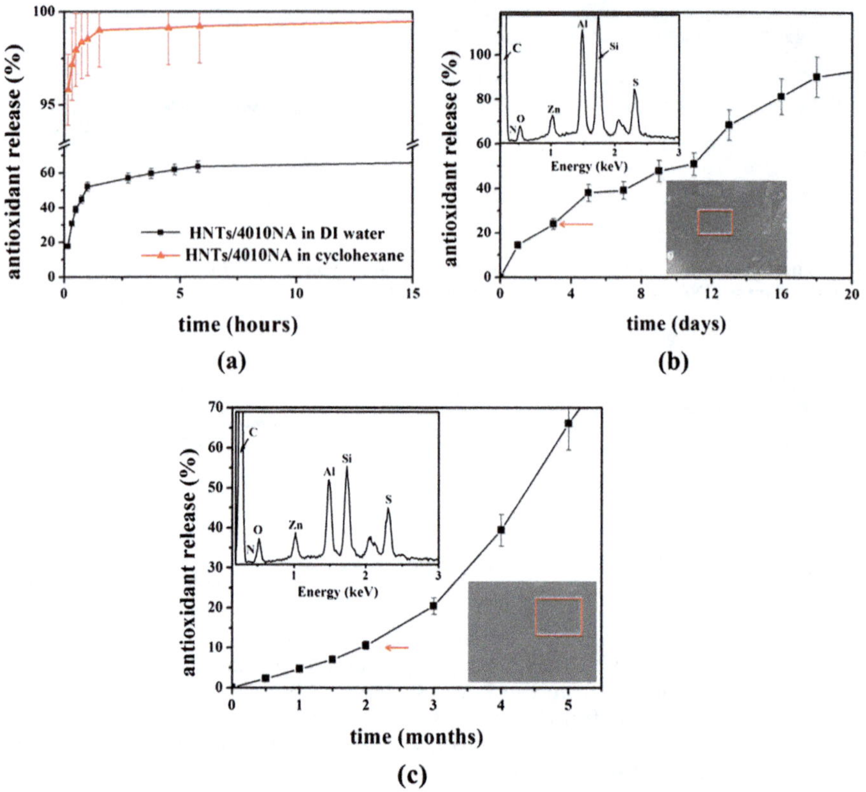

Figure 11.3 Antioxidant 4010NA release from (a) loaded halloysite in water and cyclohexane; halloysite rubber composite at (b) 90 °C and (c) 25 °C. The inset images are the SEM images and the EDS spectra of the selected area. Reprinted with permission from Y. Fu, D. Zhao, P. Yao, W. Wang, L. Zhang and Y. Lvov, *ACS Appl. Mater. Inter.*, 2015, 7, 8156 (ref. 50). Copyright (2015) American Chemical Society.

The rate of antioxidant released from HNTs in rubber composites was calculated based on the elemental amounts obtained by energy dispersive spectrometry measurement. The antioxidant release kinetics in rubber composite at 90 °C are shown in Figure 11.3b. The antioxidant-loaded HNTs showed a sustained release profile with 14.4% release after one day and 90.2% release after 18 days at 90 °C. As shown in Figure 11.3c, the rate of antioxidant release was much slower at 25 °C; about 7%, 11%, 21%, 39% and 66% after 1.5, 2, 3, 4 and 5 months, respectively. According to the extrapolation, 90% release of antioxidant was achieved within nine months. Such extended antioxidant release was based on the formation of cross-linked rubber-clogs that sealed HNT tube endings effectively.

To investigate the influence of antioxidant-loaded HNTs on the morphological, mechanical and anti-aging properties of the SBR–HNTs nanocomposites, different rubber composites were prepared. Three controlled samples, pure SBR (SBR#1), SBR with only 27 wt% HNTs (SBR#2) and SBR with only

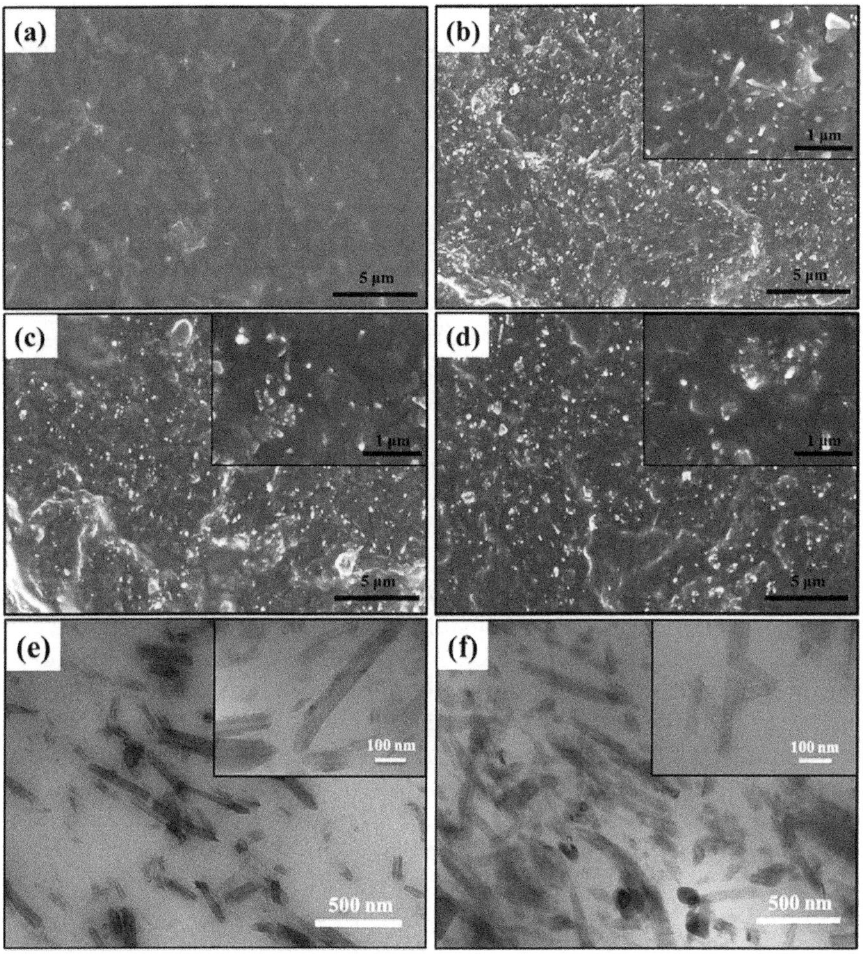

Figure 11.4 SEM images of SBR–HNTs composites with different formulas: (a) SBR#1 pure rubber, (b) SBR#2 pristine halloysite composites, (c) SBR#4 with 4010NA-loaded halloysite, and (d) SBR#5 with loaded halloysite plus admixed free antioxidant. TEM images of (e) SBR#2 and (f) SBR#5 ultrathin section surface. Reprinted with permission from Y. Fu, D. Zhao, P. Yao, W. Wang, L. Zhang and Y. Lvov, *ACS Appl. Mater. Inter.*, 2015, 7, 8156 (ref. 50). Copyright (2015) American Chemical Society.

1.0 wt% 4010NA antioxidant (SBR#3), were prepared. Two designed functional materials containing SBR filled with 27 wt% antioxidant-loaded HNTs (SBR#4) and the same material but with an extra 1.0 wt% of free antioxidant added (SBR#5) were prepared. In SBR#5, the amount of antioxidant agent dispersed in the rubber without blooming defects was tripled compared with the blooming concentration threshold of antioxidant admixed in the rubber matrix. From the SEM and TEM images of these SBR–HNTs nanocomposites (Figure 11.4), the dispersion of silane-modified HNTs was excellent and the tubular shape of HNTs embedded in the rubber matrix was preserved. Such an excellent dispersion of the modified HNTs was beneficial for the application of HNT in functional rubber composite at low cost.

The aging resistance of HNT-embedded rubber composite was studied by comparing their mechanical properties before and after thermal-oxidative aging. The tensile strength of the composite filled with pristine HNTs (SBR#2) was almost two-times that of the pure SBR due to the reinforcement by HNTs. The mechanical properties of all rubber composites after aging are shown in Figure 11.5. The SBR#5 containing 2.2 wt% antioxidant-loaded HNTs with 1.0 wt% additional antioxidant showed the best performance. Even after aging for 168 h at 90 °C, the elongation at break was almost the same while the hardness increased least. The sample (SBR#1) without antioxidant became too brittle to test its mechanical properties. This indicated the great aging-resistance of SBR#5 was achieved due to the addition of loaded HNTs for sustained release of antioxidant. HNTs–4010NA can not only play the part of reinforcing filler but also improve the aging-resistance of the designed rubber composite.

In conclusion, silanized antioxidant-loaded HNTs had favorable dispersibility in the SBR matrix, which not only improved the mechanical properties of the matrix but also achieved sustained release of antioxidant in the composite. By utilizing these modified HNTs, the antioxidant concentration in the rubber could be tripled without blooming. According to the extrapolation estimated that it would take, 90% release of the antioxidant was achieved within nine months. After thermal-oxidative aging at 90 °C for seven days, the sample with antioxidant-loaded HNTs showed well retained mechanical properties. These antioxidant-loaded HNTs may be a generic solution towards highly aging-resistant elastomers.

11.3.3.2 Halloysite Immobilized with Antioxidant and Its Rubber Composites

In another work, to strengthen the interaction between HNTs and an NR matrix, and to avoid the volatility and migration of antioxidant, MB was chemically bonded onto the surfaces of HNTs to form HNTs-s-MB and then incorporated in the NR matrix.[51] HNTs-s-MB was prepared by the reaction of MB and chlorosilane modified HNTs (m-HNTs), as shown in Figure 11.6. From the change of the characteristic absorption peaks in FTIR of HNTs before and after being modified with MB, MB was chemically bonded onto the surface of HNTs through chemical linkages by γ-chloropropyl triethoxysilane (CTS). From XPS spectra, the N1s peak was visible in HNTs-s-MB while no obvious N1s peak was detected in m-HNTs. This observation was considered by the authors as the implication for the bonding of MB onto the surface of HNTs.

The morphologies of HNTs before and after modification were investigated by SEM, as shown in Figure 11.7. It was found that the pristine HNTs

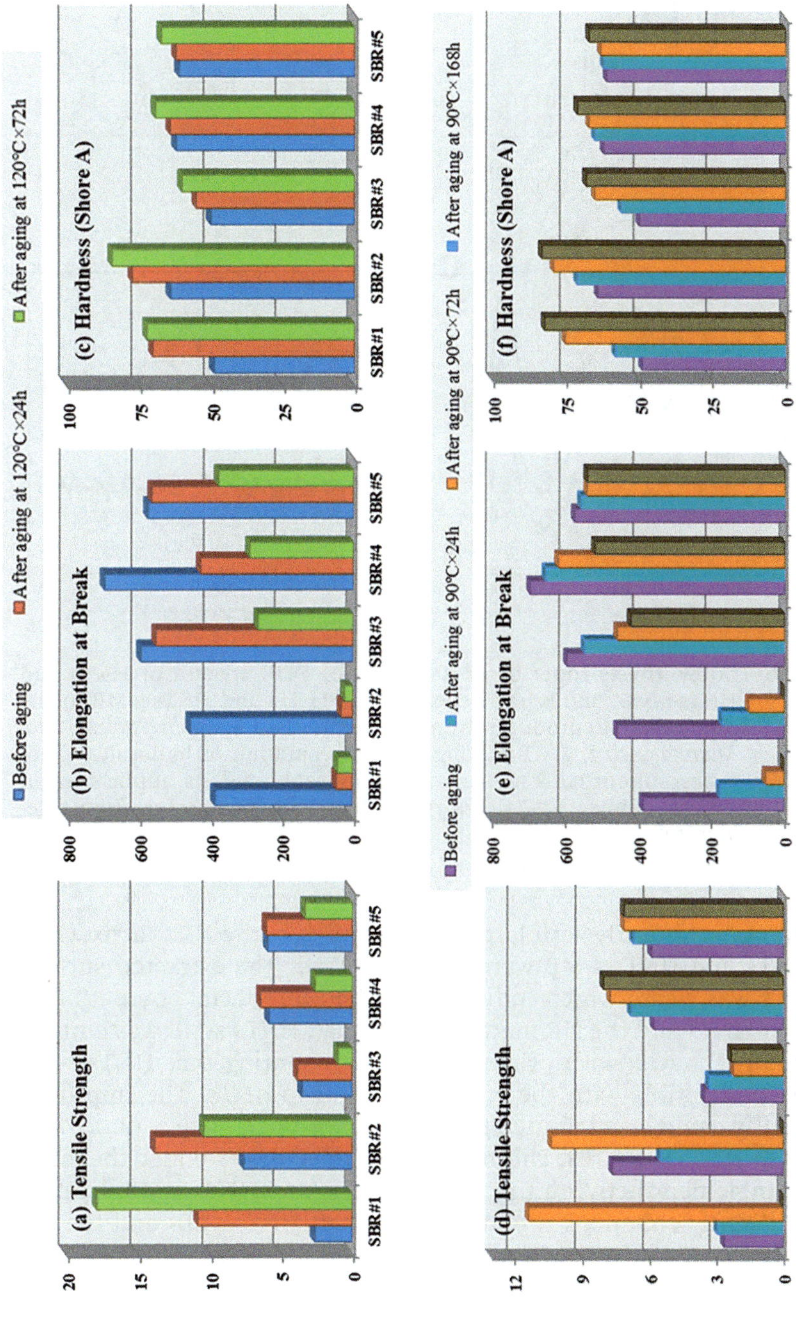

Figure 11.5 Mechanical properties of SBR–HNTs nanocomposites before and after aging at (a–c) 120 °C and (d–f) 90 °C (experimental error 5%). Reprinted with permission from Y. Fu, D. Zhao, P. Yao, W. Wang, L. Zhang and Y. Lvov, *ACS Appl. Mater. Inter.*, 2015, 7, 8156 (ref. 50). Copyright (2015) American Chemical Society.

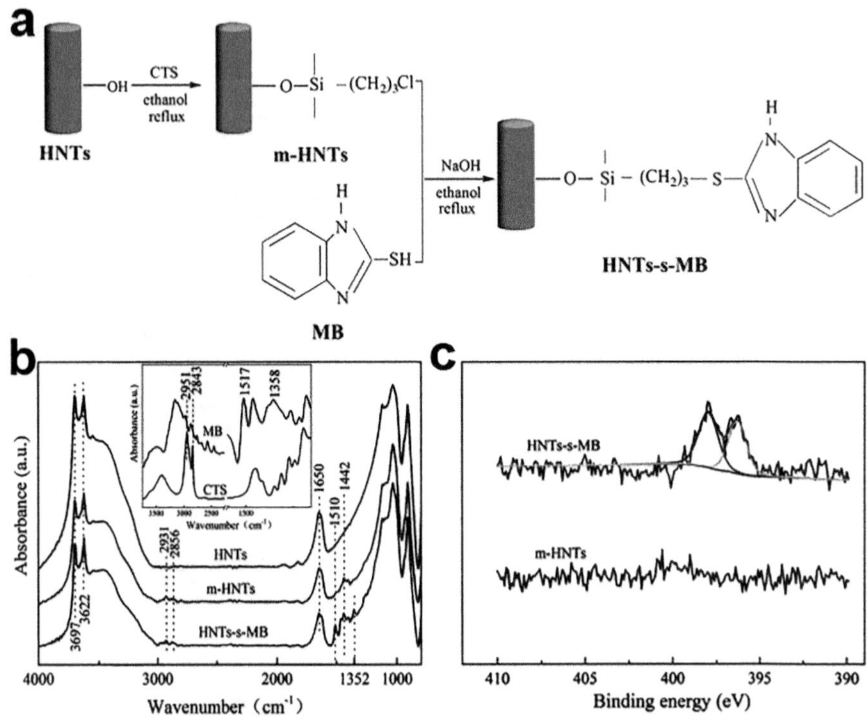

Figure 11.6 (a) Synthesis route of HNTs-s-MB, (b) FTIR spectra of HNTs and HNTs-s-MB, and (c) XPS spectra of m-HNTs and HNTs-s-MB in the N1s region. Reproduced from *Composites Part A: Applied Science and Manufacturing*, 73, B. Zhong *et al.*, Preparation of halloysite nanotubes supported 2-mercaptobenzimidazole and its application in natural rubber, 63–71, Copyright 2015 with permission from Elsevier (ref. 51).

dispersed non-uniformly with large agglomerates in the NR matrix while the m-HNTs and HNTs-s-MB were well-dispersed. The extracted surface of m-HNTs was clean, which indicated the bad interfacial compatibility between m-HNTs and the NR matrix. In contrast, HNTs-s-MB was embedded in the NR matrix with orientation, demonstrating that HNTs-s-MB was more compatible with the NR matrix than m-HNTs. The improved compatibility could be attributed to the substitution of polar monochloro groups of m-HNTs with CTS. The grafted MB molecules extended the interfacial entanglement between the rubber matrix and HNTs-s-MB, leading to a significantly enhanced interaction between HNTs-s-MB and the NR matrix.

The antioxidant ability of nanocomposites was evaluated by calculating the thermo-oxidation activation energies (E_a) from the DSC thermograms at

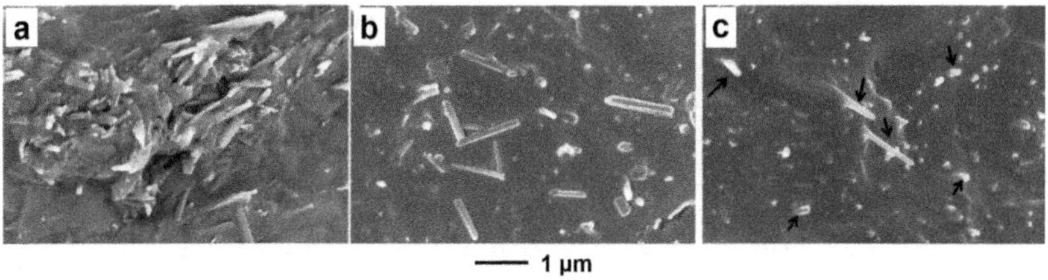

Figure 11.7 SEM photos of (a) NR–HNTs; (b) NR–m-HNTs; (c) NR–HNTs-s-MB (black arrows represent HNTs-s-MB). Reproduced from *Composites Part A: Applied Science and Manufacturing*, **73**, B. Zhong *et al.*, Preparation of halloysite nanotubes supported 2-mercaptobenzimidazole and its application in natural rubber, 63–71, Copyright 2015 with permission from Elsevier (ref. 51).

different heating rates in air flow. The E_a values of the NR nanocomposites with antioxidants were higher than those without antioxidants. Besides, the NR–HNTs-s-MB nanocomposite possessed a higher E_a value than that of the NR–m-HNTs–MB nanocomposite, indicating that the HNTs-s-MB was most responsible for the exceptional antioxidant behavior. The immobilized MB on HNTs-s-MB was hard to migrate and volatilize from the polymeric matrix even at high temperature. Moreover, the well-dispersed HNTs-s-MB promoted interfacial compatibility between the HNTs and polymer matrix. Both of the above factors contributed to the excellent antioxidant activity of the NR–HNTs-s-MB composite.

In conclusion, HNTs-s-MB could be homogeneously dispersed in the NR matrix. The strong interfacial interaction between HNTs-s-MB and NR led to the better mechanical performance of NR–HNTs-s-MB nanocomposite than that of the NR–HNTs nanocomposite. The antioxidative efficiency of NR–HNTs-s-MB composite was superior to that of the NR–m-HNTs–MB nanocomposite containing equivalent antioxidant, owing to the much lower migration and volatility of HNTs-s-MB.

11.3.3.3 The Advantages of Rubber Functionalized with Antioxidant-Loaded HNTs

Antioxidant-loaded HNTs as a filler of rubber endow the rubber with some new features. HNTs can be easily dispersed in a rubber matrix. Under the strong shearing force during the blending process, the surface-silanized HNTs exhibit much weaker inter-tubular interactions, leading to the homogeneous dispersion of HNTs in rubber. Generally speaking, strong interfacial bonding and uniform dispersion of HNTs are crucial factors to be considered for the optimization of the ultimate performance of the rubber–HNTs composites.

Blooming is usually observed when directly compounding antioxidant with rubber, leading to formation of microdefects on the surface of products, loss of antioxidant protection and contamination to the environment. In view of these problems, only about 1 wt% antioxidant can be added to a rubber formulation. The incorporation of excessive antioxidant causes a risk of blooming. Such a low load of antioxidant provides limited aging resistance which can still not inhibit blooming sufficiently. Antioxidant-loaded HNTs allow higher antioxidant loading while no blooming appears in the rubber composite. Compared with conventional antioxidant-loaded spherical fillers, like silica, the fabrication of antioxidant-loaded HNTs is more convenient and a higher antioxidant load can be achieved. HNTs also exhibit a significant reinforcing effect and higher efficiency in lowering the gas permeability for rubbers compared with conventional spherical fillers. An important advantage of tubular HNTs compared to platy clays such as montmorillonite, kaolin and bentonite which are stacked in larger crystallites, is that HNTs do not need complicated exfoliation before use. Besides, a more sustained release profile of antioxidant-loaded HNTs can be achieved. In conclusion, rubber nanocomposites with antioxidant-loaded HNTs combine the reinforcement of HNTs and the carrier function of halloysite lumens with a sustained release profile of antioxidant, leading to improved mechanical properties and aging resistance of rubber composites.

Different rubber nanocomposites with antioxidant-loaded HNTs can be fabricated through a similar method as discussed above. Such novel nanocomposites will have great potential in both the industrial and biomedical fields.

11.4 Antibacterial Rubber Composite

11.4.1 Applications of Antibacterials in Rubber Composites

Antimicrobial products have gained considerable attention from both the academic and industrial communities. Microbial infection of elastomer materials in medical apparatus, health care products, food packaging and house furnishings is getting more and more serious and thus has been brought into the focus of the public.[52] For biomedical elastomers, infections arise when bacteria attach and proliferate on surfaces of biomedical devices and implants, constituting a serious clinical problem.[53] Approximately 64% of hospital-acquired infections worldwide are attributed to attachment of viable bacteria to medical devices and implants.[54] For example, the frequency of biomaterials associated infection is more than 20% for catheters used in chronic ambulatory peritoneal dialysis.[55] Various antibacterials have been incorporated into elastomer composites to endow antibacterial function, such as inorganic substances like silver ions and various nanoparticles, and organic ones like common antibiotics and quaternary ammonium salts *etc.*

Small antibacterial molecules or nanoparticles, such as silver, gold, aluminum, titania (TiO_2), magnesium oxide (MgO), cuprous oxide (CuO), zinc oxide (ZnO), are both prominent antimicrobials. Polymer nanocomposites are considered to have enhanced barrier properties compared with pristine polymer, allowing more sustained drug release from the materials.[56] In these nanocomposite systems, the active agents are typically pre-loaded into nanoparticles (such as HNTs) and then incorporated into the polymer matrix.

NR is used extensively in various applications such as engineering tires, housewares and biomedical devices *etc.* Because NR contains protein, amino acid, carbohydrate and fat,[9] some bacteria grow easily in it, resulting in the corrosion and decomposition of the product, and possible infection to people.[10] Bacterial contamination is a major concern in NR products, especially those used in contact with humans. Incorporation of antimicrobial agents is generally the adopted method to prevent microbial contamination in NR products.[57]

Different methods have been adopted to endow antibacterial function to NR. Antibacterial agents, such as 2-hydroxypropyl-3-piperazinylquinoline-carboxylic acid methacrylate (HPQM) or silver-substituted zeolite (SSZ), were used in NR compounds.[57] Plate count-agar tests showed that diphenylguanidine at the level of 1.0 phr in vulcanized NR had the ability to kill Gram-negative *Escherichia coli* (*E. coli*) and Gram-positive *Staphylococcus aureus* (*S. aureus*) bacteria. Antibacterial group quaternary phosphonium salt was immobilized onto the chlorinated natural rubber matrix, exhibiting excellent antibacterial activity against both *E. coli* and *S. aureus*.[58]

Incorporation of metals and metal oxide nanoparticles in the NR matrix inhibited the colonization of bacteria.[59] Silver nanoparticles were synthesized by reducing silver nitrate with trisodium citrate in an aqueous medium, and then incorporated into natural rubber latex.[60] It was found that the resultant composite inhibited the growth of *E. coli*, *S. aureus* and *Staphylococcus epidermidis* in a very strong manner. Rutile-TiO_2 with median particle size of 73 nm was successfully prepared from commercial micro-TiO_2(R) by ultrasonication, and then incorporated in NR to fabricate NR–n-TiO_2(R) composites.[61] The nanocomposites showed good antibacterial properties against *E. coli* and *S. aureus* with a loading level of only 1 phr of n-TiO_2(R). ZnO-nanoparticles with an average size less than 60 nm were incorporated in natural rubber latex foam (NRLF).[62] The composite inhibited the growth of the bacteria in a very strong manner. NR loaded with charged antibacterial particles was fabricated.[63] *N*,*N*,*N*-Trimethyl chitosan (TMC) stabilized poly(methyl methacrylate) (PMMA) latex particles were synthesized by miniemulsion polymerization. TMC possesses permanent positive charges as well as greater antibacterial activity. Combining the properties of PMMA and TMC, PMMA–TMC latex nanoparticles could be used in place of inorganic lubricating powder in the fabrication of latex gloves. After immersing sulfur prevulcanized natural rubber (SPNR) film into 3 wt% of PMMA–TMC latex at pH 7, a significant amount of nanoparticles uniformly deposited onto the SPNR film and inhibited bacterial (*E. coli* and *S. aureus*) growth.

PU has an extremely broad range of applications, especially in medical devices including endotracheal tubes, vascular prostheses, components of artificial hearts, membranes for dialysis, adhesives for bone tissue and materials for dental recovering.[64] However, PU is susceptible to degradation by many types of bacteria during service and storage. It is meaningful to study the antibacterial properties of PU and its nanocomposites.

Antibacterial agents or antibacterial nanoparticles were incorporated in PU to form antibacterial PU composite. PU–β-cyclodextrin–ciprofloxacin composite films with open and interconnected macroporous pore morphology showed antimicrobial activity.[65] Silver nanoparticles (AgNPs) were incorporated into the PU matrix using different methods. Nanocomposites from a waterborne PU containing various low concentrations (15–75 ppm) of AgNPs (size approximately 5 nm) were prepared.[66] The nanocomposites showed good biocompatibility both *in vitro* and *in vivo*. Commercial catheters coated with PU–Ag 30 ppm were inserted into rat jugular veins and showed milder inflammation for PU–Ag after three months compared to non-coated catheters or pure PU-coated catheters. Anionic waterborne polyurethane (WPU)–Ag nanocomposites were prepared by *in situ* reduction of silver ions at room temperature without adding additional organic reducing/stabilizing agent.[67] The WPU–Ag nanocomposites could release silver ions and showed bacterial inhibition of 99.99% for *E. coli* and 53.97% for *S. aureus*. In another work, *in situ* light initiated synthesis of AgNP was employed for AgNP incorporation within the polymeric matrices of medical grade PU.[68] The resultant materials showed improved antibacterial and antibiofilm activity against *Pseudomonas aeruginosa* with negligible toxicity for human primary skin cells and erythrocytes. ZnO nanoparticles were also incorporated into a PU matrix, and the obtained nanocomposites showed good antibacterial activity. PU–ZnO composites with ZnO nanoparticles[69,70] or flower-like ZnO nanowhiskers[71] as reinforcement were prepared by the solution casting and evaporation method. With the combination of enhanced mechanical and antibacterial properties, the composites are potentially useful in a variety of applications.

Surface modification of material and preparation of polymeric antibacterial material with chemically anchored biocides are two efficient methods to provide material with antibacterial properties.[72,73] The surfaces of PU film were modified by the covalent immobilization of chitosan which had antibacterial activity.[74] Modification with chitosan significantly increased the antibacterial activity of PU film against Gram-positive (*S. aureus*) and Gram-negative (*P. aeruginosa*) bacteria. Antibacterial PU coatings were also prepared from quaternary ammonium salts functionalized soybean oil based polyols.[75] Due to the non-leaching nature of the bactericidal components, high biocompatibility and desired mechanical properties of the coating, these materials were considered as potential candidates for biomedical applications.

The potential benefits of nanocomposites as biomaterials include enhancement of the mechanical and barrier properties, the latter allowing more sustained release of drugs compared with other materials.[56] Polyurethane–montmorillonite (PU–MMT) nanocomposites were prepared in which the antimicrobial agent, chlorhexidine diacetate (CHX), was evaluated as an organic modifier for silicate dispersion.[76] CHX was incorporated at various concentrations through organic modification of MMT or blending directly with the bulk polymer. Intercalated and partially exfoliated nanoclays were achieved, with better dispersion occurring in the presence of free CHX within the bulk. Antibacterial activity against *S. epidermidis* was significant in materials with higher exchanged MMT and nanoclays containing free CHX.

Silicone rubber is often used in food packaging, houseware and biomedical devices that are likely to be in contact with the human body. Inflammation, infection and biofilm formation are side effects that occur during clinical use of silicon rubber. This in turn initiates the degradation of the material. Therefore, the antimicrobial performance of silicone rubber appears to be very important in preventing infection from harmful microorganisms.[77]

Conventional strategies to incorporate an antibacterial agent in a polymer matrix by solution casting or melting extrusion are simple.[78] In Taptim and Sombatsompop's work, HPQM was used as a biocide in a silicone rubber compound which showed antibacterial performance.[77] Silicone rubber compounds containing acrylated norfloxacin also showed antibacterial activity.[79] To compare the antibacterial function of different agents, silicone rubber filled with different antibacterial agents, including nano-silver colloids, SSZ and HPQM were prepared.[80] The antibacterial properties indicated that HPQM was most recommended for enhancing the antibacterial performance. To further improve the mechanical properties, silicone rubber compounds filled with different loadings of organoclay and SSZ solid fillers were prepared.[81] The antimicrobial activity of rubber added with HPQM agent and fillers increased when the contact time in the test solution was increased.

Weak adhesion between the antibacterial coatings and polymer substrates limits their clinical applications. Surface modification is an effective way to solve this intrinsic problem. In Guo's study,[82] UV irradiation was employed to activate the inert silicon rubber substrates, and then coupling agent (3-mercaptopropyl) trimethoxy silane was grafted onto the UV-activated surfaces of the substrates containing thiol groups. This process anchors Ag^+ during electroless plating which enhances the adhesion remarkably. The as-prepared Ag–polytetrafluoroethylene composite coatings possessed remarkably enhanced adhesion and superior antibacterial activity.

A hybrid composite is an efficient way to further increase the antibacterial term. A novel hydrogel–silicone composite material exhibiting sustained release of the antibiotic ciprofloxacin, which may be used for urinary tract catheters, has been reported.[83] The neat silicone elastomer was permeabilized by impregnating with a 2-hydroxyethyl methacrylate (HEMA)-based hydrogel. The hydrogel containing an active pharmaceutical ingredient provided antibacterial activity while the silicone elastomer provided sufficient mechanical stability. The antimicrobial agent, ciprofloxacin, was loaded into the polymer matrix by a post-polymerization loading procedure. Sustained release of ciprofloxacin was demonstrated, and the release of the antibacterial agent could be controlled by varying both the hydrogel content in the material and the drug concentration during loading. Devices containing 25 wt% hydrogel and loaded with ciprofloxacin displayed a strong antibacterial effect against the colonization of *S. aureus* and the formation of biofilm on the device material was inhibited for 29 days.

11.4.2 HNTs Loaded with Antibacterials

HNTs are biocompatible, as demonstrated on different cell cultures and microworms *in vitro*.[31] Surface chemistry, tubular morphology, submicron size and biocompatibility make this nanomaterial an excellent candidate for sustained release of antiseptics, drugs, and proteins.[3,29] The toxicity of HNTs *in vivo* was investigated by employing the *Caenorhabditis elegans* nematode as a model organism, and it was found that HNTs were not capable of severely damaging the nematode organism.[84] Various kinds of antibacterial drugs and metal particles were successfully loaded in HNTs which applied to different fields. HNTs loaded with simple antiseptics, such as brilliant green, providone iodine or chlorhexidine, allow antibacterial function from minutes to 5–10 hours. HNT encapsulated antibiotics (tetracycline, gentamicin, and ciprofloxacin) are even more efficient. The sustained release time is 100–200 hours, allowing them to fight even multidrug resistant bacteria.[30] Modified HNTs can be loaded with 10–30 wt% drugs and used for sustained release of chemical agents. The release time of drugs could be further extended by end-stopper or layer-by-layer encapsulation. As shown in Figure 11.8, brilliant green as an antiseptic was loaded into HNT lumens (15 nm diameter) at 15 wt% and provided a sustained release time over six hours allowing an extension of its antibacterial efficiency.[85] Formation of a benzotriazole–copper coating on HNTs allowed additional encapsulation of brilliant green and provided a more sustained release time ranging from 50 to 200 hours. The antibacterial efficiency of the brilliant green in HNTs was tested on *S. aureus* cultures and the antibacterial action extended up to 72 hours. Similar results of sustained release of amoxicillin and iodine from halloysite tubes were also demonstrated.

Antibacterials could be grafted on the surface of HNTs to realize sustained release. *N*-Halamine was grafted on HNTs and then used as an antibacterial

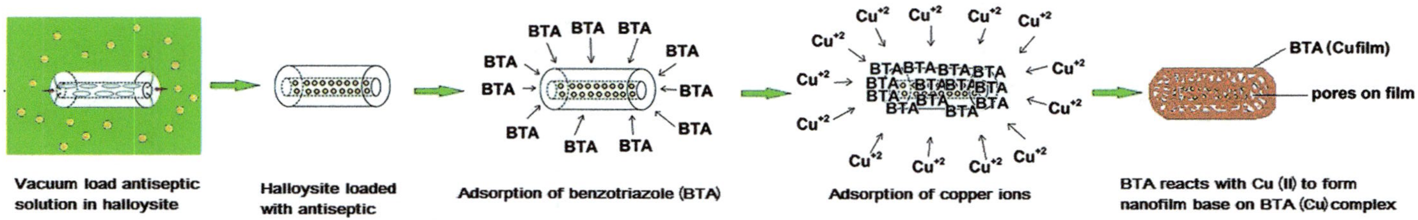

Figure 11.8 Procedure for loading brilliant green inside halloysite nanotubes and fabrication of a benzotriazole–copper film. Reproduced from ref. 85 with permission from the Royal Society of Chemistry.

agent to fabricate polyethersulfone ultrafiltration hybrid membranes.[86] The results showed that the hydrophilicity of the membranes was improved greatly after adding *N*-halamine@HNTs. The hybrid membranes showed good antibacterial activity against *E. coli*.

Deposition of metal oxides or metal nanoparticles on HNTs has been carried out to integrate antibacterial properties and the reinforcement of HNTs. AgNPs with about 5 nm diameter were homogeneously distributed and bound to the surfaces of HNTs.[87] Firstly, HNTs were modified by silane coupling agent KH-792. And then, modified HNTs were immersed in silver nitrate solution and a complex reaction between the two amino groups of KH-792 and the silver ions performed, leading to large clusters on the surfaces of HNTs. Finally, these silver containing clusters were converted into AgNPs with about 5 nm diameter by a reduction process. The powders of AgNPs–HNTs nanocomposites showed good antibacterial activities against Gram-negative bacteria (*E. coli*) and Gram-positive bacteria (*S. aureus*). Silver nanorods were also be incorporated in HNTs as antibacterial additives.[88]

Metal oxide nanoparticles (TiO_2, ZnO) with antibacterial function could also be deposited onto HNTs. For example, TiO_2 nanoparticles were deposited on the platform of HNTs by a one-step solvothermal method.[89] ZnO and TiO_2 nanoparticles were also deposited onto carbon-coated HNTs (CCH) separately.[90] HNTs were impregnated with sucrose solution. Then, CCH were produced by the carbonization of sucrose-coated HNTs in the presence of sulfuric acid. ZnO and TiO_2 nanoparticles were subsequently deposited on the surface of the CCH by sol–gel method. The aqueous solution containing the metal precursor was added, but it couldn't completely penetrate into CCH due to the carbon-coating. Thus, the inner of the HNTs was protected and the metal oxide nanoparticle deposition was realized on the outer surfaces of the HNTs. Other nanoparticles like Au nanoparticles could also be grafted on HNTs.[91]

To prepare antibacterial composites with sustained release of antibacterials, drug-loaded HNTs were then incorporated into polymer matrices. ZnO nanoparticles were deposited on the outer and inner surfaces of HNTs using a novel solvothermal method.[92] Then, these ZnO deposited HNTs were incorporated into a PLA matrix as a reinforcing filler by the solution casting method with different filler loadings (0–10 wt%). Both PLA films with ZnO or ZnO–HNTs showed exceptional antimicrobial properties against bacteria such as *E. coli* and *S. aureus*. The anti-bacterial ultrofiltration film with polyethersulfone (PES) have been prepared using such method (see more details in Chapter 10).[93,94] Alternatively, green composites composed of pectins and rosemary essential oil loaded HNTs were prepared by ball milling in the presence of water followed with solution casting.[95] A comparison of the release profile of the rosmarinic acid molecules directly dispersed in pectins and in the nano-hybrid was investigated, indicating a much slower release in the latter case.

11.4.3 Rubber Functionalized with Antibacterial-Loaded HNTs

According to the above discussion, in both the industrial and biomedical fields, the current infection prevention strategies include the use of antibiotic or antiseptic coatings on the surface of the devices and the incorporation of antibacterial metal oxides in polymer matrices.[76] However, the materials prepared by these approaches have many limitations, particularly in long-term applications. In many cases, antimicrobial agents are compounded with polymer through direct blending. However, such antimicrobial agents with low molecular weight tend to bloom to the surface of polymer products continuously which will cause inflammatory risks to the human body. Most of the antibiotics used in polymers[96] are usually effective in a short-term application.[97] Thus further research into more appropriate antibacterial materials for sustained release of antibacterial in a long-term application is highly desired. Advantageously, the antibacterial-immobilized nanocomposites may promote the mechanical properties of the final functional polymer material while endowing a sustained drug release profile.[76]

Studies into HNTs incorporated in rubber nanocomposites for sustained release of antibacterials are relatively scarce. HNTs could be easily dispersed homogeneously in various kinds of rubbers and different antibacterials could be readily loaded in HNTs or immobilized onto the surface of HNTs. Rubber composites with good mechanical properties and antibacterial function can be designed with the addition of antibacterial-loaded HNTs. Using this method, the antibacterial will be dispersed throughout the matrix, leading to sustained release of the antibacterial agent. In conclusion, the rubber–antibacterial-loaded HNTs composite represents a promising candidate for both industrial and biomedical devices to prevent bacterial colonization during long-term use. This approach may be extended for other "smart" elastomers.

11.5 Conclusions

Halloysite is a "green" and natural nanomaterial, with abundant resources and low cost. It exhibits many interesting properties, allowing versatile potential uses in a large variety of fields. With a typical hollow and tubular structure, halloysite clay tubes have been used as containers for loading and sustained release of chemical agents for application both in engineering and biomedical fields. These clay nanotubes have good dispersity in rubbers and thus enhance the mechanical properties of the composite. Furthermore, by loading active agents in the inner lumens of the tubes, additional functionality (antioxidant, antibacterial) for the composite can be achieved. For example, loaded-HNTs for sustained release of antioxidants and antibacterials can be realized and used in the rubber composites. The loaded HNTs are not only to enhance the mechanical properties of the rubber matrix, but also to achieve an ideal sustained release profile of loaded agents. This kind of composite will have great potential in both the industrial and biomedical fields. Future studies should pay more attention to the reasonable selection of the processing method, nature of matrix and types of active agents to develop novel halloysite related hybrid rubber composites with antioxidative and antibacterial abilities.

Acknowledgements

We gratefully acknowledge the National Basic Research Program of China under Grant No. 2015CB654700, and National Natural Science Foundation of China under Grant No. 51221002 and Grant No. 51320105012 for financial support.

References

1. S. Thomas and R. Stephen, *Rubber nanocomposites: preparation, properties and applications*, John Wiley & Sons, 2010, p. 8.
2. A. Das, D. Basu and G. Heinrich, in *Encyclopedia of Polymeric Nanomaterials*, ed. S. Kobayashi and K. Müllen, Springer, Berlin Heidelberg, 2015, p. 2181.
3. Y. Lvov and E. Abdullayev, *Prog. Polym. Sci.*, 2013, **38**, 1690.
4. Y. M. Lvov, D. G. Shchukin, H. Möhwald and R. R. Price, *ACS Nano*, 2008, **2**, 814.
5. S. Rooj, V. S. Raman and G. Heinrich, in *Encyclopedia of Polymeric Nanomaterials*, ed. S. Kobayashi and K. Müllen, Springer, Berlin Heidelberg, 2014, p. 1.
6. B. Gong, C. Ouyang, Y. Yuan and Q. Gao, *RSC Adv.*, 2015, **5**, 77106.
7. Z. Jia, B. Guo and D. Jia, *J. Nanosci. Nanotechnol.*, 2014, **14**, 1758.
8. V. P. Swapna, R. Stephen, T. Greeshma, C. Sharan Dev and M. S. Sreekala, *Polym. Compos.*, 2016, **37**, 602.
9. H. Ismail, S. Z. Salleh and Z. Ahmad, *J. Appl. Polym. Sci.*, 2013, **127**, 3047.
10. H. Ismail, S. Z. Salleh and Z. Ahmad, *Mater. Des.*, 2013, **50**, 790.
11. S. Rooj, A. Das, V. Thakur, R. N. Mahaling, A. K. Bhowmick and G. Heinrich, *Mater. Des.*, 2010, **31**, 2151.
12. M. Poikelispää, A. Das, W. Dierkes and J. Vuorinen, *J. Appl. Polym. Sci.*, 2013, **127**, 4688.
13. B. Guo, Y. Lei, F. Chen, X. Liu, M. Du and D. Jia, *Appl. Surf. Sci.*, 2008, **255**, 2715.
14. B. Guo, F. Chen, Y. Lei, X. Liu, J. Wan and D. Jia, *Appl. Surf. Sci.*, 2009, **255**, 7329.
15. Y. Lei, Z. Tang, L. Zhu, B. Guo and D. Jia, *Polymer*, 2011, **52**, 1337.
16. Z. Tang, Q. Wei, T. Lin, B. Guo and D. Jia, *RSC Adv.*, 2013, **3**, 17057.

17. H. Ismail, P. Pasbakhsh, M. N. A. Fauzi and A. Abu Bakar, *Polym. Test.*, 2008, **27**, 841.
18. P. Pasbakhsh, H. Ismail, M. N. A. Fauzi and A. A. Bakar, *Polym. Test.*, 2009, **28**, 548.
19. P. Pasbakhsh, H. Ismail, M. N. A. Fauzi and A. A. Bakar, *Appl. Clay Sci.*, 2010, **48**, 405.
20. P. Pasbakhsh, H. Ismail, M. N. A. Fauzi and A. A. Bakar, *J. Appl. Polym. Sci.*, 2009, **113**, 3910.
21. K. Lee, Y.-W. Chang and S. W. Kim, *J. Appl. Polym. Sci.*, 2014, **131**, 40307.
22. L. Jiang, C. Zhang, M. Liu, Z. Yang, W. W. Tjiu and T. Liu, *Compos. Sci. Technol.*, 2014, **91**, 98.
23. P. Rybiński and G. Janowska, *Thermochim. Acta*, 2012, **549**, 6.
24. P. Rybiński and G. Janowska, *Thermochim. Acta*, 2013, **557**, 24.
25. K. Senthivel, K. Manikandan and B. Prabu, *Mater. Today: Proc.*, 2015, **2**, 3627.
26. S. Rooj, A. Das and G. Heinrich, *Eur. Polym. J.*, 2011, **47**, 1746.
27. Z. Liqun and S. Rui, in *Encyclopedia of Polymeric Nanomaterials*, ed. S. Kobayashi and K. Müllen, Springer, Berlin Heidelberg, 2015, p. 172.
28. Q. Chen, L. Jin, W. D. Cook, D. Mohn, E. L. Lagerqvist, D. A. Elliott, J. M. Haynes, N. Boyd, W. J. Stark, C. W. Pouton, E. G. Stanley and A. G. Elefanty, *Soft Matter*, 2010, **6**, 4715.
29. J. Cervini-Silva, A. Nieto-Camacho, E. Palacios, J. A. Montoya, V. Gómez-Vidales and M. T. Ramírez-Apán, *Colloids Surf., B*, 2013, **111**, 651.
30. Y. Lvov, A. Aerov and R. Fakhrullin, *Adv. Colloid Interface Sci.*, 2014, **207**, 189.
31. Y. Lvov, W. Wang, L. Zhang and R. Fakhrullin, *Adv. Mater.*, 2016, **28**, 1227.
32. Q.-Z. Chen, S.-L. Liang, J. Wang and G. P. Simon, *J. Mech. Behav. Biomed. Mater.*, 2011, **4**, 1805.
33. G.-Y. Li and J. Koenig, *Rubber Chem. Technol.*, 2005, **78**, 355.
34. Q. Pan, B. Wang, Z. Chen and J. Zhao, *Mater. Des.*, 2013, **50**, 558.
35. L. Guo, H. Lei, J. Zheng, G. Huang and G. Li, *Polym. Compos.*, 2013, **34**, 1856.
36. B. Zhong, Q. Shi, Z. Jia, Y. Luo, Y. Chen and D. Jia, *Polym. Degrad. Stab.*, 2014, **110**, 260.
37. B. Zhong, Z. Jia, D. Hu, Y. Luo and D. Jia, *Composites, Part A*, 2015, **78**, 303.
38. H. Li, T. Liang, F. Li, W. Wu, X. Lai, J. Guo and X. Zeng, *Polym. Compos.*, 2015, DOI: 10.1002/pc.23688.
39. O. Mrad, J. Saunier, C. Aymes-Chodur, V. Mazel, V. Rosilio, F. Agnely, J. Vigneron, A. Etcheberry and N. Yagoubi, *Eur. Polym. J.*, 2011, **47**, 2403.
40. J. Saunier, V. Mazel, C. Aymes-Chodur and N. Yagoubi, *Int. J. Pharm.*, 2012, **437**, 89.

41. J. Saunier, J.-M. Herry, C. Marlière, M. Renault, M.-N. Bellon-Fontaine and N. Yagoubi, *Mater. Sci. Eng., C*, 2015, **56**, 522.
42. B. E. Britigan, T. L. Roeder, G. T. Rasmussen, D. M. Shasby, M. L. McCormick and C. D. Cox, *J. Clin. Invest.*, 1992, **90**, 2187.
43. K. Arunachalam and T. Parimelazhagan, *J. Ethnopharmacol.*, 2013, **145**, 139.
44. E. M. Christenson, J. M. Anderson and A. Hiltner, *J. Biomed. Mater. Res., Part A*, 2006, **76A**, 480.
45. Y. H. Kim, D. K. Han, K. D. Park and S. H. Kim, *Biomaterials*, 2003, **24**, 2213.
46. J. Cervini-Silva, A. Nieto-Camacho, E. Palacios, J. A. Montoya, V. Gomez-Vidales and M. T. Ramirez-Apan, *Colloids Surf., B*, 2013, **111**, 651.
47. V. Vergaro, Y. M. Lvov and S. Leporatti, *Macromol. Biosci.*, 2012, **12**, 1265.
48. J. Hári, Á. Gyürki, M. Sárközi, E. Földes and B. Pukánszky, *J. Colloid Interface Sci.*, 2016, **462**, 123.
49. M. H. Lee and H. J. Park, *J. Appl. Polym. Sci.*, 2015, **132**, 42771.
50. Y. Fu, D. Zhao, P. Yao, W. Wang, L. Zhang and Y. Lvov, *ACS Appl. Mater. Interfaces*, 2015, **7**, 8156.
51. B. Zhong, Z. Jia, Y. Luo, B. Guo and D. Jia, *Composites, Part A*, 2015, **73**, 63.
52. A. Muñoz-Bonilla and M. Fernández-García, *Prog. Polym. Sci.*, 2012, **37**, 281.
53. G. Subbiahdoss, R. Kuijer, D. W. Grijpma, H. C. van der Mei and H. J. Busscher, *Acta Biomater.*, 2009, **5**, 1399.
54. J. A. Lichter, K. J. Van Vlietpa and M. F. Rubner, *Macromolecules*, 2009, **42**, 8573.
55. P. H. S. Kwakman, A. A. te Velde, C. M. J. E. Vandenbroucke-Grauls, S. J. H. van Deventer and S. A. J. Zaat, *Antimicrob. Agents Chemother.*, 2006, **50**, 3977.
56. T. K. Chen, Y. I. Tien and K. H. Wei, *Polymer*, 2000, **41**, 1345.
57. K. Jai-eau, E. Wimolmala and N. Sombatsompop, *J. Vinyl Addit. Technol.*, 2013, **19**, 123.
58. T. Qiu, Q. Zeng and N. Ao, *Mater. Lett.*, 2014, **122**, 13.
59. I. Rathnayake, H. Ismail, B. Azahari, C. De Silva and N. Darsanasiri, *J. Appl. Polym. Sci.*, 2014, **131**, 40155.
60. W. G. I. U. Rathnayake, H. Ismail, A. Baharin, A. G. N. D. Darsanasiri and S. Rajapakse, *Polym. Test.*, 2012, **31**, 586.
61. B. Seentrakoon, B. Junhasavasdikul and W. Chavasiri, *Polym. Degrad. Stab.*, 2013, **98**, 566.
62. W. G. I. U. Rathnayake, H. Ismail, A. Baharin, I. M. C. C. D. Bandara and S. Rajapakse, *J. Appl. Polym. Sci.*, 2014, **131**, 1.
63. T. Arpornwichanop, D. Polpanich, R. Thiramanas, T. Suteewong and P. Tangboriboonrat, *Carbohydr. Polym.*, 2014, **109**, 1.
64. Y. Gustafsson, J. Haag, P. Jungebluth, V. Lundin, M. L. Lim, S. Baiguera, F. Ajalloueian, C. Del Gaudio, A. Bianco and G. Moll, *Biomaterials*, 2012, **33**, 8094.

65. D. Macocinschi, D. Filip, S. Vlad, C. G. Tuchilus, A. F. Cristian and M. Barboiu, *J. Mater. Chem. B*, 2014, **2**, 681.
66. S. H. Hsu, H. J. Tseng and Y. C. Lin, *Biomaterials*, 2010, **31**, 6796.
67. Y. Wattanodorn, R. Jenkan, P. Atorngitjawat and S. Wirasate, *Polym. Test.*, 2014, **40**, 163.
68. S. Saez, C. Fasciani, K. G. Stamplecoskie, L. B. Gagnon, T. F. Mah, M. L. Marin, E. I. Alarcon and J. C. Scaiano, *Photochem. Photobiol. Sci.*, 2015, **14**, 661.
69. J. H. Li, R. Y. Hong, M. Y. Li, H. Z. Li, Y. Zheng and J. Ding, *Prog. Org. Coat.*, 2009, **64**, 504.
70. A. M. El Saeed, M. A. El-Fattah and A. M. Azzam, *Dyes Pigm.*, 2015, **121**, 282.
71. X.-Y. Ma and W.-D. Zhang, *Polym. Degrad. Stab.*, 2009, **94**, 1103.
72. N. Kebir, I. Campistron, A. Laguerre, J. F. Pilard, C. Bunel and T. Jouenne, *Biomaterials*, 2007, **28**, 4200.
73. F. Mo, H. Ren, S. Chen and Z. Ge, *Mater. Lett.*, 2015, **145**, 174.
74. F. Kara, E. A. Aksoy, Z. Yuksekdag, N. Hasirci and S. Aksoy, *Carbohydr. Polym.*, 2014, **112**, 39.
75. H. Bakhshi, H. Yeganeh, S. Mehdipour-Ataei, M. A. Shokrgozar, A. Yari and S. N. Saeedi-Eslami, *Mater. Sci. Eng., C*, 2013, **33**, 153.
76. N. Fong, A. Simmons and L. A. Poole-Warren, *Acta Biomater.*, 2010, **6**, 2554.
77. K. Taptim and N. Sombatsompop, *J. Vinyl Addit. Technol.*, 2014, **20**, 49.
78. C. Y. Tang, D.-z. Chen, K. Y. Y. Chan, K. M. Chu, P. C. Ng and T. M. Yue, *Polym. Int.*, 2011, **60**, 1461.
79. H.-S. Yang and E.-S. Park, *Macromol. Mater. Eng.*, 2006, **291**, 621.
80. K. Taptim and N. Sombatsompop, *J. Vinyl Addit. Technol.*, 2013, **19**, 113.
81. K. Taptim, A. Ansarifar and N. Sombatsompop, *Polym. Eng. Sci.*, 2014, **54**, 932.
82. R. Guo, G. Yin, X. Sha, Q. Zhao, L. Wei and H. Wang, *Appl. Surf. Sci.*, 2015, **341**, 13.
83. S. L. Steffensen, M. H. Vestergaard, M. Groenning, M. Alm, H. Franzyk and H. M. Nielsen, *Eur. J. Pharm. Biopharm*, 2015, **94**, 305.
84. G. I. Fakhrullina, F. S. Akhatova, Y. M. Lvov and R. F. Fakhrullin, *Environ. Sci.: Nano*, 2015, **2**, 54.
85. W. Wei, R. Minullina, E. Abdullayev, R. Fakhrullin, D. Mills and Y. Lvov, *RSC Adv.*, 2014, **4**, 488.
86. L. Duan, W. Huang and Y. Zhang, *RSC Adv.*, 2015, **5**, 6666.
87. Y. Zhang, Y. Chen, H. Zhang, B. Zhang and J. Liu, *J. Inorg. Biochem.*, 2013, **118**, 59.
88. E. Abdullayev, K. Sakakibara, K. Okamoto, W. Wei, K. Ariga and Y. Lvov, *ACS Appl. Mater. Interfaces*, 2011, **3**, 4040.
89. R. Wang, G. Jiang, Y. Ding, Y. Wang, X. Sun, X. Wang and W. Chen, *ACS Appl. Mater. Interfaces*, 2011, **3**, 4154.
90. Y. Zhang, J. Ouyang and H. Yang, *Appl. Clay Sci.*, 2014, **95**, 252.
91. H. Zhu, M. Du, M. Zou, C. Xu and Y. Fu, *Dalton Trans.*, 2012, **41**, 10465.
92. R. T. De Silva, P. Pasbakhsh, S. M. Lee and A. Y. Kit, *Appl. Clay Sci.*, 2015, **111**, 10.

93. Y. Chen, Y. Zhang, H. Zhang, J. Liu and C. Song, *Chem. Eng. J.*, 2013, **228**, 12.
94. Q. Zhao, C. Liu, J. Liu and Y. Zhang, *RSC Adv.*, 2015, **5**, 38646.
95. G. Gorrasi, *Carbohydr. Polym.*, 2015, **127**, 47.
96. P. S. Stewart and J. William Costerton, *Lancet*, 2001, **358**, 135.
97. A. Gutowska, B. You Han, H. Jacobs, F. Mohammad, D. Mix, J. Feijen and K. Sung Wan, *J. Biomed. Mater. Res.*, 1995, **29**, 811.

CHAPTER 12

Halloysite–Dopamine Hybrid Nanotubes to Immobilize Biomacromolecules

YIFAN LI[a], YAFEI ZHAO[a], BING ZHANG*[a,b] AND RONGFENG CHEN[a]

[a]Zhengzhou University, School of Chemical Engineering and Energy, Zhengzhou 450001, P. R. China; [b]Anyang Normal University, College of Chemistry and Chemical Engineering, Anyang 455002, P. R. China
*E-mail: zhangb@zzu.edu.cn

12.1 Halloysite Nanotubes for Biomacromolecule Immobilization

Halloysite [$Al_2Si_2O_5(OH)_4{\cdot}2H_2O$] is a type of natural kaolinite mineral with typical tubular nanostructures. Owing to their environmentally friendly, cost effective, and biocompatible characteristics, halloysite nanotubes (HNTs) have been recognized as a fine lumen or support to immobilize a broad range of biologically active substances including drugs, enzymes, and DNA.[1–3] Compared with several traditional clay supports (*e.g.* intercalated kaolinite and montmorillonite) with narrow pore sizes of 0.3–2 nm, HNTs with a larger inner diameter of 5–20 nm are more suitable candidates to entrap biomacromolecules. Moreover, the exterior and interior of HNTs are composed of siloxane (Si–O–Si) groups and aluminol (Al–OH) groups, respectively, yielding a

RSC Smart Materials No. 22
Functional Polymer Composites with Nanoclays
Edited by Yuri Lvov, Baochun Guo and Rawil F. Fakhrullin

Published by the Royal Society of Chemistry, www.rsc.org

negatively charged outer surface and a positively charged inner lumen with a range of pH values (about 3–8.5).[4,5] Therefore, one biomacromolecule bearing the contrary charge (especially for enzymes with isoelectric points) is prone to bind to HNTs due to electrostatic attraction, which is expected to further support its immobilization. Shamsi and Geckeler managed to immobilize DNA onto HNTs in 2008, which is an early example of biomacromolecule immobilization using non-carbon nanotubes as substrates.[1] Zhang *et al.* reported the availability of HNTs for enzyme immobilization in 2010.[2,6] To be specific, two typical industrial enzymes (α-amylase and urease) with different sizes were physically adsorbed into the lumen of HNTs through electrostatic interaction. The immobilized enzymes retained a high fraction of their intrinsic activity after heating or long-term storage, and also exhibited good storage stability and reusability. Wang *et al.* found that HNTs could absorb more than 85% of lipase in 90 min, and approximately 80% of the catalytic activity was maintained compared with free lipase. They further examined the feasibility of the lipase-adsorbed HNTs for enriching lipase inhibitors from natural products.[7] Very recently, Wang *et al.* have reported a flower-like enzyme support which was fabricated by *in situ* growth of layered double hydroxide (LDH) nanoplates onto HNTs (Figure 12.1).[8] The as-prepared HNTs@LDHs possessed significantly higher surface area and lysozyme loading than the individual HNTs and LDHs. Surprisingly, lysozyme adsorbed onto HNTs@LDHs showed higher activity than free lysozyme, which was interpreted by the authors as a synergic combination of HNTs, LDHs, and lysozyme that enhanced the binding affinity towards *Escherichia coli* cells.

Nevertheless, it should be clarified that the immobilization of enzyme onto HNTs mainly relies on electrostatic interactions. The inorganic nature, limited functional groups and surface area, and insufficient biocompatibility of HNTs do not allow adequately high enzyme loading and strong affinity between the enzyme and the nanotubes. Since the immobilized enzyme often works under aqueous conditions at different pH values, electrostatic interactions may not be sufficiently strong to attach enzyme. On one hand, the hydrogen bonding interactions between water and enzyme molecules increase the tendency of enzyme dissolving. On the other hand, the strength of electrostatic interactions strongly depends upon the surface charges of HNTs, which are further determined by pH values. To be specific, the exterior of HNTs is known to be negatively charged at $pH > 3.0$, which means that a negatively charged enzyme might be difficult to approach to HNTs due to electrostatic repulsion. Yuan *et al.* employed gamma-aminopropyltriethoxysilane to prepare amino-modified HNTs (HNTs–NH_2), and reversed the outer surface charges of HNTs from negative to positive for further potential functionalization, although the specific applications were not included.[9] Zhang and coworkers proposed the strategy of employing halloysite–polymer hybrid nanotubes as prospective supports for enzyme immobilization. After depositing a positively charged polymer with hydroxyl or amino groups onto the outer surface of HNTs (negatively charged), the desired enzyme can then be connected to the HNTs–polymer composite *via* glutaraldehyde-enabled

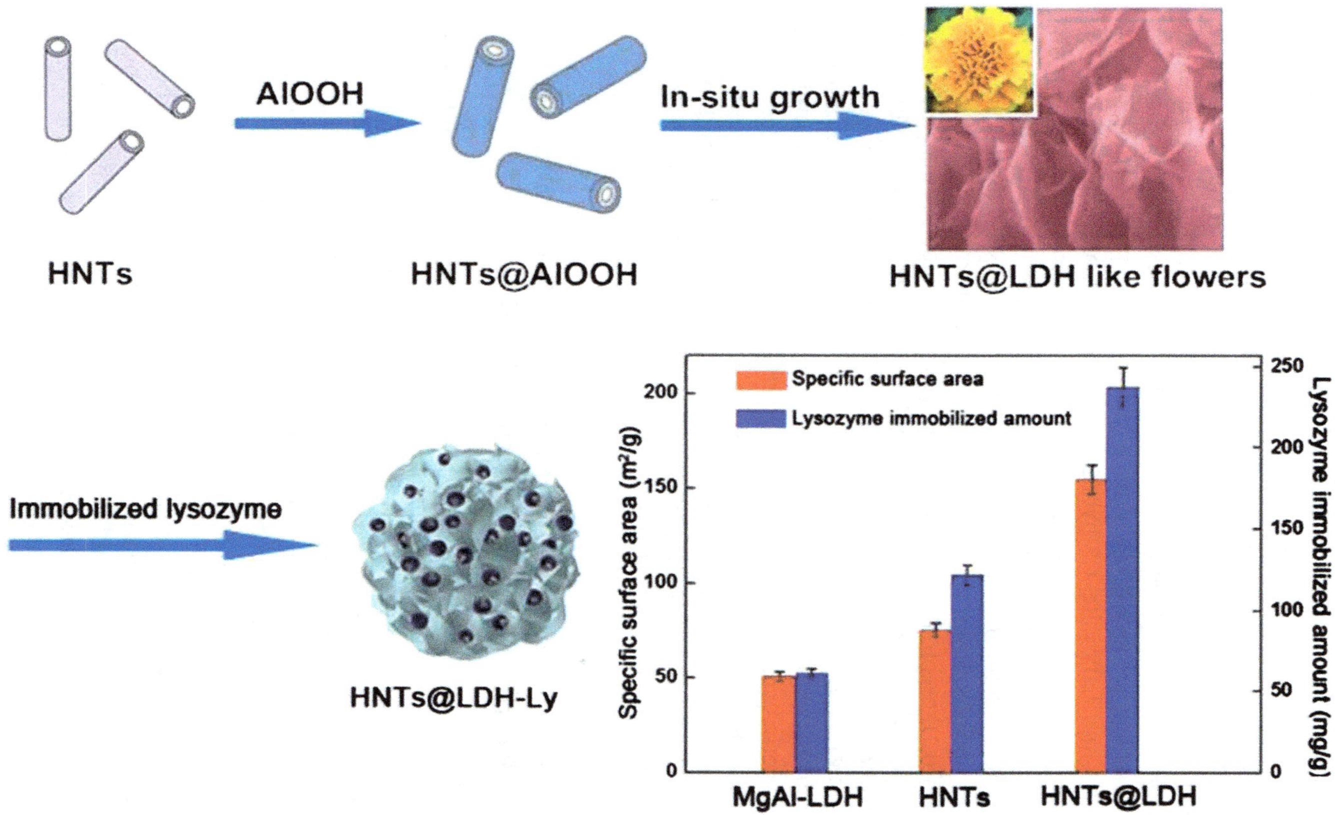

Figure 12.1 Scheme of immobilization of lysozyme on flower-like HNTs@LDH and specific surface area and lysozyme immobilized amount of different samples. Reproduced with permission from Y. Wang, C. Liu, Y. Zhang, B. Zhang, J. Liu, *ACS Sustain. Chem. Eng.*, 2015, **3**, 1183. Copyright (2015) American Chemical Society.[8]

coupling reaction.[11] For instance, when halloysite–chitosan (CS) hybrid nanotubes were prepared for horseradish peroxidase (HRP) immobilization, the maximum enzyme loading reached as large as 21.5 mg g^{-1}, compared to 3.1 mg g^{-1} of pristine HNTs. In addition, the covalently immobilized HRP did not undergo any activity loss after 35 days storage. With the encouraging results of HNTs–CS hybrid supports, porous HNTs–CS hybrid microspheres with a diameter of about 100–150 μm were also prepared by a reversed-phase suspension crosslinking technique.[12] The well-designed hybrid microspheres had a higher surface area of 114.6 m^2 g^{-1} compared to pristine halloysite of 59.6 m^2 g^{-1} and halloysite–CS hybrid of 55.2 m^2 g^{-1}, and showed a much higher loading capacity (as high as 123.1 mg g^{-1}) for laccase immobilization.

The aforementioned progress demonstrates that both physical and chemical structures significantly affect the performance of HNTs-based biomacromolecule supports. From the perspective of physical structures, fabrication of HNTs into 3D architectures proves a promising method to enlarge surface area and enhance biomacromolecule loading. On the other hand, functionalization of organic groups onto the surface of HNTs is necessary to enable covalent HNTs–biomacromolecule bonding. Furthermore, the step for biomacromolecule conjunction should be facile and mild so as to maintain the biological activity of biomacromolecules. Consequently, novel surface modification methods and new chemistries are urgently desired to introduce proper organic groups onto the surface of HNTs for easy and green biomacromolecule immobilization. In addition, methods of packing HNTs for nanostructured architectures are also to be explored.

12.2 Dopamine Chemistry for Biomacromolecule Immobilization: Nature's Choice

Attributed to the fascinating functions, structures and the formation processes of biological organisms, nature has reminded us of biomimetic solutions to sticky problems.[13–17] As a typical example, a bio-inspired surface modification strategy based on dopamine (DOPA) chemistry has been arousing renewed interest since 2007, when Messersmith and coworkers explored polydopamine as a multi-functional coating.[18] To be specific, since the marine mussel was found to secrete adhesive proteins (typically Mefp-5) so as to firmly connect its byssus onto a rocky surface, the "bioadhesion" phenomena and principles have evoked growing interest in the past two decades.[19–21] The catechol groups of DOPA residue have long been identified as the key adhesive groups of mussel's adhesive protein (Figure 12.2), while the trivial synthesis procedures of catechol-containing polymers limited their availability.[22,23] Messersmith and coworkers found that dopamine, a small-molecule catechol compound, can self-polymerize into polydopamine and simultaneously deposit on almost any type of inorganic and organic materials.[18] For example, the materials to be coated can be dipped directly into an aqueous dopamine solution at pH 8.5 (when oxygen acts as the oxidant) or in the presence of oxidants, oxidation-induced polymerization will automatically occur and the substrates become coated with a black polydopamine film. The thickness of the polydopamine layer can be precisely controlled from several nanometers to dozens of nanometers by simply tuning the coating time and the type of oxidation agent, and the film density can be tuned by varying the pH of the dopamine solution.[19,24]

More importantly, Messersmith and coworkers also found that some reactions during dopamine polymerization are useful to enable multi-functional surface modification. As shown in Figure 12.3, dopamine undergoes a series of oxidation and rearrangement reactions with resulting production

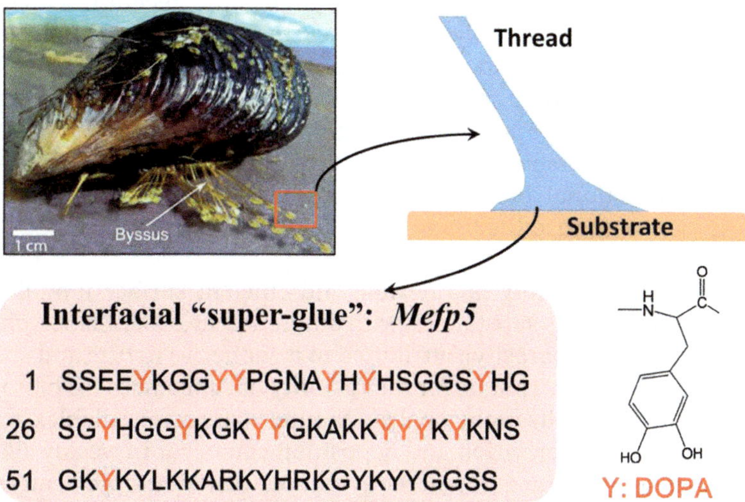

Figure 12.2 Biodistribution and amino acid composition of mussel adhesive proteins of *Mytilus edulis*.

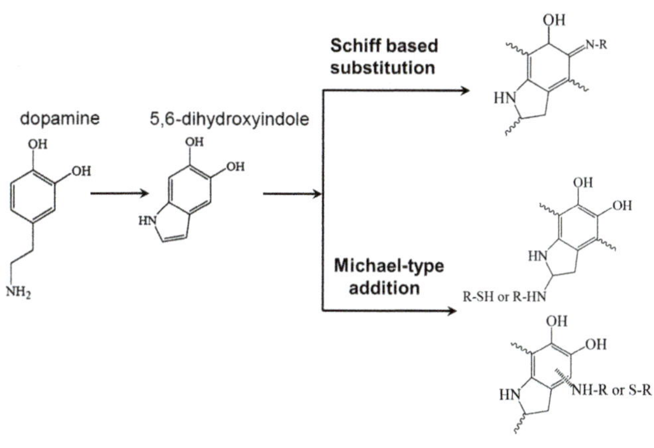

Figure 12.3 The useful dopamine polymerization routes for surface modification.

of 5,6-dihydroxylindole, a widely acknowledged intermediate during dopamine polymerization. For the following polymerization mechanisms, multiple possible routes were proposed, of which Michael addition or Schiff-based reactions were found effective to support further functionalization by thiol- or amino-containing species.[18] Waite proposed the concept of "secondary treatment" to describe the reaction platform enabled by polydopamine coating, and predicted its potential in producing various functional materials.[25] Nowadays, under the guidance of dopamine chemistry, successive polydopamine deposition and secondary treatment has become a generic, facile, and mild strategy for multi-functional coatings. Although the details of dopamine aggregation mechanisms have recently been found to be slightly different from Messersmith's hypothesis (for example, recent research data support that the so-called "polydopamine" actually comprises of dimer and trimer aggregates rather than covalently polymerized macromolecules), the mussel-inspired functional coating has proved an effective surface modification strategy for various application fields.[26,27]

In particular, biomacromolecule immobilization is a typical example where multi-functional polydopamine coatings offer significant benefits, which can be summarized as follows:

- Polydopamine coating is a flexible and efficient surface modification method for various substances (with different materials, shapes, characteristic size, and porosity), attributed to its adhesive nature and specific surface chemistry. That is, for the majority of substances for biomacromolecule immobilization, a polydopamine coating is available.
- Dopamine polymerization and deposition can occur in weakly alkaline aqueous solutions at ambient temperature, without the aid of any other additive. As such, polydopamine coating involves no complex linkers, is free of organic solvents and has good biocompatibility, making it suitable for biomaterial applications.
- The abundant functional groups (*i.e.*, catechol and amine) existing on the modified surface and the possibility of secondary treatment (based on Michael addition or Schiff-based reactions) could enhance the binding ability for various biomolecules.

Consequently, it is reasonable to speculate that dopamine chemistry is nature's choice for biomacromolecule immobilization. The first systematic study on conjugation of biomolecules to a polydopamine coating was also reported by Messersmith's group in 2009, where the chemical reactivity of polydopamine was found to be much higher than the well-known *N*-hydroxysuccinimide (NHS) after long-term preincubation. The authors found that the selectivity of the reaction with amine or imidazole functional groups of biomolecules can be tuned through pH control.[28] Jiang's group prepared polydopamine microcapsules using $CaCO_3$ as the sacrificing template for simultaneously immobilizing three enzymes within the lumen, within the shell, and outside the shell.[29] More interestingly, they found that much higher enzyme loading can be realized by co-filling enzyme and dopamine into the pores of mesoporous silica. The polydopamine deposited in the pores was found to block enzyme release due to the co-contributions from covalent binding and steric hindrance, and therefore the stability of immobilized enzyme was significantly improved.[10]

12.3 Dopamine Modified Halloysite Nanotubes for Enzyme Immobilization

Considering the intrinsic benefits of HNTs as enzyme support and the encouraging preliminary results of dopamine-enabled enzyme immobilization, researchers may wonder whether the dopamine–HNTs interactions favor the polydopamine coating process, like the cases of other substrates. On one hand, the siloxane (Si–O–Si) groups outside HNTs contain strongly electronegative oxygen atoms, which can interact with the amino groups ($-NH_2$/–NH–) and catechol groups of polydopamine *via* hydrogen bonding. On the other hand, the gibbsite-like array of aluminol (Al–OH) groups inside HNTs can interact with polydopamine *via* both hydrogen bonding and metal–organic coordination. Furthermore, the successful coating of polydopamine on HNTs has been reported in the literature.[30,31] Therefore, the versatile platform technique based on dopamine chemistry is supposed to be available for HNT modification and enzyme immobilization.[5]

12.3.1 Fabrication Procedure

According to the mentioned dopamine chemistry, the whole fabrication procedure contains only two facile and mild steps: (i) HNT functionalization: a neutral solution of dopamine hydrochloride was added into an alkaline HNT suspension followed by constant stirring, resulting in dopamine-coated HNTs (DHNTs); (ii) enzyme immobilization: DHNTs were added into a series of enzyme solutions with different concentrations followed by water-bath shaking for a certain time, and the increment of enzyme loading with its concentration became negligible until the enzyme concentration approached to a certain value. Chao *et al.* selected laccase as a model enzyme,[5] and the corresponding fabrication procedure is shown in Figure 12.4.

12.3.2 Morphologies of Dopamine Modified Halloysite Nanotubes

HNTs possessed a hollow tubular structure with both ends open, and the tube wall thicknesses of HNTs varied in the range of 10–15 nm (Figure 12.5a). For DHNTs, the tube wall became a little thicker (Figure 12.5b), implying that a nanolayer had formed on the surface of the HNTs. From Figure 12.5 it was also observed that the outer diameter of nanotubes increased by ~9 nm, while the inner diameter of nanotubes decreased by ~6 nm, compared with

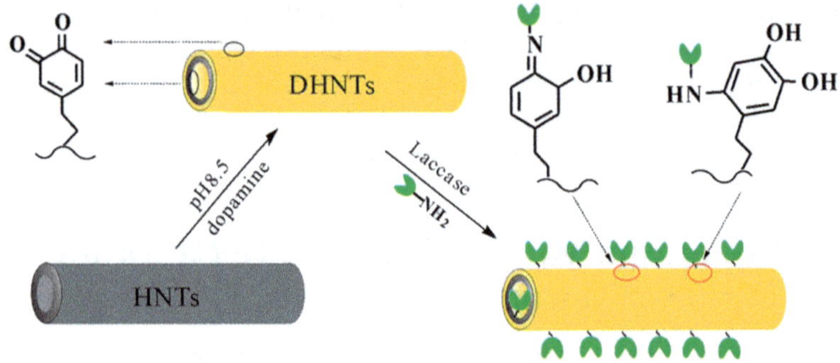

Figure 12.4 Scheme of the preparation of DHNTs and the subsequent enzyme immobilization on them. (Reproduced with permission from C. Chao, J. D. Liu, J. T. Wang, Y. W. Zhang, B. Zhang, Y. T. Zhang, X. Xiang, R. F. Chen, *ACS Appl. Mater. Interfaces*, 2013, **5**, 10 559. Copyright (2013) American Chemical Society.[5])

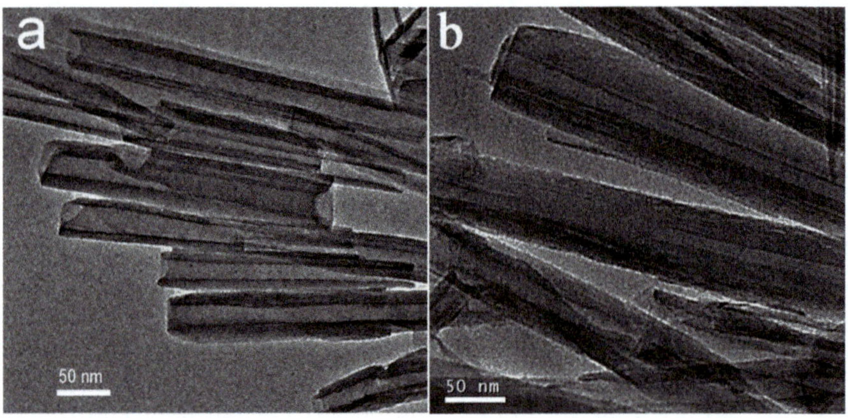

Figure 12.5 TEM images of (a) HNTs, and (b) DHNTs. (Reproduced with permission from C. Chao, J. D. Liu, J. T. Wang, Y. W. Zhang, B. Zhang, Y. T. Zhang, X. Xiang, R. F. Chen, *ACS Appl. Mater. Interfaces*, 2013, **5**, 10 559. Copyright (2013) American Chemical Society.[5])

the pristine HNTs. These results suggested that the dopamine polymer layers existed on both the inner and outer surfaces of HNTs.

12.3.3 Surface Chemical Structures of Dopamine Modified Halloysite Nanotubes

The surface chemical structures of HNTs and DHNTs are important to understand the process of polydopamine coating on HNTs. Figure 12.6a shows the presence of O, C, N, Si, and Al, confirming the presence of ultrathin layers

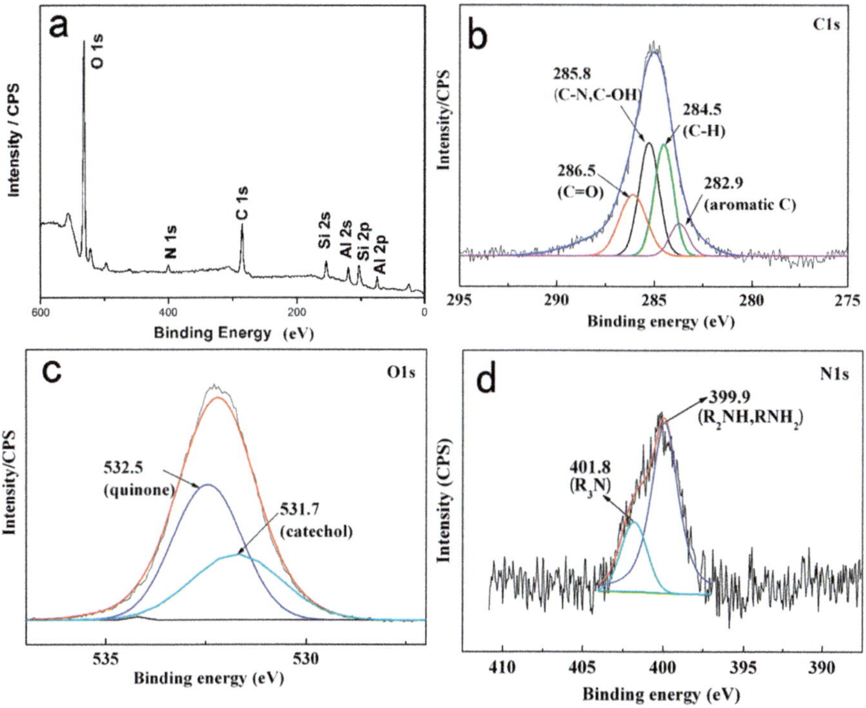

Figure 12.6 XPS analyses of DHNTs: (a) survey scan and (b) C 1s, (c) O 1s, and (d) N 1s core levels. (Reproduced with permission from C. Chao, J. D. Liu, J. T. Wang, Y. W. Zhang, B. Zhang, Y. T. Zhang, X. Xiang, R. F. Chen, *ACS Appl. Mater. Interfaces*, 2013, **5**, 10 559. Copyright (2013) American Chemical Society.[5])

on the tube surface. The complex chemical structure of polydopamine on DHNTs was evidenced in the deconvoluted core-level spectra of C 1s, O 1s, and N 1s (Figure 12.6b–d). The peaks at 286.5 eV in the C 1s spectra (Figure 12.6b) and at 531.7 eV in the O1s spectra (Figure 12.6c) demonstrated the oxidation of catechol into quinone or semi-quinone. According to the high resolution O 1s spectra, about 60% of phenolic hydroxyl groups of dopamine had been converted to carbonyl groups. Since it is commonly believed that catechol groups were oxidized to quinone form in alkaline conditions, and that the formation of polydopamine was ascribed to the crosslinking reaction between the catechol form and *o*-quinone form, the proportion of *o*-quinone form monomer implied the oxidation and polymerization degree of dopamine. Besides, the peak at 401.8 eV in the high-resolution N 1s spectra (Figure 12.6d) illustrated the presence of an R_3N group, which provided important evidence of the covalent linkage of dopamine monomers and the formation of polydopamine.

According to the surface chemical structures of HNTs and DHNTs, with the aforementioned dopamine chemistry in mind, the mechanism of polydopamine coating on HNTs could be summarized as the following points:

- Dopamine underwent a series of oxidation, cyclization, and rearrangement reactions, and resulted in some intermediate products (*e.g.* 5,6-dihydroxylindole, indoledione), which acted as diverse monomers of polydopamine.
- Covalent bonds among the monomers were evidenced, indicating the occurrence of a crosslinking reaction.
- About 60% of the catechol groups of dopamine were converted to quinone groups, which could be used to estimate the oxidation and polymerization degree of dopamine.
- The deposited dopamine could be firmly attached onto HNTs *via* hydrogen bonding, metal–organic coordination, cation–π interactions, *etc.*

12.3.4 Enzyme Immobilization on Dopamine Modified Halloysite Nanotubes

The evaluation of an immobilized enzyme mainly depends on enzyme loading capacity, activity recovery, stability, and catalytic efficiency. DHNTs were found to immobilize much more laccase than HNTs (168.8 mg g^{-1} *vs.* 11.3 mg g^{-1}), as shown in Figure 12.7. Considering that the surface area of DHNTs was found to be slightly larger than that of HNTs (63.4 m^2 g^{-1} *vs.* 59.6 m^2 g^{-1}), such a significant loading capacity enhancement could be attributed to

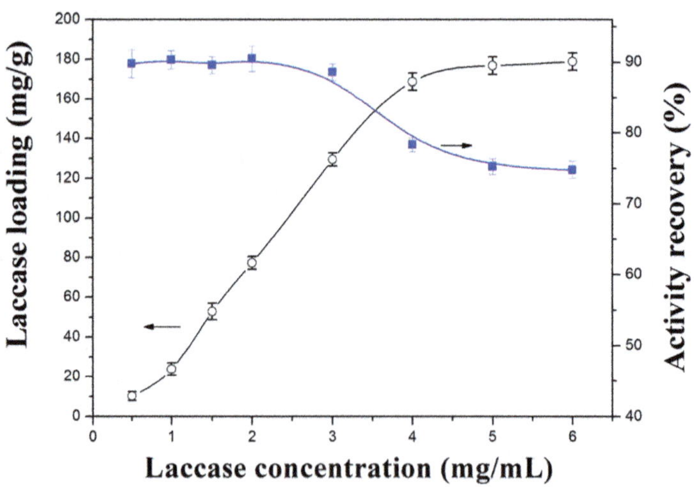

Figure 12.7 Effect of the initial laccase concentration on the laccase loading and activity recovery. (Reproduced with permission from C. Chao, J. D. Liu, J. T. Wang, Y. W. Zhang, B. Zhang, Y. T. Zhang, X. Xiang, R. F. Chen, *ACS Appl. Mater. Interfaces*, 2013, **5**, 10 559. Copyright (2013) American Chemical Society.[5])

the presence of polydopamine on the outer surface and/or the inner lumen of HNTs, which allowed secondary treatment (*via* Michael addition reaction or Schiff-based reaction) for effective laccase binding. The accurate amount of laccase loaded outside or inside the nanotubes could not be measured under the state-of-the-art characterization techniques. However, it is reasonable to estimate that the loading percentage outside the nanotubes is much higher than that inside the nanotubes, because the amount of polydopamine outside the nanotubes was observed to be much larger than that inside the nanotubes.

The activity recovery at different laccase concentration is usually calculated through dividing the activity of the immobilized laccase obtained immediately after the immobilization procedure by that of the initial laccase solutions, and the laccase activity can be measured by the previously described ABTS oxidation method.[32] As shown in Figure 12.7, the activity recovery maintained above 90% at low laccase concentration, while it gradually dropped when the laccase concentration was more than 3 mg mL^{-1}. Under a high laccase concentration up to 4.0 mg mL^{-1}, the immobilized laccase showed a rather low activity recovery down to 78%, hinting that excessive laccase loading might lead to the crowding or agglomeration of enzyme molecules onto the surface of supports, resulting in steric constraint that could hinder the availability of the active sites.

The application stability of immobilized enzyme was evaluated by the thermal stability, storage stability, and reusability. (i) Thermal stability: as shown in Figure 12.8a, the immobilized laccase exhibited a slower rate of thermal inactivation at 60 °C than that of free laccase, demonstrating a significant resistance to thermal inactivation. Notably, the immobilized one showed much higher activity than the free enzyme after a 280 min test (85% *vs.* 34%). These results suggested that the immobilized laccase could tolerate most of the unexpected temperature variations during practical

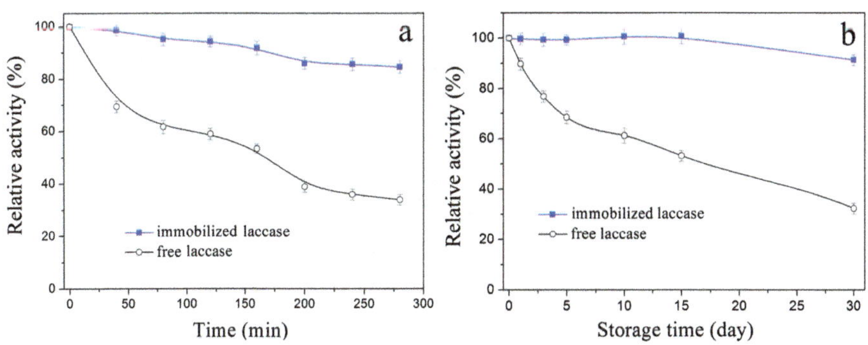

Figure 12.8 Stability of the free and immobilized laccase (4.0 mg mL^{-1}): (a) thermal stability at 60 °C and (b) storage stability at 4 °C. (Reproduced with permission from C. Chao, J. D. Liu, J. T. Wang, Y. W. Zhang, B. Zhang, Y. T. Zhang, X. Xiang, R. F. Chen, *ACS Appl. Mater. Interfaces*, 2013, 5, 10 559. Copyright (2013) American Chemical Society.[5])

operations. The enhancement in thermal stability could be attributed to the preservation of the tertiary structures and the active sites of the enzyme by the support. (ii) Storage stability: the storage stability of immobilized laccase was greatly improved compared to that of the free laccase. It can be observed from Figure 12.8b that the free laccase in a phosphate buffer (100 mM, pH 5.0) retained 32.3% activity after 30 days, while the immobilized laccase still held more than 90% activity after the same period. The improvement in storage stability resulted from the stabilization of its active conformation by multipoint-bond interactions between the adjacent support and enzyme. (iii) Reusability: as shown in Figure 12.9, the immobilized enzyme on DHNTs maintained more than 92% activity within the first five cycles. A sharp decrease of activity occurred during the sixth cycle, and then the decline trend of activity slowed down again. The activity loss might be imputed to enzyme leakage during washing and enzyme deactivation during its recycling procedures and repeated uses. Since the covalently immobilized enzyme was known to prevent itself from release, it was reasonable to assume that partial immobilized enzyme molecules were connected to the support *via* weaker interactions, *e.g.* hydrogen bonding, electrostatic interaction. Nevertheless, after 10 cycles, the immobilized enzyme retained more than 70% of its initial activity.

To test the catalytic efficiency of the immobilized laccase on DHNTs, phenolic compounds of 2,4-dichlorophen were selected as the target organic contaminant to be removed from its aqueous solution (50 mg L^{-1}), since laccase can react with dissolved phenolic compounds and form insoluble

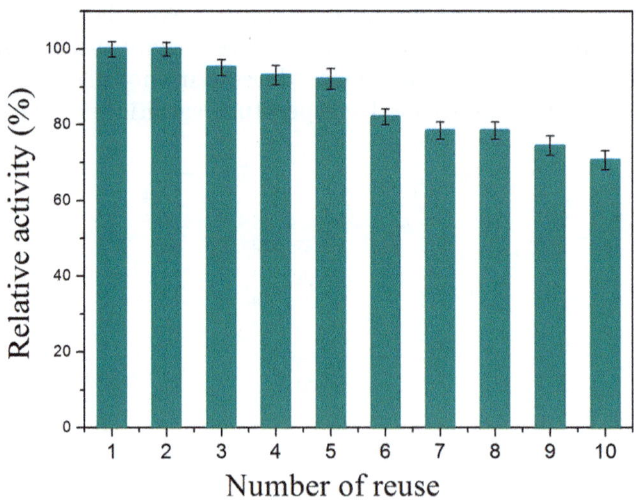

Figure 12.9 Reusability of immobilized laccase after cycles of use. (Reproduced with permission from C. Chao, J. D. Liu, J. T. Wang, Y. W. Zhang, B. Zhang, Y. T. Zhang, X. Xiang, R. F. Chen, *ACS Appl. Mater. Interfaces*, 2013, **5**, 10 559. Copyright (2013) American Chemical Society.[5])

products. As shown in Figure 12.10, a rather high removal efficiency up to about 65% had been achieved in the first two hours. After a reaction period of 2 h, the increase trend of removal efficiency *vs.* reaction time became slower and slower, demonstrating that the reaction rate decreased on account of the simultaneous consumption of the reacting species. Obviously, although the immobilized laccase displayed lower removal efficiency than free laccase (especially in the initial stage), it became closer to the value of free laccase with the increment of reaction time. At a reaction time of 20 h, the removal efficiency by free laccase reached 96.2%, only 2.1% higher than that by immobilized laccase (94.1%). These facts might be interpreted as the occurrence of activity loss of free enzyme during long reaction times.

Concluding from the fairly good stability and catalytic efficiency of immobilized enzyme on DHNTs, the design and fabrication of halloysite–dopamine hybrid nanotubes had proved effective to acquire applicable biomacromolecule supports for practical applications. The benefits of improving biomacromolecule immobilization efficiency and stability from dopamine chemistry have been identified. However, the severe agglomeration of HNTs during polydopamine coating was not solved, which restricted the available surface area of biomacromolecule supports. Further enhancement in biomacromolecule immobilization performance is to be explored by designing and tuning the physical structures of HNT aggregates.

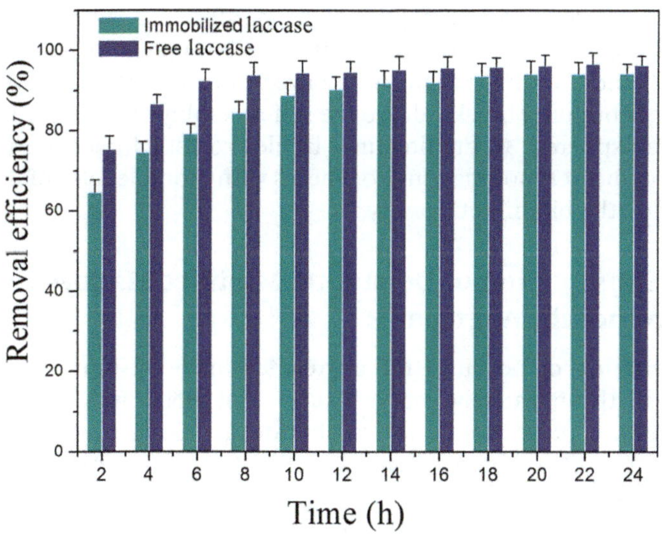

Figure 12.10 Removal efficiency of 2,4-dichlorophen by free and immobilized laccase with time. (Reproduced with permission from C. Chao, J. D. Liu, J. T. Wang, Y. W. Zhang, B. Zhang, Y. T. Zhang, X. Xiang, R. F. Chen, *ACS Appl. Mater. Interfaces*, 2013, **5**, 10559. Copyright (2013) American Chemical Society.[5])

12.4 Dopamine Modified Halloysite Nanotube Aggregates for Enzyme Immobilization

Although HNTs exhibited rather good enzyme immobilization performance after coating with polydopamine, they had a strong tendency to form bundles or agglomerates, which generally resulted in a reduction in effective specific surface area. As mentioned in Section 12.3, one effective and facile approach to address this issue is to design highly ordered HNT aggregates and thus avoid the excessively dense packing of HNTs. In Section 12.1, it has been concluded that fabrication of HNTs into a 3D porous architecture is a promising method. The hierarchical structures of a 3D porous architecture can take the advantages of both nanoscale building blocks (individual HNTs) and micro- or submicro-scale assemblies. Moreover, during the bio-catalyzed reactions, the hierarchical pores in the 3D architecture could enhance the accessibility of active sites of the biomacromolecules to be immobilized by reducing the mass-transfer barriers of reactants.

To construct a 3D porous architecture, one effective and facile approach is generally required to enable self- and/or directed assembly of nanoscale building blocks. However, few studies have been reported on the self-assembly of HNTs into microsized 3D architectures.[33] To assemble natural HNTs into hierarchical porous structures still remains a highly challenging task. In recent years, some naturally-occurring models have provided amazing and inspiring structures for researchers to design and fabricate new functional structured materials with exceptional properties and functionalities.[34,35] Inspired by the interwoven structure of the swallow nest, which is made from straw, mud, and its saliva, Zhang and coworkers chose this kind of natural nanotubes as building blocks, drove the self-assembly of them into nest-like porous microspheres (3D architecture) by electrostatic force and interfacial tension, and further modified microspheres with biomolecules of dopamine to strengthen the biomimetic entity.[36]

12.4.1 Construction of Dopamine-Modified Halloysite Nanotube Aggregates

The construction of porous HNT aggregates included four steps (Figure 12.11). First, the negatively charged surface of HNTs was modified with chitosan (CS), a polycationic polysaccharide, resulting in assembled CS–HNTs with a positive charge. Second, the aqueous suspension containing CS–HNTs was emulsified by an amphiphilic surfactant, oleic acid (OA). It is known that surfactant molecules can self-assemble into micelles above the critical micelle concentration, which is usually utilized as a soft template for synthesis of nanostructures.[37] Herein, OA formed micelles with hydrophilic and negatively charged heads outward and hydrophobic tails inward in an oil–water system. When the micelles met CS–HNTs, the attachment of individual CS–HNTs to the hydrophilic heads of OA was driven by the electrostatic force and large oil–water surface energy, which is larger than

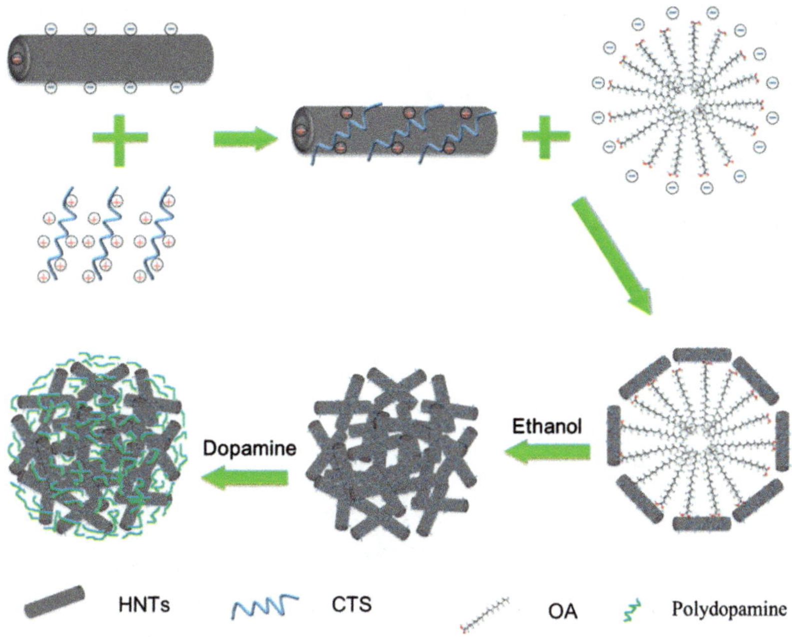

Figure 12.11 Illustration of construction of nanostructured HNT aggregates. (Reproduced with permission from C. Chao, B. Zhang, R. Zhai, X. Xiang, J. Liu, R. Chen, *ACS Sustain. Chem. Eng.*, 2014, **2**, 396. Copyright (2014) American Chemical Society.[36])

the difference between the nanotube–oil and nanotube–water surface energies.[38] In this way, OA micelles and the inner oil droplets were encapsulated by CS–HNTs aggregates. Next, the core–shell structure is transferred into an ethanol medium, in which the oil core was dissolved and the CS–HNTs aggregates in the shell further aggregated to form microspheres (designated as MS(HNTs)s). Finally, the microspheres were collected and then modified by dopamine in a basic aqueous solution, and the resultant microspheres consisting of dopamine and interwoven HNTs were designated as DMS(HNTs)s.

12.4.2 Hierarchical Structures of Dopamine-Modified Halloysite Nanotube Aggregates

The as-prepared DMS(HNTs)s were found to possess hierarchical structures, which could be divided into three aspects: (i) particle-scale structures, (ii) nanoscale structures, and (iii) molecular scale structures.

The particle-scale structures of DMS(HNTs)s were observed from optical microscope and scanning electron microscopy images, as shown in Figure 12.12 and Figure 12.13, respectively. The photograph in Figure 12.12a gave the evidence of microsphere structures, confirming the assumed assembly process in Figure 12.11. Figure 12.13a and b showed the nest-like structure

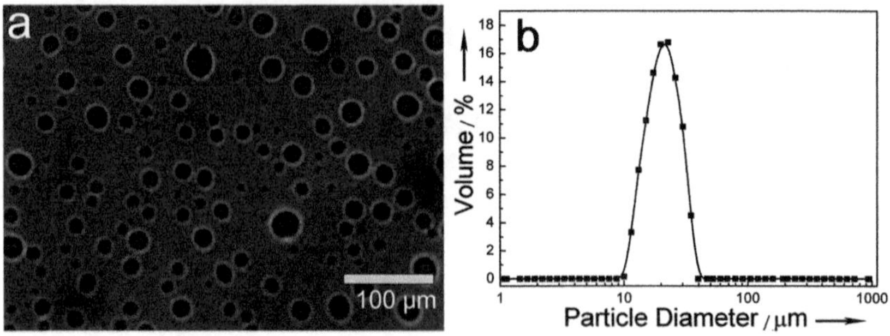

Figure 12.12 (a) Optical microscope image and (b) particle size distribution of the microspheres measured by a particle size analyzer. (Reproduced with permission from C. Chao, B. Zhang, R. Zhai, X. Xiang, J. Liu, R. Chen, *ACS Sustain. Chem. Eng.*, 2014, **2**, 396. Copyright (2014) American Chemical Society.[36])

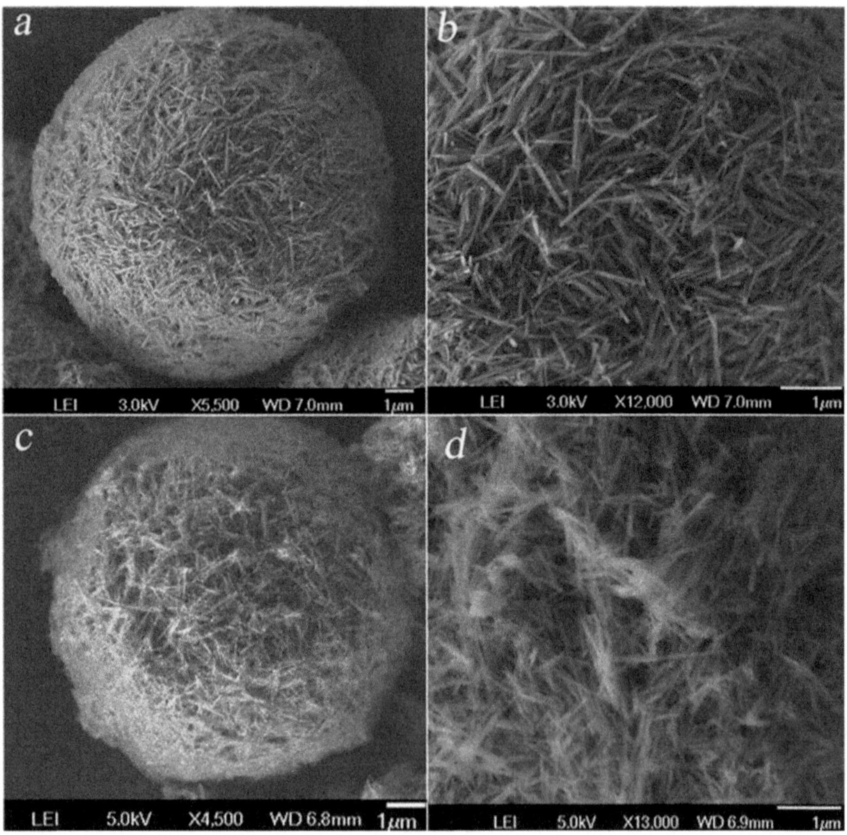

Figure 12.13 SEM photographs of the microspheres before (a) and after (b) modification of dopamine (c) and (d). (Reproduced with permission from C. Chao, B. Zhang, R. Zhai, X. Xiang, J. Liu, R. Chen, *ACS Sustain. Chem. Eng.*, 2014, **2**, 396. Copyright (2014) American Chemical Society.[36])

of MS(HNTs)s and the intrinsic tubular structure of HNTs, while in Figure 12.13c and d the microspheres were shown to be covered by cotton-like material, which was assumed to be the deposited polydopamine. The size distribution of the diameter of DMS(HNTs)s was examined using a particle size analyzer. Figure 12.12b shows a size distribution from 10 to 30 μm, and the maximum probability density was located at 22 μm. It was noted that the size of microspheres was much larger than that of an individual micelle of OA (several tens of nm).[37] Therefore, the formation of microspheres was involved in assembly from hundreds of self-assembled OA-micelle-encapsulated structures.

The nanoscale structures of DMS(HNTs)s mainly referred to the nanoporous structures. According to Table 12.1, the microspheres exhibited a higher BET surface area and larger pore volume than those of both HNTs and DHNTs, respectively, revealing the benefits from 3D structured microspheres. The pore size of DMS(HNTs)s showed a wide distribution from mesopores to macropores, demonstrating the hierarchical pore structure, and most of the pore size values lay in the range of mesopores (2–50 nm, IUPAC) (Figure 12.14). The pore distribution within the microspheres could be attributed to the interparticle space of the nanotubes as well as the contribution from the lumen space of HNTs. Especially, some of the relatively small mesopores (pore size < 20 nm) were assumed to correspond to the lumen of dopamine-coated HNTs, since the mean pore size of the pristine HNTs (~20 nm)

Table 12.1 Comparison of BET surface area and pore volume of HNTs, DHNTs, and DMS(HNTs)s.

Porous supports	HNTs	DHNTs	DMS(HNTs)s
BET surface area ($m^2\ g^{-1}$)	59.6	63.4	114.6
Pore volume ($cm^3\ g^{-1}$)	0.238	0.156	0.347

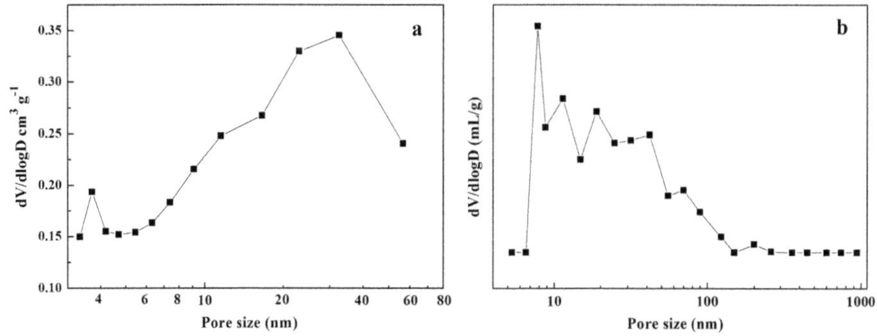

Figure 12.14 Pore size distribution of DMS(HNTs) probed by (a) BET pore size analyzer (b) mercury porosimetry. (Reproduced with permission from C. Chao, B. Zhang, R. Zhai, X. Xiang, J. Liu, R. Chen, *ACS Sustain. Chem. Eng.*, 2014, **2**, 396. Copyright (2014) American Chemical Society.[36])

would decrease after dopamine modification. For biomolecule immobilization, the hierarchical and interconnected pores in the microspheres might be more beneficial to the mass transfer of reactants and products over other narrow-pore supports (*e.g.* zeolites). In the current work, the appearance of larger pores was advantageous to enzyme immobilization or entrapment.

The molecular-scale structures of DMS(HNTs)s were characterized by XPS survey spectra. Figure 12.15a shows the additional presence of C and N, compared with the pristine HNTs. Its N/C ratio (0.084) was lower than the theoretical value of dopamine (N/C = 0.125), which might be explained by the absence of elemental N in OA, although some OA molecules had been washed out during the process of ethanol treatment (Figure 12.11). Like the case of DHNTs, the complicated nature of polydopamine on the microspheres was revealed by the different chemical states of elemental C, O, and N in deconvoluted core level spectra (Figure 12.15b–d). The four deconvoluted peaks of the C1s spectra (Figure 12.15b) agreed well with those of DHNTs, except for the peak of the C=O group (287.3 eV for DMS(HNTs)s and 286.5 eV for DHNTs). The decrease of the binding energy for the carbonyl C atom might be ascribed to the presence of acetyl group on CS and the C–O–C linkage among CS monomers. Moreover, according to the deconvoluted N1s spectra

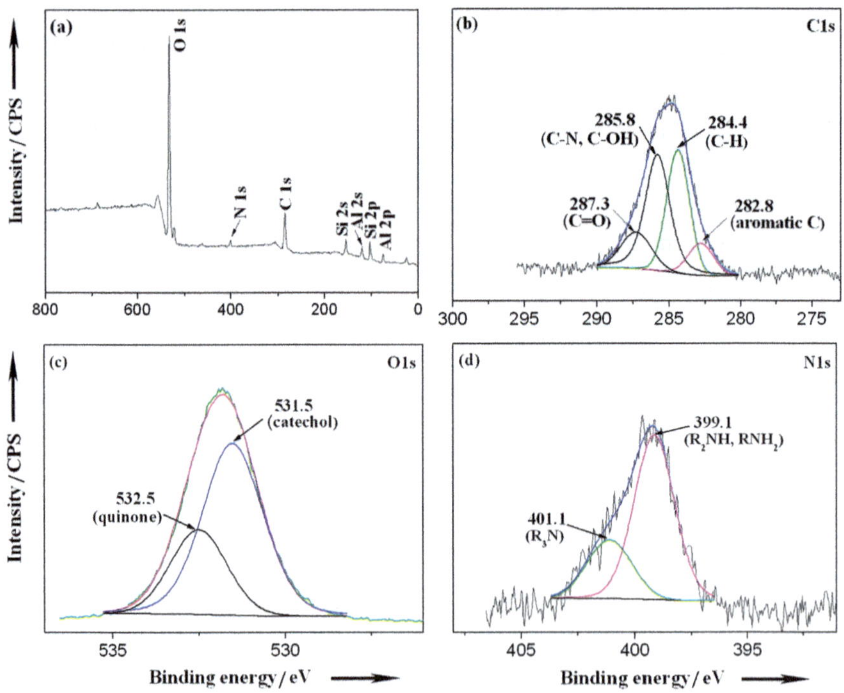

Figure 12.15 XPS analyses of DMS(HNTs) (a) survey, (b) C1s, (c) O1s, and (d) N1s core level. (Reproduced with permission from C. Chao, B. Zhang, R. Zhai, X. Xiang, J. Liu, R. Chen, *ACS Sustain. Chem. Eng.*, 2014, **2**, 396. Copyright (2014) American Chemical Society.[36])

(Figure 12.15d), the fitted peak for $-NH_2$/–NH– was located at 399.1 eV, lower than that of DHNTs (399.9 eV), because the $-NH_2$ of CS might have reacted with dopamine *via* Michael addition reaction and connected to the aromatic ring of dopamine. The fitted peak for –N= groups appeared at 401.1 eV, lower than the case of DHNTs (401.8 eV, respectively), indicating that the $-NH_2$ of CS might have also reacted with dopamine *via* Schiff-based reaction, resulting in a –C=N– bond. Consequently, evidence of both Michael addition and Schiff-based reactions could be found between CS and dopamine, confirming the surface modification mechanism proposed previously.

To sum up, the benefits from the hierarchical structures of DMS(HNTs)s encompass the following three aspects: (i) good dispersion of HNTs and easy recycling of immobilized enzymes, resulting from the micro-size nest-like aggregates (particle-scale structures); (ii) rapid mass transfer of reactants and products, resulting from the hierarchical mesopores and nanopores (nanoscale structures); (iii) stable immobilization of enzymes, resulting from the reactive groups of polydopamine (molecular scale structures).

12.4.3 Enzyme Immobilization on Dopamine Modified Halloysite Nanotube Aggregates

The laccase loading on the support was found to increase steadily with its initial concentration, while it approached the maximum value when the concentration of laccase increased up to 3.5 mg mL^{-1}. In this case, the loading on DMS(HNTs)s reached as high as 311.2 mg g^{-1}, which is 30 times higher than that on pristine HNTs (11.3 mg g^{-1}), 14 times higher than CS-modified HNTs, and twice as high as DHNTs. Furthermore, the capacity was an order of magnitude higher than most of the previously reported porous supports, as shown in Table 12.2. Although the laccase loading reported in this work could not rival that of the carbon-based mesoporous magnetic composites (491.7 mg g^{-1}), the absence of a covalent bond for the latter work could result in severe enzyme release. Moreover, DMS(HNTs)s also exhibited high loading

Table 12.2 Comparison of enzyme loading of other supports with current work.

Porous supports	Laccase loading (mg g^{-1})	Reference
Nanoporous gold	15.5	40
Chitosan	20	41
Sepabeads EC-EP3	32.6	42
Magnetic mesoporous silica spheres	82	43
Cu^{2+}-chelated magnetic mesoporous silica spheres	98.1	44
Carbon-based mesoporous magnetic composites	491.7	45
HNTs	11.3	This work
HNTs–chitosan	21.5	This work
DHNTs	168.8	This work
DMS(HNTs)	311.2	This work

ability for other enzymes (Table 12.3), indicating their generality for biomolecule immobilization.

Such a high loading capacity for DMS(HNTs)s could be interpreted from three aspects: (i) the modification of polydopamine introduced abundant functional groups to effectively bind laccase; (ii) the increased surface area of DMS(HNTs)s compared to the DHNTs (114.6 *vs.* 63.4 $m^2\ g^{-1}$) afforded more physical binding sites for enzymes; (iii) the hierarchically distributed pore size allowed low-resistance diffusion of enzyme into DMS(HNTs)s. In a word, DMS(HNTs)s integrated favorable chemical structures and physical structures for biomacromolecule immobilization.

The retention of activity in the immobilized enzymes is required for their practical application. The specific activities of the free and immobilized laccase (at an initial concentration of 3.0 mg mL^{-1}) in a phosphate buffer (pH 4.2, 30 °C) are given in Table 12.4. The laccase immobilized on the microspheres could retain a rather high specific activity of 81.8%. Further experiments indicated that the immobilized urease and HRP on DMS(HNTs)s also had a high specific activity of 86.5% and 78.6%, respectively. These facts indicated that the immobilization did not cause notable denaturation of the enzymes.

The activity of the immobilized laccase on DMS(HNTs)s was also evaluated under a variety of conditions (Figure 12.16). (i) The activity of the free and immobilized laccase as a function of pH value is shown in Figure 12.16a. Both the free and immobilized laccase exhibited a maximal activity at pH 3.2 and almost completely lost their activity above pH 6.0. Compared with free laccase, the immobilized laccase showed slightly higher adaptability to changes in pH of the medium below pH 5.0. (ii) The storage stability of free and immobilized laccase was measured in phosphate buffer (0.1 M, pH 4.2) at 4 °C for a predetermined period of time. As shown in Figure 12.16b, the activity of free laccase dropped much faster over time than that of the immobilized laccase. The immobilized laccase still kept around 75% activity in the buffer solution after 50 days. In contrast, the free laccase only maintained approximately 32% of its original activity over the same period. Nevertheless, the storage stability of the immobilized laccase on DMS(HNTs)s was a little worse than that on DHNTs, which could be interpreted that the larger

Table 12.3 Loading amount of different enzymes on DMS(HNTs).

Enzyme	Laccase	Urease	Horseradish peroxidase
Loading (mg g^{-1})	311.2	256.3	295.6

Table 12.4 Activity of free laccase and immobilized laccase on DMS(HNTs)s.

	Specific activity (U mg^{-1})	Activity recovery (%)
Free laccase	13.2	100
Immobilized laccase	10.8	81.8

pores of DMS(HNTs)s led to the leakage of the physically adsorbed laccase molecules. (iii) The thermal stability of the free and immobilized laccase was measured by handling them at a constant temperature of 60 °C in a buffer solution of pH 4.2 for a certain incubation time (Figure 12.16c). The activity of the immobilized laccase decreased much more slowly than that of the free laccase. After an incubation period of 4 h, the remaining activities of the free laccase and immobilized laccase were 32.2% and 82.6%, respectively. (iv) The reusability of immobilized laccase was estimated by measuring the relative activity after every cycle of the reaction. As shown in Figure 12.16d, more than 70% of the initial activity was retained after 10 cycles, indicating that most enzyme molecules could be immobilized firmly within the porous microspheres and not leaked from the pore channels. This result was rather good, compared with the reusability of immobilized laccase on DHNTs. Although the large pore of DMS(HNTs)s might lead to enzyme leakage, the high surface area was beneficial for a more even distribution of laccase, thus most of the intrinsic active sites on laccase remained available after being immobilized.

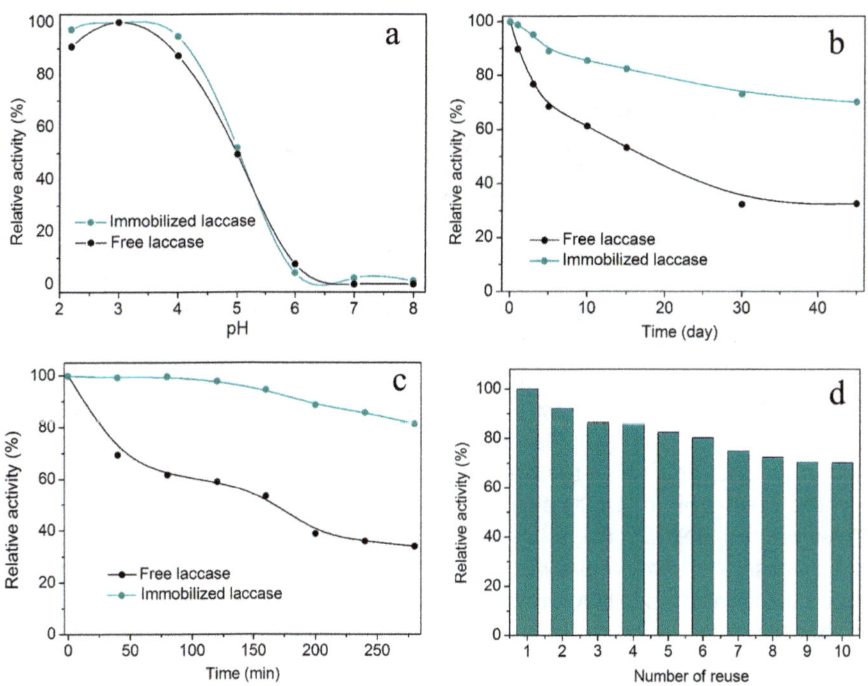

Figure 12.16 Properties of laccase. (a) Activity of free and immobilized laccase at different pH values. (b) Stability of free and immobilized laccase at 4 °C. (c) Activity of free and immobilized laccase as a function of time at 60 °C. (d) Reuse capacity of immobilized laccase. (Reproduced with permission from C. Chao, B. Zhang, R. Zhai, X. Xiang, J. Liu, R. Chen, *ACS Sustain. Chem. Eng.*, 2014, **2**, 396. Copyright (2014) American Chemical Society.[36])

In general, the biomimetic porous DMS(HNTs) was prepared and verified as an excellent support for enzyme immobilization. The major advantages of 3D structured microspheres over pristine HNTs in improving HNT dispersion and enhancing surface area were revealed. The presence of macropores reduced the inward diffusion resistance of macromolecules, while the majority mesopores served as harbors for enzymes. Consequently, the porous microspheres could provide a favorable environment for achieving high enzyme activity, significantly improved storage, and recycling stability. In addition, the preparation of microspheres was scalable and green owing to the abundant resource of HNTs and the mild procedure.

Despite these benefits, it is notable that a partial portion of immobilized biomacromolecules could not firmly attach on the supports after cycles of use or long-term storage, because only single-layer chemical adsorption was allowed on the finite surface area. One possible solution is to immobilize some biomacromolecules during the step of dopamine coating, and thus some biomacromolecules could be inserted or buried in the polydopamine layer. The ultimate solution of this issue remains enlarging the available surface area or blocking the possible leakage access of the specific biomacromolecules, which requires better designed hierarchical pores within the supports.

12.5 Other Potential Applications of Halloysite–Dopamine Hybrid Nanotubes for Biomacromolecule Immobilization

Nucleic acids represent an important class of biomacromolecules. In Section 12.1, DNA was mentioned to be successfully immobilized on HNTs.[1] In that work, DNA was assumed to wrap HNTs mainly due to the π–π interactions between the backbone of DNA and the outer surface of HNTs. The DNA-wrapped HNTs showed much higher solubility in water (10 mg mL^{-1}), and the phosphate side groups were assumed to play pivotal roles in the process of solubilization. This important finding illustrated that DNA-wrapped HNTs were excellent candidates for sustained drug delivery. For example, Lee *et al.* selected doxorubicin (DOX) as a model anticancer agent and loaded it onto DNA-wrapped HNTs.[39] The DOX-loaded DNA-wrapped HNTs show sustained DOX release over two weeks without an initial burst of DOX. However, dopamine modification of DNA-wrapped HNTs has rarely been reported yet. It is reasonable to assume that the inner space of HNTs and the functional groups can be tuned by polydopamine coating, rendering DNA-wrapped HNTs better drug release properties.

Glycosaminoglycan is another type of biomacromolecule that can be found in connective tissue of organisms. It is known to bind to some specific target proteins, and such glycosaminoglycan–receptor interactions are important for biomolecule recognition. For example, Lee *et al.* grafted partially thiolated hyaluronic acid (HA) onto a variety of polydopamine coated substrates and found that human megakaryocytic M07e did not adhere to polydopamine but did adhere to HA-grafted polydopamine surfaces, because of the expression of HA receptor CD44.[18] Consequently, immobilized glycosaminoglycan on dopamine-modified HNTs has potential applications in preventing tumor metastasis.

HNTs were also reported to immobilize macrocyclic complexes with catalytic activity, although they do not strictly belong to macromolecules. For example, Machado *et al.* managed to immobilize anionic and cationic metalloporphyrins into HNTs and investigated the catalytic activity of these novel materials in the oxidation of organic substrates, *e.g.* cyclo-octene, cyclohexane, and *n*-heptane. In the case of anionic metalloporphyrins, the aluminol groups were identified as their preferential sites on the support, which largely occurred at the ends of the tube. As for cationic metalloporphyrins, the immobilization was assumed to occur through the SiO– groups or/and on the surface of the clay. However, due to the lack of strong interactions between metalloporphyrins and HNTs, the loading amount of metalloporphyrins was reported to be very low. Bearing the mussel-inspired dopamine chemistry in mind, the interactions between metalloporphyrins and HNTs are envisaged to be enhanced *via* the mediation of polydopamine, and the loading amount of metalloporphyrins is expected to be increased. The relevant research fields remain to be explored.

12.6 Conclusion

This chapter deals with an interdisciplinary research field encompassing HNTs, marine mussel-inspired dopamine chemistry, and biomacromolecule immobilization. That is, inspired by marine mussel's bioadhesion principle, dopamine chemistry was employed for facile surface modification of HNTs and the conjunction between HNTs and biomacromolecules, especially enzymes. The roles of polydopamine modification in enhancing enzyme loading capacity, stability, reusability, and application efficiency are intensively discussed. Furthermore, the potential applications of halloysite–dopamine hybrid nanotubes for immobilizing other biomacromolecules are mentioned. Due to the simple operation and the absence of precise process control, it will be no challenge to scale up the fabrication procedures reported in this chapter. Engineers can refer to the readily reported techniques to tune the thickness, density, and functionality of the polydopamine layer.

References

1. M. H. Shamsi and K. E. Geckeler, *Nanotechnology*, 2008, **19**, 2791.
2. R. Zhai, B. Zhang, L. Liu, Y. Xie, H. Zhang and J. Liu, *Catal. Commun.*, 2010, **12**, 259.
3. E. Abdullayev and Y. Lvov, *J. Mater. Chem. B*, 2013, **1**, 2894.
4. Y. M. Lvov, D. G. Shchukin, H. Möhwald and R. R. Price, *ACS Nano*, 2008, **2**, 814.

5. C. Chao, J. D. Liu, J. T. Wang, Y. W. Zhang, B. Zhang, Y. T. Zhang, X. Xiang and R. F. Chen, *ACS Appl. Mater. Interfaces*, 2013, **5**, 10559.
6. F. Yang, B. Zhang, J. Liu, H. Zhang, M. Qiao and R. Chen, *J. Chin. Ceram. Soc.*, 2008, **36**, 862.
7. H. Wang, X. Zhao, S. Wang, S. Tao, N. Ai and Y. Wang, *J. Chromatogr. A*, 2015, **1392**, 20.
8. Y. Wang, C. Liu, Y. Zhang, B. Zhang and J. Liu, *ACS Sustainable Chem. Eng.*, 2015, **3**, 1183.
9. P. Yuan, P. D. Southon, Z. W. Liu, M. E. R. Green, J. M. Hook, S. J. Antill and C. J. Kepert, *J. Phys. Chem. C*, 2008, **112**, 15742.
10. H. L. Zhang, X. Y. Zhu, Y. G. Wu, H. Z. Song and X. W. Ba, *J. Mater. Sci.*, 2015, **50**, 4387.
11. R. Zhai, B. Zhang, Y. Wan, C. Li, J. Wang and J. Liu, *Chem. Eng. J.*, 2013, **214**, 304.
12. J. Yao, Q. Wang, Y. Wang, Y. Zhang, B. Zhang and H. Zhang, *Desalin. Water Treat.*, 2015, **55**, 1293.
13. C. Sanchez, H. Arribart and M. M. Giraud Guille, *Nat. Mater.*, 2005, **4**, 277.
14. F. Barthelat, *Philos. Trans. R. Soc., A*, 2007, **365**, 2907.
15. P.-Y. Chen, J. McKittrick and M. A. Meyers, *Prog. Mater. Sci.*, 2012, **57**, 1492.
16. J. Zhao, X. Zhao, Z. Jiang, Z. Li, X. Fan, J. Zhu, H. Wu, Y. Su, D. Yang, F. Pan and J. Shi, *Prog. Polym. Sci.*, 2014, **39**, 1668.
17. W. J. P. Barnes, *Science*, 2007, **318**, 203.
18. H. Lee, S. M. Dellatore, W. M. Miller and P. B. Messersmith, *Science*, 2007, **318**, 426.
19. B. Li, W. P. Liu, Z. Y. Jiang, X. Dong, B. Y. Wang and Y. R. Zhong, *Langmuir*, 2009, **25**, 7368.
20. J. Wang, L. Xiao, Y. Zhao, H. Wu, Z. Jiang and W. Hou, *J. Power Sources*, 2009, **192**, 336.
21. Y. Li, S. Wang, H. Wu, J. Wang and Z. Jiang, *J. Mater. Chem.*, 2012, **22**, 19617.
22. J. L. Dalsin, B. H. Hu, B. P. Lee and P. B. Messersmith, *J. Am. Chem. Soc.*, 2003, **125**, 4253.
23. H. Lee, N. F. Scherer and P. B. Messersmith, *Proc. Natl. Acad. Sci. U. S. A.*, 2006, **103**, 12999.
24. Q. Wei, F. L. Zhang, J. Li, B. J. Li and C. S. Zhao, *Polym. Chem.*, 2010, **1**, 1430.
25. J. H. Waite, *Nat. Mater.*, 2008, **7**, 8.
26. D. R. Dreyer, D. J. Miller, B. D. Freeman, D. R. Paul and C. W. Bielawski, *Langmuir*, 2012, **28**, 6428.
27. S. Hong, Y. S. Na, S. Choi, I. T. Song, W. Y. Kim and H. Lee, *Adv. Funct. Mater.*, 2012, **22**, 4711.
28. H. Lee, J. Rho and P. B. Messersmith, *Adv. Mater.*, 2009, **21**, 431.
29. L. Zhang, J. Shi, Z. Jiang, Y. Jiang, S. Qiao, J. Li, R. Wang, R. Meng, Y. Zhu and Y. Zheng, *Green Chem.*, 2011, **13**, 300.

30. H. Zhang, T. Zhang, J. Wang, F. Pei, Y. He and J. Liu, *Fuel Cells*, 2013, **13**, 1155.
31. L. Yu, Y. Zhang, B. Zhang and J. Liu, *Sci. Rep.*, 2014, **4**, 4551.
32. M. M. Bradford, *Anal. Biochem.*, 1976, **72**, 248.
33. K. Tazaki, *Clays Clay Miner.*, 2005, **53**, 224.
34. C. Ortiz and M. C. Boyce, *Science*, 2008, **319**, 1053.
35. H. Ejima, J. J. Richardson and F. Caruso, *Angew. Chem., Int. Ed.*, 2013, **52**, 3314.
36. C. Chao, B. Zhang, R. Zhai, X. Xiang, J. Liu and R. Chen, *ACS Sustainable Chem. Eng.*, 2014, **2**, 396.
37. H. Chen and J. He, *J. Phys. Chem. C*, 2008, **112**, 17540.
38. G.-H. Wang, Q. Sun, R. Zhang, W.-C. Li, X.-Q. Zhang and A.-H. Lu, *Chem. Mater.*, 2011, **23**, 4537.
39. Y. Lee, G.-E. Jung, S. J. Cho, K. E. Geckeler and H. Fuchs, *Nanoscale*, 2013, **5**, 8577.
40. H. Qiu, C. Xu, X. Huang, Y. Ding, Y. Qu and P. Gao, *J. Phys. Chem. C*, 2009, **113**, 2521.
41. J. B. Zhang, Z. Q. Xu, H. Chen and Y. R. Zong, *Biochem. Eng. J.*, 2009, **45**, 54.
42. A. Kunamneni, I. Ghazi, S. Camarero, A. Ballesteros, F. J. Plou and M. Alcalde, *Process Biochem.*, 2008, **43**, 169.
43. Y. Zhu, S. Kaskel, J. Shi, T. Wage and K. H. van Pee, *Chem. Mater.*, 2007, **19**, 6408.
44. F. Wang, C. Guo, L.-r. Yang and C.-Z. Liu, *Bioresour. Technol.*, 2010, **101**, 8931.
45. Y. Y. Liu, Z. T. Zeng, G. M. Zeng, L. Tang, Y. Pang, Z. Li, C. Liu, X. X. Lei, M. S. Wu, P. Y. Ren, Z. F. Liu, M. Chen and G. X. Xie, *Bioresour. Technol.*, 2012, **115**, 21.

CHAPTER 13

Halloysite Clay Nanotubes for Long Acting Controlled Release of Drugs and Proteins

RENATA MINULLINA[a], JOSHUA TULLY[a], RAGHUVARA YENDLURI[a] AND YURI LVOV*[a,b]

[a]Louisiana Tech University, Institute for Micromanufacturing, 911 Hergot Ave. Ruston, 71272 Louisiana, United States; [b]Kazan Federal University, Kreml uramı 18, Kazan, Republic of Tatarstan, 420008, Russian Federation
*E-mail: ylvov@latech.edu

13.1 Introduction

Halloysite is an abundantly available, naturally occurring nanomaterial that can be used for loading a wide range of biological macromolecules and other chemical agents. This versatile nanotube also offers a new means for drug encapsulation. With an inner diameter of 10–20 nm, there is ample space for loading drugs and for their sustained release (the outer diameter is 50–60 nm and length is *ca.* one micrometer).[1,2] The relatively large lumen also accommodates globular proteins with diameters of 3–6 nm. The duration of the release may significantly vary from hours to days and weeks with various modifications of the halloysite.

Being a natural mineral, halloysite is less toxic than synthetic nanomaterials. It has been shown in studies on tissues and biological cells that

RSC Smart Materials No. 22
Functional Polymer Composites with Nanoclays
Edited by Yuri Lvov, Baochun Guo and Rawil F. Fakhrullin

Published by the Royal Society of Chemistry, www.rsc.org

halloysite can be administered at a concentration of up to 0.5 mg mL^{-1} without harmful effects to cell tissue and microorganisms.[3] The properties of sustained release and biocompatibility emphasize that halloysite can be used as a tiny container for loading and sustained release of drugs and proteins. Many drugs such as resveratrol,[4] tetracycline,[2,5,6] gentamicin,[7] nifedipine, dexamethasone,[8] and doxorubicin[9] have been loaded into halloysite. However, the adsorption efficiency (mass of drug adsorbed *versus* mass of halloysite) of pristine halloysite for drugs is underwhelming. Theoretically, if one considers the average outer and inner diameters of halloysite tubes, they are capable of about 12% (w/w) efficiency. For example, the adsorption capacity of halloysite for dexamethasone, furosemide, and nifedipine is 4–6 wt% [8] and for 5-aminosalicylic acid is only about 0.4 wt%.[10] There are several strategies to increase the adsorption efficiency of halloysite and to extend release time from hours to days.

13.2 Loading and Release of Drugs with Untreated Halloysite Nanotubes

There are two major methods for adsorbing molecules in and outside of halloysite nanotubes. The first is based on loading substances by removing air trapped inside the lumen by applying a vacuum. It is postulated that this process allows for solution to flow inside the tube once the air has been displaced and capillary forces may help carry the target molecules inside the tubes. In general, the process goes as such; halloysite nanotubes are dispersed in a saturated solution of the active chemical agents (inhibitors or drugs). Next, a vacuum is repeatedly applied to the solution. Depending on the solubility of the target molecule, using a solvent such as acetone, alcohol, or hot water will help saturate the solution and increase immobilization efficiency. After the series of vacuum applications the nanotubes are washed to remove as much of the chemical from the surface as possible and dried. The dried tubes contain solidified drugs that will remain in the lumen until placed into solution to release.

This method was used with the drugs furosemide, nifedipine, and dexamethasone, and they were loaded from aqueous solutions with 0, 10, and 50% of ethanol.[8] Dexamethasone is poorly soluble in water (Table 13.1), therefore the presence of ethanol increased the solubility of the drug. Drug loading was higher at higher alcohol to water ratios: for pure water 3 wt% of the drugs were loaded; for 10% alcohol, 4 wt% were loaded; and for 50% alcohol, 5 wt%. The release study of those drugs at pH 7.4 concluded that 70 wt% of dexamethasone, 50 wt% of furosemide, and 40 wt% of nifedipine were released within the first four hours (Figure 13.1a). A linear release was observed for the drugs loaded with higher alcohol concentrations. The release was extended for a duration of 10–20 hours (Figure 13.1b).

The second method of drug loading consists of incubating halloysite overnight in a saturated solution of the desired drug with constant stirring. The

Table 13.1 Parameters essential for loading and release studies of drugs loaded in halloysite.

Drug name	Structure	Solubility in water[b] (mg mL^{-1})	pK_a[b]	Adsorption from media	Release	Reference
Tetracycline hydrochloride	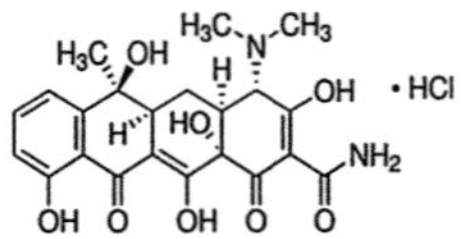	10	3.3	Water and ethanol vacuum[2] Hot water[6]	Water[2] pH 7.4[6]	2,5,6
Gentamicin	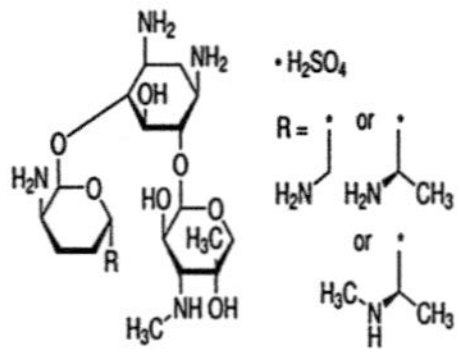	100	12.55[a] 10.18[a]	Aqueous solution 50 mg mL^{-1}	Water	7
Resveratrol	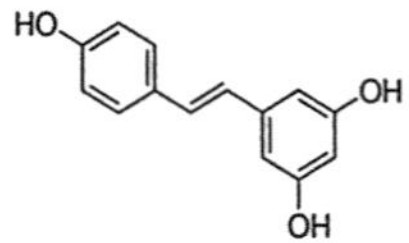	0.03	8.99 9.63 10.64	Ethanol	pH 7.4	4
Furosemide	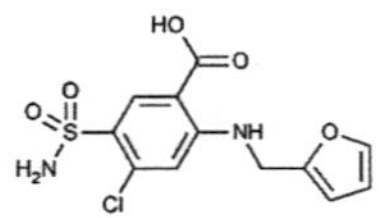	73.1 mg L^{-1}	3.8 7.5	Ethanol	pH 7.4	8
Nifedipine	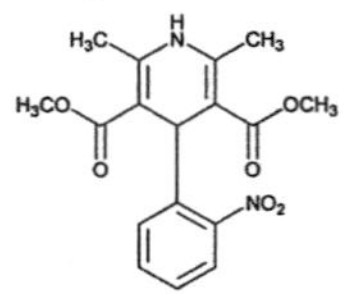	0.0177	5.33	Ethanol	pH 7.4	8
Dexamethasone	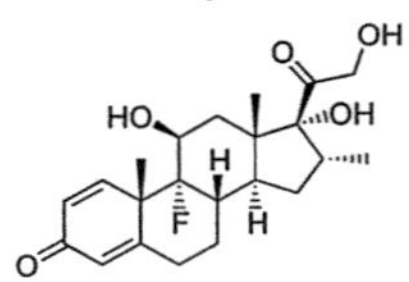	0.0505	12.42	Ethanol	pH 7.4	8

Aspirin	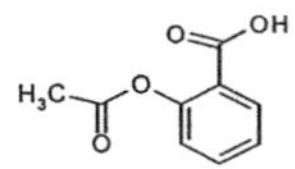	3	3.49	Ethanol 0.03 g mL^{-1} of aspirin + 2 g of HNTs; stirring 12 h	Aqueous, pH 6.8	11
5-Aminosalicylic acid	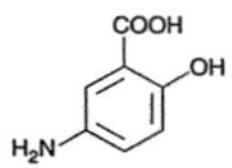	0.84	2.09, 5.26, 13.64	Aqueous 0.225 g L^{-1} at 40 ± 0.1 °C for 24 h, stirring	Water, 0.1 M HCl, pH 6.8 buffer	10
Diclofenac, sodium salt	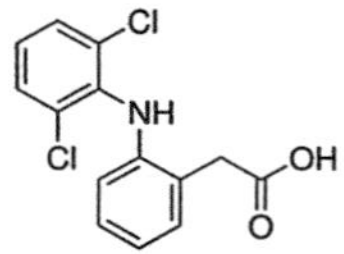	50	4.15	25 mg mL^{-1} (30 mL) + 0.5 g HNTs, stirring	pH 1.2 pH 7.4	12
Diphenhydramine hydrochloride	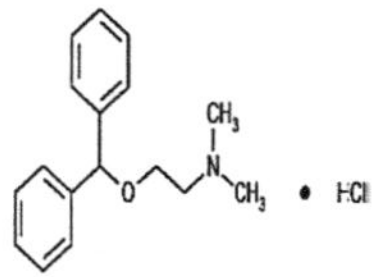	1000	8.98	29 mg mL^{-1} (30 mL) + 0.5 g HNTs, stirring	pH 1.2 pH 7.4	12
Ibuprofen	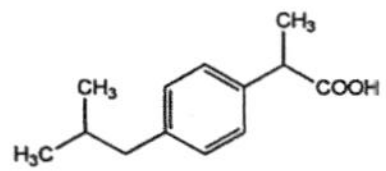	<1	4.91	0.1 g mL^{-1} in ethanol + 2 g of halloysite; stirring	—	21
Ofloxacin	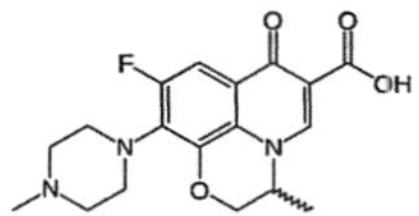	28.3	5.97 9.28	0.002 g mL^{-1} in pH 3 acetic acid + 1 g of halloysite, shaker at 35 °C	pH 1.2	15,18
Doxorubicin	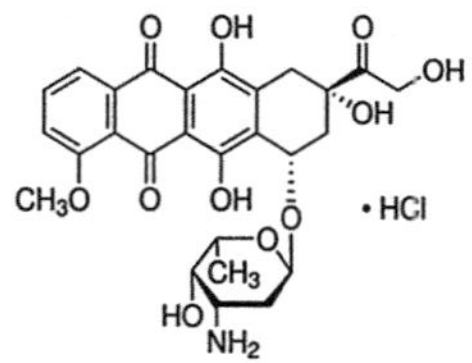	2.6	7.34 8.46 9.46	1 mL of 2 mg mL^{-1} doxorubicin solution (in water) mixed with halloysite and DNA; mixture dissolved in 30 mL water.	pH 7.4	9

[a]Data from DrugBank.
[b]Data from PubChem.

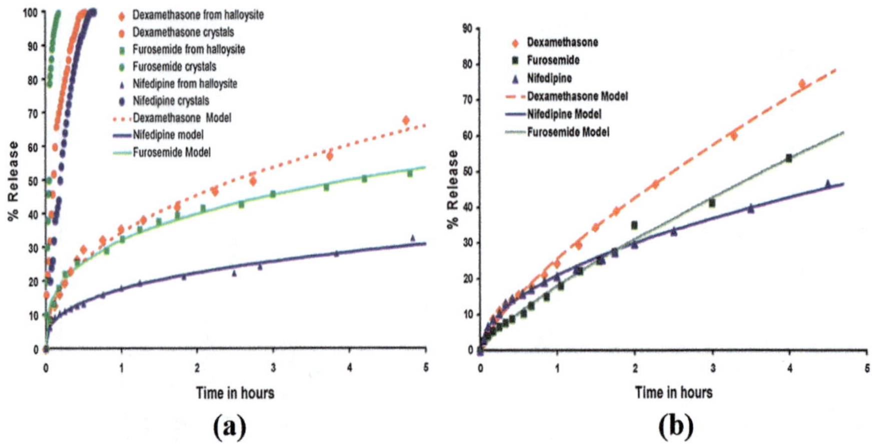

Figure 13.1 (a) Release profile of dexamethasone, furosemide, and nifedipine from microcrystals and halloysite loaded drugs using 10% ethanol–water solvent. (b) Release profile of halloysite loaded drugs using 50% ethanol. Reproduced with permission from N. G. Veerabadran, R. R. Price and Y. M. Lvov, *Nano*, 2007, **2**, 115.[8] Copyright © 2007 World Scientific Publishing.

relatively long incubation time is necessary for the system to reach equilibrium and achieve maximum drug immobilization. Various low-solubility anti-inflammatory drugs have been encapsulated using this method[10–12] (Table 13.1).

5-Amino salicylic acid was encapsulated within halloysite lumen and its release behavior was studied in pure water, 0.1 M HCl, and pH 6.8 phosphate buffer.[10] A comparison of the release profiles of the drug in the specified media are shown in Figure 13.2. There is a burst release phase in the first 10 minutes followed by extended release over 8–10 hours. In 0.1 M HCl and pH 6.8 buffer the release of the drug was slightly faster than in pure water. The release profiles show two modes of release. A fast (burst) release associated with externally adsorbed drug and molecules near the opening of the lumen occurs in the first hour. Following the burst phase, a slow release of molecules is observed. This extended release is likely due to molecules adsorbed further inside the lumen wall which tend to adsorb and quickly desorb to the oppositely charged wall. Furthermore, molecules securely inside the lumen are kept inside by the dimension lumen, in which the probability of a molecule leaving is relatively low. The release process follows first order release, which is given by eqn (13.1)

$$C_t = C_e\left[1 - e^{\left(\frac{-k_d t}{C_e}\right)}\right] \tag{13.1}$$

where C_t is drug concentration at time t; C_e is equilibrium drug concentration; k_d is specific desorption rate. Which essentially states that the release

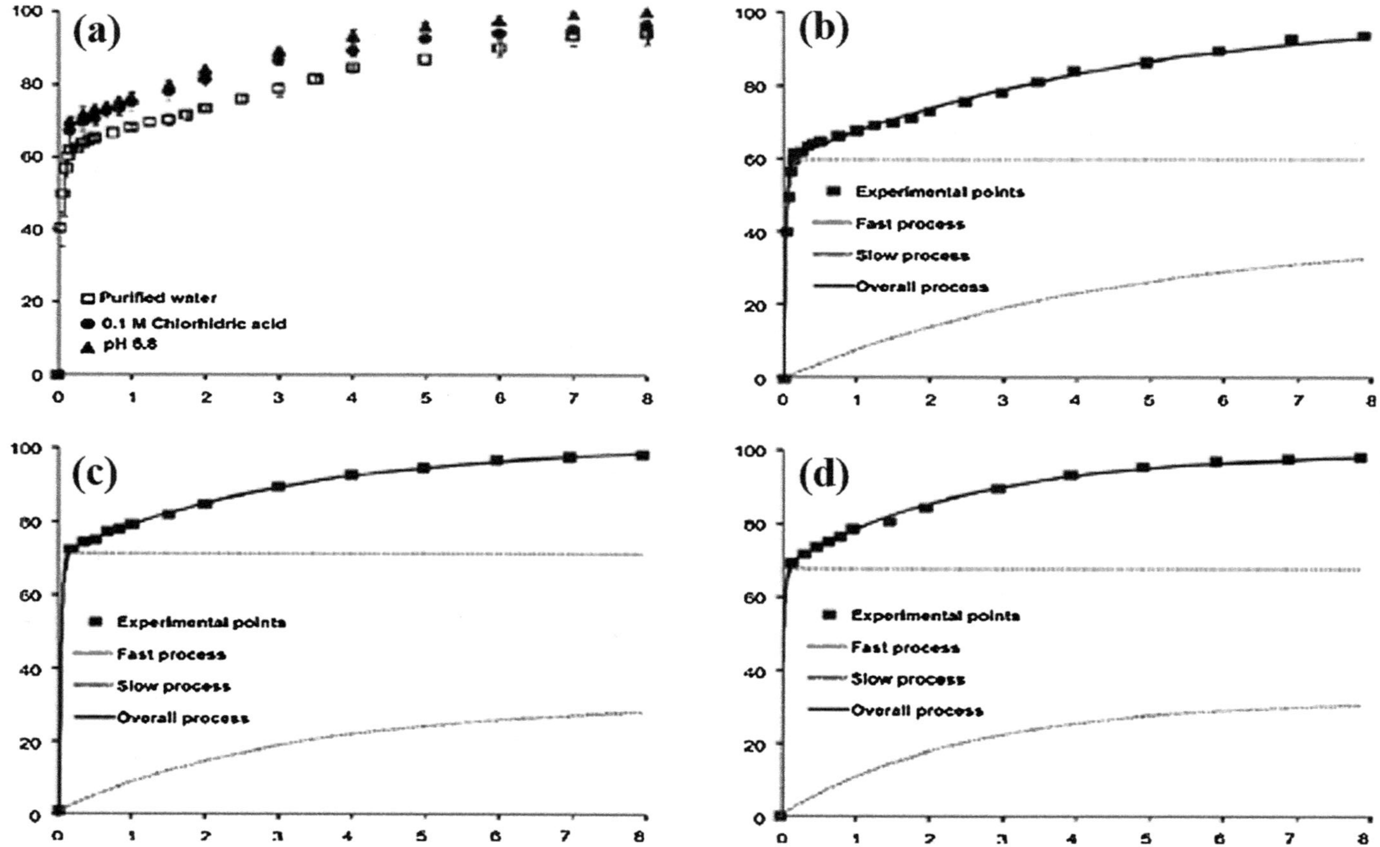

Figure 13.2 Release profiles of 5-aminosalicylic acid from halloysite nanotubes. (a) All three media, (b) purified water, (c) 0.1 M HCl solution, and (d) pH 6.8 buffer solution. Reprinted from *Colloids and Surfaces B: Biointerfaces*, **105**, C. Aguzzi *et al.*, Release kinetics of 5-aminosalicylic acid from halloysite, 75–80, Copyright 2013 with permission from Elsevier.[10]

of the drug depends only on the concentration of the immobilized drug. This means intuitively when there is a significant mass of drug available to release, desorption will occur at an accelerated rate. As the drug desorbs, there is less drug available on the tubes to desorb so the release slows down. Furthermore, when first placed into solution, there is chemical potential for desorption because of a concentration gradient leading to a flux away from the tube. As the molecules release this force is diminished and leads to a decrease in release rate.

13.3 Acid and Alkali Activation/Etching of Halloysite Nanotubes

Acid and base treatment of aluminosilicates is one method of manufacturing porous clay materials. Using this type of treatment, one can selectively increase the diameter of the lumen of halloysite and, if desired, introduce tiny pores into the nanotube's walls. The selectivity of etching is due to the different Lewis acid–base chemistry of sulfuric acid with aluminum oxide and silicon dioxide.[13] Treating halloysite nanotubes with 3 M sulfuric acid at 90 °C for eight hours significantly reduces the amount of aluminum in the sample (from 12.8 to 1.5%).[14] Sulfuric acid diffuses into the pores of halloysite and reacts with the alumina (and partially with the silicon dioxide) resulting in aluminum sulfate according to reaction (13.2). This process produces tubes of the same external diameter but larger lumen. Over time the products of the reaction diffuse out of the lumen (Figure 13.3a). For long term treatment, the resulting tubes look like a network of spherical nanoparticles organized in a rod shaped structure (Figure 13.3b).

$$Al_2O_3 + 3H_2SO_4 \rightarrow Al_2(SO_4)_3 + 3H_2O \tag{13.2}$$

The same phenomenon, of organized spherical particles, was observed by Abdullayev and coauthors[13] even after removing 100% of the alumina layer. The authors also managed to find a combination of conditions where the tubular structure of halloysite was preserved but the lumen was enlarged (Figure 13.3c). In particular, the loading efficiency for benzotriazole increased four times by selectively etching 60% alumina within the tubes' lumens. The specific surface area of halloysite increased over six times, from 50 to 250 $m^2\ g^{-1}$, after acid treatment.

The loading and release of ofloxacin from calcinated and hydrochloric acid-treated halloysite in acidic conditions simulating gastric fluid were studied.[15] The pK_{a_1} and pK_{a_2} of ofloxacin are 5.8 and 8.4, respectively[16,17] (Table 13.1), which means that there are cationic, neutral, and anionic species in the used solutions (loading was performed as follows: ofloxacin was dissolved in pH 3 acetic acid to a final concentration of 0.002 g mL^{-1} + 1 g of halloysite, shaker at 35 °C). The surface area, as determined by BET, increased with the concentration of HCl. This is indicative of the formation of pores and/or an increase in

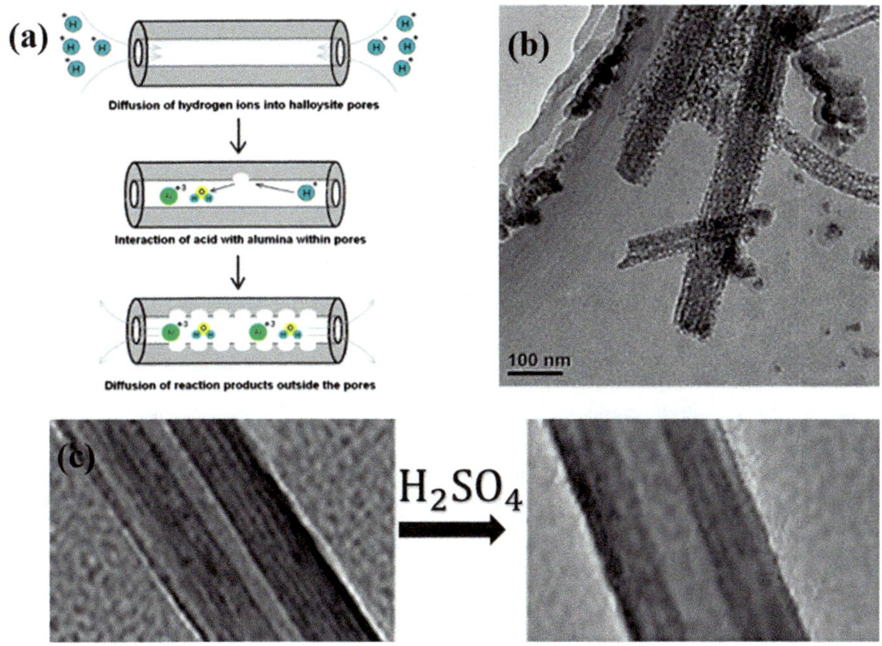

Figure 13.3 (a) Schematic representation of acid treatment of halloysite nanotubes. Reprinted with permission from E. Abdullayev, A. Joshi, W. Wei, Y. Zhao and Y. Lvov, *ACS Nano*, 2012, **6**, 7216.[13] Copyright (2012) American Chemical Society. (b) TEM microphotography of acid-treated halloysite nanotubes (for 8 h 3 M H_2SO_4). Reprinted from *Colloids and Surfaces S: Physicochemical and Engineering Aspects*, **396**, A.-B. Zhang *et al.*, Effects of acid treatment on the physic chemical and pore characteristics of halloysite, 182–188, Copyright 2012 with permission from Elsevier.[14] (c) TEM of an enlarged halloysite lumen after treatment with sulfuric acid.

the size of the lumen. *In vitro* release of ofloxacin from acid-treated halloysite is shown in Figure 13.4. The cumulative release at pH 1.2 shows significant differences as the concentration of HCl used for treatment increased. The cumulative release of ofloxacin from untreated halloysite is about 57% after 12 hours. For the halloysite sample treated with 2 M HCl, a slight increase in the cumulative release to 64% is observed. Further increases in HCl concentrations result in an evident increase in the cumulative release and almost all the adsorbed ofloxacin can be released with 12 M HCl-treated halloysite after 12 hours.

The cumulative release increases gradually as the calcining temperature is increased from 25 to 400 °C. The release of ofloxacin is slow and the cumulative release is less than 70% after 12 hours. However, the release reaches 100% after 12 hours from the halloysite calcined at 500 °C. Although the cumulative percentage of ofloxacin released is increasing with temperature, the amount of released ofloxacin is in fact decreasing with increasing temperature owing to the low adsorption capacity of heat-treated halloysite.

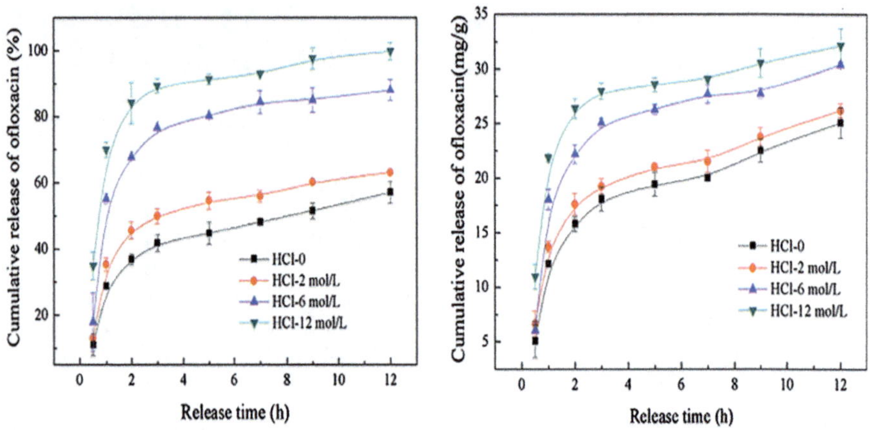

Figure 13.4 *In vitro* release of ofloxacin from acid processed halloysite nanotubes at pH 1.2. Reprinted from *Colloids and Surfaces B: Biointerfaces*, **113**, Q. Wang *et al.*, Adsorption and release of ofloxacin from acid- and heat-treated halloysite, 51–58, Copyright 2014 with permission from Elsevier.[15]

Alkali treatment of halloysite has also been performed.[18] The silica layer located on the outside of the halloysite nanotubes reacts with sodium hydroxide following reaction (13.3). It has been shown that treatment of raw halloysite with 6 M sodium hydroxide increases the BET surface area and pore volume from 32.8 $m^2\,g^{-1}$ and 0.09 $cm^3\,g^{-1}$ to 39.2 $m^2\,g^{-1}$ and 0.12 $cm^3\,g^{-1}$ respectively, indicating the formation of a more porous material.

$$SiO_2 + 2NaOH \rightarrow Na_2SiO_3 + H_2O \tag{13.3}$$

These modified halloysite tubes were also used to study ofloxacin adsorption and desorption. In contrast to hydrochloric acid treatment, a decrease in release rate was observed as the concentration of NaOH used for halloysite treatment was increased (Figure 13.5).

Therefore, with moderate acid treatment one may increase the volume of the lumen and harsh treatments produce a kind of perforated nanotube enriched with silica. Reacting the nanotubes with bases also etches alumina, essentially increasing the general porosity of the nanotubes. Calcination decreases hydrophobicity of the nanotubes and may enhance etching efficiency.

13.4 Modification of the Inner and Outer Surfaces of Halloysite

The exterior silicate surface of halloysite interacts with molecules through hydrogen bonding or van der Waals forces[19] while the lumen, which is positive below pH 9, enhances the loading of negatively charged molecules.[1] However, the surface of the tube can also be modified using various silicate specific chemistries to change the properties of the outer surface.[20]

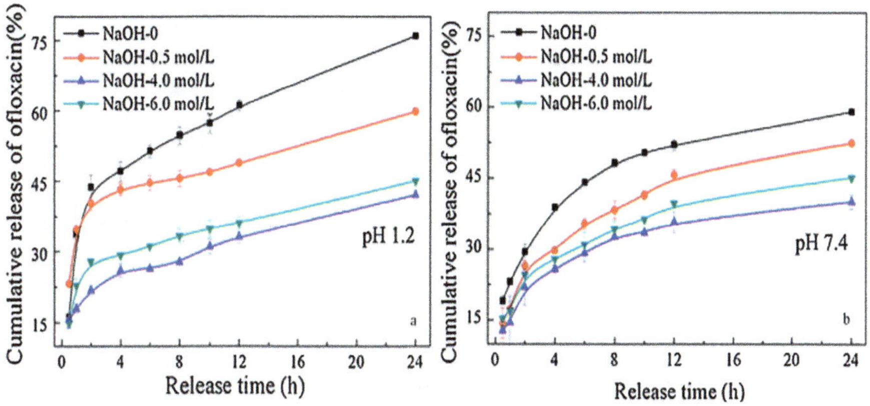

Figure 13.5 Release of ofloxacin from alkali-treated halloysite at various concentrations of sodium hydroxide at pH 1.2 and 7.4. Reprinted from *Applied Surface Science*, **287**, Q. Wang *et al.*, Alkali activation of halloysite for adsorption and release of ofloxacin, 54–61, Copyright 2013 with permission from Elsevier.[18]

Grafting molecules to the inner/outer surface of the tubes is a promising tool for increasing loading efficiency. The technique often used is called silanization, and halloysite is a viable candidate for organosilane modification such as with 3-aminopropyltriethoxysilane (APTES). The disadvantage to this technique is that it is not selective and will modify both the inner and outer surfaces of halloysite. It has been shown that non-selective functionalization of halloysite with APTES increased the loading of the anti-inflammatory drug ibuprofen in halloysite[21] by 25% (this is two-times larger than the lumen volume, indicating that a large portion of the drug adsorption occurred at the outer tube surface). The authors believe that the increased loading capacity of the modified halloysite is related to two primary mechanisms. First, the aminopropyl groups added to the surface create an electrostatic interaction between halloysite and ibuprofen molecules, which enhances the ibuprofen loading in halloysite. Second, the ibuprofen loading in APTES-grafted halloysite is promoted by the thermal pretreatment of halloysite, which regulates the APTES grafting by reducing the water content in halloysite and preserves the lumen space for ibuprofen loading. The schematic representation of the process is shown in Figure 13.6.

The same grafting technique was applied to halloysite nanotubes for aspirin loading. The results indicated that the modification of halloysite nanotubes increased the amount of loaded aspirin from 3.8 to 12.0 wt%.[11,22]

The external surface of halloysite was modified with positively charged triazolium salts such as methyl and butyl triazolium[22] (Figure 13.7). The triazolium modified halloysite nanotubes were used to study the adsorption and release of curcumin. The solubility of curcumin in water is 0.6 μg mL^{-1} (Table 13.1) and it degrades rapidly under neutral and alkaline conditions,

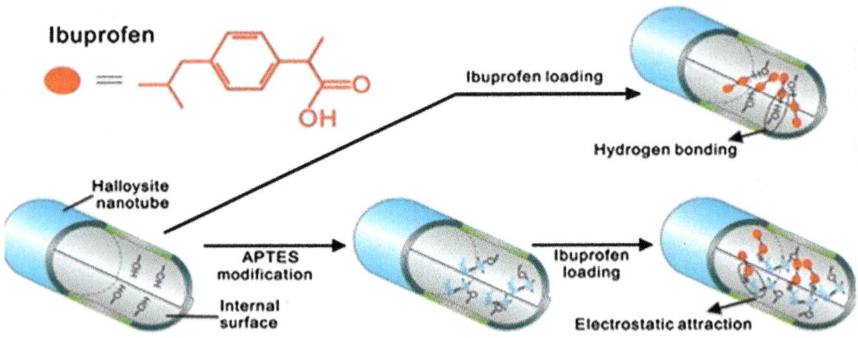

Figure 13.6 Loading process of ibuprofen into halloysite. Reprinted from Microporous and Mesoporous Materials, 179, D. Tan *et al.*, Natural halloysite nanotubes as mesporous ccarriers for the loading of ibuprofen, 89–98, Copyright 2012 with permission from Elsevier.[21]

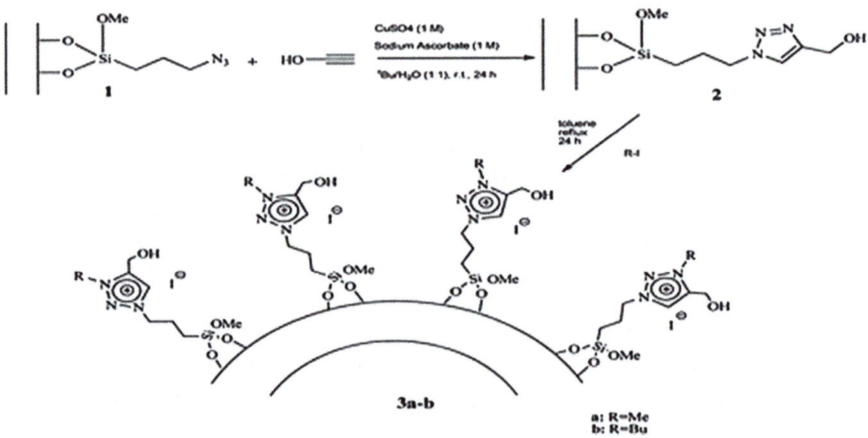

Figure 13.7 Scheme of modification of halloysite with triazolium derivatives. Reprinted from International Journal of Pharmaceutics, 475, S. Riela *et al.*, Development and characterization of co-loaded curcumin/triazole-halloysite systems and evaluation of their potential anticancer activity, 613–623, Copyright 2014 with permission from Elsevier.[22]

with a half-life shorter than 10 min. Release studies were conducted in solutions of pH 1, 5, and 8, and show a significant difference depending on the pH used. The first factor to take into account is the solubility of curcumin as a function of pH. Curcumin molecules ($pK_{a_1} = 8.38$, $pK_{a_2} = 9.9$, and $pK_{a_3} = 10.5$) can exist as cationic, neutral, or anionic species in acidic, neutral, or alkaline solutions, respectively.[23] Therefore, curcumin will readily diffuse out of the lumen of halloysite at pH 1 in its cationic form, resulting in a higher release as compared with that at pH 5.0 and 8.0.

Non-selective halloysite modifications allow for increased adsorption of drug molecules but this type of modification neglects the major advantage of

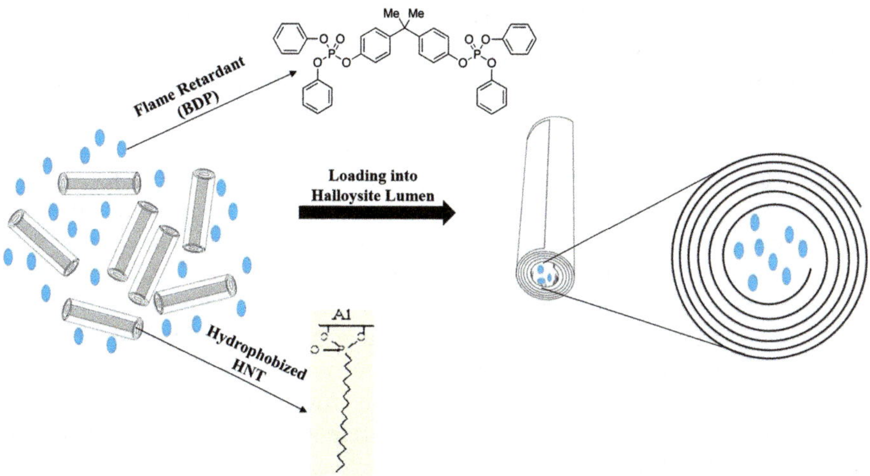

Figure 13.8 Schematic representation of loading bisphenol A–bis(diphenyl phosphate) into ODP-modified halloysite. Reproduced with permission from ref. 27. Copyright (2013) by Japanese Chemical Society.

the nanotubes, a selective adsorption and controlled release by the nanoscaled inner pore. Next, we will consider a non-covalent but selective modification of the inner/outer tube surfaces.

The lumen of halloysite has been selectively modified with octadecylphosphonic acid (ODP) preferably reacting with Al_2O_3 rather than with SiO_2 resulting in the formation of inorganic micelle-like tubes with inner aliphatic chains and outer hydrophilic silicate shells.[24] This made the inner core of the tubes hydrophobic and promoted adsorption of non-polar molecules inside the tubes. To characterize specific adsorption and release properties of clay nanotubes with a hydrophobized interior, the authors used hydrophobic and hydrophilic ferrocene derivatives which were loaded into the halloysite. Ferrocene exhibits biological activity and has been used as a potential anticancer agent.[25,26] For pristine halloysite, only a small initial burst release is observed due to dissolution of ferrocene from the halloysite surface. ODP modified halloysite releases four times more ferrocene in the same period of time, indicating its higher adsorption capacity.

In a similar approach, the inner lumen of halloysite was treated with octadecylphosphonic acid to enhance internal tube loading of water-insoluble flame retardant bisphenol A-bis(diphenyl phosphate) (Figure 13.8) and this processing resulted in linear 10 hour release kinetics.[27]

13.5 End Stoppers and Encapsulation

Another way to increase time of release of active molecules from halloysite is to fabricate "stoppers" at the ends of the nanotubes or by adding a coating around the tubes which clogs the lumen openings, further slowing diffusion out of the tubes. These stoppers can be divided into two categories: polymeric coatings and metal chelate networks.

Stoppers can be created through complexation of these anionic nanotubes with polycations. In this case, the tube openings are clogged with bulk polymers, slowing diffusion. It is better to use low molecular weight polymers to manufacture fine stopper formation on each end to avoid aggregation.

The highly water soluble diltiazem chloride and propranolol chloride were loaded in halloysite then the loaded tubes were non-selectively covered with poly-cyanoacrylates.[28] As one can see in Figure 13.9a, such treatment significantly slows down the release of diltiazem. The minor toxicity of cyanoacrylates is the obvious disadvantage of using this class of polymers as a potential stopper. Therefore, the authors also tested biocompatible polycations such as chitosan and polyethyleneimine. It has been shown that this treatment reduced the initial burst release of the drugs mentioned above (Figure 13.9b).

Gentamicin loaded halloysite doped into PMMA bone cement released the antibiotics over 400 hours. The sluggish release is due to limited diffusion of the drug from nanotubes embedded in the bone cement matrix.[7]

For the first time, highly specific formation of copper(II) stoppers on halloysite tubes were observed when studying benzotriazole, a corrosion inhibitor for copper.[29] Benzotriazole was released from the tubes and reacted with copper ions in the vicinity of the tube-ends. The complex of copper ions and benzotriazole has been described in ref. 30. Benzotriazole forms insoluble two dimensional networks in water by complexation with copper. The structure of the complex clogging the tube-ends is shown in Figure 13.10.

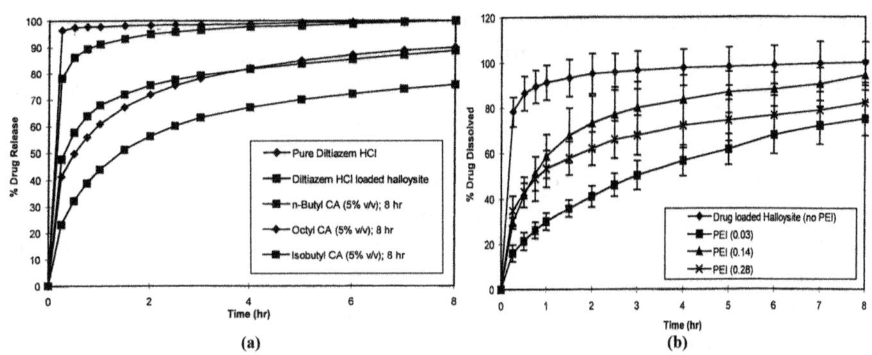

Figure 13.9 (a) The release profile of diltiazem chloride from the halloysite nanotubes, coated with poly-cyanoacrylate (CA) at pH 6.8 for 8 h at 37 °C. (b) The release profile of diltiazem chloride from the halloysite nanotubes, coated with PEI (the number in parentheses indicates the PEI/halloysite ratio) at pH 6.8 for 8 h at 37 °C. Reprinted from *International Journal of Pharmaceutics*, **253**, S. R. Levis and P. B. Deasy, Use of coated microtubular halloysite for the sustained release of diliazem hydrochloride and propranolol hydrochloride, 145–157, Copyright 2003 with permission from Elsevier.[28]

The method described above was also used for stopper formation on halloysite nanotubes loaded with the antiseptic brilliant green.[31] The release profile of brilliant green from nanotubes with stoppers clearly indicates slower, sustained release as compared to the rate of release from pristine halloysite (Figure 13.11). Seventy percent of brilliant green was released from halloysite within three hours, whereas, the microcrystalline powder of brilliant green was completely dissolved in 15 minutes. In total, brilliant green was slowly released from the nanotubes over a period of 30 hours when no more detectable amount of antiseptic was present.

It is also possible to coat the entire tube with a very thin multilayer of alternating polycations and polyanions with the layer-by-layer method, as was first demonstrated for extending dexamethasone release from encapsulated clay nanotubes.[32]

Resveratrol, an antioxidant from red wine, was also loaded into halloysite nanotubes. The tubes were then wrapped using the layer-by-layer technique

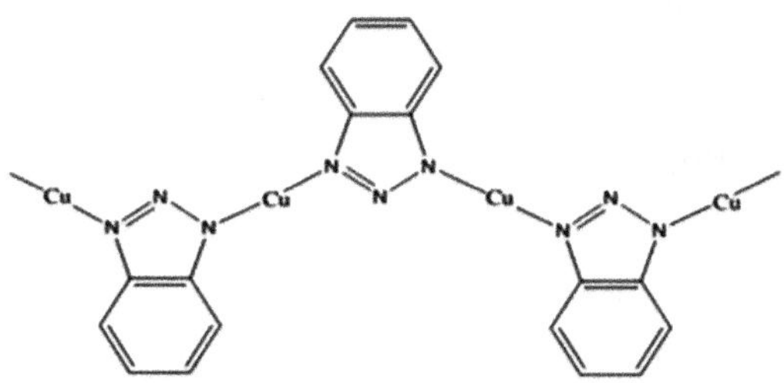

Figure 13.10 Structure of copper benzotriazole complex.[30]

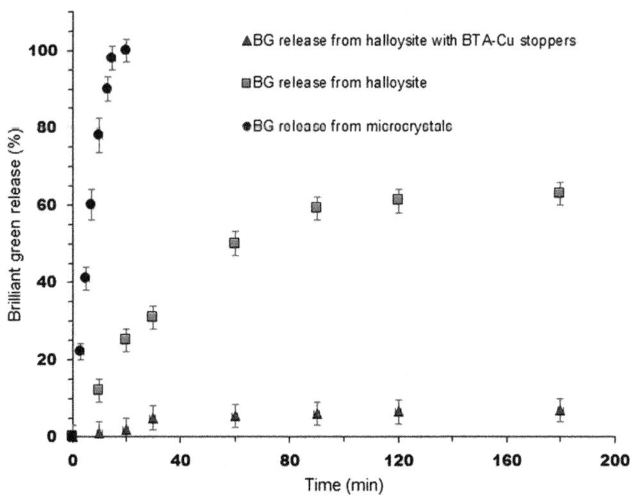

Figure 13.11 Brilliant green (BG) release from microcrystals powder (black), pristine halloysite nanotubes (red) and halloysite with copper-benzotriazole (BTA-Cu) stoppers (dark blue).

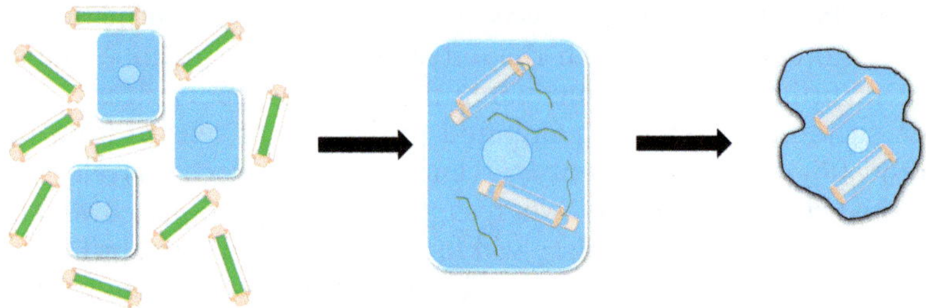

Figure 13.12 Scheme of intracellular delivery of brilliant green loaded halloysite with dextrin stopper.

to provide control over the release.[4] The authors used two combinations of polyelectrolytes, synthetic polymers PEI-PSS–PAH and natural polymers protamine–dextran sulfate. It has been shown that nanotubes loaded with resveratrol and covered with PEI-PSS–PAH did not exhibit a cytotoxic effect towards a breast cancer cell line (MCF-7) whereas loaded tubes covered with protamine–dextran sulfate triggered apoptosis of the cancer cells. This result indicates that the protamine/dextran sulfate tube coating is biodegradable by intracellular enzymes.

A biodegradable dextrin-based stopper was also assembled on halloysite nanotubes loaded with brilliant green.[33] The stopper was created by non-specific dextrin adsorption onto halloysite which gave a thicker coating at the tube-ends. Earlier, it was shown that, apart from its antibacterial properties, brilliant green is capable of suppressing mitochondria in cancer cells.[34] The authors demonstrated that the dextrin stoppers can be cleaved by intercellular enzymes which facilitate the release of the drug within the cells, leading to the death of the cancer cells (Figure 13.12). The efficacy of this formulation was tested using two types of cancer cells: human lung cancer cells (A549) and hepatoma cells.

13.6 Antibacterial Halloysite Nanocomposites

Nanocomposites of halloysite with silver,[35] gold,[36] platinum,[37] cobalt,[38] and iron oxide[39] nanoparticles have been synthetized. In a simple approach, one can just carry out the synthesis reaction in the presence of halloysite and the clay nanotube surface will be seeded with 5–6 nm diameter nanoparticles. If the reagents are selectively loaded into halloysite, then nanoparticles, such as nanorods, may be grown within the tube interior[35] (Figure 13.13d).

Silver nanorods were synthesized inside the halloysite lumen.[35] The resulting composites were admixed into oil-based paint and the paint was applied on aluminum foil. Then, the painted pieces of foil were arranged on the surface of agar plates seeded with *Staphylococcus aureus* and *Escherichia coli* bacteria. This experiment demonstrated the inhibition of bacterial growth around the foil containing paint with silver halloysite nanocomposites.

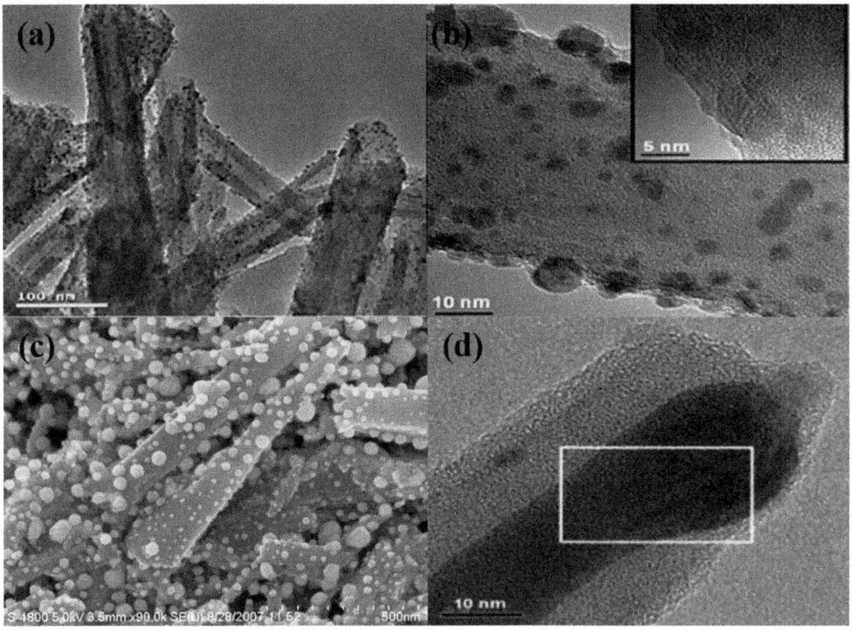

Figure 13.13 Halloysite silver nanocomposites (a), (b) TEM microphotography of halloysite seeded with silver nanoparticles on the outer surface. Reprinted from Journal of Inorganic Biochemistry, 118, Y. Zhang *et al.*, Potent antibacterial activity of a novel silver nanoparticle-halloysite nanotube nanocomposite powder, 59–64, Copyright 2013 with permission from Elsevier.[40] (c) SEM microphotography of halloysite seeded with silver nanoparticles. (d) Silver nanorod inside of the halloysite lumen. Reproduced with permission from E. Abdullayev, K. Sakakibara, K. Okamoto, W. Wei, K. Ariga and Y. Lvov, *ACS Appl. Mater. Interfaces*, 2011, 3, 4040. Copyright (2012) American Chemical Society.[35]

In another approach, the outer surface of halloysite was silanized, then the modified tubes were mixed with silver nitrate solution. The precursor molecules adsorbed to the surface of the tubes and were reduced with sodium borate, resulting in the formation of silver nanoparticle clusters on the outer surface of halloysite (Figure 13.13a–c). The final product exhibited significant antibacterial properties against Gram-positive and Gram-negative microorganisms.[40]

Polyethersulfone membrane composites for water purification doped with 3 wt% halloysite and loaded with copper ions were prepared and their antibacterial properties were determined. First, halloysite nanotubes were modified with *N*-β-(aminoethyl)-γ-aminopropyltrimethoxy silane (AEAPTMS) and mixed with copper chloride solution resulting in deposition of copper ions on the surface of the nanotubes. The results indicated that the hybrid membranes had significant inhibition on the growth of both Gram-negative and Gram-positive bacteria.[41]

An interesting application of halloysites is their inclusion in electrospun microfibers.[42] Such polymeric fibers are typically of 400–500 nm in diameter, which forces longer halloysite tubes (around 1 μm in length) to orient along the fibers making them stronger. The addition of 10 vol% of halloysite loaded with drugs into the polycaprolactone fibers allowed for extended drug release (from 10–20 hours typical for polymeric electrospun fibers to hundreds of hours for drug loaded halloysite fibers). Guided tissue regeneration has become a standard in therapy; however, infections are a common cause of failure in clinical applications. A nano-micro composite electrospun membrane with sustained three-week drug delivery was developed using metronidazole-loaded halloysite clay nanotubes doped into poly(caprolactone)–gelatin fibers. This will allow proper infection protection.

In another approach, electrospun poly (lactic-*co*-glycolic) acid nanofibers doped with halloysite nanotubes were prepared and their biocompatibility was tested. The hemolysis and anticoagulation assay results indicated that 5 wt% halloysite nanotubes admixed into the fibers did not display hemotoxicity.[43] Similar fibers were prepared with tetracycline hydrochloride loaded halloysite nanotubes and demonstrated sustained release of the antibiotic for 42 days. These fibers also displayed antibacterial activity towards *S. aureus*. It has been shown that the composite membranes are safe for fibroblast cells.[6,44]

13.7 Protein and DNA Loading

Given the ample amount of research concerning halloysite and synthetic polymers it is a natural extension to consider the interaction of halloysite with biopolymers. Indeed, when surveying research with halloysite, one will find work that has been done with halloysite and various biocompatible polymers, but there is little on halloysite and protein composites. Primarily, it is interesting to consider composites consisting of enzymes and halloysite; however, non-enzymatic proteins may also be important considering the applications of proteins as hormones, anti-toxins, structural support, or any of the variety of roles proteins serve. As for practical applications of protein–halloysite composites, the immediate use for halloysite is as an immobilization substrate.

One of the difficulties of directly applying enzymes in industrial or analytical settings is that most are water-soluble and will be washed away with the solution carrying the analyte. The solution to this issue is to immobilize the enzyme using chemisorption or physisorption to an insoluble porous substrate, over which the analyte is passed and the results of the catalyzed reaction are assessed. One of the industrial uses includes enzymatic conversion of glucose to fructose. These types of immobilized proteins can also be used for automatic assays such as in handheld blood sugar monitors. The downside to immobilized enzymes is that, unlike with inorganic catalysts, they tend to degrade at a much faster rate.

In comparison to *de facto* halloysite–polymer composites, in which halloysite is a minor constituent of the composite, often composing only 5–10 wt%

of the composite, in protein immobilization halloysite is the primary constituent. Conceptually, proteins are added to the halloysite instead of nanotubes being added to the protein, as illustrated in Figure 13.14. This role maybe reversed if one considers structural proteins such as fibrin; however, it would be expected that halloysite would behave the same as with synthetic polymers.

Given this shift in ideology, research concerning protein immobilization takes a slightly different approach to the application of halloysite. We are no longer as concerned with the mechanical properties of the resulting composite as we are with the biochemical effects of immobilization with halloysite. These effects include enhanced (or reduced) temperature and pH stability. It is also interesting to study the ability to recover the composite using filtration or centrifugation and reuse the composite for an assay and even to study the temporal stability of the enzyme. Ideally, and it is generally acknowledged that, the four parameters mentioned – pH stability, thermal stability, reusability, and temporal stability – are all enhanced by proper immobilization, but this enhancement comes at a cost.

The first and the most significant disadvantage may be the loss of specific activity or in other terms catalytic efficiency. That is to say, immobilization often significantly decreases the rate of catalysis per unit mass. There are various reasons this loss occurs. If the adsorption occurs by physisorption then the enzyme may denature upon adsorption. Another possibility is that the protein adsorbs preferentially with its active site towards the substrate. This will be especially true if the adsorbent has hydrophobic patches, as many enzymes' active sites are hydrophobic pockets that allow the enzyme control over the aqueous environment of the pocket. If adsorption occurs by chemisorption then the chemical bonds formed may disrupt the hydrogen bonding network that assists the protein in proper folding. There is some debate as to which factors contribute the most energy to the folding of the protein. It would seem that hydrophobic interactions contribute the most (though some may say van der Waals contribute more), and the folding is more or less "completed" by the hydrogen bonding energy. Therefore, it may be appropriate to assume that denaturation occurs mostly due to disruption of the proteins' hydrophobic interior. And finally, any surface enzyme immobilization decreases a substrate delivery rate at least two-times.

When discussing the adsorption of molecules onto or inside halloysite it is assumed that physisorption is the primary mechanism driving the

Figure 13.14 Schematic representation of protein fate after interaction with halloysite nanotubes.

immobilization of molecules inside or on the surface of halloysite. Specifically, it is the electrostatic interactions of the two oppositely charged surfaces of halloysite with the adsorbate. We see evidence of this with organic molecules such as dyes, in which positively charged dyes desorb much more quickly and fully than negatively charged dyes.[45] It is postulated that negatively charged dyes adsorb to the lumen of the nanotube and desorb slower than those on the surface. Recent computational studies showed that molecules inside the lumen are constrained due to the miniscule cross-sectional area of the lumen.[46] Furthermore, because of the small volume in which the molecules are constrained they are more likely to re-adsorb to the wall of the lumen than molecules on the surface are to re-adsorb to the outer surface.

There exists a small selection of research papers that address halloysite protein composites with a focus on adsorption inside the lumen. This important distinction is what differentiates work with halloysite nanotubes from the plethora of publications on protein adsorption on the outer surface of platy clays, such as kaolin, montmorillonite, LAPONITE®, and bentonite. Generally, halloysite–protein papers address manufacturing composites in one of two ways: specific loading of proteins inside the lumen or conjugation with polymers adsorbed to the surface of the nanotubes. Works by Shchukin *et al.*,[47] Zhai *et al.*,[48] and Zheng *et al.*[49] pursue the first method, while others such as Chao[50] and Yao[51] prefer functionalizing the surface of the tube before covalently bonding the proteins. There is merit to both approaches, however adsorbing polymers to the surface of the nanotubes seems unnecessary. There is plentiful research on cross-linking polymers and proteins to form structured composites with a variety of properties that depend upon the constituent materials. Therefore, this approach has a sense of redundancy but as with many novel approaches there may be specific applications where one may find the addition of polymer useful.

It is interesting to assess protein adsorption without the addition of polymers. Indeed, halloysite is a suitable substrate for immobilization but there are some issues that need to be addressed. The first and perhaps the most important is the quantification of protein immobilized on/inside the tubes. This quantity is computed by measuring the concentration of the washing/filtered solution after performing the immobilization procedure, which consisted of five hours of incubation at room temperature.[48] This solution was measured using the standard Bradford quantification technique. However, there is no mention of the protein used as a standard for calibration. If using a BSA standard (as is typical), both the original α-amylase solution (15 mg mL^{-1}) and the subsequent solutions after immobilization would need to be measured. Simply assuming the quantities obtained from the immobilization procedure would provide misleading results of the actual immobilized proteins. Besides this point, another criticism would be that it is a bold suggestion to assume that simply because the post-immobilization solutions did not contain some amount of protein that it must exist on the tube alone. Proteins adsorb readily onto almost any surface not designed to resist non-specific adsorption. In spite of these critiques, the paper was successful in showing that protein adsorbed on halloysite can be easily reused by filtration and that a considerable number of cycles can be achieved using a simple immobilization process.

This paper[48] can be considered a proof of concept, showing that halloysite–enzyme composites can be created and that the composite is reusable along with showing increases in enzyme stability. In the paper though, there is some uneasiness concerning the location of the protein, specifically if the protein resides inside, outside, or on both surfaces of the tubes. It is not a leap of faith to assume proteins adsorb on the outer surface of halloysite, as it is chemically equivalent to glass. Most likely both non-specific and electrostatic interactions are the driving forces behind adsorption on the surface of the tubes. The more curious question is if the protein resides inside the nanotubes. The increase in temperature or temporal stability can be easily attributed to adsorption to the silica surface. It would seem intuitive though that the proteins should be even further shielded from detrimental environmental conditions if they resided in the inner lumen. Essentially, limited diffusion of molecules inside the tubes should prevent inhibitors, oxidizers, and other damaging chemicals from impacting the functionality of the protein along with the structural stability gained by immobilization.

The similarities between Shchukin's and Zheng's papers are uncanny. Both demonstrated that an insoluble product is formed inside the nanotubes by immobilizing enzymes inside the tubes and reacting the products of catalysis with specific chemicals. The first author used urease and calcium salts to form calcium carbonate whereas the latter also chose urease but used iron salts (iron chloride and iron sulfate) to form iron oxide. The assumption of their work was that proteins must reside inside the tube and as the enzyme catalyzes the conversion of urea to ammonia the products formed due to the increase of pH inside the tubes forms insoluble nanoconfined products, as seen in Figure 13.15. TEM images of the tubes show that there are products of the desired reactions mostly inside the tubes. The method these authors used is quite intelligent. It is difficult to image proteins using electron microscopy because of their light elemental composition and low electron density as compared with alumosilicate clay. Our attempts to stain proteins with heavy metal ions for electron microscopy imaging have not been much help in imaging proteins inside the tubes' lumen.

The downside to this approach is that one can argue that the enzyme could exist anywhere in solution and the products of the catalyzed reaction may diffuse freely into and out of the tubes. This may be a reasonable critique because the molecules used are small and there is little to prevent them from haphazardly diffusing in and out of the lumen. It would be interesting to compare the images after the catalyzed reaction to images produced without enzymes, but still yielding the same products. This may be achieved by incubating tubes with salts then adding them to solutions with the appropriate pH.

According to Tully *et al.* the loading efficiency for proteins with halloysite is approximately 5–7% depending on the protein, as determined by TGA analysis. Of this 5–7%, when placed into solution for release studies only one

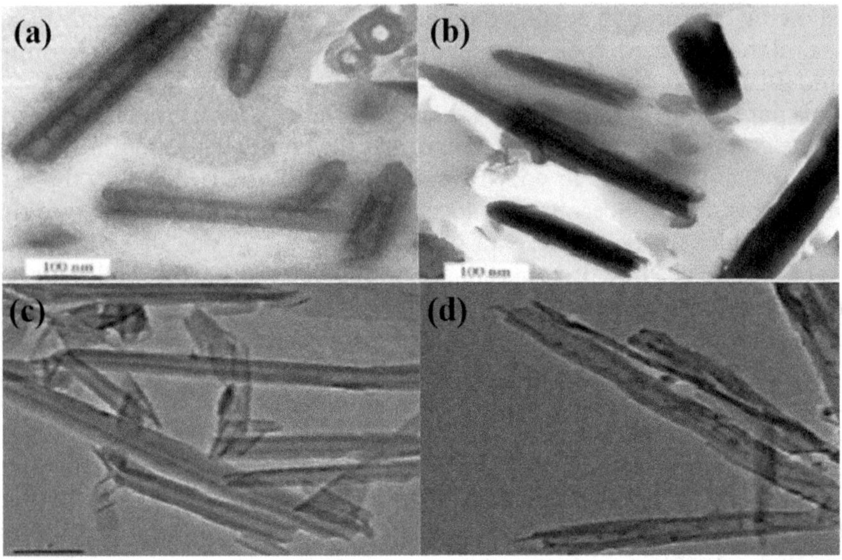

Figure 13.15 TEM images of halloysite before (a), (c) and after (b), (d) the catalyzed reactions. (b) The result of reaction of urease with urea in presence of $CaCl_2$ to generate $CaCO_3$. (d) The result of urease catalyzing urea into CO_2 and NH_3 and the resulting basic solution reacting with $FeSO_4$ and $FeCl_3$ to generate Fe_3O_4. (a) and (b). Reproduced from ref. 47 with permission from John Wiley and Sons. Copyright © 2005 WILEY-VCH Verlag GmbH & Co. KGaA, Weinheim; (c) and (d) Reproduced from Materials Chemistry and Physics, 151, P. Zheng *et al.*, Selective fabrication of iron oxide particles in halloysite lumen, 14–17, Copyright 2015 with permission from Elsevier.[49]

third of the protein will desorb while the remaining fraction stays immobilized virtually permanently. The stationary fraction may serve as a biocatalytical reactor, in which substrates diffuse inside the tubes where the enzyme is protected, the catalyzed reaction occurs, and the products diffuse out of the lumen. These masses come from what would be considered a "standard environment" where the protein was adsorbed to the tubes in water and released in water.

However, to illustrate an earlier supposition of electrostatic interaction driving the adsorption it was shown in the same paper that varying the pH of the solution during immobilization, which affects only the magnitude of the charge of the tube not the sign, protein adsorption is heavily impacted by the charge of both the protein and tube, as seen in Figure 13.16. The concept is as follows, proteins cannot be composed of charges of a single sign; however, the net charge of the protein will either be positive, negative, or neutral. The pH at which the protein is neutrally charged is called the isoelectric point. At a pH above the isoelectric point proteins are negatively charged while below the isoelectric point they are positively charged. By varying the charge of the protein, as seen in Figure 13.16a, proteins that are positively charged adsorb

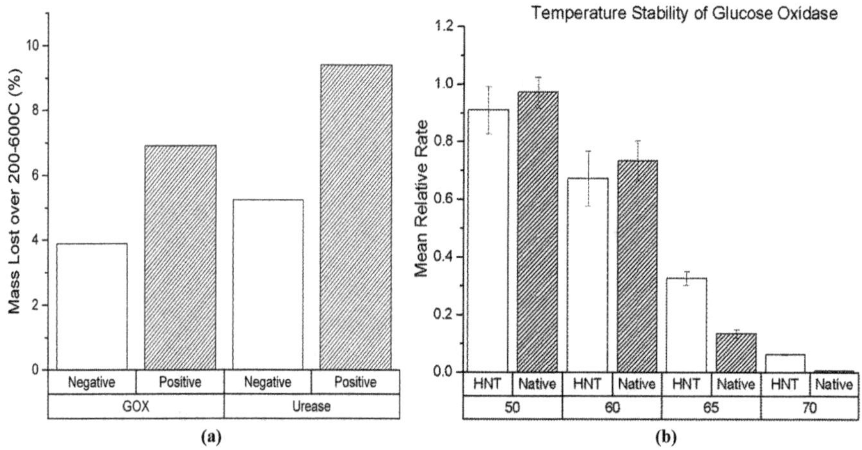

Figure 13.16 (a) Proteins were adsorbed at pH 2 and 10, which affects the charge of the protein as indicated by the labels positive and negative, respectively. From this chart, we see that negatively charged proteins adsorbed in amounts 40% less than that of positively charged proteins. (b) Immobilized (HNT) and native glucose oxidase activity were measured after 15 minutes of heat treatment as specified by the bottom axis. Activities are given relative to samples that were stored at room temperature and did not undergo heat treatment. Immobilized glucose oxidase retains enzymatic activity after heat treatment at temperatures which completely denature the native protein. Reproduced with permission from J. Tully, R. Yendluri and Y. Lvov, *Biomacromolecules*, 2016, **17**, 615. Copyright (2016) American Chemical Society.[53]

in much higher quantities (~40%) than negatively charged proteins. This is indicative that by controlling the pH of the solution during immobilization one can "coerce" the protein to adsorb predominantly inside the tubes (when the protein is negatively charged) or predominantly on the outside (when the protein is positively charged).

An increase in the thermal stability of glucose oxidase was also observed after immobilizing the protein on halloysite. In Figure 13.16b, immobilization doesn't seem to provide a benefit to the thermal stability of the protein and in fact it seems to be detrimental. However, as the temperature increases it becomes apparent that immobilization greatly extends the maximum temperature at which glucose oxidase can be used.

Another promising but underdeveloped field of research is loading DNA into halloysite clay nanotubes.[9,52] The authors of these papers assumed, because of the increase in negativity of the tubes' charge, that the DNA was adsorbed to the surface of the tube. It is more probable that the anionic DNA adsorbed inside the cationic lumen, which would also result in an increase of the magnitude of the tube's negative ξ-potential. These DNA–clay composites may be used for intra-cellular delivery. It is obligatory however to disclose that, when discussing clay nanotube drug formulations, we do not expect intravenous injection to become a possibility. This is largely because these aluminosilicate tubes are not biodegradable. As opposed to internal applications, it is more realistic to consider external medical applications (creams, wound healing tissue, implants) or products for the modification of cell cultures and bacteria.

13.8 Conclusion

The advantage of halloysite to other clays is due to its unique tubule morphology. As a scrolled derivative of the platy kaolin, it retains the same chemical structure, which is ideal. However, the position of charges of the lumen and surface enable applications unimaginable to kaolin. The lumen of halloysite is 15–20 nm in diameter, which slows the diffusion of the tube-loaded molecules. The positive charge of the lumen also allows for selective adsorption and functionalization due to electrostatic interaction. These columbic interactions may also play a role in the extended release of negatively charged molecules from the tubes. The surface of halloysite may be functionalized, allowing further application in non-traditional environments such as hydrophobic polymeric matrices.

When embedded into a polymer, halloysite enhances the mechanical properties of the composite by acting like a skeleton of the body. Functional molecules in the nanotubes imbue the composite with the desired functionality such as anti-corrosion, catalysis, or antibiotics. In another ideology, halloysite acts as an immobilization substrate that, due to its structure, provides a sort of refuge to proteins and other molecules.

References

1. Y. Lvov and E. Abdullayev, *Prog. Polym. Sci.*, 2013, **38**, 1690.
2. R. Price, B. Gaber and Y. Lvov, *J. Microencapsulation*, 2001, **18**, 713.
3. V. Vergaro, E. Abdullayev, Y. M. Lvov, A. Zeitoun, R. Cingolani, R. Rinaldi and S. Leporatti, *Biomacromolecules*, 2010, **11**, 820.
4. V. Vergaro, Y. M. Lvov and S. Leporatti, *Macromol. Biosci.*, 2012, **12**, 1265.
5. C. J. Ward, S. Song and E. W. Davis, *J. Nanosci. Nanotechnol.*, 2010, **10**, 6641.
6. R. Qi, R. Guo, F. Zheng, H. Liu, J. Yu and X. Shi, *Colloids Surf., B*, 2013, **110**, 148.
7. W. Wei, E. Abdullayev, A. Hollister, D. Mills and Y. M. Lvov, *Macromol. Mater. Eng.*, 2012, **297**, 645.
8. N. G. Veerabadran, R. R. Price and Y. M. Lvov, *Nano*, 2007, **2**, 115.
9. Y. Lee, G.-E. Jung, S. J. Cho, K. E. Geckeler and H. Fuchs, *Nanoscale*, 2013, **5**, 8577.
10. C. Aguzzi, C. Viseras, P. Cerezo, I. Salcedo, R. Sánchez-Espejo and C. Valenzuela, *Colloids Surf., B*, 2013, **105**, 75.
11. H. Lun, J. Ouyang and H. Yang, *RSC Adv.*, 2014, **4**, 44197.

12. H. Hemmatpour, V. Haddadi-Asl and H. Roghani-Mamaqani, *Polymer*, 2015, **65**, 143.
13. E. Abdullayev, A. Joshi, W. Wei, Y. Zhao and Y. Lvov, *ACS Nano*, 2012, **6**, 7216.
14. A.-B. Zhang, L. Pan, H.-Y. Zhang, S.-T. Liu, Y. Ye, M.-S. Xia and X.-G. Chen, *Colloids Surf., A*, 2012, **396**, 182.
15. Q. Wang, J. Zhang, Y. Zheng and A. Wang, *Colloids Surf., B*, 2014, **113**, 51.
16. A. A. Elbashir, B. Saad, A. S. Mohamed Ali, M. I. Saleh and H. Y. Aboul-Enein, *J. Liq. Chromatogr. Relat. Technol.*, 2007, **31**, 348.
17. S. A. Zaidi, K. M. Han, S. S. Kim, D. G. Hwang and W. J. Cheong, *J. Sep. Sci.*, 2009, **32**, 996.
18. Q. Wang, J. Zhang and A. Wang, *Appl. Surf. Sci.*, 2013, **287**, 54.
19. C. Aguzzi, P. Cerezo, C. Viseras and C. Caramella, *Appl. Clay Sci.*, 2007, **36**, 22.
20. P. Yuan, P. D. Southon, Z. Liu and C. J. Kepert, *Nanotechnology*, 2012, **23**, 375705.
21. D. Tan, P. Yuan, F. Annabi-Bergaya, H. Yu, D. Liu, H. Liu and H. He, *Microporous Mesoporous Mater.*, 2013, **179**, 89.
22. S. Riela, M. Massaro, C. G. Colletti, A. Bommarito, C. Giordano, S. Milioto, R. Noto, P. Poma and G. Lazzara, *Int. J. Pharm.*, 2014, **475**, 613.
23. M. Bernabé-Pineda, M. a. T. Ramírez-Silva, M. Romero-Romo, E. González-Vergara and A. Rojas-Hernández, *Spectrochim. Acta, Part A*, 2004, **60**, 1091.
24. W. O. Yah, A. Takahara and Y. M. Lvov, *J. Am. Chem. Soc.*, 2012, **134**, 1853.
25. C. Ornelas, *New J. Chem.*, 2011, **35**, 1973.
26. B. Webber, A. Serafin, J. Milchie, C. Van Rensburg, J. Swarts and L. Bohm, *Anticancer Res.*, 2004, **24**, 763.
27. H. Jing, Y. Higaki, W. Ma, H. Wu, W. O. Yah, H. Otsuka, Y. M. Lvov and A. Takahara, *Chem. Lett.*, 2013, **42**, 121.
28. S. Levis and P. Deasy, *Int. J. Pharm.*, 2003, **253**, 145.
29. E. Abdullayev, R. Price, D. Shchukin and Y. Lvov, *ACS Appl. Mater. Interfaces*, 2009, **1**, 1437.
30. M. Finšgar and I. Milošev, *Corros. Sci.*, 2010, **52**, 2737.
31. W. Wei, R. Minullina, E. Abdullayev, R. Fakhrullin, D. Mills and Y. Lvov, *RSC Adv.*, 2014, **4**, 488.
32. N. G. Veerabadran, D. Mongayt, V. Torchilin, R. R. Price and Y. M. Lvov, *Macromol. Rapid Commun.*, 2009, **30**, 99.
33. M. R. Dzamukova, E. A. Naumenko, Y. M. Lvov and R. F. Fakhrullin, *Sci. Rep.*, 2015, **5**, 1–11.
34. X. Zhang, Y. Zheng, L. E. Fried, Y. Du, S. J. Montano, A. Sohn, B. Lefkove, L. Holmgren, J. L. Arbiser and A. Holmgren, *Free Radical Biol. Med.*, 2011, **50**, 811.
35. E. Abdullayev, K. Sakakibara, K. Okamoto, W. Wei, K. Ariga and Y. Lvov, *ACS Appl. Mater. Interfaces*, 2011, **3**, 4040.
36. H. Zhu, M. Du, M. Zou, C. Xu and Y. Fu, *Dalton Trans.*, 2012, **41**, 10465.

37. D. Zhang, W. Huo, J. Wang, T. Li, X. Cheng, J. Li and A. Zhang, *J. Appl. Polym. Sci.*, 2012, **126**, 1580.
38. Y. Zhang and H. Yang, *Phys. Chem. Miner.*, 2012, **39**, 789.
39. Y. Xie, D. Qian, D. Wu and X. Ma, *Chem. Eng. J.*, 2011, **168**, 959.
40. Y. Zhang, Y. Chen, H. Zhang, B. Zhang and J. Liu, *J. Inorg. Biochem.*, 2013, **118**, 59.
41. Y. Chen, Y. Zhang, J. Liu, H. Zhang and K. Wang, *Chem. Eng. J.*, 2012, **210**, 298.
42. J. Xue, Y. Niu, M. Gong, R. Shi, D. Chen, L. Zhang and Y. Lvov, *ACS Nano*, 2015, **9**, 1600.
43. Y. Zhao, S. Wang, Q. Guo, M. Shen and X. Shi, *J. Appl. Polym. Sci.*, 2013, **127**, 4825.
44. R. Qi, X. Cao, M. Shen, R. Guo, J. Yu and X. Shi, *J. Biomater. Sci., Polym. Ed.*, 2012, **23**, 299.
45. Y. Zhao, E. Abdullayev, A. Vasiliev and Y. Lvov, *J. Colloid Interface Sci.*, 2013, **406**, 121.
46. D. N. Elumalai, Y. Lvov and P. Derosa, *J. Encapsulation Adsorpt. Sci.*, 2015, **5**, 74.
47. D. G. Shchukin, G. B. Sukhorukov, R. R. Price and Y. M. Lvov, *Small*, 2005, **1**, 510.
48. R. Zhai, B. Zhang, L. Liu, Y. Xie, H. Zhang and J. Liu, *Catal. Commun.*, 2010, **12**, 259.
49. P. Zheng, Y. Du and X. Ma, *Mater. Chem. Phys.*, 2015, **151**, 14.
50. C. Chao, B. Zhang, R. Zhai, X. Xiang, J. Liu and R. Chen, *ACS Sustainable Chem. Eng.*, 2014, **2**, 396.
51. J. Yao, Q. Wang, Y. Wang, Y. Zhang, B. Zhang and H. Zhang, *Desalin. Water Treat.*, 2014, 1.
52. M. H. Shamsi and K. E. Geckeler, *Nanotechnology*, 2008, **19**, 075604.
53. J. Tully, R. Yendluri and Y. Lvov, *Biomacromolecules*, 2016, **17**, 615.

CHAPTER 14

Biocompatible Electrospun Polymer–Halloysite Nanofibers for Sustained Release

YILI ZHAO[†a,b], ZHE WANG[†a], MINGWU SHEN[a] AND XIANGYANG SHI*[a,b]

[a]State Key Laboratory for Modification of Chemical Fibers and Polymer Materials, College of Chemistry, Chemical Engineering and Biotechnology, Donghua University, Shanghai 201620, People's Republic of China; [b]Key Laboratory of Textile Science & Technology, Ministry of Education, College of Textiles, Donghua University, Shanghai 201620, People's Republic of China

*E-mail: xyshius@yahoo.com

14.1 Introduction

Electrospinning has seen growing interest as a technique of fabricating nanofibers for applications in tissue engineering and drug delivery.[1] The micro-nano scale organization and high porosity of electrospun membranes, which are similar to the natural extracellular matrix (ECM), are favorable for adhesion and proliferation of cells and decrease in the immune response.[2] In addition, controlled local drug delivery can be realized by electrospun fibers.

[†]These authors equally contributed to this chapter.

RSC Smart Materials No. 22
Functional Polymer Composites with Nanoclays
Edited by Yuri Lvov, Baochun Guo and Rawil F. Fakhrullin

Published by the Royal Society of Chemistry, www.rsc.org

Drug-loaded nanofibers can be formed easily by electrospinning polymer solution and drug together or immersing the electrospun nanofibers into drug solution. However, the burst release effect at the initial period is often unavoidable. Therefore, coaxial and emulsion electrospinning approaches have been utilized to form core–shell nanofibers with drug enclosed in the core region of the fibers.[3–5] In this case, the burst release effect can be avoided, the diffusion rate of the therapeutic agent is fairly stable, and the drug could be released in a sustained manner. Nevertheless, the stringent specification of the electrospinning equipment in the coaxial method and the poor biocompatibility of emulsifiers in the emulsion method limit their biomedical applications. More recently, the fabrication of "nano in nano" composites for drug delivery has witnessed a growing interest. These "nano in nano" composites consist of biodegradable drug-loaded polymer nanoparticles or biodegradable drug-loaded inorganic nanoparticles (*e.g.*, halloysite nanotubes (HNTs), LAPONITE® (LAP), nano-hydroxyapatite (n-HA), mesoporous silica (MMS), glass, *etc.*) incorporated into polymer nanofibers.[6–14] The formed composite nanofibers can efficiently modulate the drug release profile, but an extra step to prepare drug-loaded polymer nanoparticles or drug-loaded inorganic nanoparticles is required prior to the fabrication of the nanofibers. Among these nanoparticles, the natural clay halloysite is distinct.

HNTs are a kind of nanotubular kaolinite clay.[15] Their basic structure and properties have been described in Chapter 5. Potential applications of HNTs as a drug delivery vehicle have been reported in the literature[16–21] and the drug release rate could be retarded by coating polymers onto the drug-loaded HNTs. Also, HNTs are reported to be biocompatible,[22,23] suggesting that they are promising drug containers. In general, the drug-loaded HNTs are presenting as a powder or suspension, which may limit their application in the development of drug-loaded devices. It is expected that by the combination of the drug loading capability of HNTs and the eletrospinning technology, a new drug delivery system with sustained release property could be developed for various applications such as tissue engineering and pharmaceutical sciences.

In this chapter, we review recent progress toward the development of electrospun polymer–halloysite nanofibers for sustained release and their biomedical applications.

14.2 Electrospinning Technology

Electrospinning, first reported in 1934, has been used for more than 80 years. The term "electrospinning", derived from "electrostatic spinning", was coined relatively recently. Since the 1980s and especially in recent years, the electrospinning process has regained more attention, probably due in part to a surge of interest in nanotechnology, as ultrafine fibers or fibrous structures of various polymers with diameters in the submicron/nanometer range can be easily fabricated *via* electrospinning.[24–26]

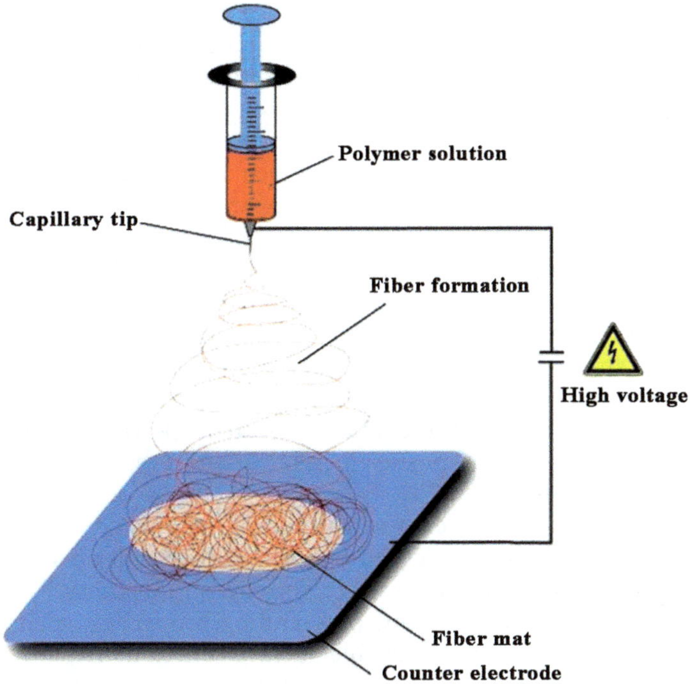

Figure 14.1 A laboratory setup for an electrospinning experiment with a perpendicular arrangement of the electrodes.[27] (Reprinted from ref. 27 with permission from John Wiley and Sons. Copyright © 2007 Wiley-VCH Verlag GmbH & Co. KGaA, Weinheim.)

A schematic diagram demonstrating the process of electrospinning to form polymer nanofibers is shown in Figure 14.1.[27] There are basically three components: a high voltage supplier, a capillary tube with a pipette or needle of small diameter, and a metal collector. A high voltage is used to create an electrically charged jet of polymer solution or melt out of the pipette. Before reaching the collector, the solution jet evaporates or solidifies, and is collected as an interconnected web of small fibers. One electrode is placed into the spinning solution/melt and the other attached to the collector. In most cases, the collector is simply grounded. The electric field is applied across the end of the capillary tube that contains the fluid held by its surface tension. This induces a charge on the surface of the liquid. Mutual charge repulsion and the contraction of the surface charges to the counter electrode create a force directly opposite to the surface tension. As the intensity of the electric field is increased, the hemispherical surface of the fluid at the tip of the capillary tube elongates to form a conical shape known as the Taylor cone. Further increasing the electric field, a critical value is attained with which the repulsive electrostatic force overcomes the surface tension and the charged jet of the fluid is ejected from the tip of the Taylor cone. The discharged polymer solution jet undergoes an instability and elongation process, which allows the jet to become very long and thin. Meanwhile, the solvent evaporates, leaving behind a charged polymer fiber. In the case of the melt, the discharged jet solidifies when it travels in the air stream.[25,26,28–33]

Electrospinning appears to be affected by the following parameters and variables: (1) system parameters such as molecular weight, molecular weight distribution and architecture (branched, linear, *etc.*) of the polymer, and polymer solution properties (viscosity, conductivity, dielectric constant, and surface tension, charge carried by the spinning jet) and (2) process parameters such as electric potential, flow rate, distance between the capillary and collector, ambient parameters (temperature, humidity, and air velocity in the chamber) and finally motion of the collector.[34] By appropriately varying one or more of the above parameters, nanofibers can be successfully electrospun from a rich variety of materials that include synthetic polymers, biopolymers (*e.g.*, DNA and protein), composites, ceramics, and even small molecules such as phospholipids.[25,26,28]

Electrospinning has attracted much attention not only because of its versatility in spinning a wide variety of polymeric fibers but also because of its consistency in producing fibers with submicron thickness.[35,36] Electrospun nanofibers have shown many outstanding properties, such as large surface area, high aspect ratio, flexible surface functionality, and tunable surface topology. Among many possible applications of electrospun fibers, such as composite materials reinforcement, sensors, filtration, catalysis, protective clothing, solar cells, optoelectronics, and biomedical sciences (including wound dressing and scaffolds for tissue engineering, implants, membranes, and drug delivery systems), the uses of them for tissue engineering and drug delivery have become the most interesting two areas.[37] Tissue engineering is an interdisciplinary field that applies the principles of chemistry, physics, materials science, engineering, cell biology, and medicine to develop biological substitutes that restore, maintain or improve tissue/organ functions.[38] Electrospun nanofibers quite mimic the ECM[39] and are able to enhance cell adhesion, migration, proliferation, and differentiation.[40]

Besides tissue engineering applications, electrospun fibers have also been used for topical drug delivery. Localized therapy of skin and wounds using appropriate electrospun fibers as delivery vehicles could significantly reduce the systemic absorption of the drugs and also provide localized therapeutic effect at lower drug concentrations. Some of the unique features of electrospun fibers as drug delivery vehicles are the ability to incorporate a wide range of drugs in an amorphous or solid solution form within the fibers, the high surface area of the fibers leading to efficient drug release, the interconnecting porous structure with high permeability, and the ease of fabrication of the delivery vehicle in the required architectures/forms. Currently, polymeric nanofibers for drug delivery are under extensive investigation for a variety of other biomedical applications, such as post-surgical adhesion prevention membranes.[41–43]

14.2.1 Traditional Electrospun Nanofibers for Drug Delivery

The traditional method to prepare drug-carrying fibers is to blend drugs into the polymer materials in solution, followed by electrospinning to form drug-loaded ultrafine fibers.[44–46] In this way, drugs could easily be located on the fiber surface because of the rapid evaporation of the solvent. Therefore, fibrous mats prepared by the blending method often result in a burst release effect in the initial stage of drug release.[45,47] In some cases, the initial burst release is necessary to eliminate the intruding bacteria before they begin to proliferate in the case of antibiotics.[48] For example, bio-absorbable polylactic acid (PLA) nanofibers were used to load an antibiotic drug Mefoxin, and the efficiency of this nanofibrous membrane was superior to that of bulk film.[49] Cui *et al.* prepared drug-loaded electrospun nanofibers using the blending method and then investigated the effect of fiber diameter and drug loading on the release behavior. They found that electrospun fibers with a larger diameter exhibited a longer period of nearly zero order release, and higher drug encapsulation resulted in a more significant burst release.[50] For example, electrospun fibers with 5.0% of drug entrapment and with a diameter of 212 nm, 551 nm, and 1.31 μm all exhibited a linear release behavior during the first stage. Nearly zero order release profiles were observed during the first 24, 48, and 96 h, and the total amount of release was 80%, 50%, and 80%, respectively, during this stage. Besides the fiber diameter, the fiber morphology, drug content, and fiber degradation properties may also have an effect on local drug delivery. For potential use in topical drug administration, Verreck *et al.* investigated poorly water-soluble drugs loaded in a water-insoluble, non-biodegradable polymer.[51] The collected non-woven fibers were shown to release the drugs at various rates and profiles based on the nanofiber morphology and drug content. Cui *et al.* synthesized acid-labile polymers by introducing acetal groups and *ortho* ester groups into the biodegradable backbone of poly(DL-lactide)–poly(ethylene glycol) (PELA), respectively, and investigated the degradation and drug release behaviors responding to the local pathological pH environment.[52] An *in vitro* release study showed that the total amount of drug released from acid-labile polymeric fibers was accelerated due to the pH-induced structural and morphological changes of the fibrous mats and the degradation of matrix polymers, and the burst release was significantly higher for polymers with higher contents of acid-labile segments.[53]

14.2.2 Coaxial and Emulsion Electrospinning Nanofibers for Drug Delivery

The coaxial electrospinning method has been used to prepare a fibrous drug delivery system with a core–shell structure, in which two miscible or immiscible components can be conveniently electrospun into a composite fiber with a core encapsulated inside a shell. In this way, drugs or their mixture can be enclosed within the polymeric outer shell to form a core–shell type drug delivery device.[54,55] In coaxial electrospinning, bioactive agents or drugs encapsulated within electrospun nanofibers show steady release characteristics. A model drug, tetracycline hydrochloride (TCH), encapsulated into poly (L-lactic acid) (PLLA) fibers has been used to fabricated core–shell fibers through coaxial electrospinning.[56] Through an *in vitro* release study, they found that the threads made from the TCH-loaded PLLA core–shell fibers could suppress the initial burst release and provide a sustained drug release profile. Mickova *et al.* embedded the enzyme-loaded liposomes inside polyvinyl alcohol-core–poly-ε-caprolactone-shell nanofibers, and results indicated that the system showed a slower initial release and also preserved the enzymatic activity.[3] In another study, paclitaxel (PTX)-loaded poly(lactic acid-*co*-ε-caprolactone) (75 : 25) (P(LLA-CL) (core/shell)) was electrospun into fibers using 2,2,2-trifluoroethanol as the solvent. Results indicated that PTX could be released from the fibers in a short burst period over 24 h, followed by very slow release over the following 60 days.[4]

Emulsion electrospinning has also been used to generate core–shell nanofibers for drug delivery applications. For instance, Qi *et al.* obtained core–shell structured ultrafine fibers of bovine serum albumin (BSA)/PLLA through emulsion electrospinning.[57] They found that the formed core–shell nanofibers may serve as a promising scaffold for sustainable, controllable, and effective release of bioactive proteins for tissue engineering and other applications. Yang *et al.* showed that BSA-loaded PLA core–shell nanofibers *via* emulsion electrospinning could reduce the initial drug burst release and provide sustained therapeutic activity.[58] In another study, Rhodamine B and BSA were successfully incorporated into the P(LLA-CL) nanofibers by means of emulsion electrospinning. *In vitro* dual drug release behaviors revealed that the incorporated drug and/or proteins in composite fibrous mats could be released in a controlled manner by adjusting the processes of emulsion preparation.[59]

Both the coaxial electrospinning and emulsion electrospinning method can be used to make core–shell nanofibers. By using these two methods, the initial burst drug release can be suppressed. Also, drugs or biologically active objects encapsulated inside can be protected from harsh solvents. Factors that influence the sustained drug delivery of core–shell fibers fabricated by coaxial electrospinning or emulsion electrospinning are nearly the same, such as the types of polymers used, and the porosity or the cracks on the shell structure.[60]

14.2.3 Novel Electrospinning Nanofibers for Drug Delivery

An ordinary single fluid electrospinning method permits direct integration of drug molecules within nanofibers by simply electrospinning the drug and polymer mixture solution or post-adsorption of drugs onto/within the nanofibers.[42,45,61] However, such a method often displays a burst release profile, which is not desirable in most cases. Emulsion and coaxial electrospinning are two improved techniques to be used for drug delivery applications, which are able to mitigate the burst release of the drug to some extent.[57,62,63] In both methods, the drugs can be incorporated into the core region of the nanofibers to form a so called "core-sheath" structure.[64] Nevertheless, there are still some issues and challenges in the emulsion and coaxial electrospinning techniques. The emulsifier used in emulsion electrospinning may be cytotoxic, and coaxial electrospinning may need substantial optimization of the electrospinning parameters. In another aspect, both the traditional and coaxial or emulsion electrospun nanofibers suffer from a problem of weak mechanical durability, which is not desirable for practical biomedical applications. Therefore, development of other nanofiber systems that can overcome the burst release of the encapsulated drugs and have improved mechanical durability still remains a great challenge.

More recently, fabrication of "nano in nano" composites, which consist of biodegradable drug-loaded polymer nanoparticles or biodegradable drug-loaded inorganic nanoparticles incorporated into polymer nanofibers, has received a great deal of attention for applications in drug delivery.[6–14] Zheng *et al.* demonstrated that amoxicillin (AMX) can be successfully adsorbed onto the n-HA surface and the formed AMX–n-HA–poly(lactic-*co*-glycolic acid) (PLGA) composite nanofibers display a uniform and smooth morphology with improved mechanical durability. Cell viability assays and cell morphology observations reveal that the formed AMX–n-HA–PLGA composite nanofibers are cytocompatible. Importantly, the loaded AMX within the n-HA–PLGA hybrid nanofibers shows a sustained release profile and a non-compromised activity to inhibit the growth of a model bacterium, *Staphylococcus aureus*.[8] Hu *et al.* fabricated electrospun composite fibrous membranes based on drug-loaded modified MMS and PLLA. The electrospun PLLA–MMS–IBU (IBU denotes ibuprofen) composite fibrous membranes showed significantly lower initial burst release (6% release in the first 12 h) compared with that of electrospun PLLA–IBU fibrous membranes (46% release in the first 12 h) in *in vitro* release tests. Moreover, the IBU release from PLLA–MMS–IBU was significantly longer than that from PLLA–IBU (100 *vs.* 20 days). In animal studies, both PLLA–IBU and PLLA–MMS–IBU showed improved anti-adhesion properties and anti-inflammatory effects compared with PLLA fibrous membrane alone at four weeks post implantation.[12] Kouhi *et al.* demonstrated that poly (ε-caprolactone) (PCL) nanofibers containing bioactive glass (BG) nanoparticles and simvastatin drug could be produced by electrospinning. *In vitro* degradation tests indicated that the presence of BG led to a faster degradation of the nanofibrous webs, and drug release studies performed in phosphate buffered saline (PBS) at pH 7.4 showed that the drug release rate was affected by the BG concentration.[13]

Moreover, other nanoclay materials such as LAP can also be used as drug carriers. For example, Wang *et al.* encapsulated a chemotherapeutic drug doxorubicin (DOX) into the interlayer space of LAP through an ionic exchange process.[65] The formed LAP–DOX nanodisks demonstrated a pH-dependent *in vitro* drug release profile and higher therapeutic efficacy in inhibiting the growth of cancer cells. Li *et al.* further analyzed the long-term *in vivo* antitumor efficacy and toxicology of the prepared LAP–DOX complexes using a tumor-bearing mouse model.[66] In order to further increase the targeted delivery of drugs, folic acid or lactobionic acid was covalently conjugated onto the surface of LAP nanodisks.[67,68]

In addition, the clay materials LAP and attapulgite (ATT) can also be applied for the fabrication of "nano in nano" composites for osteogenic differentiation. The LAP or ATT incorporated electrospun PLGA nanofibers display uniform fibrous morphology, improved mechanical durability, increased surface hydrophilicity and protein adsorption capacity, good hemocompatibility, and superior cytocompatibility. Most importantly, the doped LAP or ATT within the PLGA nanofibers is able to induce the osteoblastic differentiation of human mesenchymal stem cells (hMSCs) in growth medium without any inducing factors.[69,70] In a recent study, Wang *et al.* encapsulated an AMX drug within LAP, followed by electrospinning of the mixture of LAP–AMX complexes and PLGA to form "nano in nano" composite nanofibers that display sustained release of the AMX drug and improved mechanical durability.[69]

14.3 Electrospun Halloysite–Polymer Nanofibers

HNTs have been added to biodegradable polymers such as PLA[71] and PLGA[7,72] to form composite nanofibers. A recent study by Zhao *et al.* showed that HNTs were well distributed in the PLGA nanofibers in a coaxial manner (Figure 14.2), and the incorporated HNTs did not significantly change the morphology of the PLGA nanofibers. Importantly, the HNT-doped PLGA nanofibers display enhanced mechanical properties and non-compromised hemocompatibility.[73] Moreover, the electrospun HNTs–PLGA nanofibrous mats have

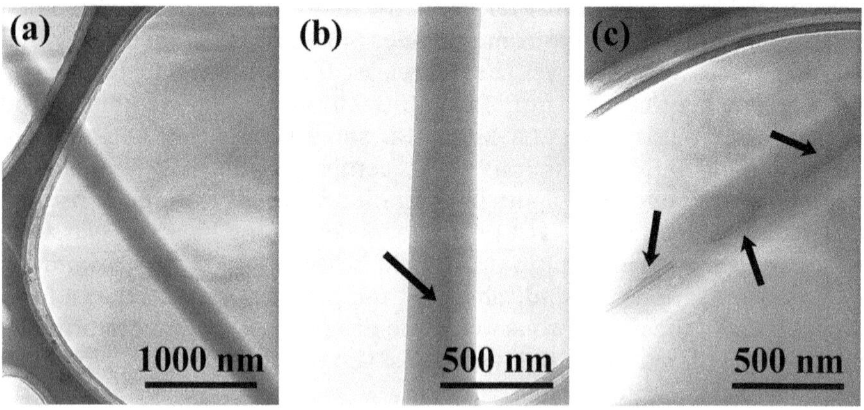

Figure 14.2 TEM micrographs of (a) PLGA nanofibers, (b) HNT-doped PLGA nanofibers with 5% HNTs relative to PLGA, and (c) CNT-doped PLGA nanofibers with 5% CNTs relative to PLGA. The arrows indicate the nanotubes embedded within the PLGA nanofibers.[73] (Reprinted from ref. 73 with permission from John Wiley and Sons. Copyright © 2012 Wiley Periodicals, Inc.)

excellent biocompatibility.[7,72] An MTT proliferation assay showed that the incorporation of HNTs did not significantly influence the cell proliferation or viability compared to the pure PLGA fibers. SEM observations of the cell morphology showed that the HNTs–PLGA nanofibrous scaffolds could provide a three-dimensional structure with interconnected pores for cell attachment and migration. The fibroblast cells cultured on the nanocomposite scaffolds displayed a phenotypic shape, suggesting that the cells could penetrate and migrate within the scaffolds similar to native ECM.

As a biocompatible material, HNTs have their own strong ability to absorb a variety of drug molecules. By incorporating drug-loaded HNTs within polymer nanofibers to form "nano in nano" composite nanofibers, one can design a desired drug delivery system that possesses many advantages such as improved mechanical properties, enhanced cellular compatibility, or significantly reduced burst release of the incorporated drugs.

14.4 Properties of Electrospun Drug-Loaded Halloysite–Polymer Nanofibers

14.4.1 Mechanical Reinforcement

Mechanical properties are of paramount importance for the fabricated nanofibers to be applied in tissue engineering. The natural tubular morphology, nanoscale diameter, and high aspect ratio of HNTs make them particularly attractive for reinforcing in polymer nanocomposites. The improved mechanical properties are always attributed to the fact that the external load is efficiently transferred from the polymer fiber matrix to the HNTs.[7,74,75] In a recent study, Qi *et al.* used water as the medium to load TCH into HNTs with an optimized TCH loading percentage of 42.65%. Then the TCH-loaded HNTs were blended with PLGA solution for subsequent electrospinning to obtain the TCH–HNTs–PLGA composite nanofibers.[74] As can be seen from Figure 14.3, the incorporation of TCH-loaded HNTs within PLGA nanofibers does not significantly compromise the uniform fibrous structure of the PLGA nanofibers. More importantly, as can be seen from Figure 14.4, compared with PLGA fibrous mats, the breaking strength, Young's modulus, and failure strain significantly increase with the addition of only a small amount of HNTs (1 wt% relative to PLGA).[76] This enhancement may be ascribed to the alignment of HNTs in the PLGA fibers. The strength and modulus of HNT-doped polymer nanofibers are also related to the nanotube loading percentage, dispersion, and interfacial interactions in the systems.[77]

14.4.2 Sustained Drug Release Effect

Due to the fact that the drug molecules have to be first released from the HNTs to the polymer fiber matrix, and then released from the polymer fiber matrix to the releasing medium, the composite HNTs–PLGA nanofibers show an improved sustained release profile with a release rate much slower than

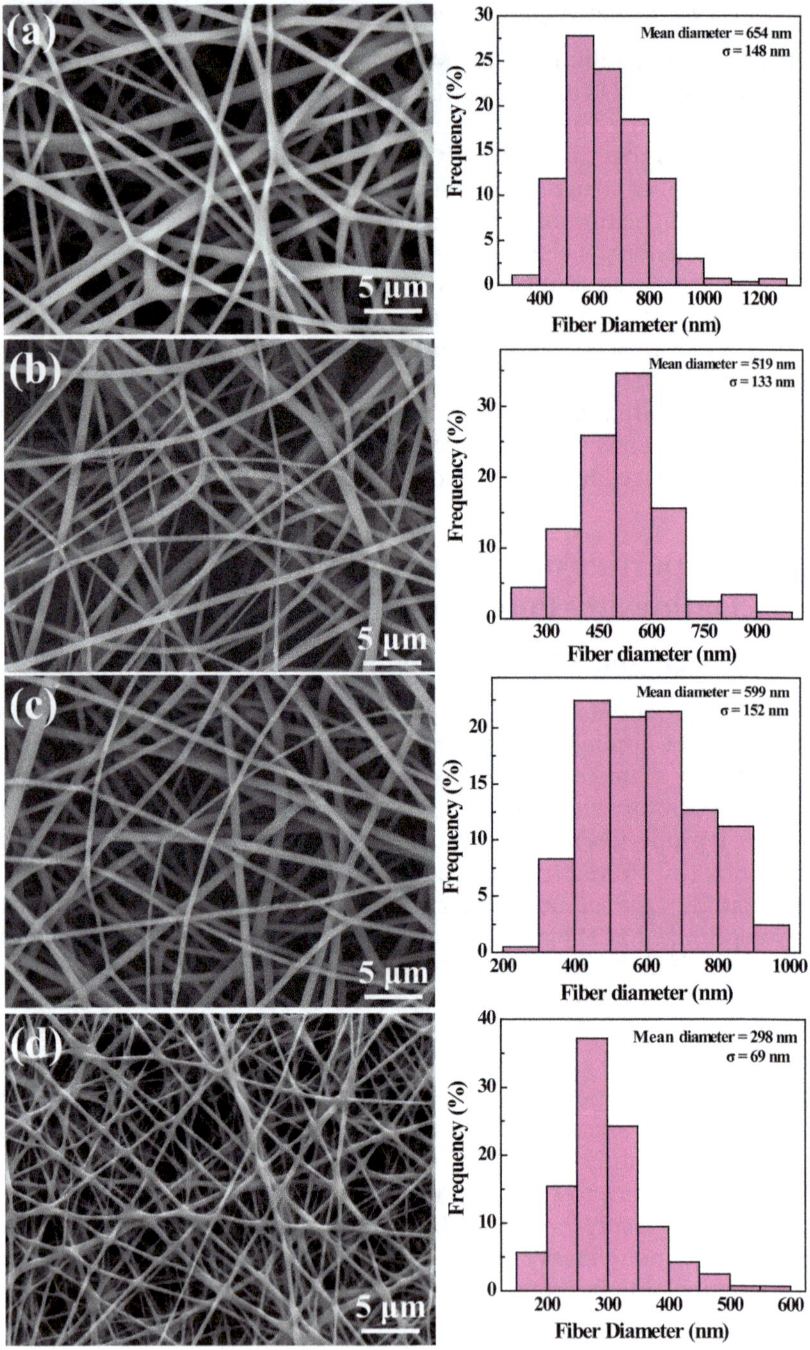

Figure 14.3 SEM micrographs of the electrospun (a) PLGA, (b) TH-1–PLGA, (c) TH-2–PLGA, and (d) TCH–PLGA nanofibers, and the corresponding fiber diameter distribution histograms.[74] (Reprinted from *Colloids and Surfaces B: Biointerfaces*, **110**, R. Qi *et al.*, Controlled release and antibacterial activity of antibiotic-loaded electrospun halloysite/poly(lactic-*co*-glycolic acid) composite nanofibers, 148–155. Copyright 2013, with permission from Elsevier.[74])

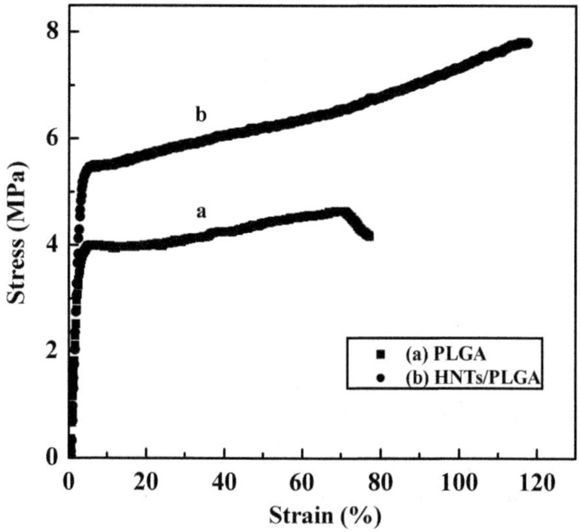

Figure 14.4 Strain–stress curves of electrospun (a) PLGA and (b) HNTs–PLGA (1 wt% HNTs relative to PLGA) nanofibrous mats.[76] (Reprinted from ref. 76 with permission from the Royal Society of Chemistry.)

those of the TCH–PLGA nanofibers and TCH–HNTs powder.[74–76] Both electrospun TCH–PLGA nanofibers and TCH–HNTs powders exhibit an obvious initial burst release. Around 83.8% and 89.4% of TCH were released from the TCH–PLGA nanofibers and the TCH–HNTs powders, respectively within the first day. And then a plateau was reached after two days with a release percentage of >90% for both formulations. In sharp contrast, the medicated TCH–HNTs–PLGA composite nanofibers showed a relatively sustained TCH release profile. Only about 18.6% of TCH was released from the TCH–HNTs–PLGA composite nanofibers within the first day, and after 42 days, about 77.6% of the TCH drug was released. The TCH release percentage from the TCH–HNTs–PLGA composite nanofibers at 42 days was much lower than those of the TCH–PLGA nanofibers and TCH–HNTs powder within the first day.[74] In another study, Xue *et al.* prepared PCL–gelatin (PG) solution with a concentration of 6 wt% by dissolving PCL and gelatin in trifluoroethanol under magnetic stirring for 24 h at room temperature.[75] Then four different kinds of electrospinning solution were prepared by adding 20 wt% HNTs, metronidazole (MNA), MNA-loaded HNTs, or MNA-loaded HNTs mixed with free MNA into the PG solution. The electrospun fibrous membranes with different compositions were labeled as PG–HNT, PG–MNA, PG–HNT-MNA, and PG-MNA–HNT-MNA, respectively. For the PG–HNT-MNA nanofibers, the same sustained release effect as TCH–HNTs–PLGA composite nanofibers can be observed. More importantly, for the PG-MNA–HNT-MNA nanofibers, the drug can be loaded both in the fiber matrix and in the HNTs, and the release kinetics represent a combination of the two processes. Firstly, about 75 wt% MNA was released from the fibrous membrane within four days, and then

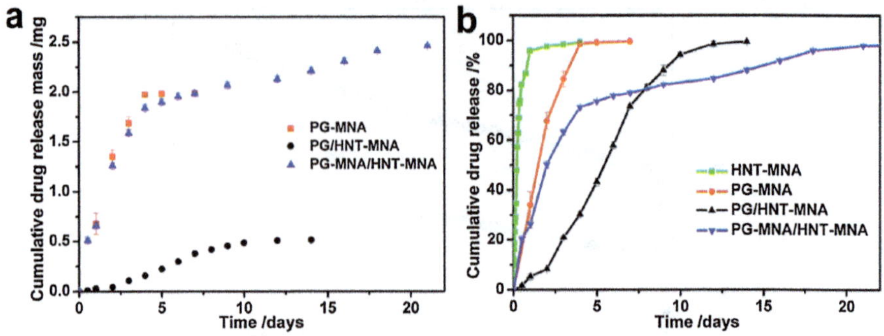

Figure 14.5 Cumulative drug release profiles: (a) released metronidazole (MNA) mass and (b) released percent for electrospun drug-loaded membranes at different soaking times in PBS buffer.[75] (Reprinted with permission from J. J. Xue, Y. Z. Niu, M. Gong, R. Shi, D. F. Chen, L. Q. Zhang and Y. Lvov, *ACS Nano*, 2015, **9**, 1600.[75] Copyright (2015) American Chemical Society.)

the remaining 25% MNA was released in a sustained manner for over three weeks due to the location of the drugs inside the nanotubes. After 10 days, the drug release rate was still high enough to prevent bacterial growth. This formulation combined 5 wt% drug loading into the nanotubes with an additional 20 wt% loaded into the polymeric fiber matrix, which is an essential improvement for the membrane antibacterial efficiency due to the release rate that can be extended up to three weeks (Figure 14.5).[75] In the first week after the membrane was implanted into a tissue defect site, prevention of the high-incidence of infection and inflammation is required. Therefore, a high initial release rate of antibiotics in the first few days is desirable for eliminating the intruding bacteria. For the few bacteria that survive in the initial stage, a continued drug release for two weeks is necessary to inhibit the occurrence of latent infection.[48]

14.4.3 Drug Activity Maintenance

TCH has a broad antibiotic spectrum that includes both Gram-negative and Gram-positive bacteria. The bacterial inhibition percentages after incubation of the bacterial suspension with different samples are shown in Figure 14.6.[74] It is clear that the samples without TCH encapsulation (HNT powder, and PLGA and HNTs–PLGA nanofibers) do not display any inhibitory effect on the bacterial growth. In contrast, free TCH, TCH–HNTs powders, and drug-loaded nanofibrous mats of TH-1–PLGA (TCH–HNTs–PLGA nanofibers with 1% TCH) and TCH–PLGA are all able to inhibit the bacterial growth. Free TCH and TCH–HNTs can effectively inhibit the growth of *S. aureus* by 81.1% and 80.3%, respectively. The approximately similar bacterial inhibition effect for both the TCH–HNTs powder and free TCH is likely due to the efficient exposure of drug molecules to the bacterial suspension. Moreover, the electrospun TH-1–PLGA and TCH–PLGA nanofibrous mats also display effective antibacterial activities with bacterial inhibition percentages of

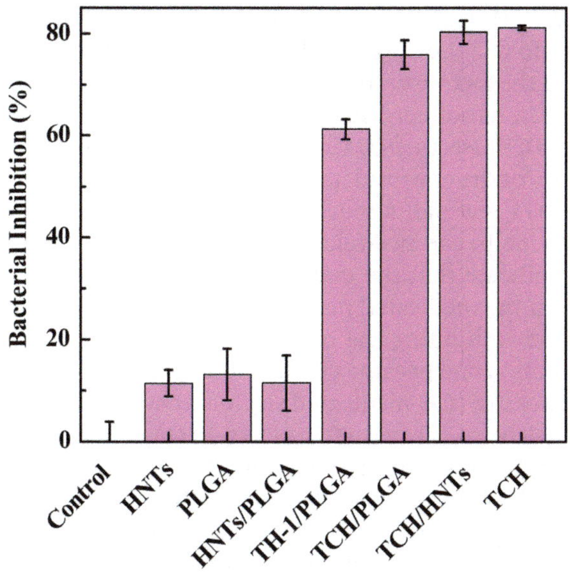

Figure 14.6 Inhibition of bacterial (*S. aureus*) growth after 24 h of incubation with free TCH, HNTs powders, TCH–HNTs powders, and PLGA, HNTs–PLGA, TH-1–PLGA, and TCH–PLGA nanofibers. The concentration of free TCH and TCH in medicated samples was 20 μg mL^{-1}.[74] (Reprinted from *Colloids and Surfaces B: Biointerfaces*, **110**, R. Qi *et al.*, Controlled release and antibacterial activity of antibiotic-loaded electrospun halloysite/poly(lactic-*co*-glycolic acid) composite nanofibers, 148–155, Copyright 2013, with permission from Elsevier.[74])

61.3% and 75.9%, respectively. The antibacterial effect of TH-1–PLGA was lower than that of TCH–PLGA nanofibers. This could be due to the fact that the release rate of TCH from the composite HNTs–PLGA nanofibers is significantly slower than that of the TCH–PLGA nanofibers, and hence the actual concentration of TCH exposed to the bacterial suspension is much lower than that of the TCH–PLGA nanofibers. These results confirm that the encapsulation of TCH in composite HNTs–PLGA nanofibers has no adverse effect on the structure of drugs, and the TCH released from the electrospun mats retains the inherent antibacterial function.

The bacterial inhibition efficacy of the TCH-medicated nanofibrous mats was also investigated using a modified Kirby–Bauer method.[74] PLGA, HNTs–PLGA, TH-1–PLGA, TH-2–PLGA (TCH–HNTs–PLGA nanofibers with 2% TCH), and TCH–PLGA nanofibrous mats were pasted onto the agar plate. After 4 h of incubation, the fibrous mats were removed and the bacterial suspension was spread onto the regions where the different fibrous mats were pasted. It is clear that after incubation of the *S. aureus* for 24 h, 48 h, and 60 h, respectively, the agar regions with TCH-medicated scaffolds pasted display visible bacterial inhibition rings. In contrast, TCH-free PLGA and HNTs–PLGA mats do not show an obvious bacterial inhibition effect. The bacterial inhibition ring for the TCH-medicated scaffold is bigger than the scaffold size due to the diffusion of the released TCH drug onto the agar plate. And the inhibition zone of the TCH–PLGA scaffold is the biggest, which should be attributed to a faster drug release rate and the fact more TCH is released from TCH–PLGA than from the TCH–HNTs–PLGA nanofibers. It is worth noting that the slow release of the TCH from the TCH–HNTs–PLGA nanofibers does not compromise the antibacterial efficacy of the encapsulated TCH drug. The size of the inhibition ring of the TH-2–PLGA scaffold is much larger than that of the TH-1–PLGA mat, which is likely due to the two-fold TCH loading in the TH-2–PLGA scaffold. These results indicate that the encapsulated antibiotic drug TCH is able to be released from the medicated nanofibrous mats and diffuses onto the agar, thereby effectively inhibiting the bacterial growth. In addition, although the bacterial inhibition zone became smaller at longer incubation times, the antibacterial ability of the TCH-medicated scaffolds could still be observed after 60 h of incubation. The medicated TCH–HNTs–PLGA composite nanofibers have great potential to be used in functional wound dressing and tissue engineering.

For the PG–HNT-MNA and PG-MNA–HNT-MNA nanofibers, the MNA maintained its antibacterial activity well and the antibacterial activity was consistent with the release rate from different materials.[75] HNTs were incorporated into PCL scaffolds for applications in wound healing. An antiseptic, brilliant green, was loaded and its release was analyzed. Results indicated that the release of brilliant green from electrospun composite could be extended to five days, which is contrary to 30 h for the unmodified HNTs. About 20% of the initial drug burst is associated with the externally adsorbed molecules. Figure 14.7 depicts the inhibition zone of bacteria growth around an HNTs–PCL scaffold square, indicating its good antibacterial properties.[78]

Figure 14.7 Zone of inhibition of *S. aureus* growth around empty halloysite–PCL scaffolds and the one loaded with brilliant green after seven days exposition in lysogeny/agar broth.[78] (Reprinted from *Progress in Polymer Science*, **38**, Y. Lvov and E. Abdullayev, Functional polymer-clay nanotube composites with sustained release of chemical agents, 1690–1719, Copyright 2013, with permission from Elsevier.[78])

14.4.4 Biocompatibility and Biodegradability

Drug-loaded HNTs–polymer nanofibers display good biocompatibility and biodegradability, which is desirable for their biomedical applications. Xue *et al.* reported that the MNA-loaded HNT fibrous membranes did not show cytotoxicity. The cells incubated for 72 h grew well and showed spindle, triangular, and quadrangular shapes. The influence of drugs on the biocompatibility of the membranes was also investigated by studying the proliferation of the cells. For all of the tested membranes, the number of cells increased continuously during seven days of incubation, indicating that the membranes were nontoxic and support cell proliferation (Figure 14.8).[75] As shown in SEM and confocal laser scanning microscopy (CLSM) micrographs, after five days of seeding, the cells reach 50–60% confluency with stretching shape and demonstrate extension and spreading cytoskeleton. Specific structures, filopodia, and lamellipodia, related to the cell motility of fibroblasts can be observed.

In another study, Qi *et al.* also demonstrated that the TCH–HNTs–PLGA nanofibrous mats had good cytocompatibility.[7,74,76] As assessed by the protein adsorption assay, the protein adsorption onto PLGA scaffolds doped with 1, 3, and 5% HNTs was 1.9–2.1 times that of pure PLGA scaffolds, clearly indicating that the incorporation of HNTs favors protein adsorption, which is beneficial for enhanced cell growth and proliferation.[7]

Therefore, even at a high loading content of drugs, the developed HNT-doped nanofibers do not have any negative effects on the cell viability and proliferation. Thus, the HNT-based fiber drug delivery system could be used as an efficient anti-infection membrane with a long-lasting drug release profile.

14.5 Conclusion and Outlook

In summary, this chapter reviews the electrospun polymer–HNTs nanofibers-based drug delivery system that can be used for efficient drug loading and sustained release. The incorporation of drug-loaded HNTs within polymer nanofibers does not significantly compromise the uniform fibrous structure of the polymer nanofibers, and instead significantly improves their mechanical properties. The developed "nano in nano" composite fiber system could release the drug in a sustained manner, appreciably weakening the initial burst release characteristics. Therefore, the developed electrospun polymer–HNTs composite nanofibrous mats may be used in various therapeutic applications requiring extended therapeutic efficacy, including pathologies demanding chronic drug treatments, wound healing, prevention of postsurgical adhesions, and tissue engineering applications.

The research area of electrospun HNTs–polymer hybrid nanofiber-based drug delivery systems still remains open since a large amount of issues regarding their fabrication and applications remains to be addressed. For

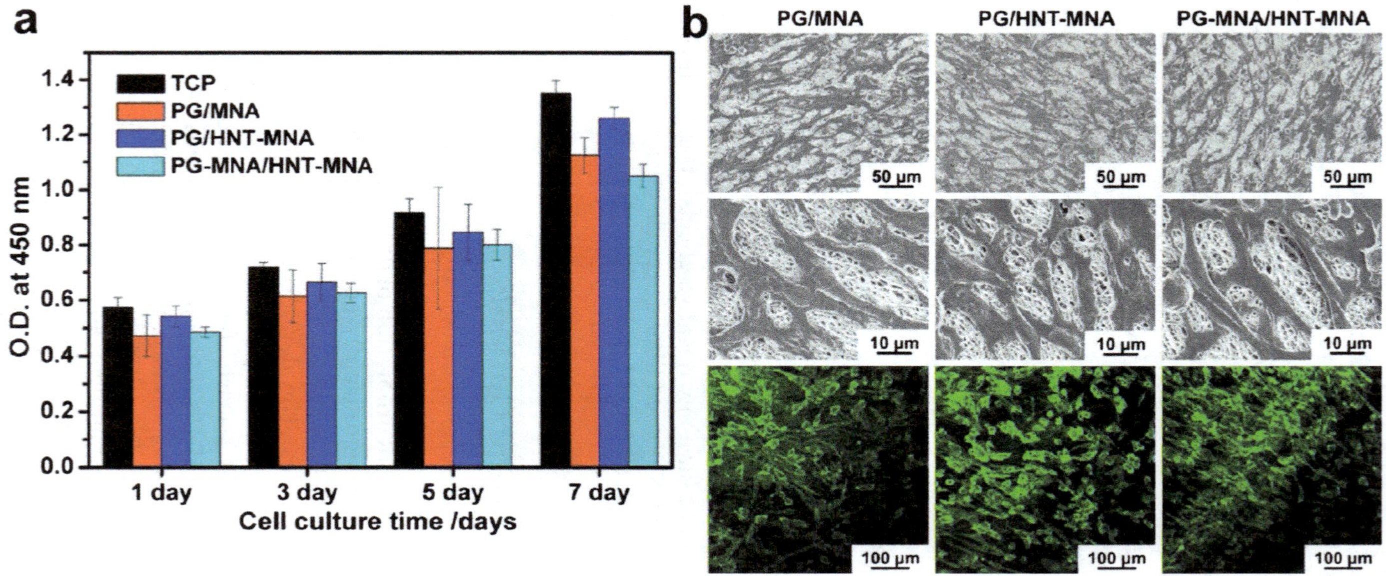

Figure 14.8 *In vitro* biocompatibility of the electrospun drug-loaded nanomicro fiber membranes: (a) O.D. of L929 cells proliferated for 1, 3, 5, and 7 days on the membranes tested using the CCK-8 assay; (b) SEM micrographs and confocal fluorescent images of L929 fibroblasts proliferated on the membranes after five days.[75] (Reprinted with permission from J. J. Xue, Y. Z. Niu, M. Gong, R. Shi, D. F. Chen, L. Q. Zhang and Y. Lvov, *ACS Nano*, 2015, **9**, 1600.[75] Copyright (2015) American Chemical Society.)

instance, the HNTs–polymer electrospun nanofibers may be processed to have different shapes, morphology, or yarn-shaped fibers to meet different requirements. In addition, different polymer systems having different properties may be used to form hybrid nanofibers with different properties. In addition, beyond the use of HNTs, there are some other nano-clay materials (ATT, sepiolite, montmorillonite, *etc.*) that can also be used to form electrospun organic–inorganic composite nanofibers for drug release and other biomedical applications. Furthermore, current efforts to use HNTs–polymer hybrid nanofibers for drug delivery applications are still limited to antibacterial drugs. Other types of drugs and other *in vitro*/*in vivo* models should be developed to test the therapeutic efficacy of the HNTs–polymer hybrid drug delivery system.

Abbreviations

ECM	Extracellular matrix
HNTs	Halloysite nanotubes
LAP	LAPONITE®
n-HA	nano-Hydroxyapatite
MMS	Mesoporous silica
PLA	Polylactic acid
PELA	Poly(DL-lactide)-poly(ethylene glycol)
TCH	Tetracycline hydrochloride
PLLA	Poly (L-lactic acid)
BSA	Bovine serum albumin
AMX	Amoxicillin
PLGA	Poly(lactic-*co*-glycolic acid)
IBU	Ibuprofen
PCL	Poly (ε-caprolactone)
BG	Bioactive glass
DOX	Doxorubicin
ATT	Attapulgite
hMSCs	human Mesenchymal stem cells
PG	Poly(ε-caprolactone)–gelatin
MNA	Metronidazole
TH-1–PLGA	TCH–HNT–PLGA nanofibers with 1% TCH
TH-2–PLGA	TCH–HNT–PLGA nanofibers with 2% TCH

Acknowledgements

This research is financially supported by the Program for Professor of Special Appointment (Eastern Scholar) at Shanghai Institutions of Higher Learning and the Key Laboratory of Textile Science & Technology, Ministry of Education, "111 Project", B07024. M. Shen thanks the Fundamental Research Funds for the Central Universities. Y. Zhao thanks the Chinese Universities Scientific Fund (101-06-0019014).

References

1. T. J. Sill and H. A. Von Recum, *Biomaterials*, 2008, **29**, 1989.
2. W. Y. Liu, S. Thomopoulos and Y. N. Xia, *Adv. Healthcare Mater.*, 2012, **1**, 10.
3. A. Mickova, M. Buzgo, O. Benada, M. Rampichova, Z. Fisar, E. Filova, M. Tesarova, D. Lukas and E. Amler, *Biomacromolecules*, 2012, **13**, 952.
4. H. H. Huang, C. L. He, H. S. Wang and X. M. Mo, *J. Biomed. Mater. Res., Part A*, 2009, **90**, 1243.
5. Z. M. Huang, C. L. He, A. Z. Yang, Y. Z. Zhang, X. J. Han, J. L. Yin and Q. S. Wu, *J. Biomed. Mater. Res., Part A*, 2006, **77**, 169.
6. M. Beck-Broichsitter, M. Thieme, J. Nguyen, T. Schmehl, T. Gessler, W. Seeger, S. Agarwal, A. Greiner and T. Kissel, *Macromol. Biosci.*, 2010, **10**, 1527.
7. R. L. Qi, X. Y. Cao, M. W. Shen, R. Guo, J. Y. Yu and X. Y. Shi, *J. Biomater. Sci., Polym. Ed.*, 2012, **23**, 299.
8. F. Y. Zheng, S. G. Wang, S. H. Wen, M. W. Shen, M. F. Zhu and X. Y. Shi, *Biomaterials*, 2013, **34**, 1402.
9. F. Y. Zheng, S. G. Wang, M. W. Shen, M. F. Zhu and X. Y. Shi, *Polym. Chem.*, 2013, **4**, 933.
10. Z. Y. Hou, X. J. Li, C. X. Li, Y. L. Dai, P. A. Ma, X. Zhang, X. J. Kang, Z. Y. Cheng and J. Lin, *Langmuir*, 2013, **29**, 9473.
11. E. Y. Yan, S. Fan, X. Q. Li, C. Wang, Z. Y. Sun, L. Ni and D. Q. Zhang, *Mater. Sci. Eng., C*, 2013, **33**, 461.
12. C. M. Hu, S. Liu, Y. Zhang, B. Li, H. L. Yang, C. Y. Fan and W. G. Cui, *Acta Biomater.*, 2013, **9**, 7381.
13. M. Kouhi, M. Morshed, J. Varshosaz and M. H. Fathi, *Chem. Eng. J.*, 2013, **228**, 1057.
14. H. J. Haroosh, Y. Dong and G. D. Ingram, *J. Polym. Sci., Part B: Polym. Phys.*, 2013, **51**, 1607.
15. S. B. Hendricks, *Nature*, 1938, **142**, 38.
16. H. M. Kelly, P. B. Deasy, E. Ziaka and N. Claffey, *Int. J. Pharm.*, 2004, **274**, 167.
17. Y. M. Lvov, D. G. Shchukin, H. Mohwald and R. R. Price, *ACS Nano*, 2008, **2**, 814.
18. R. Price, B. P. Gaber and Y. R. Lvov, *J. Microencapsulation*, 2001, **18**, 713.
19. N. G. Veerabadran, D. Mongayt, V. Torchilin, R. R. Price and Y. M. Lvov, *Macromol. Rapid Commun.*, 2009, **30**, 99.
20. N. G. Veerabadran, R. R. Price and Y. M. Lvov, *Nano*, 2007, **2**, 115.
21. S. R. Levis and P. B. Deasy, *Int. J. Pharm.*, 2003, **253**, 145.
22. D. S. Kommireddy, I. Ichinose, Y. M. Lvov and D. K. Mills, *J. Biomed. Nanotechnol.*, 2005, **1**, 286.
23. V. Vergaro, E. Abdullayev, Y. M. Lvov, A. Zeitoun, R. Cingolani, R. Rinaldi and S. Leporatti, *Biomacromolecules*, 2010, **11**, 820.
24. F. Anton, *U.S. Patent*, 1,975,504, 1934.
25. J. Doshi and D. H. Reneker, *J. Electrost.*, 1995, **35**(2-3), 151.
26. Z. M. Huang, Y. Z. Zhang, M. Kotaki and S. Ramakrishna, *Compos. Sci. Technol.*, 2003, **63**, 2223.

27. A. Greiner and J. H. Wendorff, *Angew. Chem., Int. Ed.*, 2007, **46**, 5670.
28. M. G. McKee, J. M. Layman, M. P. Cashion and T. E. Long, *Science*, 2006, **311**, 353.
29. F. Ignatious and J. Baldoni, *U.S. Patent*, 10,181,640, 2001.
30. Y. Dzenis, *Science*, 2004, **304**, 1917.
31. I. S. Chronakis, *J. Mater. Process. Technol.*, 2005, **167**, 283.
32. S. Agarwal, J. H. Wendorff and A. Greiner, *Polymer*, 2008, **49**, 5603.
33. D. G. Yu, X. X. Shen, C. Branford-White, K. White, L. M. Zhu and S. W. A. Bligh, *Nanotechnology*, 2009, **20**, 055104.
34. W. G. Cui, X. H. Li, S. B. Zhou and J. Weng, *J. Appl. Polym. Sci.*, 2007, **103**, 3105.
35. A. Frenot and I. S. Chronakis, *Curr. Opin. Colloid Interface Sci.*, 2003, **8**, 64.
36. T. Subbiah, G. S. Bhat, R. W. Tock, S. Parameswaran and S. S. Ramkumar, *J. Appl. Polym. Sci.*, 2005, **96**, 557.
37. M. Y. Li, M. J. Mondrinos, M. R. Gandhi, F. K. Ko, A. S. Weiss and P. I. Lelkes, *Biomaterials*, 2005, **26**, 5999.
38. R. Langer, J. P. Vacanti, C. A. Vacanti, A. Atala, L. E. Freed and G. Vunjak-Novakovic, *Tissue Eng.*, 1995, **1**, 151.
39. R. G. Flemming, C. J. Murphy, G. A. Abrams, S. L. Goodman and P. F. Nealey, *Biomaterials*, 1999, **20**, 573.
40. T. G. Kim and T. G. Park, *Tissue Eng.*, 2006, **12**, 221.
41. X. H. Zong, S. Li, E. Chen, B. Garlick, K. Kim, D. F. Fang, J. Chiu, T. Zimmerman, C. Brathwaite and B. S. Hsiao, *Ann. Surg.*, 2004, **240**, 910.
42. N. Bölgen, I. Vargel, P. Korkusuz, Y. Z. Menceloğlu and E. Pişkin, *J. Biomed. Mater. Res., Part B*, 2007, **81**, 530.
43. Z. W. Ma, M. Kotaki and S. Ramakrishna, *J. Membr. Sci.*, 2005, **265**, 115.
44. S. Tungprapa, I. Jangchud and P. Supaphol, *Polymer*, 2007, **48**, 5030.
45. E. R. Kenawy, G. L. Bowlin, K. Mansfield, J. Layman, D. G. Simpson, E. H. Sanders and G. E. Wnek, *J. Controlled Release*, 2002, **81**, 57.
46. P. Taepaiboon, U. Rungsardthong and P. Supaphol, *Nanotechnology*, 2006, **17**, 2317.
47. J. Zeng, X. Y. Xu, X. S. Chen, Q. Z. Liang, X. C. Bian, L. X. Yang and X. B. Jing, *J. Controlled Release*, 2003, **92**, 227.
48. K. Kim, Y. K. Luu, C. Chang, D. F. Fang, B. S. Hsiao, B. Chu and M. Hadjiargyrou, *J. Controlled Release*, 2004, **98**, 47.
49. Y. Z. Zhang, C. T. Lim, S. Ramakrishna and Z. M. Huang, *J. Mater. Sci.: Mater. Med.*, 2005, **16**, 933.
50. W. G. Cui, X. H. Li, X. L. Zhu, G. Yu, S. B. Zhou and J. Weng, *Biomacromolecules*, 2006, **7**, 1623.
51. G. Verreck, I. Chun, J. Rosenblatt, J. Peeters, A. Van Dijck, J. Mensch, M. Noppe and M. E. Brewster, *J. Controlled Release*, 2003, **92**, 349.
52. W. G. Cui, M. B. Qi, X. H. Li, S. Z. Huang, S. B. Zhou and J. Weng, *Int. J. Pharm.*, 2008, **361**, 47.
53. M. B. Qi, X. H. Li, Y. Yang and S. B. Zhou, *Eur. J. Pharm. Biopharm.*, 2008, **70**, 445.
54. M. Bognitzki, H. Hou, M. Ishaque, T. Frese, M. Hellwig, C. Schwarte, A. Schaper, J. H. Wendorff and A. Greiner, *Adv. Mater.*, 2000, **12**, 637.

55. Z. C. Sun, E. Zussman, A. L. Yarin, J. H. Wendorff and A. Greiner, *Adv. Mater.*, 2003, **15**, 1929.
56. C. L. He, Z. M. Huang and X. J. Han, *J. Biomed. Mater. Res., Part A*, 2009, **89**, 80.
57. H. X. Qi, P. Hu, J. Xu and A. J. Wang, *Biomacromolecules*, 2006, **7**, 2327.
58. Y. Yang, X. H. Li, W. G. Cui, S. B. Zhou, R. Tan and C. Y. Wang, *J. Biomed. Mater. Res., Part A*, 2008, **86**, 374–385.
59. Y. Su, X. Q. Li, S. P. Liu, X. M. Mo and R. Seeram, *Colloids Surf., B*, 2009, **73**, 376–381.
60. A. L. Yarin, *Polym. Adv. Technol.*, 2011, **22**, 310–317.
61. H. S. Yoo, T. G. Kim and T. G. Park, *Adv. Drug Delivery Rev.*, 2009, **61**, 1033.
62. C. Wang, K. W. Yan, Y. D. Lin and P. C. H. Hsieh, *Macromolecules*, 2010, **43**, 6389.
63. X. L. Xu, X. S. Chen, P. A. Ma, X. R. Wang and X. B. Jing, *Eur. J. Pharm. Biopharm.*, 2008, **70**, 165.
64. A. K. Moghe and B. S. Gupta, *Polym. Rev.*, 2008, **48**, 353.
65. S. G. Wang, Y. L. Wu, R. Guo, Y. P. Huang, S. H. Wen, M. W. Shen, J. H. Wang and X. Y. Shi, *Langmuir*, 2013, **29**, 5030.
66. K. Li, S. G. Wang, S. H. Wen, Y. Q. Tang, J. P. Li, X. Y. Shi and Q. H. Zhao, *ACS Appl. Mater. Interfaces*, 2014, **6**, 12328.
67. G. X. Chen, D. Li, J. C. Li, X. Y. Cao, J. H. Wang, X. Y. Shi and R. Guo, *New J. Chem.*, 2015, **39**, 2847.
68. Y. L. Wu, R. Guo, S. H. Wen, M. W. Shen, M. F. Zhu, J. H. Wang and X. Y. Shi, *J. Mater. Chem. B*, 2014, **2**, 7410.
69. S. G. Wang, R. Castro, X. An, C. L. Song, Y. Luo, M. W. Shen, H. Tomas, M. F. Zhu and X. Y. Shi, *J. Mater. Chem.*, 2012, **22**, 23357.
70. Z. Wang, Y. L. Zhao, Y. Luo, S. G. Wang, M. W. Shen, H. Tomas, M. F. Zhu and X. Y. Shi, *RSC Adv.*, 2015, **5**, 2383.
71. H. J. Haroosh, Y. Dong, D. S. Chaudhary, G. D. Ingram and S. Yusa, *Appl. Phys. A: Mater. Sci. Process.*, 2013, **110**, 433.
72. R. L. Qi, M. W. Shen, X. Y. Cao, R. Guo, X. J. Tian, J. Y. Yu and X. Y. Shi, *Analyst*, 2011, **136**, 2897.
73. Y. L. Zhao, S. G. Wang, Q. S. Guo, M. W. Shen and X. Y. Shi, *J. Appl. Polym. Sci.*, 2013, **127**, 4825.
74. R. L. Qi, R. Guo, F. Y. Zheng, H. Liu, J. Y. Yu and X. Y. Shi, *Colloids Surf., B*, 2013, **110**, 148.
75. J. J. Xue, Y. Z. Niu, M. Gong, R. Shi, D. F. Chen, L. Q. Zhang and Y. Lvov, *ACS Nano*, 2015, **9**, 1600.
76. R. L. Qi, R. Guo, M. W. Shen, X. Y. Cao, L. Q. Zhang, J. J. Xu, J. Y. Yu and X. Y. Shi, *J. Mater. Chem.*, 2010, **20**, 10622.
77. M. X. Liu, Z. X. Jia, D. M. Jia and C. R. Zhou, *Prog. Polym. Sci.*, 2014, **39**, 1498.
78. Y. Lvov and E. Abdullayev, *Prog. Polym. Sci.*, 2013, **38**, 1690.

CHAPTER 15

Toxicological Evaluation of Clay Nanomaterials and Polymer–Clay Nanocomposites

EKATERINA A. NAUMENKO[a] AND RAWIL F. FAKHRULLIN*[a]

[a]Kazan Federal University, Kreml uramı 18, Kazan, Republic of Tatarstan, 420008, Russian Federation
*E-mail: kazanbio@gmail.com

15.1 Introduction

In recent years, scientists worldwide have focused on fabrication of synthetic and natural polymer composites doped with nanomaterials (*i.e.* particles having at least one dimension less than 100 nm) to improve the functional properties of the resulting nanocomposites.[1] Polymer nanocomposites are crucially important in applications such as fabrication of microelectronic devices; piezoelectric actuators and sensors; organic solar cells; microelectronics and microsystems technologies, as well as in fabrication of various biomedical devices, to mention just a few.[2] In the polymer industry, mixing of polymers with inorganic additives traditionally takes place to improve the polymer's properties (*i.e.* mechanical, chemical, thermal, *etc.*) and to reduce the costs of the final product. Polymers doped with inorganic micro- and nanoparticles are typically classified as *microcomposites*, indicating the dimensions of the inorganic particles added as fillers into

RSC Smart Materials No. 22
Functional Polymer Composites with Nanoclays
Edited by Yuri Lvov, Baochun Guo and Rawil F. Fakhrullin

Published by the Royal Society of Chemistry, www.rsc.org

the polymer matrix. New composites are constantly being developed worldwide to address current challenges in the polymer composite industry.[3] If the dimension of the fillers used for production of functional composite is less than 100 nm, such composites are called *nanocomposites*. Nanoparticles used typically exhibit specific properties unlike larger particles of the same chemical structure.

Based on morphology, clay nanoparticles can be divided into two groups: (1) layered or platelet nanoparticles (montmorillonite, bentonite, kaolinite, *etc.*) (Figure 15.1A–C) and (2) tubular nanoparticles [halloysite (Figure 15.1D) and imogolite (Figure 4.1 in Chapter 4)]. Here, we overview selected literature reports aimed at investigating the toxicity of certain nanoclays and nanoclay-doped polymer composites, focusing on the ones used in current technological applications or suggested as potent biomedical nanomaterials.

Among the platelet nanoparticles, natural clay montmorillonite is a popular filler for nanocomposite fabrication used extensively in industrial applications. Montmorillonite is a layered alumina silicate of the smectite minerals having an octahedral laminated sheet of aluminium hydroxide positioned between tetrahedral silicate layers.[4] Montmorillonite demonstrates

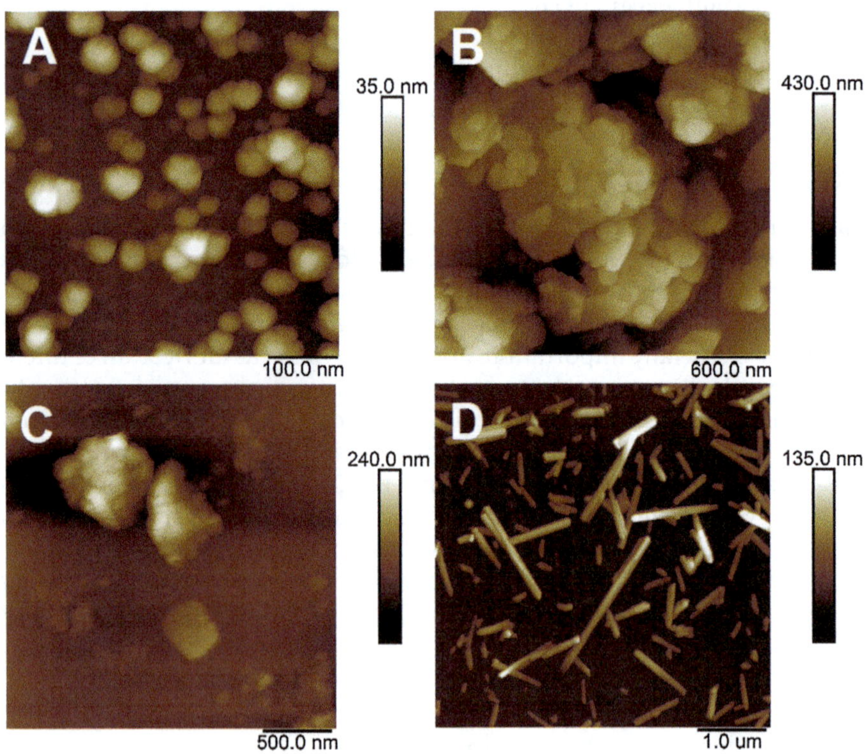

Figure 15.1 Atomic force microscopy images depicting platelet and tubular nanoclays. A – Bentonite, B – kaolinite, C – montmorillonite, D – halloysite.

outstanding cation exchange properties and could form many nanocomposites with different organic compounds within the interlayer spacing of *ca.* 0.9 to 1.2 nm. Montmorillonite is typically added into polymer matrices at 1–5 wt%. Nanocomposites based on polymers mixed with layered silicates have gained enormous attention from researchers due to the undoubted advantages associated with the use of layered silicates as fillers in polymer composites. Clay materials are available in large amounts and at relatively low costs. However, chemical surface modification of these materials is normally required for their better compatibility with the organic polymer matrices. The incorporation of the layered silicate materials into the polymer composites originates from 1950, when Carter *et al.* used organically modified layered silicates for reinforcement of elastomers.[5] After that, many other studies confirming the great potential of layered silicates in the fabrication of composites were reported.[6,7]

Halloysite[8–11] is another aluminosilicate with a hollow tubular structure originating from the strain-induced rolling of kaolin sheets (for its structure and properties, see Chapter 5 for more details). Halloysite tubes can be loaded with different active agents (*i.e.* drugs, antiseptics, antibiotics, antibacterial compounds, *etc.*, see Chapters 7 and 13 for more details) within the inner lumen, followed by the retention of the active compounds and their slow release.[10,12–21] Numerous studies report the production of halloysite-based functional nanocontainers for industrial and biomedical applications. The different chemical structure and opposite charge of the inner lumen and outer surfaces in halloysite tubes allow for the selective and predetermined modification and labelling of nanotubes. Importantly, the inner lumen diameter of halloysite tubes allows for fitting in a number of (bio)macromolecules, including proteins, nucleic acids and antibiotics. As a result, halloysite was found to be an inexpensive nanocontainer for encapsulation of drugs and functional reagents. This possibility was first demonstrated by the teams of Price, Lvov, and Kelly;[10,12,13] then their pioneering reports have inspired others, resulting in almost exponential growth of publications reporting fabrication of halloysite-based materials. For instance, halloysite nanotubes were used to supplement polymer constructs for bone regeneration[13,22–24] and for fabrication of wound-healing materials.[25] The addition of halloysite to polymers has shown significant improvement in their mechanical and thermal properties.[1,26]

Before any practical application of nanocomposite materials (especially in biomedical areas or in the food industry), it is essential to examine the toxic effects of these materials. Biocompatibility is one of the main requirements for the safe use of products based on nanoclay particles for medical formulations and household devices and materials. The toxicological evaluation of nanocomposites can be assessed using both *in vitro* and *in vivo* approaches. The *in vitro* evaluation of toxicological profiles of different clays and derived nanocomposites has shown that the toxicity level depends on the modifiers' nature, the experimental model selected for the study, the exposure times and the concentrations of nanoparticles.

15.2 Evaluation of Toxicity of Nanoclays

15.2.1 Nanoclay Toxicity Towards Unicellular Prokaryotic and Eukaryotic Organisms

Microorganisms have been extensively employed in the investigation of the toxicity of nanoclays, due to the relatively cheap and easy experimental techniques associated with microbial toxicity testing. As often occurs with the toxicity of nanomaterials, literature reports are somewhat contradicting. For example, most of the recent reports suggest that halloysite nanoclay shows a very low toxic potential towards bacteria;[27] as a result, it cannot be used as an antimicrobial agent *per se*, but should be loaded with an appropriate antimicrobial inorganic[28] or organic[29] modifier. However, Chen *et al.* found that pristine (not loaded with any other compounds) halloysite exhibited higher toxicity to Gram-negative bacteria (*Escherichia coli*) and Gram-positive bacteria (*Staphylococcus aureus*) compared with amine- and protein-functionalized clay nanotubes.[28] The authors have postulated that this effect is caused by direct contact with the cell walls of microorganisms and formation of reactive oxygen species. Choi *et al.* reported the reduction of *E. coli* cell viability caused by pristine halloysite in a range of concentrations (0.05 to 1 mg mL^{-1}), apparently depending on nanotube surface modification.[30] The authors have examined the cytotoxicity of native and modified halloysite using a wide range of methods from cell growth monitoring to protein expression analysis. Halloysite nanotubes were modified by amine functionalization (NH_2-HNTs) or bovine serum albumin (BSA-HNTs) coating, while the pristine nanotubes were used as control samples. Interestingly, native and NH_2-modified nanotubes have exhibited increased cell death levels at concentrations ≥0.5 mg mL^{-1}. The major biological parameters (cellular viability, metabolic activity and DNA replication) decreased with increasing concentrations of native and NH_2-modified halloysite. For BSA-coated nanotubes, the cytotoxicity was less pronounced, exhibiting the same or similar levels of cell proliferation and cell viability as the untreated cells. Moreover, even higher levels of metabolic activity and plasmid DNA replication than those detected in cells treated with native halloysite and NH_2-HNTs were detected. A high ranaspumin-2 protein level was observed in bacteria grown in media supplemented with BSA-HNTs (100, 83 and 80% of yield at 0.01, 0.05 and 0.1 mg mL^{-1}, respectively). Apparently, the bactericidal effects of halloysite correlated well with nanotube concentration and the surface modification of nanotubes with BSA, which seems to be an effective strategy to reduce the cytotoxicity up to 0.1 mg mL^{-1} of nanotubes.

Eukaryotic microorganisms were also subjected to halloysite[31] during the fabrication of hybrid microparticles using isolated yeast cells as templates for immobilization of nanoclay *via* layer-by-layer assembly. Neither pristine halloysite nor polyelectrolyte functionalized nanotubes significantly affected the viability of halloysite-coated yeast cells, as demonstrated by cell growth monitoring (Figure 15.2).

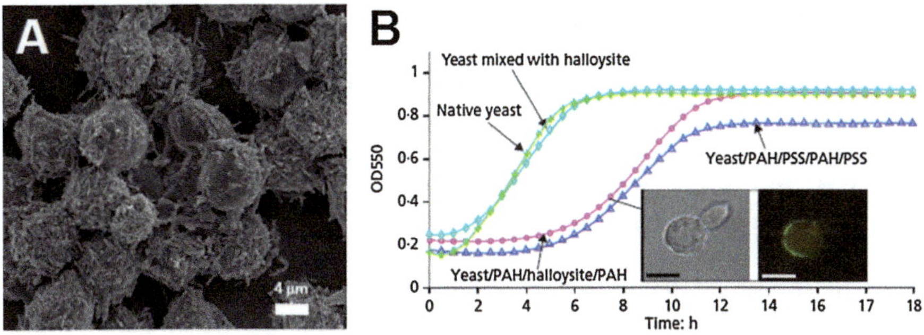

Figure 15.2 Scanning electron microscopy microphotographs of layer-by-layer deposited polyelectrolyte–halloysite nanotube-coated yeast cells (A) and growth curves of native and halloysite-coated cells (B). Reprinted from ref. 31 with permission from The Royal Society of Chemistry.

The negative effect of a nanocomposite coating on cell walls was manifested in the increased delay in budding, most likely due to the time required for the breaking of the halloysite shells by the growing bud. In a more recent paper, Cavallaro *et al.*[32] employed both bacterial cells (*E. coli*) and yeast cells (*Saccharomyces cerevisiae*) to investigate the potential toxic effects of HNTs. Halloysite within the concentration range of 0.5–2.5 mg mL^{-1} does not significantly affect the proliferation of bacterial and yeast cells. The slight reduction in growth at higher concentrations is explained by the mechanistic effects on cells, possibly during the adsorption to cell walls. Viability tests and fluorescence microscopy were employed to demonstrate the high viability levels of cells subjected to halloysite nanotubes. These results are consistent with a previous report on modification of the surface of yeast cells *S. cerevisiae* with halloysite using polyelectrolyte multilayers,[31] where 99% of yeast cells mixed with HNTs retained viability. Li *et al.* have employed several strains of *Salmonella typhimurium* bacteria to determine the genotoxicity of montmorillonite using the Ames test.[33] For all five *S. typhimurium* strains investigated, no mutations were found for up to 1 mg mL^{-1} of montmorillonite, indicating the low genotoxicity of nanoclays.

Clay nanomaterials are frequently being compared with carbon nanomaterials, including evaluation of the toxicities of both. Recently, Taylor *et al.*[34] have investigated the impact of trace concentrations of clay nanomaterials and multi-walled carbon nanotubes (MWCNT) using Gram-negative bacteria. Several *S. typhimurium* and *E. coli* mutant and wild type strains were tested, including oxyR and rpoS mutants, which are susceptible to oxidative stress. The inhibition of the oxyR mutants aerobically grown under illumination in the presence of montmorillonite nanoparticles was demonstrated, whereas no toxic effects were observed in the dark or during anaerobic growth. A slight toxic effect was detected for *E. coli* rpoS mutants cultured with montmorillonite and MWCNT only if grown aerobically under illumination. There was no effect on bacterial growth in the presence of halloysite or if *S. typhimurium* rpoS mutants were used. These observations suggest that clay nanoparticles (montmorillonite and halloysite) as well as MWCNT do not exert a mutagenic effect and do not have a general toxic effect among the bacterial species tested. In another study, the toxicity of four different plate-like and tubular nanoclays (bentonite, montmorillonite, kaolin and halloysite) towards a protozoan microorganism *Parmecium caudatum* was investigated in comparison with graphene oxide (GO) nanoflakes.[35] GO was used as a nanomaterial having a similar size to plate-like nanoclays, but a different chemical structure and already established toxicity,[36] along with monodisperse silica nanoparticles (120 nm). Different physiological parameters, such as growth rate, reproduction, nutrition, oxidative stress induction and chemotaxis of cells in contact with nanoclays, GO and silica nanoparticles were investigated. Interestingly, the protozoans demonstrated positive chemotaxis (~70%) towards halloysite, kaolin and montmorillonite, whereas silica and bentonite induced overall negative chemotaxis (~80%) at 10 mg mL^{-1}. Noteworthy, GO repelled protozoans at a much lower concentration (1 mg mL^{-1}). Clay nanoparticles affected food vacuole formation and digestion; among the investigated clays halloysite exhibited the lowest reduction of food vacuoles (Figure 15.3).

The biosafety of the examined nanoparticles can be demonstrated within the following sequence (from relatively non-toxic to toxic nanoparticles): halloysite – kaolin – montmorillonite – silica – bentonite – graphene oxide. Among the analyzed nanoclay particles, halloysite was most biocompatible for *P. caudatum* and the safe concentration for these ciliate organisms was up to 10 mg mL^{-1}, which is 10-fold more than the previously established safe concentration for mammalian cell cultures,[11] suggesting that an additional protective mechanism in exists ciliate organisms.

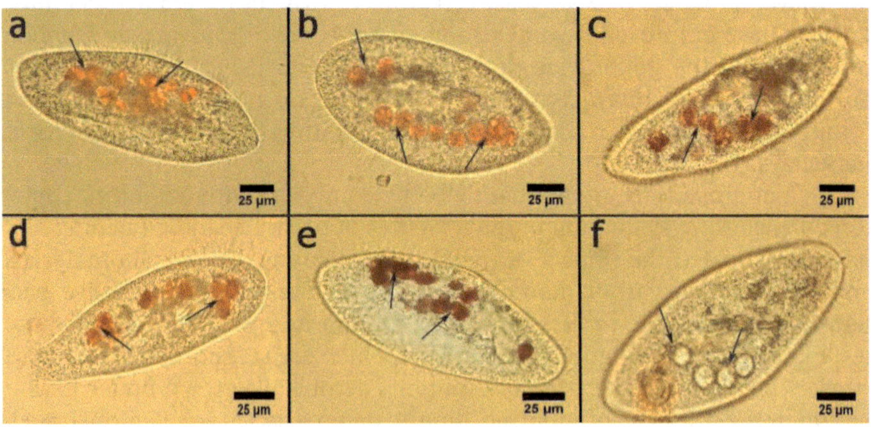

Figure 15.3 Congo Red-stained food vacuoles in *P. caudatum*: (a) intact cell; and cells in contact with (b) halloysite; (c) kaolin; (d) montmorillonite; (e) bentonite; (f) silica. Food vacuoles are indicated by arrows. Reprinted from ref. 35 with permission from The Royal Society of Chemistry.

15.2.2 Toxicological Evaluation of Nanoclays Using Mammalian Cell Cultures

The toxic effects of nanoclays and nanoclay-doped polymer composites are normally tested using human cells as suitable models to predict the influence of nanoclays on the human organism. A wide range of nanoclay particles (both with and without chemical surface modification) has been studied by different research groups worldwide. Human cell lines apparently represent the most frequently used experimental model in toxicological studies of nanoclays. Among human model cells, intestine and liver cells are mostly used, followed by lung and skin cells. The selection of these types of cells reflects the growing interest in the use of these clay materials for the fabrication of packages in the food industry.

Murphy *et al.*[37] have investigated the cytotoxicity of unmodified montmorillonite, bentonite, kaolinite and erionite using human umbilical vein endothelial (HUVEC), N1E-115 neuroblastoma and ROC-1 oligodendroglial cells. The clays were added in concentrations from 0.01 to 0.1 mg mL^{-1}; no lysis in ROC-I and NlE-115 cells was detected. Moreover, this range of clay concentrations did not induce a dose-dependent increase in fatty acid levels. In addition, there were no significant increases in lactate dehydrogenase activity in neuronal cells. Oppositely, adding montmorillonite, kaolinite and bentonite to HUVECs caused a dose-dependent increase of fatty acid concentrations and induced cell lysis. The authors postulate that the cytotoxicity of clays containing an aluminium silicate towards HUVECs may be behind the blood–brain barrier destruction in the affected areas, allowing entry of the clay particle into the brain. A similar work by Li *et al.*[33] with Chinese hamster ovary (CHO) cells exposed to montmorillonite demonstrated low cytotoxicity towards the tested cells at concentrations below 1 mg mL^{-1} after 12 h of incubation and a dose-dependent effect after 24 h of incubation using lactate dehydrogenase (LDH) release assays. Genotoxic effects were investigated, indicating no significant DNA damage caused by exfoliated montmorillonite. Baek *et al.*[38] reported a more pronounced toxic effect of montmorillonite using intestinal cells (INT-407) as a model. Montmorillonite exposure significantly inhibited cell proliferation (>100 μg mL^{-1}) after 24–72 h of incubation. However, the membrane damage determined by measuring the LDH release level was evident only at 1 mg mL^{-1} after 48–72 h and reactive oxygen species generation was found in the cells incubated with >50 μg mL^{-1} after 48–72 h. No reactive oxygen species production occurred in montmorillonite-treated cells within 24 h at concentrations below 500 μg mL^{-1}. These results suggest that montmorillonite may induce oxidative stress after relatively long exposure in concentrations above 500 μg mL^{-1}.

Sharma *et al.* have reported that chemical modification of montmorillonite with organic compounds modulates the toxicity of the resulting clay. For example, CloisitesNa® (CNa®) did not show any negative effect in colon cancer cells (Caco-2), although Cloisites30B (C30B) induced approximately 40% cytotoxicity at a concentration 226 mg mL^{-1} after 24 h of exposure.[39] Lordan *et al.*[40] observed a lower cytotoxicity in the case of incubation of CNa® with hepatoma cells (HepG2) than that after exposure to Cloisite®93A (C93A) for 24 h. Active bentonite (treated with H_2SO_4) toxicity was compared with that of native bentonite; the former was found to be more toxic towards human B lymphoblast cells. Houtman *et al.*[41] employed the MTS assay, IL-6 leakage, protein content identification, DNA damage assay and reactive oxygen species determination for cytotoxicity assessment of three modified montmorillonite clays (commercially available Cloisite®20A, Clay 1 and Clay 2). Two human cell lines, Caco-2 and HepG2 (hepatoma cells), were used in this study. All three clays have been chemically modified with quaternary ammonium salts. The results obtained in this study demonstrate that the main time-dependent cytotoxic effects were induced by Clay 2. Toxicity assessment of Cloisite®20A and Clay 1 revealed no cytotoxic and genotoxic effects at the concentrations studied. Clay 1 and Clay 2 at a concentration of 8 μg mL^{-1} in Caco-2 cells induced a low decrease of viability. Oppositely, in HepG2 8 μg mL^{-1} of Clay 1 induced a significant decrease of protein content, but 8 μg mL^{-1} Clay 2 did not induce any toxic effects. Clay 2 induced DNA strand breaks in both tested cell lines and increased DNA strand breaks in HepG2 cells time-dependently.

Two types of cells, human breast cancer cells (MCF-7) and human epithelial adenocarcinoma cells (HeLa), were used by Vergaro *et al.*[11] to determine the toxicity and cellular uptake of halloysite by adding cell cultivation growth media supplemented with increasing concentrations of both untreated and functionalized nanotubes. Fluorescently-labelled halloysite allowed visualization of the distribution of nanotubes in cytoplasm using confocal microscopy. The distribution of nanotubes inside the MCF-7 cells is demonstrated in Figure 15.4.

MCF-7 cells accumulated halloysite in cytoplasm with concentrating around the cells' nuclei; however, proliferation was unaffected. A similar pattern of nanotube distribution has been observed in HeLa cells. Cellular viability was evaluated using an enzymatic activity test (MTT assay) which has shown that the cells were viable up to 70% when the halloysite concentration was lower than 0.075 mg mL^{-1}. At higher concentrations (from 0.1 to 1 mg mL^{-1}) halloysite induced significant cell death. The intracellular distribution of halloysite coated with dextrin to form end stoppers[42] was demonstrated recently. Apparently, different cells exhibit a different uptake rate of halloysite from growth media which depends on cell type (Figure 15.5). Two cell lines (Hep3B and A549) have been employed, confirming that dextrin-coated halloysite penetrated predominantly A549 cell membranes and concentrated mostly around the nuclei, while Hep3B cells exhibited a low uptake level and distribution of nanotubes among the microvilli network. Dextrin-coated halloysite exhibited extremely low levels of toxicity towards human cells in concentrations up to 100 μg mL^{-1}. It should be noted that differential uptake of nanotubes may be utilized in selective killing of cells by drug-loaded nanotubes towards the cells exhibiting a higher uptake rate.

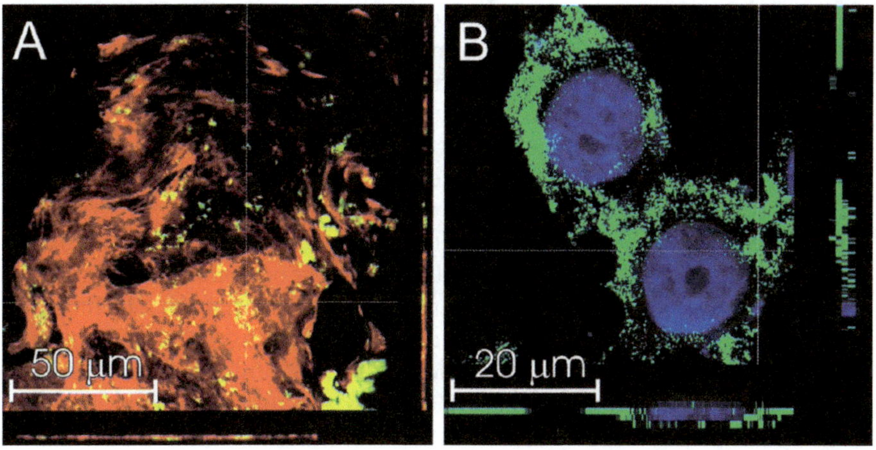

Figure 15.4 A – Intracellular uptake of halloysite by MCF-7 cells (yellow) and phalloidin-TRITC visualization on the cytoskeleton (red). B – Confocal image of halloysite (functionalized by aminopropyltriethosilane (APTES)) intracellular uptake by HeLa cells (halloysite + APTES (green) and Hoechst-stained HeLa nuclei (blue) overlaid images). Reprinted with permission from V. Vergaro, E. Abdullayev, Y. M. Lvov, A. Zeitoun, R. Cingolani, R. Rinaldi and S. Leporatti, *Biomacromolecules*, 2010, **11**, 820.[11] Copyright (2010) American Chemical Society.

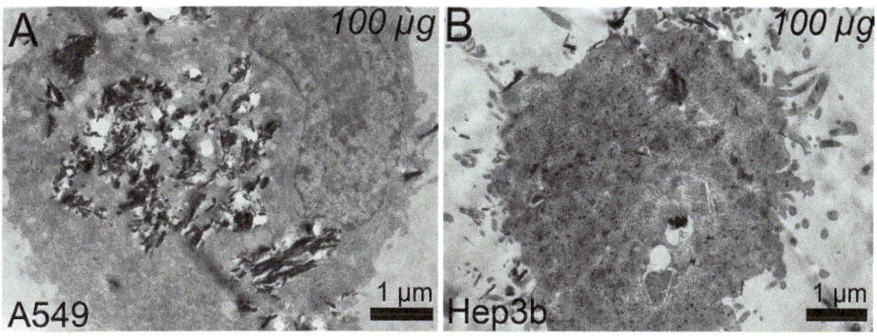

Figure 15.5 Transmission electron microscopy images of A549 (A) and Hep3B (B) cells incubated with dextrin-coated halloysite nanocontainers, demonstrating the differential uptake by various cell cultures (both cells types were administered 100 µg per 100 000 cells).

Several attempts have been made to elucidate the potential mechanisms of halloysite action towards human cells. In particular, Lai *et al.*[43] have performed a proteomic profiling of intestinal cell co-culture human colorectal adenocarcinoma cells–human colon epithelial cells (Caco-2–HT29-MTX cells) exposed to halloysite. The authors observed exposure-specific changes in expression among 4081 proteins, which have shown pro-inflammatory effects of halloysite exposure in concentrations lower than 1 mg mL^{-1}, while the cells exposed to extremely high concentrations (~100 mg mL^{-1}) demonstrated significant changes in protein expression. These observations indicate that halloysite is not toxic at moderate levels of exposure; moreover, halloysite stimulates cell growth and proliferation.

Liu *et al.*[44] studied the enzymatic activity of caspase 3 and a range of apoptotic/necrotic cells using flow cytometry in two different cell lines, mouse fibroblasts NIH3T3 and human embryonic kidney cells (HEK293) exposed to unmodified montmorillonite and oligo(Styrene-*Co*-Acrylonitrile)-modified montmorillonite at 1 mg mL^{-1}. Modified montmorillonite did not induce apoptosis unlike the native montmorillonite in both treated cell lines, but all tested clays significantly induced cell death and caspase 3 activity. Janer *et al.*[45] obtained similar results employing HepG2 and SKMEL28 cell lines. They used pristine montmorillonite along with montmorillonite modified with quaternary ammonium modifiers and observed an increase in the caspase 3/7 activity after 4.5 h in the HepG2 cell line and in both cell lines after 48 h of exposure. Using flow cytometry for determination of apoptotic cells, the authors demonstrated a seven-fold increase of the number of HepG2 cells stained with Annexin V exposed to 100 mg mL^{-1} montmorillonite. Oppositely, Lordan *et al.*[40] observed no increase in caspase 3/7 activity in HepG2 cells treated with unmodified and modified montmorillonite. Maisanaba *et al.*[46] also did not detect any variation in caspase 3/7 activity in the HepG2 cell line exposed to Cloisite 30B (another modified montmorillonite clay) for 24 and 48 h.

15.2.3 Nanoclay Toxicity Evaluation *In vivo* Using Multicellular Eukaryotic Organisms

The number of *in vivo* experimental toxicity studies involving clay nanoparticles in the scientific literature is relatively low in comparison with the *in vitro* studies.[47] Recently, a free living nematode *Caenorhabditis elegans* was employed for the investigation of the effects of halloysite nanoclays.[48] Halloysite was delivered into the organism using so-called "nanobaits"; in other words, the authors immobilized halloysite on *E. coli* cells (the normal nematode food source) by the sequential layer-by-layer deposition of (poly) allylamine (polycation) and (poly)styrenesulfonate (polyanion), and halloysite was sandwiched between the polymers. The resulting charge of the nanotube-coated cells was negative, which is similar to that of intact bacterial cells. The immobilization on *E. coli* cells was monitored using dark field microscopy. The uptake of halloysite did not affect the alimentary system of the animals (though the nanotubes were located exclusively in the intestines), and the reproductive organs (ovaries, uterus and spermatheca) were free of halloysite (Figure 15.6).

These results suggest that clay nanotubes have no profound toxic effects on *C. elegans* nematodes, unlike other nanomaterials, such as single-walled[49] or multi-walled[50] carbon nanotubes, graphene oxide,[51] TiO_2 nanoparticles[51] or platinum nanoparticles.[52] The minor toxic effects of halloysite in

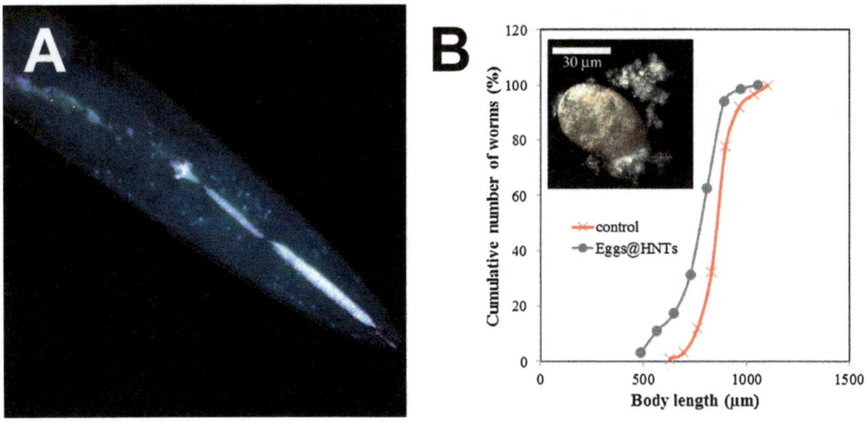

Figure 15.6 Halloysite nanotube interaction with *C. elegans* microworms. A – Merged enhanced dark field and fluorescence image HNTs in foregut, B – the cumulative curves of body length of nematodes hatched from the HNT-coated eggs; inset shows an EDF microscopy image of HNT-coated egg. Reprinted from ref. 48 with permission from The Royal Society of Chemistry.

comparison with other nanomaterials can be explained by the relatively low uptake of nanotubes by intestinal cells and a very limited transport to other tissues and organs of nematodes.

For the evaluation of toxicity of nanoclays in mammals, oral administration of nanomaterials was predominantly used. The report by Li *et al.*[33] confirms low (if any) acute oral toxicity of montmorillonite in rats. Exfoliated montmorillonite was fed to rats, with no toxic effects detected (no significant changes in the body weight or food consumption of the treated group compared to the control group). In this study the median lethal dose (LD_{50}) for acute oral toxicity of MMT was identified at >5700 mg kg^{-1} for rats, which is unlikely to occur *via* normal exposure. A study by Baek *et al.*[38] has also demonstrated the absence of any remarkable toxicity in mice orally exposed to a single montmorillonite dose of 1000 mg kg^{-1}; the LD_{50} value in mice was 41 000 mg kg^{-1}. The absence of toxicity, or very low toxicity, was also reported by Wiles *et al.*[53] in rats exposed to montmorillonite during pregnancy. The authors report the absence of any significant differences between the animals and postulated an extremely low level of teratogenicity. However, Patterson and Staszak have found developmental effects manifesting in maternal anemia and reduction in the birth weight of newborn rats after exposure to 20% kaolin in food during gestation.[54] Their results clearly indicate that more efforts should be focused on evaluation of toxicity of nanoclay materials to unquestionably determine the safe exposure parameters.

Apart from experimental results, ethological observations suggest that normally animals consume clays in nature; the process is termed geophagia.[55]

It is assumed that animals can consume clay for detoxification of toxic compounds present in the diet, for treatment of gastrointestinal upsets, such as diarrhea, for supplementation of the body with minerals or for decreasing of hyperacidity in the digestive tract.[56] These facts also indicate the low toxicity of nanoclays.

15.3 Toxicity of Nanoclay-Doped Polymer Nanocomposites

Nanocomposites can be fabricated using different polymers and nanomaterials and both can induce toxic effects. Obviously, for biomedical and food industry applications safe and inert polymer nanocomposites must be primarily used. However, occasionally the biocompatibility of the combination of a polymer and nanoclay components is different from that of each of the initial materials used. Consequently, the toxicological evaluation should include the *in vitro* (evaluation using isolated cells or tissues) and *in vivo* (the use of whole animal models) toxicity estimation using a wide range of models (Figure 15.7).

Despite the attention on nanocomposite formation and associated practical applications, so far only a limited number of papers report results evaluating the toxicological properties of polymer–clay nanocomposites. Apparently, tissue engineering scaffolds are the materials which humans are most likely

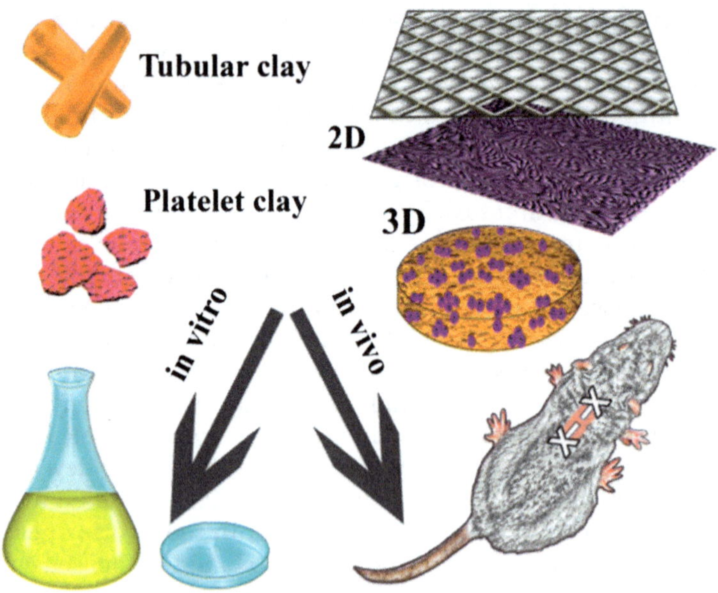

Figure 15.7 A sketch demonstrating the *in vitro* and *in vivo* models for nanomaterials and nanocomposite toxicity evaluation.

to be exposed to, therefore they have been extensively studied in terms of their toxicity. Tissue engineering scaffolds are three-dimensional support structures typically produced using (bio)polymers and specifically designed to provide a friendly environment and support for cultivated human cells. Tissue engineering aims at the fabrication of human organs *de novo*; as a result, much effort is given to envisaging novel strategies for fabrication of versatile and biocompatible scaffolds. These scaffolds should exhibit exceptional structural and functional properties, namely cell adhesion "friendly" surface properties, shape memory, high porosity, biocompatibility and mechanical strength, among many others. Within recent decades, polymer scaffolds have been fabricated and applied in tissue engineering and regenerative medicine. Nanomaterials have been regarded as an effective means to modify the properties of polymer scaffolds, improving considerably a number of practically-relevant parameters if compared with the pristine polymer scaffolds. Obviously, the ultimate goal of tissue engineering scaffold design is the development of versatile and preferably universal technology allowing for scalable and reproducible fabrication of biocompatible supports for cell cultivation *in vitro* and subsequent implantation *in vivo*. Although there are other clay-doped nanocomposites apart from tissue engineering scaffolds, the results on their toxicity are characteristic and can be extrapolated to other systems.

15.3.1 *In vitro* Study of Polymer–Nanoclay Composites on Mammalian Cell Lines

Before starting the toxicological study of nanocomposites, it is essential to understand the kinetics of nanoclay–polymer material decomposition and the rate of clay particle release to biological tissue. Toxicological evaluation of halloysite nanoclay has shown that the material is not toxic within moderate concentrations up to 0.5 mg mL^{-1} and over reasonable exposure periods for a wide range of organisms and cells. Liu *et al.* have performed a toxicological evaluation of a halloysite-doped polymer scaffold and demonstrated that addition of nanotubes did not reduce the biocompatibility of the biopolymer matrix.[57,58] As confirmed by cell growth monitoring, adding up to 10% of halloysite as a nanofiller to the polymer for scaffold fabrication did not affect the cell proliferation (Figure 15.8).

Moreover, the authors have shown that mixing halloysite with chitosan leads to a significant enhancement of mechanical properties (compressive strength, compressive modulus and thermal stability) compared with that of the pure chitosan scaffold. Porous tissue engineering constructs built using chitosan doped with halloysite demonstrated relatively low toxicity and high cytocompatibility for fibroblasts. No significant negative effects on attachment, development and cytoskeleton formation were observed in a comparative study between growth on tissue culture treated plastic, pure chitosan and chitosan–HNTs nanocomposite (Figure 15.9).

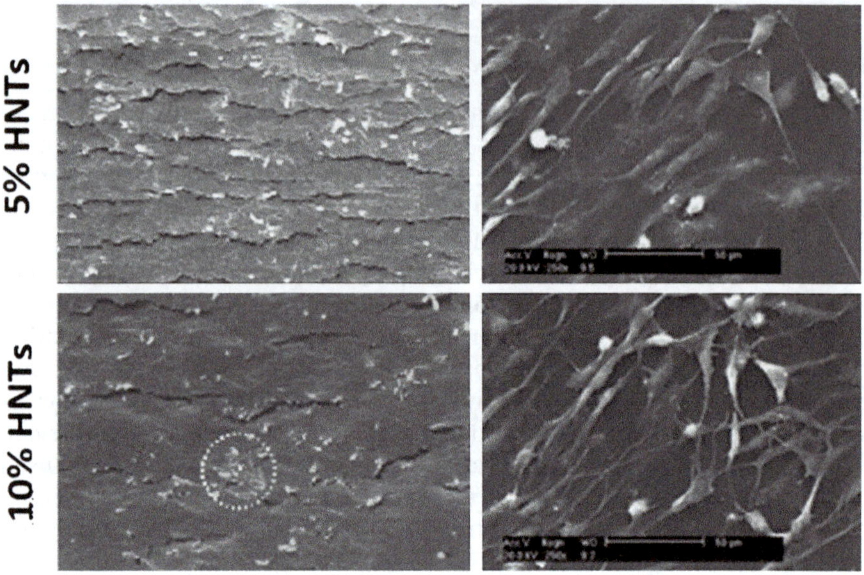

Figure 15.8 Scanning electron microscopy images of chitosan–halloysite nanocomposite with 5% and 10% halloysite and distribution of fibroblasts cultured on nanocomposites. Reprinted from *International Journal of Biological Macromolecules*, **51**, M. Liu *et al.*, Chitosan–halloysite nanotubes bionanocomposites: Structure, mechanical properties and biocompatibility, Copyright 2012 with permission from Elsevier.[57]

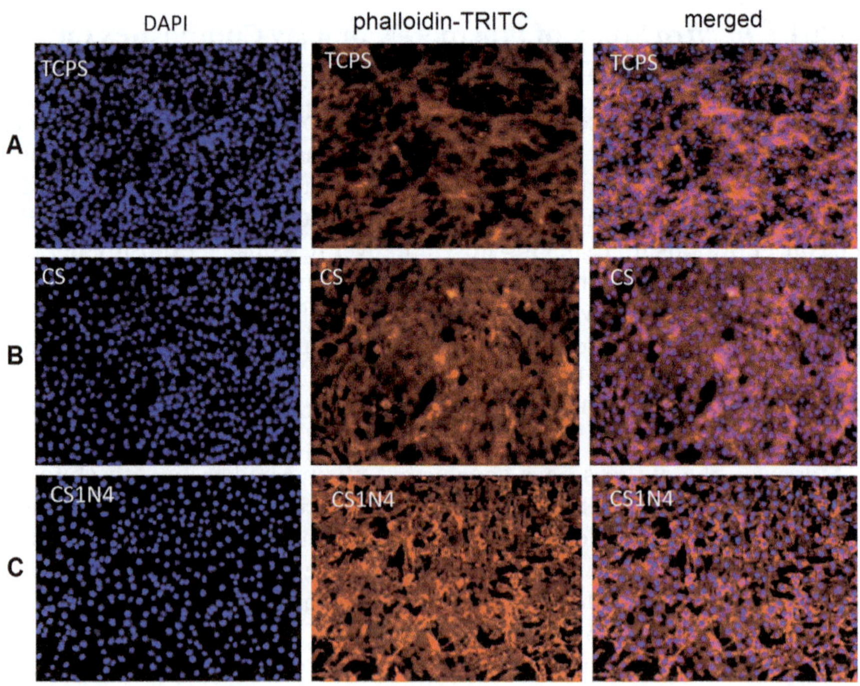

Figure 15.9 Confocal microscopy images of cytoskeleton organization in NIH3T3 cells grown on tissue culture-treated polystyrene (TCPS) (A), pure chitosan scaffold (B) and chitosan–halloysite nanocomposites (C) after three days culture. Reprinted from ref. 58 with permission from The Royal Society of Chemistry.

Another investigation has demonstrated the growth stimulation of fibroblasts derived from dental pulp by halloysite-doped polymer dental implants.[59] Kong *et al.*[60,61] prepared chitosan–hydroxyapatite nanocomposite scaffolds, which exhibited pronounced porosity (typically stimulating cell growth) and improved biocompatibility. A higher proliferation of cells was observed in the case of nanoclay-doped scaffolds if compared with pure chitosan scaffolds. Katti and co-workers[62] described the fabrication and characterization of chitosan–montmorillonite–hydroxyapatite scaffolds for engineering of bone tissues. Remarkably, the osteoblasts (bone cells) proliferated better on the composite chitosan–montmorillonite–hydroxyapatite scaffold rather than on HAP–chitosan scaffold. Barua *et al.* demonstrated the absence of toxicity of thermosensitive hyperbranched epoxy nanocomposites doped with *Homalomena aromatica* rhizome oil-modified bentonite and organically modified montmorillonite using two types of primary human cells: liver hepatocytes and cardiac cells.[63]

So far, most of the published reports demonstrate the good biocompatibility and lack of toxicity of clay–polymer nanocomposites for a wide range of cell cultures *in vitro*. Composite materials demonstrated improved surface characteristics, thermal and mechanical properties and promoted cell proliferation even better than the pristine polymers without nanoclay filling.

15.3.2 Evaluation of Implanted Clay-Doped Nanocomposite Toxicity *In vivo*

Doping tissue engineering scaffolds with nanoclays is one of the emerging directions in tissue engineering, for example exceptional improvement of scaffolds' mechanical properties, such as impact, scratch hardness, tensile strength and elongation at break, was demonstrated in hyperbranched epoxy nanocomposites doped with nanoclays.[63] Normally, most of the polymer nanoclay composite tissue engineering constructs have demonstrated high biocompatibility *in vitro*, exhibiting good cell growth rates (see Section 15.3.1). However, the investigation of biocompatibility and structural stability of tissue engineering scaffolds employing laboratory animals is crucially important before the practical application of polymer–clay scaffolds in human patients. One needs to bear in mind that the data available so far are certainly fragmented, which outlines the need to perform a systematic study of the effects of nanoclays for biomedical polymer composite doping. On the other hand, results obtained for nanoclay-doped tissue engineering scaffolds can, to a certain extent, be extrapolated to other nanoclay-doped polymer composites.

So far, several studies have reported the results of *in vivo* study of nanocomposite scaffolds fabricated from biological or synthetic polymers doped with clay nanomaterials. In general, these studies have confirmed the biocompatibility and safety of clay-doped polymer scaffolds for tissue engineering.

A detailed study of biocompatibility of clay–polymer nanocomposites subcutaneously implanted into rats was performed;[63] the authors investigated the histopathological features of brain, heart, liver and skin samples taken from animals exposed to bentonite or montmorillonite-containing scaffolds. Implanted scaffolds supported the proliferation of dermatocytes, importantly, no acute toxic effects were detected in the organs assessed. Moreover, Hsu *et al.* reported that montmorillonite-doped chitosan (a polysaccharide extracted from crustaceans) nanocomposite three-dimensional constructs even promoted a higher *in vivo* cellular compatibility compared with pure chitosan constructs.[64] The montmorillonite–chitosan nanocomposite also demonstrated improved dynamic tensile properties, higher antimicrobial activity, enhanced cell proliferation and a faster degradation rate than the pure chitosan scaffolds. Remarkably, the collagen capsules developed around the implantation site were thinner in the case of montmorillonite-doped scaffolds (Figure 15.10), which also indicates the high biocompatibility of the clay-doped scaffolds.

Recently, implantable porous 3D scaffolds built from biopolymers (a mixture of chitosan, agarose and gelatin) doped with nanoclay particles were fabricated.[65] In this case, the addition of 3–6 wt% of halloysite nanotubes led to significant improvement of the mechanical stability and wettability of the scaffolds obtained. The biocompatibility of these scaffolds was assessed using a combination of scanning electron microscopy, confocal microscopy and viability tests based on the enzymatic activity of cells. Two types of human cells were employed to investigate the distribution of nanoclay in cells and tissues and the *in situ* viability and proliferative activity of cells grown on 3D scaffolds. Cells confluently colonized the walls

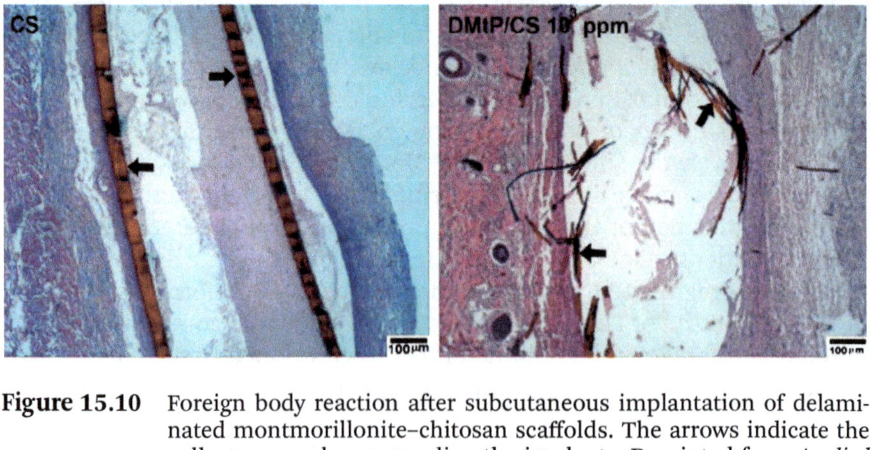

Figure 15.10 Foreign body reaction after subcutaneous implantation of delaminated montmorillonite–chitosan scaffolds. The arrows indicate the collagen capsule surrounding the implants. Reprinted from *Applied Clay Science*, 56, S.-h. Hsu *et al.*, Biocompatibility and antimicrobial evaluation of montmorillonite/chitosan nanocomposites, Copyright 2012 with permission from Elsevier.[64]

of the microscopic pores in scaffolds, retaining the characteristic cellular morphology. The scaffolds did not affect cytoskeleton formation during cell growth, which secures the attachment and distribution of the cells on the scaffold surface. In addition, *in vivo* experiments using the subcutaneous implantation of scaffolds into rats were performed. The implanted area (presumably containing nanoclay leaked from the scaffold residuals) was examined histologically; almost total resorption of material during the six weeks after implantation was detected. Moreover, the clay-doped scaffolds induced the blood vessel formation and blood flow restoration that was demonstrated (Figure 15.11). Dark field microscopy was employed to demonstrate the distribution of halloysite tubes within the implantation area. The reason for the such low toxicity of the scaffolds is apparently due to the use of freeze-drying technology without using any cross-linkers, which can be toxic for cells, whereas the mechanical properties were secured by the halloysite doped into the scaffolds.

In addition to tissue engineering scaffolds, the biocompatibility of polymer–nanoclay composites for drug encapsulation and delivery was also investigated. Veterinarian formulations are deemed to be good candidates for evaluation of biocompatibility. Kelly *et al.* studied the biocompatibility of a polymer–nanoclay-based dental drug delivery composite for the treatment of periodontitis *in vivo*.[13] The nanocomposite was developed using two principal components, namely nanoclay (halloysite) loaded with antibiotic (tetracycline) was coated with biopolymer chitosan to further slow the release of antibiotic for up to 50 days. Then the halloysite-based carrier was added to the thermoresponsive polymer composite (poloxamer 407, octyl cyanoacrylate and polyethylene glycol). This thermosensitive polymer doped with antibiotic-loaded halloysite has demonstrated good antibacterial activity based on effective tetracycline release for up to six weeks *in vivo* in dogs, with no negative side effects. Furthermore, Pinto *et al.*[66] demonstrated the safety and improved drug release properties (prolonged drug release for 30 days) of biodegradable polyurethane–montmorillonite nanocomposite with a dispersed drug triamcinolone acetonide in comparison with formulation lacking montmorillonite clay.

Apart from direct biomedical application, nanoclay-doped polymer composites are used in the packaging industry, *i.e.* for food packaging. As a consequence, food consumers may contact nanoclay particles. To investigate the possible toxicity of nanoclays in food packages, Maisanaba *et al.*[67] exposed rats to extracted obtain from packaging built from a poly(lactic) acid–montmorillonite nanocomposite. The animals were subjected to skin application of the samples to imitate the exposure scenario for consumer exposure while using the food package nanocomposite material. No toxic effects were reported after the detailed histopathological study (combined with optical and electron microscopy) of lung, heart, spleen, kidney, liver, intestine of rats as well as in the evaluation of oxidative stress biomarkers.

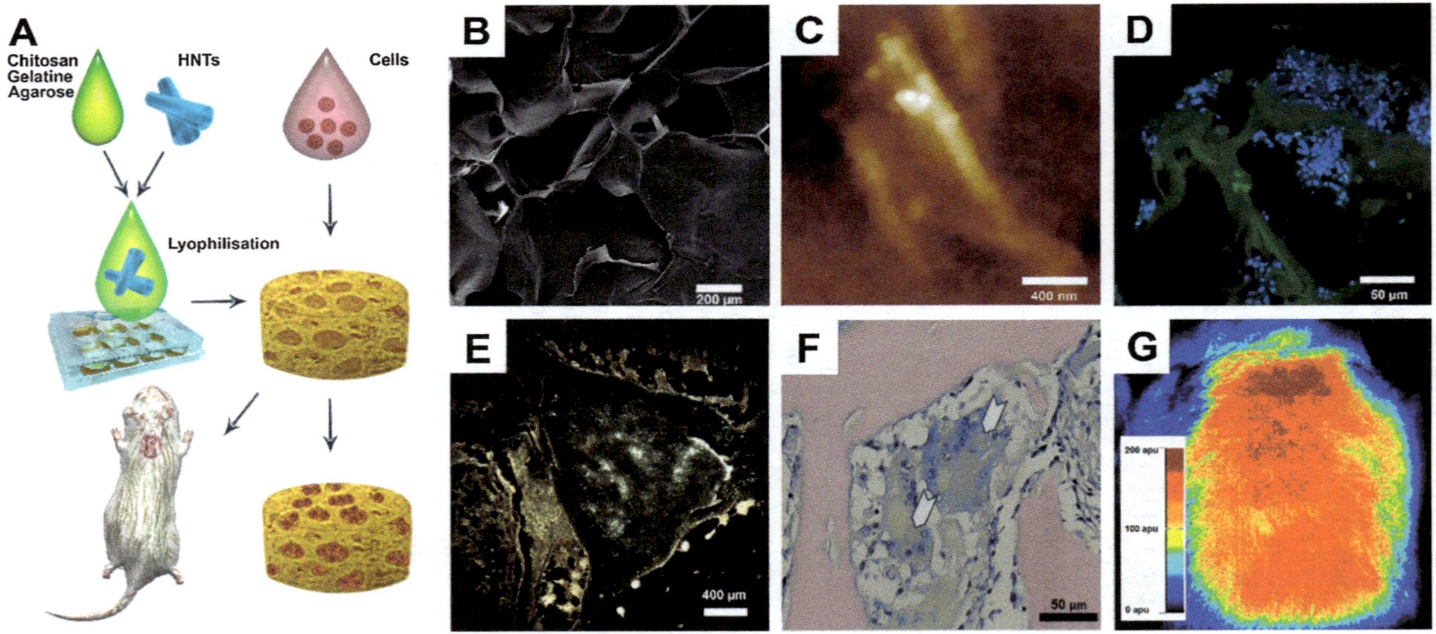

Figure 15.11 Halloysite-doped polymer composite scaffold biocompatibility evaluation: A – scheme demonstrating the fabrication of halloysite-doped scaffolds and the subsequent biocompatibility assays; B – scanning electron microscopy image of chitosan–agarose–gelatine–HNTs (3 wt%) scaffold without cells; C – atomic force microscopy image of chitosan–agarose–gelatine–HNTs (3 wt%) scaffold (note the separate HNTs); D – confocal image of A549 cells distribution on porous 3D scaffold; E – EDF microscopy image of peri-implant subcutaneous area three weeks after transplantation. Note the moderate release of halloysite nanotubes into surrounding tissue; F – hematoxylin–eosin staining of chitosan–agarose–gelatine–HNTs 6 wt% scaffold (three weeks after subcutaneous implantation); G – Laser Doppler Instrument-image of implantation area demonstrating the full restoration of blood supply six weeks after subcutaneous implantation of the porous 3D scaffold. Reprinted from ref. 65 with permission from The Royal Society of Chemistry.

15.4 Conclusion

Nanoclay particles and clay-doped polymer composite have been subjected to toxicological evaluation to find out the limits of their practical use. Based on the already published results, one may conclude that, in general, clay nanoparticles are considerably less toxic than other nanomaterials playing the same role in nanocomposite fabrication. However, we are still far from having unambiguous nanosafety data on clay nanomaterials, therefore future research in this area is deemed to be relevant.

Acknowledgements

This study was supported by Russian Science Foundation grant No 14-14-00924. Authors thank Ms. A. Dubkova for technical help.

References

1. M. L. Du, B. C. Guo and D. M. Jia, *Polym. Int.*, 2010, **59**, 574.
2. T. Hanemann and D. V. Szabó, *Materials*, 2010, **3**, 3468.
3. R. Rothon, *Particulate-Filled Polymer Composites*, Longman, Harlow, UK, 1995.
4. V. Mittal, *Materials*, 2009, **2**, 992.
5. L. W. Carter, J. G. Hendricks and D. S. Bolley, US Pat., No. 2,531,396, 1950.
6. K. Yano, A. Usuki, A. Okada, T. Kurauchi and O. Kamigaito, *J. Polym. Sci., Part A: Polym. Chem.*, 1993, **31**, 2493.
7. Y. Kojima, K. Fukumori, A. Usuki, A. Okada and T. Kurauchi, *J. Mater. Sci. Lett.*, 1993, **12**, 889.
8. E. Joussein, S. Petit, J. Churchman, B. Theng, D. Righi and B. Delvaux, *Clay Miner.*, 2005, **40**, 383.
9. G. Tari, I. Bobos, C. Gomes and J. Ferreira, *J. Colloid Interface Sci.*, 1999, **210**, 360.
10. R. Price, B. Gaber and Y. Lvov, *J. Microencapsulation*, 2001, **18**, 713.
11. V. Vergaro, E. Abdullayev, Y. M. Lvov, A. Zeitoun, R. Cingolani, R. Rinaldi and S. Leporatti, *Biomacromolecules*, 2010, **11**, 820.
12. D. Shchukin, R. Price, G. Sukhorukov and Y. Lvov, *Small*, 2005, **1**, 510.
13. H. Kelly, P. Deasy, E. Ziaka and N. Claffey, *Int. J. Pharm.*, 2004, **274**, 167.
14. N. Veerabadran, R. Price and Y. Lvov, *Nano*, 2007, **2**, 215.
15. Y. Lvov, D. Shchukin, H. Mohwald and R. Price, *ACS Nano*, 2008, **2**, 814.
16. E. Abdullayev, R. Price, D. Shchukin and Y. Lvov, *ACS Appl. Mater. Interfaces*, 2009, **2**, 1437.
17. N. Veerabadran, D. Mongayt, V. Torchilin, R. Price and Y. Lvov, *Macromol. Rapid Commun.*, 2009, **30**, 94.
18. D. Kommireddy, I. Ichinose, Y. Lvov and D. Mills, *J. Biomed. Nanotechnol.*, 2005, **1**, 286.
19. Y. Lvov, K. Ariga, I. Ichinose and T. Kunitake, *Langmuir*, 1996, **12**, 3038.

20. M. T. Viseras, C. Aguzzi, P. Cerezo, G. Cultrone and C. Viseras, *J. Microencapsulation*, 2009, **26**, 279.
21. M. Liu, B. Guo, Q. Zou, M. Du and D. Jia, *Nanotechnology*, 2008, **19**, 205709.
22. W. Zhou, B. Guo, M. Liu, R. Liao, A. Rabie and D. Jia, *J. Biomed. Mater. Res., Part A*, 2009, **2**, 1456.
23. P. Yuan, P. Southon, Z. Liu, M. Green, J. Hook, S. Antill and C. Kepert, *J. Phys. Chem. C*, 2008, **112**, 15742.
24. P. Liu and M. Zhao, *Appl. Surf. Sci.*, 2009, **255**, 3989.
25. M. Dai, X. L. Zheng, X. Xu, X. Y. Kong, X. Y. Li, G. Guo, F. Luo, X. Zhao, Y. Q. Wei and Z. Qian, *J. Biomed. Biotechnol.*, 2009, **2009**, 595126.
26. Y. P. Ye, H. B. Chen, J. S. Wu and L. Ye, *Polymer*, 2007, **48**, 6426.
27. Y. Zhang, Y. Chen, H. Zhang, B. Zhang and J. Liu, *J. Inorg. Biochem.*, 2013, **118**, 59.
28. Y. Chen, Y. Zhang, J. Liu, H. Zhang and K. Wang, *Chem. Eng. J.*, 2012, **210**, 298.
29. W. Wei, R. Minullina, E. Abdullayev, R. Fakhrullin, D. Mills and Y. Lvov, *RSC Adv.*, 2014, **4**, 488.
30. H.-J. Choi, T. J. Stazak Jr and C. Montemagno, *J. Nanopart. Res.*, 2013, **15**, 1.
31. S. A. Konnova, I. R. Sharipova, T. A. Demina, Y. N. Osin, D. R. Yarullina, O. N. Ilinskaya, Y. M. Lvov and R. F. Fakhrullin, *Chem. Commun.*, 2013, **49**, 4208.
32. G. Cavallaro, G. Lazzara, S. Konnova, R. Fakhrullin and Y. Lvov, *Green Mater.*, 2014, **2**, 232.
33. P.-R. Li, J.-C. Wei, Y.-F. Chiu, H.-L. Su, F.-C. Peng and J.-J. Lin, *ACS Appl. Mater. Interfaces*, 2010, **2**, 1608.
34. A. A. Taylor, G. M. Aron, G. W. Beall, N. Dharmasiri, Y. Zhang and R. J. C. McLean, *Environ. Toxicol.*, 2014, **29**, 961.
35. M. Kryuchkova, A. Danilushkina, Y. Lvov and R. Fakhrullin, *Environ. Sci.: Nano*, 2016, **3**, 442–452.
36. X. Guo and N. Mei, *J. Food Drug Anal.*, 2014, **22**, 105.
37. E. J. Murphy, E. Roberts and L. A. Horrocks, *Neuroscience*, 1993, **55**, 597.
38. M. Baek, A. J. Lee and S. J. Choi, *Mol. Cell. Toxicol.*, 2012, **8**, 95.
39. A. K. Sharma, B. Schmidt, H. Frandsen, N. R. Jacobsen, E. H. Larsen and M. L. Binderup, *Mutat. Res.*, 2010, **700**, 18.
40. S. Lordan, J. E. Kennedy and C. L. Higginbotham, *J. Appl. Toxicol.*, 2011, **31**, 27.
41. J. Houtman, S. Maisanaba, M. Puerto, D. Gutiérrez-Praena, M. Jordá, S. Aucejo and A. Jos, *Appl. Clay Sci.*, 2014, **90**, 150.
42. M. R. Dzamukova, E. A. Naumenko, Y. M. Lvov and R. F. Fakhrullin, *Sci. Rep.*, 2015, **5**, 10560.
43. X. Lai, M. Agarwal, Y. Lvov, C. Pachpande, K. Varahramyan and F. Witzmann, *J. Appl. Toxicol.*, 2013, **33**, 1316.
44. Q. Liu, Y. Liu, S. Xiang, X. Mo, S. Su and J. Zhang, *Appl. Clay Sci.*, 2011, **51**, 214.

45. G. Janer, E. Fernández-Rosas, E. Masdel Molino, D. González-Gálvez, G. Vilar, C. López-Iglesias, V. Ermini and S. Vázquez-Campos, *Nanotoxicology*, 2014, **8**, 279.
46. S. Maisanaba, M. Puerto, S. Pichardo, M. Jordá, F. J. Moreno, S. Aucejo and A. Jos, *Food Chem. Toxicol.*, 2013, **57**, 266.
47. S. Maisanaba, S. Pichardo, M. Puerto, D. Gutiérrez-Praena, A. M. Cameán and A. Jos, *Environ. Res.*, 2015, **138**, 233.
48. G. I. Fakhrullina, F. S. Akhatova, Y. M. Lvov and R. F. Fakhrullin, *Environ. Sci.: Nano*, 2015, **2**, 54.
49. P.-H. Chen, K.-M. Hsiao and C.-C. Chou, *Biomaterials*, 2013, **34**, 5661.
50. A. Nouara, Q. Wu, Y. Li, M. Tang, H. Wang, Y. Zhao and D. Wang, *Nanoscale*, 2013, **5**, 6088.
51. Q. Wu, Y. Zhao, G. Zhao and D. Wang, *Nanomedicine*, 2014, **10**, 1263.
52. J. Kim, T. Shirasawa and Y. Miyamoto, *Biomaterials*, 2010, **31**, 5849–5854.
53. M. W. Wiles, H. J. Huebner, E. Afriyie-Gyawu, R. J. Taylor, G. R. Bratton and T. D. Phillips, *J. Toxicol. Environ. Health, Part A*, 2004, **67**, 863.
54. E. C. Patterson and D. J. Staszak, *J. Nutr.*, 1977, **107**, 2020.
55. R. Slamova, M. Trckova, H. Vondruskova, Z. Zraly and I. Pavlik, *Appl. Clay Sci.*, 2011, **51**, 395.
56. M. J. Wilson, *J. Chem. Ecol.*, 2003, **29**, 1525.
57. M. Liu, Y. Zhang, C. Wu, S. Xiong and C. Zhou, *Int. J. Biol. Macromol.*, 2012, **51**, 566.
58. M. Liu, C. Wu, Y. Jiao, S. Xiong and C. Zhou, *J. Mater. Chem. B*, 2013, **1**, 2078.
59. M. C. Bottino, G. H. Yassen, J. A. Platt, N. Labban, L. J. Windsor, K. J. Spolnik and A. H. Bressiani, *J. Tissue Eng. Regener. Med.*, 2013, **9**, E116.
60. L. Kong, Y. Gao, W. Cao, Y. Gong, N. Zhao and X. Zhang, *J. Biomed. Mater. Res.*, 2005, **5A**, 275.
61. L. Kong, Y. Gao, W. Cao, Y. Gong, N. Zhao and X. Zhang, *Eur. Polym. J.*, 2006, **42**, 3171.
62. K. S. Katti, D. R. Katti and R. Dash, *Biomed. Mater.*, 2008, **3**, 034122.
63. S. Barua, N. Dutta, S. Karmakar, P. Chattopadhyay, L. Aidew, A. K. Buragohain and N. Karak, *Biomed. Mater.*, 2014, **9**, 025006.
64. S.-h. Hsu, M.-C. Wan and J.-J. Lin, *Appl. Clay Sci.*, 2012, **56**, 53.
65. E. A. Naumenko, I. D. Guryanov, R. Yendluri, Y. M. Lvov and R. F. Fakhrullin, *Nanoscale*, 2016, **8**, 7257–7271.
66. F. C. H. Pinto, A. Silva-Cunha, G. A. Pianetti, E. Ayres, R. L. Orefice and G. R. Da Silva, *J. Nanomater.*, 2011, **2011**, 528628.
67. S. Maisanaba, D. Gutiérrez-Praena, M. Puerto, M. Llana-Ruiz-Cabello, S. Pichardo, R. Moyano, A. Blanco, M. Jordá-Beneyto and A. Jos, *J. Toxicol. Environ. Health, Part A*, 2014, **77**, 731.

Subject Index